Biology

FOR THE IB DIPLOMA

SECOND EDITION

Biology

FOR THE IB DIPLOMA

SECOND EDITION

C.J. Clegg

HODDER
EDUCATION
AN HACHETTE UK COMPANY

This material has been developed independently by the publisher and the content is in no way connected with, nor endorsed by, the International Baccalaureate Organization.

Orders: please contact Bookpoint Ltd, 130 Milton Park, Abingdon, Oxon OX14 4SB. Telephone: (44) 01235 827720. Fax: (44) 01235 400454. Lines are open from 9.00-5.00, Monday to Saturday, with a 24 hour message answering service. You can also order through our website www.hoddereducation.com

© C.J. Clegg 2014

First edition published in 2007

by Hodder Education

An Hachette UK Company

Carmelite House, 50 Victoria Embankment

London EC4Y 0DZ

This second edition published 2014

Impression number   5 4

Year   2019

Cover photo © Patryk Kosmider - Fotolia

Illustrations by Barking Dog, Aptara, Inc., and Oxford Designers & Illustrators

Typeset in 10/12 Goudy Oldstyle Std by Aptara, Inc.

Printed in India

A catalogue record for this title is available from the British Library

ISBN: 978 1471 828997

Contents

Options
Available on the website accompanying this book: www.hoddereducation.com/IBextras

Option A

Appendices and chapter summaries

Available on the website accompanying this book: www.hoddereducation.com/IBextras
Appendix 1: Background chemistry for biologists
Appendix 2: Investigations, data handling and statistics
Appendix 3: Defining ethics and making ethical decisions
Chapter summaries 1–11

Answers to self-assessment questions in Chapters 1–11

Answers to the self-assessment questions in the Options Chapters 12–15 are available on the website accompanying this book: www.hoddereducation.com/IBextras
Answers to the examination questions at the end of each chapter are also on the website.

Glossary

Index

Introduction

Welcome to the second edition of *Biology for the IB Diploma*, updated and designed to meet the criteria of the 2014 International Baccalaureate (IB) Diploma Programme Biology Guide. The structure and content of this second edition follow the structure and content of the IB Biology Subject Guide.

■ Using this book

Special features of the chapters of Biology for the IB Diploma include:

- Each chapter begins with **Essential ideas** that summarize the concepts on which the chapter is based.

- **Applications** in the Guide are integrated within the main content and are used to illustrate the various **Understandings** listed in the Guide.

- **Skills** are highlighted with this icon. Students are expected to be able to show these skills in the examination, so we have explicitly pointed these out when they are mentioned in the Guide.

Nature of Science

- The **Nature of Science (NoS)** is a theme that runs throughout the course, and can be examined in Biology papers. It explores the scientific process itself, and how science is represented and understood by the general public. It also examines the way in which science is the basis for technological developments and how these new technologies, in turn, drive developments in science.

- **International mindedness** explores how the exchange of information and ideas across national boundaries has been essential to the progress of science, and illustrates the international aspects of Biology.

- **Self-assessment questions (SAQs)** are phrased so as to assist comprehension and recall, but also to help familiarize students with the assessment implications of the command terms. Answers to all SAQs are given, either in this book or on the accompanying website.

- Links to the interdisciplinary **Theory of Knowledge (TOK)** element of the IB Diploma course are made at appropriate places in most chapters.

- Links to relevant material available on the website that accompanies this book (www.hoddereducation.com/IBextras) are highlighted with this icon.

- At the end of each chapter, there is a selection of **examination questions**. Some are questions taken from past exam papers, others are exam-style questions written for this book. Answers are available on the accompanying website.

The Options (Chapters 12–15) are available on the website accompanying this book, as are useful appendices and additional student support for Chapters 1–11, including further opportunities to practise data response questions: www.hoddereducation.com/IBextras

■ Author's acknowledgements

I am indebted to IB teachers who have welcomed me into their departments and have updated me on the delivery of the IB Diploma programme, both in the UK and in Hong Kong, and to the international students who have discussed their recent experiences of the course with me.

In the production of this second edition, I have had the benefit of detailed feedback on the approach and content of each chapter during its creation, by Dr Andrew Davis, Head of Environmental Science at St Edward's School. His insights and observations, based in part on his experiences of delivering the current course, and his interpretation of the new syllabus, have been invaluable.

Mrs Lynda Brooks, the Librarian of the Linnean Society, London, UK, provided invaluable help in the issue of the reclassification of the figwort family.

The following components of this edition have been authored by:

Dr Andrew Davis, Head of Environmental Science, St Edward's School, Oxford, UK:

- Option C (Chapter 14) Ecology and conservation.
- The data in 'Practical ecology: Testing for associations between species' (pages 187–91).

Mrs Lucy Baddeley, Biology Department, St Edward's School, Oxford, UK:

- Use of the Gene Bank database to determine differences in base sequence of a gene in two species (pages 137–38).

Mr Luis F. Ceballos Posada, Head of the Biology Department, Changchun American International School (CAIS), Changchun, Jilin Province, China:

- Option B (Chapter 13) Section 13.5 Bioinformatics.

Finally, I am indebted to the International publishing team at Hodder Education, ably led by So-Shan Au, International Publisher, Asia and IB, whose skill and patience have brought together the text and illustrations as I have wished, and I am most grateful to them.

Dr Chris Clegg
Salisbury
Wiltshire
UK
June 2014

Acknowledgements

The Publishers would like to thank the following for permission to reproduce copyright material. Every effort has been made to trace all copyright holders, but if any have been inadvertently overlooked the Publishers will be pleased to make the necessary arrangements at the first opportunity.

■ Photo credits

p.4 *l* © J.C. Revy, ISM/Science Photo Library; **p.4** *r* © Gene Cox; **p.5** *bl* © macropixel – Fotolia.com; **p.5** *br* © vlad61_61 – Fotolia.com; **p.10** *t* © Gene Cox; **p.10** *b* © Power and Syred/Science Photo Library; **p.11** *bl* © George Chapman/Visuals Unlimited/Getty Images (modified photograph); **p.11** *br* © ScienceFoto/Getty Images; **p.13** © Gene Cox; **p.17** © Sinclair Stammers/Science Photo Library; **p.18** *t* © Power and Syred/Science Photo Library; **p.18** *bl* © Phototake Inc./Alamy; **p.18** *br* © Biophoto Associates/Science Photo Library; **p.20** © Kevin S. Mackenzie, Technician, School of Medical Science, Aberdeen University, Foresterhill, Aberdeen; **p.22** *t* © Don W Fawcett/Science Photo Library; **p.22** *br* © CNRI/Science Photo Library; **p.23** © Dr Kari Lounatmaa/Science Photo Library; **p.24** *t* © Medimage/Science Photo Library; **p.24** *b* © Omikron/Science Photo Library; **p.25** *t* © Carolina Biological Supply Co/Visuals Unlimited, Inc./Science Photo Library; **p.27** *tl* © David T. Moran, Ph.D., Visual Histology; **p.27** *tr* © Dr Jeremy Burgess/Science Photo Library; **p.27** *b* © Steve Gschmeissner/Science Photo Library **p.29** © Kwangshin Kim/Science Photo Library; **p.30** © CNRI/Science Photo Library; **p.33** *l* © NIBSC/Science Photo Library; **p.33** *r* © Don W Fawcett/Science Photo Library; **p.54** *tl* © Michael Abbey/Science Photo Library; **p.54** *tr* © Michael Abbey/Science Photo Library; **p.54** *c* © Michael Abbey/Science Photo Library; **p.54** *br* © Michael Abbey/Science Photo Library; **p.54** *bl* © Michael Abbey/Science Photo Library; **p.56** © Science Photo Library; **p.57** © Gene Cox; **p.61** © Don Fawcett-Keith Porter/Photo Researchers, inc. **p.71** © Rossi Paolo – Fotolia.com; **p.72** © Biophoto Associates/Science Photo Library; **p.77** *both* Andrew Lambert/Photography/Science Photo Library; **p.79** © Biophoto Associates/Science Photo Library; **p.80** *cl* © Andrew Lambert Photography/Science Photo Library; **p.80** *cr* © Andrew Lambert Photography/Science Photo Library; **p.80** *b* © Phototake Inc./Alamy; **p.104** © Cordelia Molloy/Science Photo Library; **p.108** *tr* © A. Barrignton Brown/Science Photo Library; **p.108** *c* © Science Photo Library; **p.108** *cr* © Science Photo Library; **p.116** *l* © Martyn F. Chillmaid/Science Photo Library; **p.116** *r* © Reuters/Corbis; **p.136** © TEK Image/Science Photo Library; **p.140** © Leonard Lessin/Science Photo Library; **p.146** *r* © Hattie Young/Science Photo Library; **p.146** *l* © U.S. Department of Energy Human Genome Program; **p.148** © Oxford Scientific/Getty Images **p.149** © Science Photo Library; **p.157** *t* © Dr P Marazzi/Science Photo Library; **p.157** *b* © Chris Bjornberg/Science Photo Library; **p.160** © Adam Hart-Davis/Science Photo Library; **p.163** © Igor Kostin/Sygma/Corbis; **p.170** © Dr Gopal Murti/Science Photo Library; **p.173** *l* © Hiroya Minakuchi/Minden Pictures/FLPA; **p.173** *r* © Makato Iwafuji/Eurelios/Science Photo Library; **p.178** © David Parker/Science Photo Library; **p.180** © Roslyn Institute, The University of Edinburgh; **p.182** © National Human Genome Research Institute **p.186** © CuboImages srl/Alamy; **p.187** © Biophoto Associates/Science Photo Library; **p.189** *l* © C.J. Clegg; **p.189** *c* © Richard Becker/Alamy; **p.189** *r* © D. Hurst/Alamy; **p.207** *l* © Nigel Dickenson/Still Pictures/Robert Harding; **p.212** *tr* © Ken Lucas, VISUALS UNLIMITED/ SCIENCE PHOTO LIBRARY; **p.212** *bl* © Sally A. Morgan; Ecoscene/Corbis; **p.212** *br* © Daniel Eskridge/Stocktrek Images/Getty Images; **p.213** *cl* © Dave Watts/Alamy; **p.213** *bl* © Jon Durrant/Alamy; **p.213** *br* © Nigel Cattlin/Alamy; **p.218** *l* © Owen Franken/Corbis; **p.218** *r* © NHPA/Photoshot; **p.225** *tl* © idp wildlife collection/Alamy; **p.225** *tc* © Lip Kee/http://www.flickr.com/photos/lipkee/5657636385/sizes/m/in/pool-42637302@N00/ (https://creativecommons.org/licenses/by-sa/2.0/); **p.225** *tr* © FLPA/Alamy; **p.225** *cl* © DDniki – Fotolia.com; **p.225** *cr* © Gerry Ellis/Minden Pictures/FLPA; **p.244** *tl* © Steve Taylor ARPS/Alamy; **p.244** *tr* © imageBROKER/Alamy; **p.244** *cl* © C.J. Clegg; **p.244** *cr* © Premium Stock Photography GmbH/Alamy; **p.244** *bl* © Chris Howes/Wild Places Photography/Alamy; **p.244** *br* © C.J. Clegg; **p.246** *tl* © Simon

Colmer/Alamy; **p.246** *tr* © Mira/Alamy; **p.246** *cl* © John Tiddy (MYN) Nature Picture Library/ Corbis; **p.246** *cr* © Vangert–Fotolia.com; **p.246** *bl* © izmargad–Fotolia.com; **p.246** *br* © David Colo/Alamy; **p.252** © Dr Keith Wheeler/Science Photo Library; **p.256** © Gene Cox; **p.257** © Gene Cox; **p.260** © Gene Cox; **p.267** © Everett Collection Historical/Alamy; **p.268** © CNRI/ Science Photo Library; **p.270** © ISM/Science Photo Library; **p.273** © CNRI/Science Photo Library; **p.274** © St Mary's Hospital Medical School/Science Photo Library; **p.278** © Thomas Deerinck, NCMIR/Science Photo Library; **p.284** © Gene Cox; **p.289** © Moredum Animal Health Ltd/Science Photo Library; **p.290** © Biophoto Associates/Science Photo Library; **p.297** © Prof S. Cinti/Science Photo Library; **p.300** *tl* © Heather Angel/Natural Visions; **p.300** *tc* © NHPA/Photoshot; **p.300** *tr* © Tomas Friedmann/Science Photo Library; **p.300** *bl* © Heather Angel/Natural Visions; **p.300** *br* © NHPA/Photoshot; **p.301** © Nickel ElectroLtd; **p.302** © Gene Cox; **p.303** © Saturn Stills/Science Photo Library; **p.316** © Science Picture Co/Science Picture Co; **p.317** Biophoto Associate Science Source; **p.318** © Biozentrum, University of Basel/Science Photo Library; **p.334** *tl* Courtesy of Fvoigtsh/Wikipedia Commons (http:// creativecommons.org/licenses/by-sa/3.0/deed.en); **p.334** *tr* © Laguna Design/Science Photo Library; **p.336** © Biophoto Associates/Science Photo Library; **p.342** © Steve Gschmeissner/ Science Photo Library; **p.358** © CNRI/Science Photo Library; **p.359** © Mariam Ghochani and Terrence G. Frey, San Diego State University; **p.360** © Dr Kenneth R. Miller/Science Photo Library; **p.370** © Dr Kari Lounatmaa/Science Photo Library; **p.374** © Gene Cox; **p.375** © Gene Cox; **p.378** © Dr David Furness, Keele University/Science Photo Library; **p.379** *both* © Gene Cox; **p.382** *tr* © Gene Cox; **p.382** *cl* © Gene Cox; **p.386** *l* © C.J. Clegg; **p.386** *r* © Gene Cox; **p.387** *l* © Scott Camazine/Alamy; **p.387** *r* © Wildlife GmbH/Alamy; **p.388** © Biophoto Associates/Science Photo Library; **p.390** *r* © Eye of Science/Science Photo Library; **p.395** *l* © Gene Cox; **p.395** *l* © Gene Cox; **p.401** *r* © Dr Jeremy Burgess/Science Photo Library; **p.401** *l* © Nigel Cattlin/Alamy; **p.402** *t* © NHPA/Photoshot; **p.402** *bl* © Bill Ross/Corbis; **p.402** *br* © Jane Sugarman/Science Photo Library; **p.403** © C.J. Clegg; **p.404** *tl* © Siloto/Alamy; **p.404** *tc* © Steve Byland – Fotolia.com; **p.404** *tr* © Rolf Nussbaumer Photography/Alamy; **p.416** © Carolina BiologicaL Supply Co/Visuals Unlimited, Inc./Science Photo Library; **p.422** © David Q. Cavagnaro/Photolibrary/Getty Images; **p.424** © NHPA/Photoshot; **p.433** *tr* © tbkmedia. de/Alamy; **p.433** *c* © Konrad Wothe/Minden Pictures/Getty Images; **p.434** *tl* © ams images/ Alamy; **p.434** *tr* © Eric Brasseur/Photonica/Getty Images; **p.434** *b* © imageBROKER/Alamy; **p.446** © Steve Gschmeissner/Science Photo Library; **p.452** *l* © Bettmann/Corbis; **p.452** *r* © Jean-Loup Charmet/Science Photo Library; **p.461** © Gene Cox; **p.462** *tl* © P. Navarro, R. Bick, B. Poindexter, UT Medical SchooL/Science Photo Library; **p.462** *c* © Mark Rothery (http:// www.mrothery.co.uk/); **p.463** *t* © Mark Rothery (http://www.mrothery.co.uk/); **p.463** *b* © Mark Rothery (http://www.mrothery.co.uk/); **p.467** © Alaska Stock/Alamy; **p.470** © age fotostock/ SuperStock; **p.471** © Steve Gschmeissner/Science Photo Library; **p.484** *tl* © Astrid & Hans-Frieder Michler/Science Photo Library; **p.484** *br* © Gene Cox; **p.487** © Jean Claude Revy-A. Goujeon, ISM/Science Photo Library; **p.490** © Edelmann/Science Photo Library; **p.492** © Dr G Moscoso/Science Photo Library; **p.496** © Mark Rothery (http://www.mrothery.co.uk/)

■ Text and artwork credits

p.14 From *The Life of Mammals*, John Zachary Young, Oxford University Press, 1957, reproduced by permission of the publisher; **p.74** Dr Peter H. Gleick, Pacific Institute, California; **p.87** Adapted from *Report of the WHO Study Group*, Andrew Edmonson and David Druce, Oxford University Press, 1996, reproduced by permission of the publisher; **p.138** From National Center for Biotechnology Information websites, www.ncbi.nlm.nih.gov/gene and www.ncbi. nlm.nih.gov/Blast.cgi; **p.147** *Effect of maternal age on chromosome abnormalities*, from PubMed Central, International Journal of Women's Health, 2013 and www.ncbi.nlm.nih.gov/pmc/ articles/PMC3581291/figure/f1-ijwh-5-065, reproduced by permission of Dove Medical Press; **p.202** Adapted from 'Overview of the Carbon Cycle From the Systems Perspective', Timothy Bralower and David Bice, from www.e-education.psu.edu/earth103/print/book/export/html/692, © D. Bice; **p.203** From *Biology for the IB Diploma*, C.J. Clegg, Hodder Education, 2007, © The Intergovernmental Panel on Climate Change; **p.207** 'Yo! Amigo!' greenhouse effect cartoon from *San Jose Mercury News*, reproduced by permission of Scott Willis; **p.215** C.J. Clegg, based on data from *How & Why Species Multiply:The Radiation of Darwin's Finches*, Peter R.Grant and B. Rosemary Grant, Princeton University Press, 2008; **p.223** *op.cit* C.J. Clegg; **p.228** From Lipid structure of cell membranes in the three Domains, http://en.wikipedia.org/wiki/Archaea; **p.245** From 'Whatever happened to the Scrophulariaceae', *Fremontia*, 17:30.2, adapted by Richard G. Olmstead; **p.267** Adapted by Colin Clegg, from *Exercise Physiology and Functional Anatomy*, revised by Stephen Ingham, Feltham Press, 1995; **p.269** Adapted from *Principles of Anatomy and Physiology*, 9th edition, Gerard Tortora and Sandra Grabowski, John Wiley & Sons, 1999; **p.271** *World Health and Disease*, Alastair Gray, Open University Press, 1985, reproduced by permission of Professor Alastair Gray; **p.275** 'The number of antibodies developed annually', from 'Clinical Infectuous Diseases', *New Scientist*, Volume 38, 29 September 2007, reproduced by permission of Tribune Content Agency; reused from *Cambridge International AS & A Level Biology*, C.J. Clegg, Hodder Education, 2014, © Crown copyright; **p.279** From 'The immunopathogenesis of human immunodeficiency virus infection', G. Pantaleo, C. Graziosi and A.S. Fauci, *The New England Journal of Medicine*, 325(5),1993; **p.321** Adapted from *Life Chemical and Molecular Biology*, W.R. Pickering, E.J. Wood and C. Smith, Portland Press, 1997; **p.322** From Eukaryotic genes, http://upload.wikimedia.org/wikipedia/commons/0/07/Gene.png; **p.357** From *Sea of Cortez: A Leisurely Journal of Travel and Research*, John Steinbeck and Edward F. Ricketts, Penguin Books, 2009; **p.379** C.J. Clegg, based on a figure from *Anatomy & Activities of Plants*; **p.437** From *Evolution and Pollution (Studies in Biology, No.130)*, A.D. Bradshaw and T. McNeilly, Hodder Arnold, 1981; **p.451** Adapted from 'Tubercolosis: the global challenge', *Biological Science Review 8*, S.J.G. Kavanagh and D.W. Denning, Hodder Arnold, 1981; **p.460** C.J. Clegg, modified from a figure from *Introduction to Advanced Biology*, Simpson et.al, Hodder Education, 2000

■ Examination questions credits

The publishers would like to thank the International Baccalaureate Organization for permission to reproduce its intellectual property.

Cell biology

ESSENTIAL IDEAS

- The evolution of multicellular organisms allowed cell specialization and cell replacement.
- Eukaryotes have a much more complex cell structure than prokaryotes.
- The structure of biological membranes makes them fluid and dynamic.
- Membranes control the composition of cells by active and passive transport.
- There is an unbroken chain of life, from the first cells on Earth to all cells in organisms alive today.
- Cell division is essential but must be controlled.

1.1 Introduction to cells – *the evolution of multicellular organisms allowed cell specialization and cell replacement*

The cell is the basic unit of living matter – the smallest part of an organism which we can say is alive. It is cells that carry out the essential processes of life. We think of them as self-contained units of structure and function.

Cells are extremely small – most are only visible as distinct structures when we use a **microscope** (although a few types of cell are just large enough to be seen by the naked eye).

Observations of cells were first reported over 300 years ago, following the early development of microscopes (Figure 1.2, page 3). Today we use a compound light microscope to investigate cell structure – perhaps you are already familiar with the light microscope as a piece of laboratory equipment. You may have used one to view living cells, such as the single-celled animal, *Amoeba*, shown in Figure 1.1.

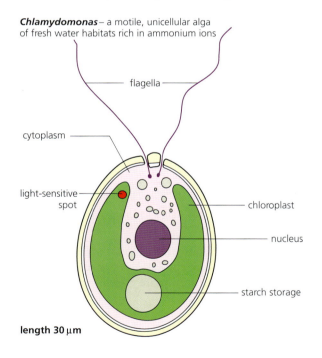

Chlamydomonas – a motile, unicellular alga of fresh water habitats rich in ammonium ions

- flagella
- cytoplasm
- light-sensitive spot
- chloroplast
- nucleus
- starch storage

length 30 μm

Amoeba – a protozoan of freshwater habitats
- cytoplasm
- endoplasm
- clear ectoplasm
- pseudopodia
- plasma membrane
- contractile vacuole
- nucleus
- food vacuoles

length 400 μm

Escherichia coli – a bacterium found in the intestines of animals, e.g. humans
- cell wall (polysaccharide + amino acids)
- plasma membrane
- cytoplasm
- plasmid
- pili
- circular DNA
- ribosomes

length 2.0 μm

■ **Figure 1.1** Introducing unicellular organization

■ Unicellular and multicellular organisms

Some organisms are made of a single cell and are known as **unicellular**. Examples of unicellular organisms are introduced in Figure 1.1. In fact, there are vast numbers of different unicellular organisms in the living world, many with very long evolutionary histories.

Other organisms are made of many cells and are known as **multicellular** organisms. Examples of multicellular organisms are the mammals and flowering plants. Much of the biology in this book is about multicellular organisms, including humans, and the processes that go on in these organisms. But remember, single-celled organisms carry out all the essential functions of life too, within the confines of a single cell.

1   State the essential processes characteristic of living things.

■ The features of cells

A cell consists of a **nucleus** surrounded by cytoplasm, contained within the cell membrane. The nucleus is the structure that controls and directs the activities of the cell. The **cytoplasm** is the site of the chemical reactions of life, which we call 'metabolism'. The cell membrane, known as the **plasma membrane**, is the barrier controlling entry to and exit from the cytoplasm.

Newly formed cells grow and enlarge. A growing cell can normally divide into two cells. Cell division is very often restricted to unspecialized cells, before they become modified for a particular task.

Cells may develop and specialize in their structure and in the functions that they carry out. A common outcome of this is that many fully specialized cells are no longer able to divide, for example. But as a consequence of specialization, **cells show great variety in shape and structure**. This variety in structure reflects the evolutionary adaptations of cells to different environments, and to different specialized functions – for example, within multicellular organisms.

■ Cell theory – a summary statement

The **cell theory** – the statement that cells are the unit of structure and function in living things – contains three very basic ideas:

- Cells are the building blocks of structure in living things.
- Cells are the smallest unit of life.
- Cells are derived from other cells (pre-existing cells) by division.

Today, we can confidently add two concepts to the theory:

- Cells contain a blueprint (information) for their growth, development and behaviour.
- Cells are the site of all the chemical reactions of life (metabolism).

■ Cell size

Since cells are so small, we need appropriate units to measure them. The **metre** (symbol **m**) is the standard unit of length used in science (it is an internationally agreed unit, or **SI unit**).

1 metre (m)	= 1000 millimetres (mm)
1 mm	= 1000 micrometres (μm) (or microns)
1 μm	= 1000 nanometres (nm)

■ **Table 1.1** Units of length used in microscopy

Look at Table 1.1, showing the subdivisions of the metre that are used to measure cells and their contents. These units are listed in descending order of size. You will see that each subdivision is one thousandth of the unit above it. The smallest units are probably quite new to you; they may take some getting used to.

So, the dimensions of cells are expressed in the unit called a **micrometre** or micron (**μm**). Notice, this unit is one thousandth (10^{-3}) of a millimetre. This gives us a clear idea about how small cells are when compared to the millimetre, which you can see on a standard ruler.

Bacteria are really small, typically 0.5–10 μm in size, whereas the cells of plants and animals are often in the range 50–150 μm or larger. In fact, the lengths of the unicells shown in Figure 1.1 are approximately:

2   Calculate:
   a how many cells of 100 μm diameter will fit side by side along a millimetre line
   b the magnification of the image of *Escherichia coli* in Figure 1.1.

- *Chlamydomonas*   30 μm
- *Escherichia coli*   2 μm
- *Amoeba*   400 μm (but its shape and, therefore, length vary greatly).

The origins of cell theory

Many biologists contributed to the development of the cell theory. This concept evolved gradually in Western Europe during the nineteenth century, as a result of the steadily accelerating pace of developments in **microscopy** and **biochemistry**. You can see a summary of the earliest steps in Figure 1.2.

■ **Figure 1.2**
Early steps in the development of the cell theory

Hooke's microscope, and a drawing of the cells he observed

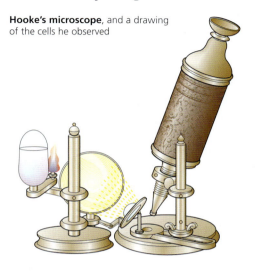

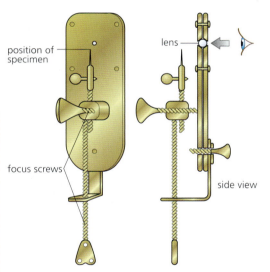

van Leeuwenhoek's microscope

Robert Hooke (1662), an expert mechanic and one of the founders of the Royal Society in London, was fascinated by microscopy. He devised a compound microscope, and used it to observe the structure of cork. He described and drew cork cells, and also measured them. He was the first to use the term 'cells'.

Anthonie van Leeuwenhoek (1680) was born in Delft. Despite no formal training in science, he developed a hobby of making lenses, which he mounted in metal plates to form simple microscopes. Magnifications of ×240 were achieved, and he observed blood cells, sperms, protozoa with cilia, and even bacteria (among many other types of cells). His results were reported to the Royal Society, and he was elected a fellow.

Robert Brown (1831), a Scottish botanist, observed and named the cell nucleus. He also observed the random movements of tiny particles (pollen grains, in his case) when suspended in water (Brownian movement).

Matthias Schleiden (1838) and Theodor Schwann (1839), German biologists, established cells as the natural unit of form and function in living things: 'Cells are organisms, and entire animals and plants are aggregates of these organisms arranged to definite laws.'

Rudolf Virchow (1856), a German pathologist, established the idea that cells arise only by division of existing cells.

Louis Pasteur (1862), a brilliant French microbiologist, established that life does not spontaneously generate. The bacteria that 'appear' in broth are microbes freely circulating in the air, which contaminate exposed matter.

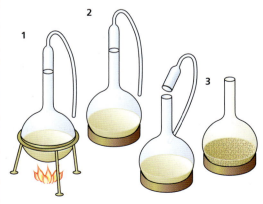

Pasteur's experiment, in which broth was sterilized (**1**), and then either exposed to air (**3**) or protected from air-borne spores in a swan-necked flask (**2**). Only the broth in **3** became contaminated with bacteria.

■ Introducing animal and plant cells

■ **Figure 1.3**
Animal and plant cells from multicellular organisms

No 'typical' cell exists – there is a very great deal of variety among cells. However, we shall see that most cells have features in common. Viewed using a compound microscope, the *initial* appearance of a cell is of a simple sac of fluid material, bound by a membrane, and containing a nucleus. *Look at the cells in Figure 1.3.*

Canadian pondweed (*Elodea*) grows submerged in fresh water

human

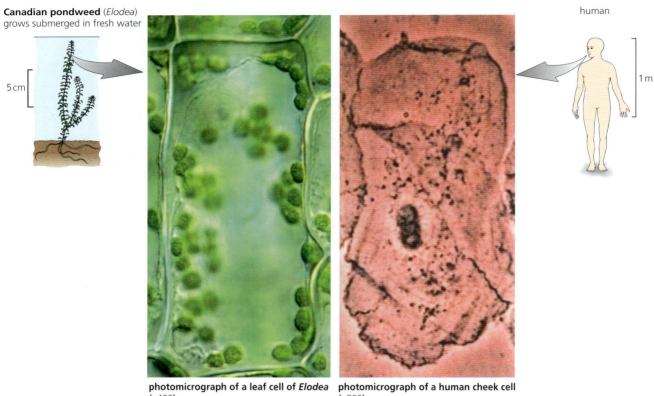

photomicrograph of a leaf cell of *Elodea* (×400)

photomicrograph of a human cheek cell (×800)

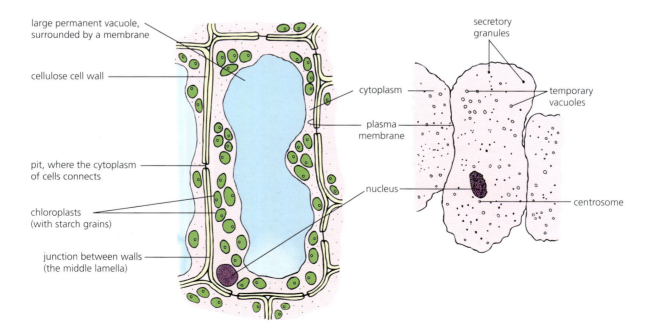

large permanent vacuole, surrounded by a membrane

cellulose cell wall

pit, where the cytoplasm of cells connects

chloroplasts (with starch grains)

junction between walls (the middle lamella)

secretory granules

cytoplasm

plasma membrane

temporary vacuoles

nucleus

centrosome

Animal and plant cells have at least three structures in common. These are their **cytoplasm** with its **nucleus**, surrounded by a **plasma membrane**. In addition, there are many tiny structures in the cytoplasm, called **organelles**, most of them common to both animal and plant cells. An organelle is a discrete structure within a cell, having a specific function. Organelles are all too small to be seen at this magnification. We will learn about the structure of organelles using the electron microscope (page 17).

There are some important basic differences between plant and animal cells (Table 1.2). For example, there is a tough, slightly elastic **cell wall**, made largely of cellulose, present around plant cells (page 4). Cell walls are absent from animal cells.

A **vacuole** is a fluid-filled space within the cytoplasm, surrounded by a single membrane. Plant cells frequently have a large, permanent vacuole present. By contrast, animal cells may have small vacuoles, but these are mostly temporary.

Green plant cells also contain organelles called **chloroplasts** in their cytoplasm. These are not found in animal cells. The chloroplasts are the sites where green plant cells manufacture food molecules by a process known as photosynthesis.

The **centrosome**, an organelle that lies close to the nucleus in animal cells (Figure 1.22), is not present in plants. This tiny organelle is involved in nuclear division in animal cells.

Finally, the **storage carbohydrate** (energy store) differs, too. Animal cells may store glycogen (page 79); plant cells normally store starch.

The profoundly different ways that unicellular organisms may differ are illustrated in Figure 1.5.

■ **Table 1.2**
Differences between plant and animal cells

Plant cells	Feature	Animal cells
cellulose cell wall present	**cell wall**	no cellulose cell walls
many cells contain chloroplasts; site of photosynthesis	**chloroplasts**	no chloroplasts; animal cells cannot photosynthesize
large, fluid-filled vacuole typically present	**permanent vacuole**	no large permanent vacuoles
no centrosome	**centrosome**	a centrosome present outside the nucleus
starch	**carbohydrate storage product**	glycogen

TOK Link

Living and non-living

You are familiar with the characteristics of living things (question 1). How could these be used to **explain** to non-biologists why a copper sulfate crystal growing in a solution of copper sulfate (or stalactites and stalagmites growing in a cave) are not living, yet corals are?

Flooded limestone cave where stalactites have formed in the roof and stalagmites on the floor beneath

Corals are formed by sedentary animals called polyps that secrete a calcareous shell around themselves

■ **Figure 1.4**

Paramecium – a large protozoan (about 600 μm), common in freshwater ponds.

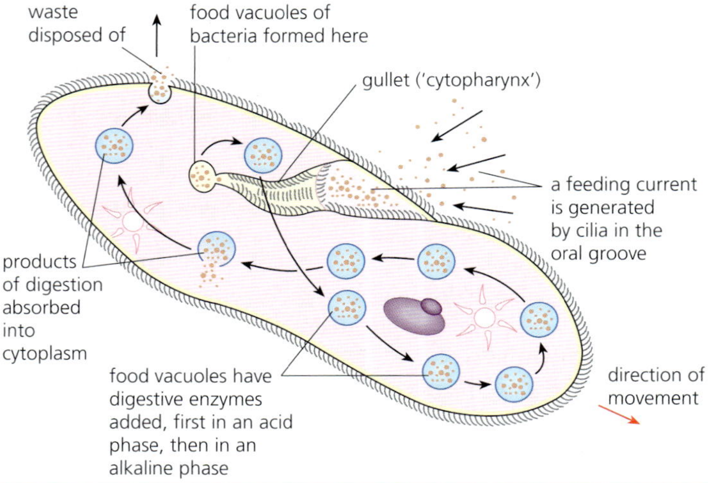

Chlorella – a small alga (about 20 μm), abundant in freshwater ponds where its presence colours the water green.

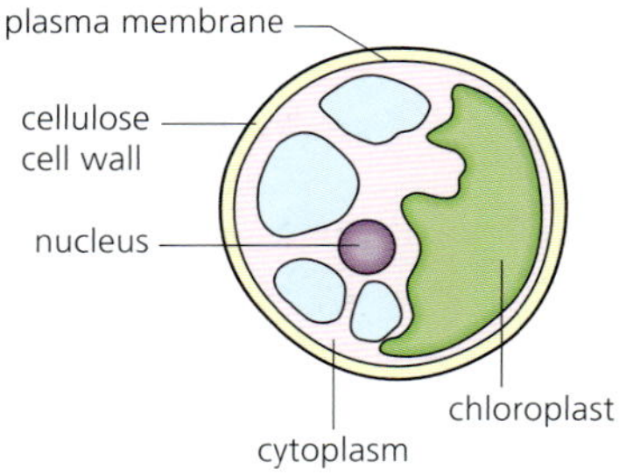

waste disposed of

food vacuoles of bacteria formed here

gullet ('cytopharynx')

a feeding current is generated by cilia in the oral groove

products of digestion absorbed into cytoplasm

food vacuoles have digestive enzymes added, first in an acid phase, then in an alkaline phase

direction of movement

plasma membrane

cellulose cell wall

nucleus

chloroplast

cytoplasm

Paramecium		*Chlorella*
A 'particle feeder', it takes in small floating unicellular organisms into food vacuoles in the cytoplasm where the contents are digested and the products absorbed.	**nutrition**	Manufactures sugars by photosynthesis in the light, using carbon dioxide and water (in a way that is almost identical to photosynthesis in flowering plants).
Respires aerobically, transferring energy to maintain cell functions.	**respiration**	Respires aerobically, transferring energy to maintain cell functions.
Obtains the biochemicals it requires for metabolism by digestion of food particles. Energy transferred by respiration makes this possible.	**metabolism**	Manufactures all biochemicals it requires for metabolism using sugars (from photosynthesis) and ions (such as nitrates) from the surrounding water. Energy transferred by respiration makes this possible.
Loss of waste products (mainly CO_2 and NH_3) over the entire cell surface.	**excretion**	Loss of waste products (mainly CO_2) over the entire cell surface.
Commonly, reproduction occurs by nuclear division followed by a transverse constriction of the cytoplasm.	**reproduction**	Periodically the cell contents divide into four autospores that each forms a cell wall around themselves. Eventually these are released by breakdown of the mother-cell wall.
Swims rapidly through the water, rotating as it goes.	**movement/locomotion**	A stationary cell.
Food vacuoles within can be seen being carried around the cytoplasm.		Cytoplasm within stream around within the plasma membrane.
Typically detects favourable food particles in the water and moves towards them.	**sensitivity**	Typically responds to the absence of light by nuclear followed by cell division.
Small cells grow to full size prior to cell division (dividing into two cells).	**growth/development**	Small cells grow to full size prior to cell division into autospores.

■ **Figure 1.5** Investigating the functions of life in unicellular organisms

■ Examining cells, and recording structure and size

We use microscopes to magnify the cells of biological specimens in order to view them. Figure 1.6 shows two types of light microscope.

In the simple microscope (**hand lens**), a single biconvex lens is supported in a frame so that the instrument can be held very close to the eye. Today, a hand lens is mostly used to observe external structure, although some of the earliest detailed observations of living cells were made with single-lens instruments.

In the **compound microscope**, light rays are focused by the **condenser** on to a specimen on a microscope slide on the stage of the microscope. Light transmitted through the specimen is then focused by two sets of lenses (hence the name 'compound microscope'). The **objective lens** forms an image (in the microscope tube) which is then further magnified by the **eyepiece lens**, producing a greatly enlarged image.

Biological material to be examined by compound microscopy must be sufficiently transparent for light rays to pass through. When bulky tissues and parts of organs are to be examined, thin sections are cut. Thin sections are largely colourless.

■ **Table 1.3**
The skills of light microscopy

You need to master and be able to demonstrate these aspects of good practice
Knowledge of the parts of your microscope, and care of the instrument – its light source, lenses and focusing mechanisms.
Use in low-power magnification first, using prepared slides and temporary mounts.
Switching to high-power magnification, maintaining focus and examining different parts of the image.
Types of microscope slides and the preparation of temporary mounts, both stained and unstained.
Getting started: With a slide, a drop of water and a cover slip, you can trap tiny air bubbles under the cover slip. Now try examining one of these air bubbles under low-power magnification and then its meniscus under high power.

■ **Figure 1.6**
Light microscopy

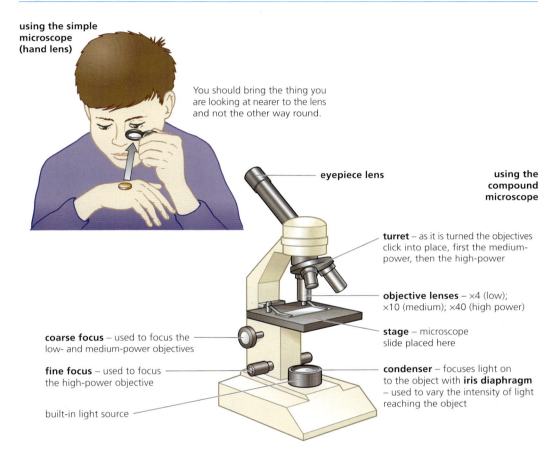

using the simple microscope (hand lens)

You should bring the thing you are looking at nearer to the lens and not the other way round.

eyepiece lens

using the compound microscope

turret – as it is turned the objectives click into place, first the medium-power, then the high-power

objective lenses – ×4 (low); ×10 (medium); ×40 (high power)

stage – microscope slide placed here

coarse focus – used to focus the low- and medium-power objectives

fine focus – used to focus the high-power objective

condenser – focuses light on to the object with **iris diaphragm** – used to vary the intensity of light reaching the object

built-in light source

■ Recording observations

Images of cells and tissues may be further magnified, displayed, projected and saved for printing by the technique of **digital microscopy** (Figure 1.7). A digital microscope is used or, alternatively, an appropriate video camera is connected by microscope coupler or eyepiece adaptor that replaces the standard microscope eyepiece. Images are displayed via video recorder, TV monitor or computer, and may be printed out by the latter.

Alternatively, a record of what you see via the compound microscope may be recorded by drawings of various types (Figure 1.9). For a clear, simple drawing:

■ use a sharp HB pencil and a clean eraser

■ use unlined paper and a separate sheet for each specimen you record

■ draw clear, sharp outlines, avoiding shading or colouring (density of structures may be represented by degrees of stippling)

■ label each drawing with appropriate information, such as the species, conditions (living or stained; if stained, note which stain was used) and type of section (transverse section, TS, or longitudinal section, LS)

■ label your drawing fully, with labels well clear of the structures shown, remembering that label lines should not cross

■ annotate (add notes about function, role and development) if appropriate

■ include a statement of the magnification under which the specimen has been observed.

■ **Figure 1.7**
Digital microscopy
in action

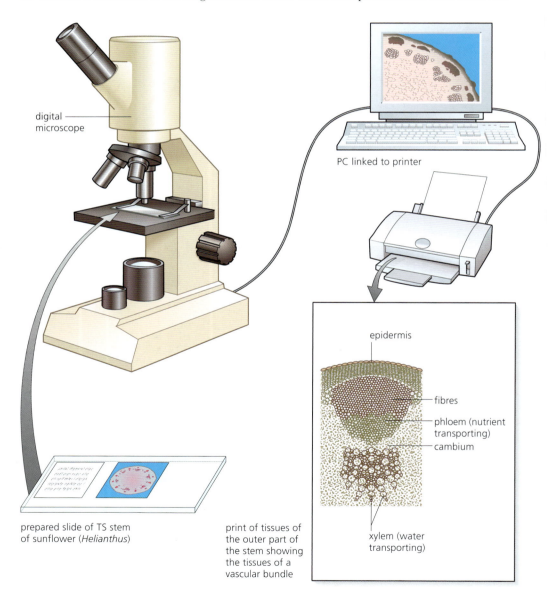

digital
microscope

PC linked to printer

prepared slide of TS stem
of sunflower (*Helianthus*)

print of tissues of
the outer part of
the stem showing
the tissues of a
vascular bundle

epidermis

fibres

phloem (nutrient
transporting)

cambium

xylem (water
transporting)

■ Measuring microscopic objects

The size of a cell can be measured under the microscope. A transparent scale, called a **graticule**, is mounted in the eyepiece at the focal plane (there is a ledge for it to rest on). In this position, when the object under observation is in focus, so too is the scale. The size (for example, length or diameter) of the object may then be recorded in arbitrary units. Next, the graticule scale is calibrated using a **stage micrometer** – in effect, a tiny, transparent ruler, which is placed on the microscope stage in place of the slide and then observed. With the eyepiece and stage micrometer scales superimposed, the true dimensions of the object can be estimated in micrometres. Figure 1.8 shows how this is done.

■ **Figure 1.8**
Measuring the
size of cells

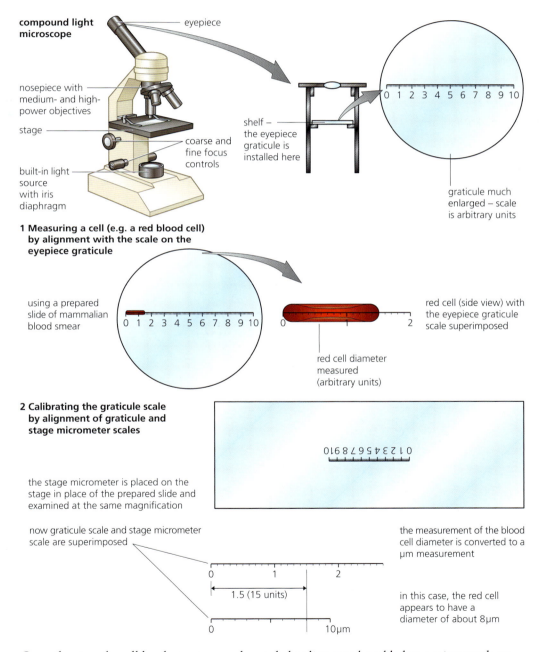

Once the size of a cell has been measured, a scale bar line may be added to a micrograph or drawing to record the actual size of the structure, as illustrated in Figure 1.10.

■ **Figure 1.9**
Recording cell
structure by drawing

view (phase contrast) of the layer of the cells (epithelium) lining the stomach wall

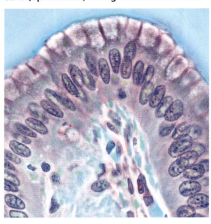

The lining of the stomach consists of columnar epithelium. All cells secrete mucus copiously.

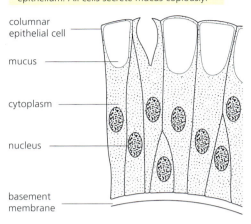

columnar
epithelial cell

mucus

cytoplasm

nucleus

basement
membrane

interpretive drawing

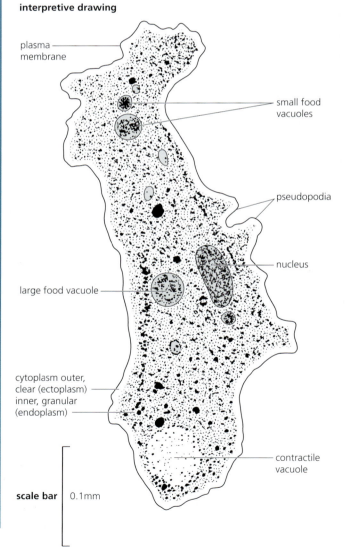

plasma
membrane

small food
vacuoles

pseudopodia

nucleus

large food vacuole

cytoplasm outer,
clear (ectoplasm)
inner, granular
(endoplasm)

contractile
vacuole

scale bar | 0.1mm

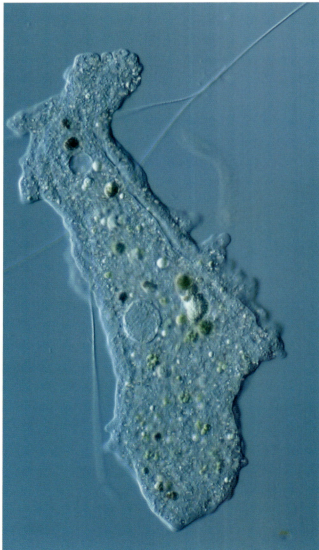

photomicrograph of *Amoeba proteus* (living specimen) – phase contrast microscopy

■ **Figure 1.10** Recording size by means of scale bars

3 Using the scale bar given in Figure 1.10, **calculate** the maximum observed length of the *Amoeba* cell.

Magnification and resolution of an image

Magnification is the number of times larger an image is than the specimen. The magnification obtained with a compound microscope depends on which of the lenses you use. For example, using a ×10 eyepiece and a ×10 objective lens (medium power), the image is magnified ×100 (10 × 10). When you switch to the ×40 objective (high power) with the same eyepiece lens, then the magnification becomes ×400 (10 × 40). These are the most likely orders of magnification you will use in your laboratory work.

Actually, there is **no limit to magnification**. For example, if a magnified image is photographed, then further enlargement can be made photographically. This is what may happen with photomicrographs shown in books and articles. Magnification is given by the formula:

$$\text{magnification} = \frac{\text{size of image}}{\text{size of specimen}}$$

So, for a particular plant cell of 150 μm diameter, photographed with a microscope and then enlarged photographically, the magnification in a print showing the cell at 15 cm diameter (150 000 μm) is:

$$\frac{150\ 000}{150} = 1000$$

If a further enlargement is made to show the same cell at 30 cm diameter (300 000 μm), then the magnification is

$$\frac{300\ 000}{150} = 2000$$

4 **Calculate** what magnification occurs with a ×6 eyepiece and a ×10 objective.

In this case, the image size has been doubled, but **the detail will be no greater**. You will not be able to see, for example, details of cell membrane structure, however much the image is enlarged. This is because the layers making up a cell's membrane are too thin to be seen as separate structures using the light microscope (Figure 1.11).

■ **Figure 1.11** Magnification without resolution

chloroplast enlarged (×6000)
a) from a transmission electron micrograph

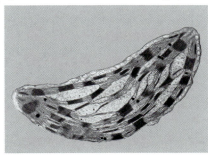

b) from a photomicrograph obtained by light microscopy

The **resolution** (resolving power) of a microscope is its ability to separate small objects which are very close together. If two separate objects cannot be resolved, they are seen as one object. Merely enlarging them does not separate them. Resolution is a property of lenses that is quite different from their magnification – and is more important.

Resolution is determined by the wavelength of light. Light is composed of relatively long wavelengths, whereas shorter wavelengths give better resolution. For the light microscope, the limit of resolution is about 0.2 μm. This means two objects less than 0.2 μm apart may be seen as one object.

Calculating **linear magnification** and **actual size** of images and objects is detailed in *Appendix 2: Investigations, data handling and statistics* which is available on the accompanying website.

■ Cell size and cell growth

The materials required for growth and maintenance of a cell enter through the outermost layer of the cytoplasm, a membrane called the **plasma membrane**. Similarly, waste products must leave the cell through the plasma membrane.

The rates at which materials can enter and leave a cell depend on the surface area of that cell, but the rates at which materials are used and waste products are produced depend on the amount of cytoplasm present within the cell. Similarly, heat transfer between the cytoplasm and environment of the cell is determined by surface area.

Surface area:volume ratios and cell size

As the cell grows and increases in size, an important difference develops between the surface area available for exchange and the volume of the cytoplasm in which the chemical reactions of life occur. The volume increases faster than the surface area; the **surface area:volume ratio** falls (SA:V, Figure 1.12). So, with increasing size of a cell, less and less of the cytoplasm has access to the cell surface for exchange of gases, supply of nutrients and loss of waste products.

■ **Figure 1.12** The effect of increasing size on the surface area:volume ratio

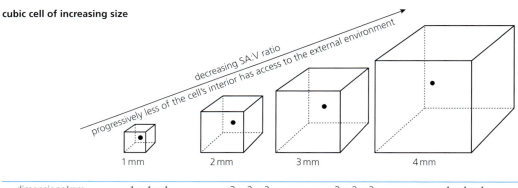

cubic cell of increasing size

decreasing SA:V ratio

progressively less of the cell's interior has access to the external environment

	1 mm	2 mm	3 mm	4 mm
dimensions/mm	1 × 1 × 1	2 × 2 × 2	3 × 3 × 3	4 × 4 × 4
surface area/mm^2	6	24	54	96
volume/mm^3	1	8	27	64
surface area:volume ratio	6:1 = 6/1 = 6	24:8 = 24/8 = 3	54:27 = 54/27 = 2	96:64 = 96/64 = 1.5

5 Consider imaginary cubic 'cells' with sides 1, 2, 4 and 6 mm.
  a **Calculate** the volume, surface area and ratio of surface area to volume.
  b **State** the effect on the SA:V ratio of a cell as it increases in size.
  c **Explain** the effect of increasing cell size on the efficiency of diffusion in the removal of waste products from cell cytoplasm.

Put another way, we can say that the smaller the cell is, the more quickly and easily materials can be exchanged between its cytoplasm and environment. One consequence of this is that cells cannot grow larger indefinitely. When a maximum size is reached, cell growth stops. The cell may then divide. The process of cell division is discussed later (page 51).

Metabolism and cell size

The extent of chemical reactions that make up the metabolism of a cell is not directly related to the surface area of the cell, but is related to the amount of cytoplasm, expressed as the cell mass. In summary, we can say that the rate of metabolism of a cell is a function of its mass, whereas the rate of exchange of materials and heat energy that metabolism generates is a function of the cell's surface area. Metabolism is the subject of later chapters (pages 63 and 345).

■ Multicellular organisms – specialization and division of labour

We have seen that unicellular organisms, though structurally simple, carry out all the functions and activities of life within a single cell. The cell feeds, respires, excretes, is sensitive to internal and external conditions (and may respond to them), may move, and eventually divides or reproduces.

By contrast, the majority of multicellular organisms – like the mammals and flowering plants – are made of cells, most of which are highly **specialized** to perform a particular role or function (Figures 1.13 and 1.14). Specialized cells are organized into tissues and organs. A **tissue** is a group of similar cells that are specialized to perform a particular function, such as heart muscle tissue of a mammal. An **organ** is a collection of different tissues which performs a specialized function, such as the heart of a mammal. So, the tissues and organs of multicellular organisms consist of specialized cells.

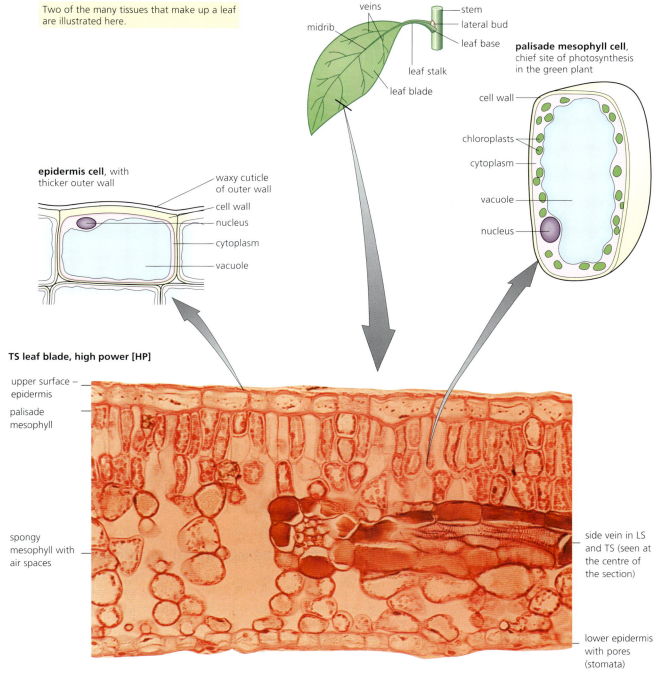

Two of the many tissues that make up a leaf are illustrated here.

veins

midrib

stem

lateral bud

leaf base

leaf stalk

leaf blade

palisade mesophyll cell, chief site of photosynthesis in the green plant

cell wall

chloroplasts

cytoplasm

vacuole

nucleus

epidermis cell, with thicker outer wall

waxy cuticle of outer wall

cell wall

nucleus

cytoplasm

vacuole

TS leaf blade, high power [HP]

upper surface – epidermis

palisade mesophyll

spongy mesophyll with air spaces

side vein in LS and TS (seen at the centre of the section)

lower epidermis with pores (stomata)

■ **Figure 1.13** Tissues of a leaf

■ **Figure 1.14**
Tissues of part of the mammalian gut

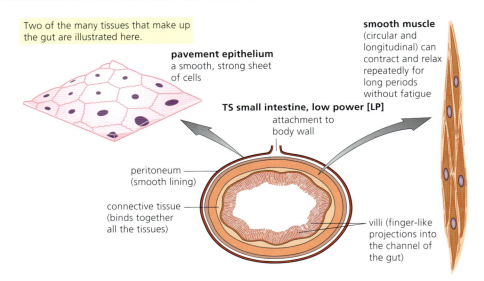

Two of the many tissues that make up the gut are illustrated here.

pavement epithelium
a smooth, strong sheet of cells

smooth muscle
(circular and longitudinal) can contract and relax repeatedly for long periods without fatigue

TS small intestine, low power [LP]
attachment to body wall

peritoneum (smooth lining)

connective tissue (binds together all the tissues)

villi (finger-like projections into the channel of the gut)

Control of cell specialization

We have seen that the nucleus of each cell is the structure that controls and directs the activities of the cell. The information required for this exists in the form of a nucleic acid, DNA. The nucleus of a cell contains the DNA in thread-like **chromosomes**, which are linear sequences of **genes** (page 131). Genes control the development of each cell within the mature organism. We can define a gene in different ways, including:

■ a specific region of a chromosome which is capable of determining the development of a specific characteristic of an organism

■ a specific length of the DNA double helix (hundreds or thousands of base pairs long) which codes for a protein.

So, when a cell is becoming specialized – we say the cell is differentiating – some of its genes are being activated and expressed. These genes determine how the cell develops. In the next chapter we explore both what happens during gene expression and the mechanism by which a cell's chemical reactions are controlled. For the moment, we can just note that the nucleus of each cell contains all the information required to make each type of cell present within the whole organism, only a selected part of which information is needed in any one cell and tissue. Which genes are activated and how a cell specializes are controlled by the immediate environment of the differentiating cell, and its position in the developing organism.

The cost of specialization

Specialized cells are efficient at carrying out their particular function, such as transport, support or protection. We say the resulting differences between cells are due to **division of labour**. By specialization, increased efficiency is achieved, but at a price. The specialized cells are now totally dependent on the activities of other cells. For example, in animals, nerve cells are adapted for the transport of nerve impulses, but depend on red blood cells for oxygen, and on heart muscle cells to pump the blood. This modification of cell structure to support differing functions is another reason why no 'typical' cell really exists.

Nature of Science | **Looking for trends and discrepancies**

■ Non-cellular organization – an exceptional condition

Although most organisms conform to the cell theory, there are exceptions. In addition to the familiar unicellular and multicellular organization of living things, there are a few multinucleate organs and organisms that are not divided into separate cells. This type of organization is called acellular. An example of an **acellular organism** is the pin mould *Rhizopus*, in which the 'plant' body consists of fine, thread-like structures called hyphae. An example of an **acellular organ** is the striped muscle fibres that make up the skeletal muscles of mammals (Figure 1.15). The internodal cells of the giant alga *Nitella* are also multinucleate.

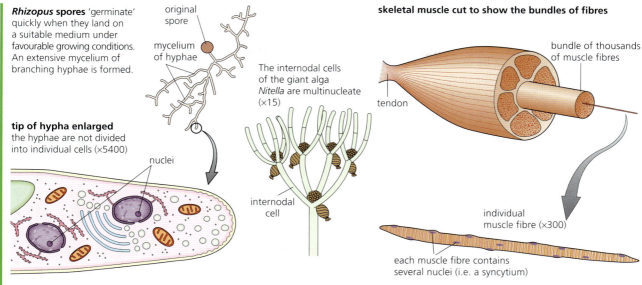

■ **Figure 1.15** Acellular organization in *Rhizopus,* in skeletal muscle fibres and in *Nitella*

■ The life history of the cell and the nature of stem cells

Multicellular organisms begin life as a single cell, which grows and divides, forming very many cells, and these eventually form the adult organism (Figure 1.16). So, cells arise by division of existing cells. The time between one cell division and the next is known as the **cell cycle**.

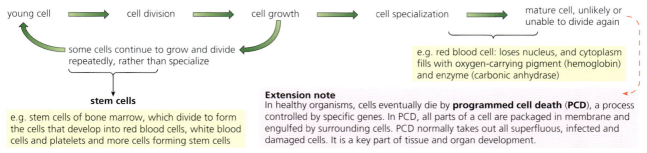

■ **Figure 1.16** The life history of a cell and the role of stem cells

In the development of a new organism the first step is one of continual cell division to produce a tiny ball of cells. All these cells are capable of further divisions, and they are known as **embryonic stem cells**.

A **stem cell** is a cell that has the capacity for repeated cell division while maintaining an undifferentiated state (self-renewal), and the subsequent capacity to differentiate into mature cell types (potency). Stem cells are the building blocks of life; they divide and form cells that develop into the range of mature cells of the organism. Stem cells are found in all multicellular organisms.

At the next stage of embryological development most cells lose the ability to divide as they develop into the tissues and organs that make up the organism, such as blood, nerves, liver, brain and many others. However, a very few cells within these tissues do retain many of the properties of embryonic stem cells, and these are called **adult stem cells**. Table 1.4 compares embryonic and adult stem cells.

■ **Table 1.4**
Differences between embryonic and adult stem cells

Embryonic stem (ES) cells	Adult stem cells
These are undifferentiated cells capable of continual cell division and of developing into all the cell types of an adult organism.	Undifferentiated cells capable of cell divisions, these give rise to a limited range of cells within a tissue, for example blood stem cells give rise to red and white blood cells and platelets only.
These make up the bulk of the embryo as it commences development.	Occurring in the growing and adult body, within most organs, they replace dead or damaged cells, such as in bone marrow, brain and liver.

Are there roles for isolated stem cells in medical therapies?

If stem cells can be isolated in large numbers and maintained in viable cell cultures, they have uses in medical therapies to replace or repair damaged organs. To do this, isolated stem cells must be manipulated under reproducible conditions so that they:

■ continue to divide in a sterile cell-culture environment (relatively large volumes of tissue are required)

■ differentiate into specific, desired blood cell types, like heart muscle

■ survive in a patient's body after they have been introduced

■ integrate into a particular tissue type in the patient's body

■ function correctly in the body for the remainder of the patient's life

■ do not trigger any harmful reactions within the tissues of the patient's body.

Medical conditions have been identified in which stem cell technologies may have the potential to bring relief or cure, a few examples of which are listed in Table 1.5.

■ **Table 1.5**
Examples of diseases that may be treated by stem cell technology

In **Stargardt's disease** there is a breakdown of light-sensitive cells in the retina in the area where fine focusing occurs. Peripheral vision is not initially affected, but blindness is a typical outcome in most cases. It is an inherited condition, due to a mutation of a gene associated with the processing of vitamin A in the eye. Currently, embryonic stem cells are being used in human clinical trials to regenerate damaged light-sensitive cells. Stem cells have been tested in animal models, resulting in 100% improvement in some cases.

Parkinson's disease arises from the death of neurons (nerve cells) in the part of the mid-brain that controls subconscious muscle activities by means of a neurotransmitter substance called dopamine. Movement disorders result, with tremors in the hands, limb rigidity, slowness of movements and impaired balance.

Cardiac muscle damage (death of muscle fibres) can be due to **myocardial infarction**, or heart attack, and is caused by major interruption to the blood supply to areas of cardiac muscle.

Type 1 diabetes arises when the β-cells of the pancreas are destroyed by the body's immune system and a severe lack of insulin results. Insulin (a hormone) normally maintains the blood's glucose concentration at about $90\,mg/100\,cm^3$. In diabetics, the level of blood glucose is not controlled and generally becomes permanently raised. Glucose is regularly excreted in the urine.

Nature of Science

Ethical implications of research

■ Where do therapeutic stem cells come from?

Stem cells may be obtained in several different ways.

1 Embryonic stem (ES) cells may come from the '**spare' embryos** produced by the infertility clinics that treat infertile couples, *provided this is allowed in law and agreed by the parents.* Today, this remains controversial – the chief objection is that the embryo's 'life' is destroyed in the process of gathering stem cells.

2 Blood extracted from the umbilical cord at the time of birth (**cord blood**) contains cells indistinguishable from the ES cells obtained as described above. Samples of cord blood (typically $40–100\,cm^3$) are collected, the stem cells are harvested, and then multiplied by sterile cell-culture technique to yield sufficient ES cells for practical purposes. Since 100 million babies are born each year, this source should surely grow to be significant.

3 Sources of **adult stem cells** are also sought. They have been identified in many organs and tissues including the brain, bone marrow, skin and liver, although present there in tiny quantities and in a non-dividing state. These stem cells are naturally activated by damage or disease in the organ where they occur. The stem cells that generate blood cells are obtained from bone marrow and are already used in treatments.

■ Ethical implications in stem cell research

Stem cell research generates ethical issues. **Ethics** are the moral principles that we feel ought to influence the conduct of a society. The field of ethics is concerned with how we decide what is right and what is wrong. Today, developments in science and technology influence many aspects of people's lives and often raise ethical issues. Stem cell research is just one case in point.

 You can learn more about ethics and how ethical decisions are made in *Appendix 3: Defining ethics and making ethical decisions* on the accompanying website.

6 **Identify** the points you feel are important in **support** of and in **opposition** to the harvest and use of ES cells in medical therapies.

■ Keeping in touch with developments

ES cell techniques are controversial and experimental, and new therapeutic developments and challenges arise all the time as research progresses in many countries. You can keep in touch with developments in this (and other) aspects of modern biology by reference to journals such as *Biological Sciences Review* (www.bsr.manchester.ac.uk) and *Scientific American* (www.sciam.com). Other sources, including the *BioNews* website (www.bionews.org.uk), may be accessed using an internet search engine.

1.2 Ultrastructure of cells *– eukaryotes have a much more complex cell structure than prokaryotes*

■ Electron microscopy – the discovery of cell ultrastructure

Microscopes were invented simultaneously in different parts of the world at a time when information travelled slowly. Modern-day advances in microscopy and communications have allowed for improvements in the ability to investigate and collaborate, enriching scientific endeavour.

The electron microscope uses electrons to make a magnified image in much the same way as the optical microscope uses light. However, because an electron beam has a much shorter wavelength, its resolving power is much greater. When the electron microscope is used with biological materials, the limit of resolution is about 5 nm. (The size of nanometres is given in Table 1.1, page 2.)

Only with the electron microscope can the detailed structures of the cell organelles be observed. This is why the electron microscope is used to resolve the fine detail of the contents of cells, the organelles and cell membranes, collectively known as **cell ultrastructure**. It is difficult to exaggerate the importance of electron microscopy in providing our detailed knowledge of cells.

In the electron microscope, the electron beam is generated by an **electron gun**, and focusing is by **electromagnets**, rather than by glass lenses. We cannot see electrons, so the electron beam is focused onto **a fluorescent screen** for viewing, or onto a **photographic plate** for permanent recording (Figure 1.17).

In **transmission electron microscopy**, the electron beam is passed through an extremely thin section of material. Membranes and other structures are stained with heavy metal ions, making them electron-opaque so they stand out as dark areas in the image.

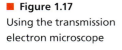
■ **Figure 1.17**
Using the transmission electron microscope

Electron microscopes have a greater resolving power than light microscopes. Their application to biology has established the presence and structure of all the cell organelles.

In **scanning electron microscopy**, a narrow electron beam is scanned back and forth across the surface of the specimen. Electrons that are reflected or emitted from this surface are detected and converted into a three-dimensional image (Figure 1.18, Figure 6.40, page 290 and Figure 9.6, page 378).

■ **Figure 1.18**
A scanning electron micrograph

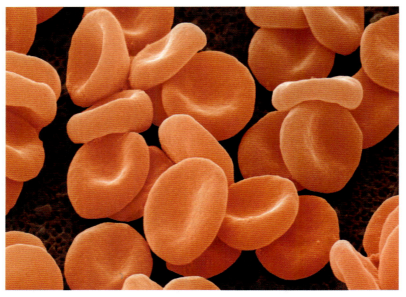

red blood cells
(5.7 μm in diameter)

Freeze etching

In an **alternative method of preparation**, biological material is *instantly* frozen solid in liquid nitrogen. At atmospheric pressure this liquid is at –196°C. At this temperature living materials do not change shape as the water present in them solidifies instantly.

This solidified tissue is then broken up in a vacuum, and the exposed surfaces are allowed to lose some of their ice; the surface is described as 'etched'.

Finally, a carbon replica (a form of 'mask') of this exposed surface is made and coated with heavy metal to strengthen it. The mask of the surface is then examined in the electron microscope. The resulting electron micrograph is described as being produced by **freeze-etching**.

A comparison of a cell nucleus observed by both transmission electron microscopy and by freeze etching is shown in Figure 1.19. *Look at these images carefully.* The picture we get of nucleus structure is consistent; we can be confident that our views of cell structure obtained by electron microscopy are realistic.

observed as thin section · the nucleus of a liver cell · replica of freeze-etched surface

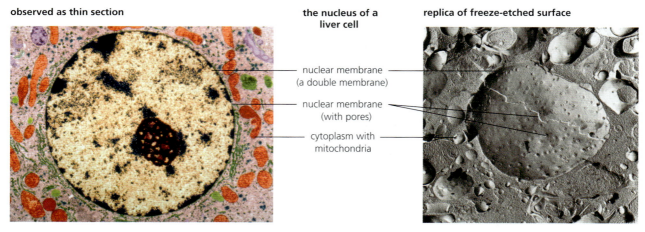

nuclear membrane
(a double membrane)

nuclear membrane
(with pores)

cytoplasm with
mitochondria

■ **Figure 1.19** Electron micrographs from thin-sectioned and freeze-etched material

7 Distinguish between resolution and magnification.

Nature of Science   **Developments in scientific research follow improvements in apparatus**

■ The impact of electron microscopy on cell biology

The presence and structure of organelles

The nucleus is the largest substructure (organelle) of a cell and may be observed with the light microscope. However, most organelles cannot be viewed by light microscopy and none is large enough for internal details to be seen. It is by means of the electron microscope that we have learnt about the fine details of cell structure. We now think of the eukaryotic cell as a bag of organelles suspended in a fluid matrix, contained within a special membrane, the plasma membrane.

■ Prokaryotic and eukaryotic organization

Living things have traditionally been divided into two major groupings: animals and plants. However, the range of biological organization is more diverse than this. The use of the electron microscope in biology has led to the discovery of two types of cellular organization, based on the presence or absence of a nucleus.

Cells of plants, animals, fungi and protoctista have cells with a large, obvious nucleus. The surrounding cytoplasm contains many different membranous organelles. These types of cells are called **eukaryotic cells** (meaning a 'good nucleus').

On the other hand, bacteria contain no true nucleus and their cytoplasm does not have the organelles of eukaryotes. These are called **prokaryotic cells** (meaning 'before the nucleus').

This distinction between prokaryotic and eukaryotic cells is a fundamental division and is more significant than the differences between plants and animals. We will shortly return to examine the detailed structure of the prokaryotic cell, choosing a bacterium as our example (page 28). First, we need to look into the main organelles in eukaryotic cells.

The ultrastructure of the eukaryotic cell

In the living cell there is a fluid around the organelles. This is a watery (aqueous) solution of chemicals, called the **cytosol**. The chemicals in the cytosol are substances formed and used in the chemical reactions of life. All the reactions of life are known collectively as **metabolism**, and the chemicals are known as metabolites.

Cytosol and organelles are contained within the **plasma membrane**. This membrane is clearly a barrier of sorts. It must be crossed by all the metabolites that move between the cytosol and the environment of the cell. We will return to the structure of cell membrane and how molecules enter and leave cells. We next consider the structure and function of the **organelles**. The ultrastructure of a mammalian liver cell is shown in Figure 1.20. The interpretive drawing is an illustration of the application of the rules relating to observing microscopic structure (page 8).

Our knowledge of organelles has been built up by examining electron micrographs of many different cells. The outcome, a detailed picture of the ultrastructure of animal and plant cells, is represented diagrammatically in a generalized cell in Figure 1.21.

electron micrograph of liver cells (×17 500)

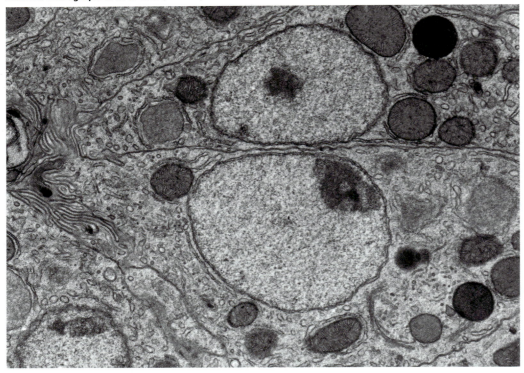

interpretive drawing

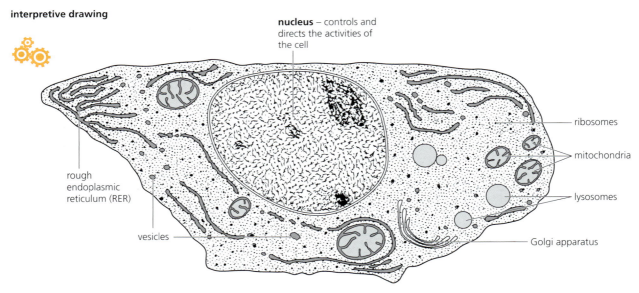

nucleus – controls and directs the activities of the cell

ribosomes

mitochondria

lysosomes

Golgi apparatus

rough endoplasmic reticulum (RER)

vesicles

■ **Figure 1.20** Electron micrograph of a mammalian liver cell with interpretive drawing (×17 500)

Introducing the organelles

1 Nucleus

The appearance of the nucleus in electron micrographs is shown in Figure 1.19 (page 18). The nucleus is the largest organelle in the eukaryotic cell, typically 10–20 µm in diameter. It is surrounded by a double-layered membrane, the **nuclear envelope**. This contains many **pores**. These pores are tiny, about 100 nm in diameter. However, the pores are so numerous that they make up about one third of the nuclear membrane's surface area. This suggests that communications between nucleus and cytoplasm are important.

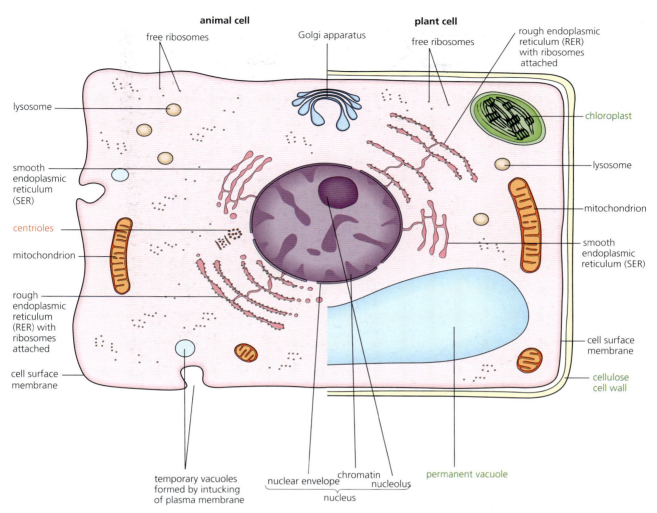

animal cell

free ribosomes

Golgi apparatus

plant cell

free ribosomes

rough endoplasmic reticulum (RER) with ribosomes attached

lysosome

smooth endoplasmic reticulum (SER)

centrioles

mitochondrion

rough endoplasmic reticulum (RER) with ribosomes attached

cell surface membrane

chloroplast

lysosome

mitochondrion

smooth endoplasmic reticulum (SER)

cell surface membrane

cellulose cell wall

temporary vacuoles formed by intucking of plasma membrane

nuclear envelope

chromatin

nucleolus

permanent vacuole

nucleus

■ **Figure 1.21** The ultrastructure of the eukaryotic animal and plant cell

The nucleus contains the **chromosomes**. These thread-like structures are visible at the time the nucleus divides (page 142). At other times, the chromosomes appear as a diffuse network called **chromatin**.

One or more **nucleoli** are present in the nucleus, too. A nucleolus is a tiny, rounded, darkly-staining body. It is the site where ribosomes (see below) are synthesized. Chromatin, chromosomes and the nucleolus are visible only if stained with certain dyes. The everyday role of the nucleus in cell management, and its behaviour when the cell divides, are the subject of Section 1.6 (page 51).

Here, we can note that most cells contain one nucleus but there are interesting exceptions. For example, both the red blood cells of mammals (page 256) and the sieve tube element of the phloem of flowering plants (page 388) are without a nucleus. Both lose their nucleus as they mature.

2 Centrioles

A centriole is a tiny organelle consisting of nine paired **microtubules** (Figure 1.22), arranged in a short, hollow cylinder. In animal cells, two centrioles occur at right-angles, just outside the nucleus, forming the **centrosome**. Before an animal cell divides, the centrioles replicate, and their role is to grow the spindle fibres – the spindle is the structure responsible for movement of chromosomes during nuclear division.

Microtubules themselves are straight, unbranched hollow cylinders, only 25 nm wide. The cells of all eukaryotes, whether plants or animals, have a well-organized system of these microtubules which shape and support the cytoplasm. Microtubules are involved in movement of other cell components within the cytoplasm too, acting to guide and direct them. The network of microtubules is made of a globular protein called tubulin. This is built up and broken down in the cell as the microtubule framework is required in different places for different tasks.

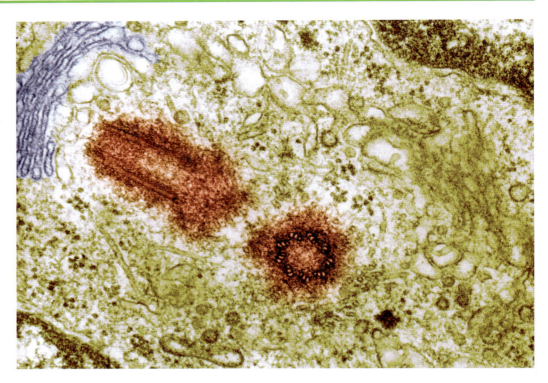

3 Mitochondria

Mitochondria appear mostly as rod-shaped or cylindrical organelles in electron micrographs (Figure 1.23). Occasionally their shape is more variable. They are relatively large organelles, typically 0.5–1.5 µm wide, and 3.0–10.0 µm long. Mitochondria are found in all cells and are usually present in very large numbers. Metabolically very active cells contain thousands of them in their cytoplasm – for example, muscle fibres and hormone-secreting cells.

The mitochondrion also has a double membrane. The outer membrane is a smooth boundary, the inner membrane is folded to form **cristae**. The interior of the mitochondrion contains an aqueous solution of metabolites and enzymes. This is called the **matrix**. The mitochondrion is the site of the aerobic stages of respiration (page 118).

■ **Figure 1.23**
The mitochondrion

a mitochondrion, cut open
to show the inner membrane and cristae

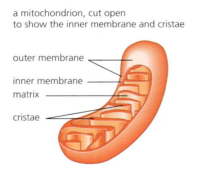

outer membrane
inner membrane
matrix
cristae

In the mitochondrion many of the enxymes of respiration are housed, and the 'energy currency' molecules adenosine triphosphate (ATP) are formed.

TEM of a thin section of a mitochondrion

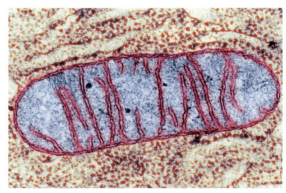

4 Chloroplasts

Chloroplasts are large organelles, typically biconvex in shape, about 4–10 µm long and 2–3 µm wide. They occur in green plants, where most occur in the mesophyll cells of leaves. A mesophyll cell may be packed with 50 or more chloroplasts. Photosynthesis is the process that occurs in chloroplasts. Photosynthesis is covered in Chapter 2.

Look at the chloroplasts in the electron micrograph in Figure 1.24. Each chloroplast has a double membrane. The outer layer of the membrane is a continuous boundary, but the inner layer is tucked to form a system of branching membranes called lamellae or **thylakoids**. In the interior of the chloroplast, the thylakoids are arranged in flattened circular piles called **grana** (singular **granum**). These look a little like a stack of coins. It is here that the chlorophylls and other pigments are located. There are a large number of grana present. Between them, the branching membranes are very loosely arranged in an aqueous matrix, usually containing small starch grains. This part of the chloroplast is called the **stroma**.

Chloroplasts are one of a larger group of organelles called **plastids**. Plastids are found in many plant cells but never in animals. The other members of the plastid family are **leucoplasts** (colourless plastids) in which starch is stored, and **chromoplasts** (coloured plastids), containing non-photosynthetic pigments such as carotene, and occurring in flower petals and the root tissue of carrots.

■ **Figure 1.24**
The chloroplast

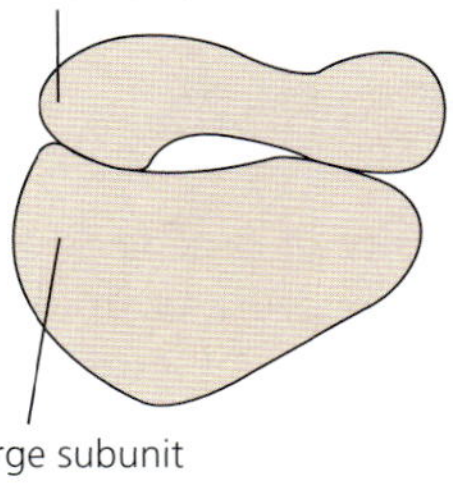

5 Ribosomes

small subunit

large subunit

■ **Figure 1.25**
The ribosome

Ribosomes are tiny structures, approximately 25 nm in diameter. They are built of two subunits and do not have membranes as part of their structures. Chemically, they consist of protein and a nucleic acid known as RNA. Ribosomes are found free in the cytoplasm and bound to endoplasmic reticulum (rough endoplasmic reticulum – RER, see below). They also occur within the mitochondria and in the chloroplasts. The sizes of tiny objects like the ribosomes are recorded in Svedberg units (S). This is a measure of their rate of sedimentation in centrifugation, rather than of their actual size. Ribosomes of mitochondria and chloroplasts are slightly smaller (70S) than those in the rest of the cell (80S). We will return to this issue later (page 51).

Ribosomes are the sites where proteins are made in cells. The structure of a ribosome is shown in Figure 1.25. Many different types of cell contain vast numbers of ribosomes. Some of the cell proteins produced in the ribosomes have structural roles. Collagen is an example (page 92). Most cell proteins are enzymes. These are biological catalysts. They cause the reactions of metabolism to occur quickly under the conditions found within the cytoplasm.

8 **Explain** why the nucleus in a human cheek cell (Figure 1.3, page 4) may be viewed by light microscopy in an appropriately stained cell, but the ribosomes cannot.

6 Endoplasmic reticulum

The endoplasmic reticulum consists of a network of folded membranes formed into sheets, tubes or sacs that are extensively interconnected. Endoplasmic reticulum 'buds off' from the outer membrane of the nuclear envelope, to which it may remain attached. The cytoplasm of metabolically active cells is commonly packed with endoplasmic reticulum. In Figure 1.26 we can see there are two distinct types of endoplasmic reticulum.

■ **Rough endoplasmic reticulum (RER)** has ribosomes attached. At its margin, vesicles are formed from swellings. A vesicle is a small, spherical organelle bounded by a single membrane, which becomes pinched off as it separates. These tiny sacs are then used to store and transport substances around the cell. For example, RER is the site of synthesis of proteins that are 'packaged' in the vesicles and then typically discharged from the cell. Digestive enzymes are discharged in this way.

SER and RER in cytoplasm, showing origin from outer membrane of nucleus

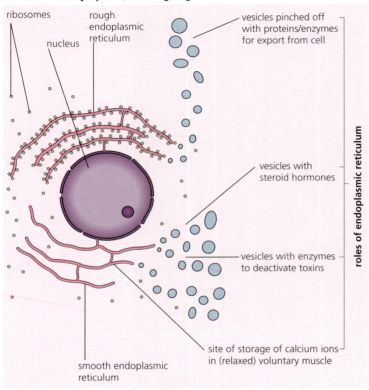

electron micrograph of RER

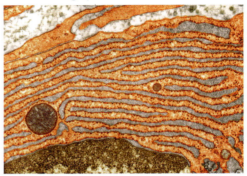

electron micrograph of SER

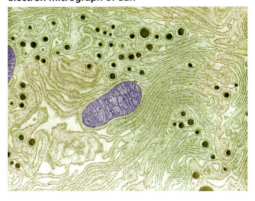

■ **Figure 1.26** Endoplasmic reticulum, rough (RER) and smooth (SER)

■ **Smooth endoplasmic reticulum (SER)** has no ribosomes. SER is the site of synthesis of substances needed by cells. For example, SER is important in the manufacture of lipids. In the cytoplasm of voluntary muscle fibres, a special form of SER is the site of storage of calcium ions which have an important role in the contraction of muscle fibres.

7 Golgi apparatus

The Golgi apparatus consists of a stack-like collection of flattened membranous sacs (Figure 1.27). One side of the stack of membranes is formed by the fusion of membranes of vesicles from SER. At the opposite side of the stack, vesicles are formed from swellings at the margins that, again, become pinched off.

The Golgi apparatus occurs in all cells, but it is especially prominent in metabolically active cells – for example, secretory cells. It is the site of synthesis of specific biochemicals, such as hormones and enzymes. These are then packaged into vesicles. In animal cells these vesicles may form lysosomes. Those in plant cells may contain polysaccharides for cell wall formation (page 78).

electron micrograph of Golgi apparatus, in section

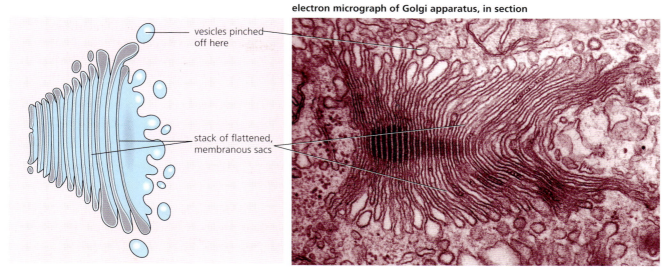

vesicles pinched off here

stack of flattened, membranous sacs

■ **Figure 1.27** The Golgi apparatus

8 Lysosomes

Lysosomes are tiny spherical vesicles bound by a single membrane (Figure 1.28). They contain a concentrated mixture of 'digestive' enzymes. These are correctly known as hydrolytic enzymes. They are produced in the Golgi apparatus or by the rough ER.

■ **Figure 1.28** Lysosomes

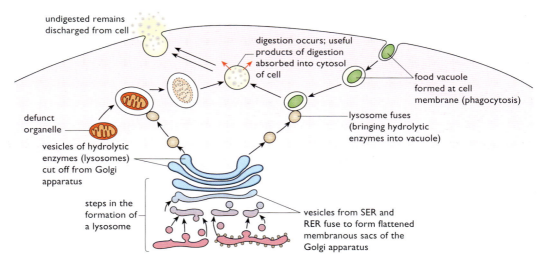

undigested remains discharged from cell

digestion occurs; useful products of digestion absorbed into cytosol of cell

food vacuole formed at cell membrane (phagocytosis)

defunct organelle

vesicles of hydrolytic enzymes (lysosomes) cut off from Golgi apparatus

lysosome fuses (bringing hydrolytic enzymes into vacuole)

steps in the formation of a lysosome

vesicles from SER and RER fuse to form flattened membranous sacs of the Golgi apparatus

Lysosomes are involved in the breakdown of the contents of 'food' vacuoles. For example, harmful bacteria that invade the body are taken up into tiny vacuoles (they are engulfed) by special white cells called macrophages. Macrophages are part of the body's defence system (Chapter 6).

Any foreign matter or food particles taken up into these vacuoles are then broken down. This occurs when lysosomes fuse with the vacuole. The products of digestion then escape into the liquid of the cytoplasm. Lysosomes will also destroy damaged organelles in this way.

When an organism dies, it is the hydrolytic enzymes in the lysosomes of the cells that escape into the cytoplasm and cause self-digestion, known as autolysis.

9 Plasma membrane – the cell surface membrane

The plasma membrane is an extremely thin structure – less than 10 nm thick. It consists of a lipid bilayer in which proteins are embedded. This membrane has a number of roles. Firstly, it retains the fluid cytosol. The cell surface membrane also forms the barrier across which all substances entering and leaving the cell must pass. In addition, it is where the cell is identified by surrounding cells.

The detailed structure and function of the cell surface membrane is the subject of Section 1.3 (page 30).

10 Cilia and flagella

Cilia and flagella are organelles that project from the surface of certain cells. Structurally, cilia and flagella are almost identical, and both can move.

Cilia occur in large numbers on certain cells, such as the ciliated lining (epithelium) of the air tubes serving the lungs (bronchi), where they cause the movement of mucus across the cell surface. It is the cilia of this 'bronchial tree' that cigarette smoke destroys over time. Flagella occur singly, typically on small, motile cells, such as sperm, or they may occur in pairs.

Cells may have extracellular components

We have noted that the contents of cells are contained within the plasma membrane. However, cells may secrete material outside the plasma membrane; for example, plant cells have an external wall, and many animal cells secrete glycoproteins.

The plant cell and its wall

The plant cell differs from an animal cell in that it is surrounded by a wall. This wall is completely external to the cell; it is not an organelle. Plant cell walls are primarily constructed of cellulose – a polysaccharide and an extremely strong material. Cellulose molecules are very long, and are arranged in bundles called microfibrils (Figure 2.18, page 79).

Cell walls make the boundaries of plant cells easy to see when plant tissues are examined by microscopy. The presence of this strong structure allows the plant cell to develop high internal pressure due to water uptake, without danger of the cell bursting. This is a major difference between the cell water relations of plants and animals.

Extracellular glycoproteins around animal cells

9 **Outline** how the electron microscope has increased our knowledge of cell structure.

Many animal cells are able to adhere to one other. This property enables cells to form compact tissues and organs. Other animal cells occur in simple sheets or layers, attached to a basement membrane below them. These cases of adhesion are brought about by glycoproteins that the cells have secreted. Glycoproteins are large molecules of protein to which large sugar molecules (called oligosaccharides) are attached.

■ Analysing transmission electron micrographs of cells

1 Comparing the organelles of cells

Examine the electron micrographs of specialized animal and plant cells in Figure 1.29, and then answer question 10.

10 **List** the organelles common to the animal and plant cells illustrated in Figure 1.29. **Annotate** your list by recording the principal role or function of these structures.

   List separately any organelles you observe to be present only in the plant cell.

11 **Draw** and **label** a representation of the electron micrograph of the palisade mesophyll cell in Figure 1.29, using the interpretive drawing of an animal cell in Figure 1.20 as a model, and following the guidelines on biological drawing on page 8.

electron micrograph of an exocrine gland cell of the mammalian pancreas

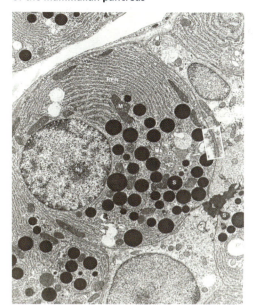

electron micrograph of a palisade mesophyll cell

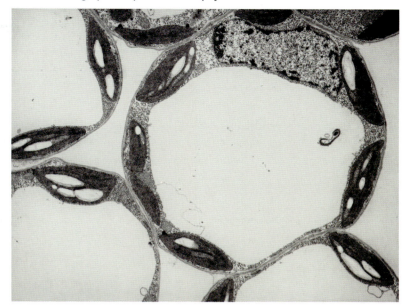

■ **Figure 1.29** Electron micrographs of named animal and plant cells

 2 Deducing the function of specialized cells

The organelles present in a specialized cell, and their relative numbers, may suggest a specialized role for that cell within the multicellular organism in which it occurs. With this in mind, examine the cell illustrated in Figure 1.30, and then answer question 12.

■ **Figure 1.30**
Electron micrograph of a specialized cell (×4500)

12 Identify the features of structure in the cell in Figure 1.30 – its shape, size and the organelles present. On the basis of these observations **deduce** the specialized role of the cell, giving your reasons.

The ultrastructure of prokaryotic cells

We have seen that the use of the electron microscope in biology led to the discovery of eukaryotic and prokaryotic cell structure (page 19). Bacteria and cyanobacteria are prokaryotes. The generalized structure of a bacterium is shown in Figure 1.31. The distinctive features of the prokaryotes are:

■ they are exceedingly small – about the size of individual organelles found in the cells of eukaryotes

■ they contain no true nucleus but have a single, circular chromosome in the cytoplasm, referred to as a nucleoid

■ their cytoplasm does not have the organelles of eukaryotes.

■ **Figure 1.31**
The structure of
a bacterium

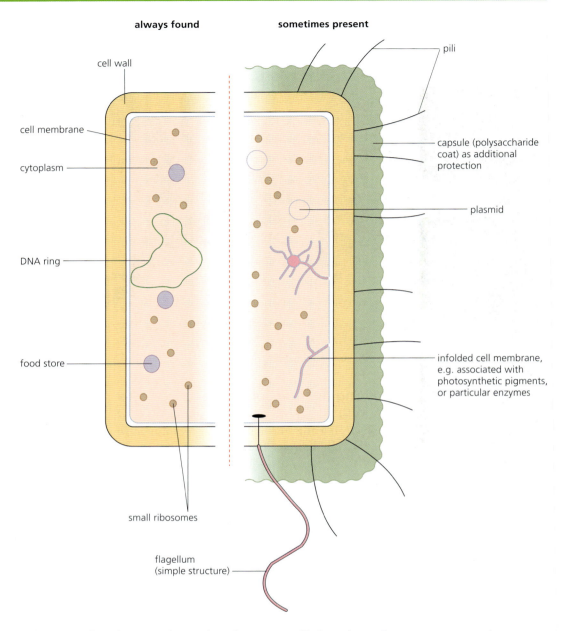

always found

sometimes present

cell wall

cell membrane

cytoplasm

DNA ring

food store

small ribosomes

flagellum
(simple structure)

pili

capsule (polysaccharide
coat) as additional
protection

plasmid

infolded cell membrane,
e.g. associated with
photosynthetic pigments,
or particular enzymes

Drawing the ultrastructure of prokaryotic cells based on electron micrographs

In Figure 1.32, the ultrastructure of *Eschericha coli* is shown. *E. coli* is a common bacterium of the human gut – it occurs in huge numbers in the lower intestine of humans and other endothermic (once known as 'warm-blooded') vertebrates, such as the mammals. It is a major component of the faeces of these animals.

This tiny organism was named by a bacteriologist, Professor T. Escherich, in 1885. Notice the scale bar in Figure 1.32. This bacterium is typically about 1–3 μm in length – about the size of a mitochondrion in a eukaryotic cell.

The functions of each of the structures present – cell wall, plasma membrane, cytoplasm, **pili**, **flagella**, **ribosomes** and **nucleoid**, are included as annotations to their labels.

You can practise the skill of drawing the ultrastructure of a eukaryotic cell, using the electron micrograph shown in Figure 1.32.

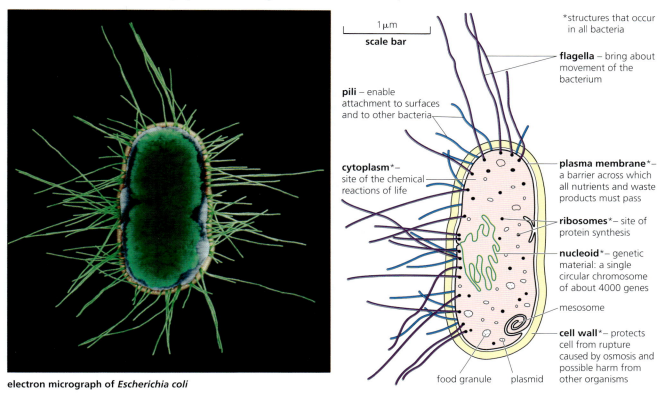

1 μm
scale bar

*structures that occur in all bacteria

flagella – bring about movement of the bacterium

pili – enable attachment to surfaces and to other bacteria

plasma membrane*– a barrier across which all nutrients and waste products must pass

cytoplasm*– site of the chemical reactions of life

ribosomes*– site of protein synthesis

nucleoid*– genetic material: a single circular chromosome of about 4000 genes

mesosome

cell wall*– protects cell from rupture caused by osmosis and possible harm from other organisms

food granule   plasmid

electron micrograph of *Escherichia coli*

■ **Figure 1.32** The structure of *Escherichia coli*, together with an interpretive drawing

13 Calculate the approximate magnification of the image of *E. coli* in Figure 1.32.

■ **Table 1.6** Prokaryote and eukaryote cells compared

Prokaryotic and eukaryotic cells compared

By contrasting Figures 1.32 and 1.20 we can see that there are fundamental differences between prokaryotes and eukaryotes, both in cell size and cell complexity. In Table 1.6, prokaryotic and eukaryotic cells are compared.

Prokaryotes	Eukaryotes
e.g. bacteria, cyanobacteria	e.g. mammals, green plants, fungi
cells are extremely small, typically about 1–5 μm in diameter	cells are larger, typically 50–150 μm
nucleus absent; circular DNA helix in the cytoplasm, DNA not supported by histone proteins	nucleus has distinct nuclear membrane (with pores), with chromosomes of linear DNA helix supported by histone protein
cell wall present (peptidoglycan – long molecules of chains of amino acids and sugars)	cell wall present in plants (largely of cellulose) and fungi (largely of the polysaccharide chitin)
few organelles; membranous structures absent	many organelles bounded by double membrane (e.g. chloroplast, mitochondria, nucleus) or single membrane (e.g. Golgi apparatus, lysosome, vacuole, endoplasmic reticulum)
proteins synthesized in small ribosomes (70S)	proteins synthesized in large ribosomes (80S)
some cells have simple flagella	some cells have cilia or flagella, 200 nm in diameter
some can fix atmospheric nitrogen gas for use in the production of amino acids for protein synthesis	none can metabolize atmospheric nitrogen gas but, instead, require nitrogen already combined in molecules in order to make proteins from amino acids (page 89)

■ Cell division and reproduction in bacteria – the cell cycle

Bacterial cells grow to full size and then divide in two by a process called **binary fission**. The complete cycle of growth, from new cell to the point of division, may take as little as 20 minutes, provided the necessary conditions are maintained. *E. coli* is one of many species that can reproduce at this rate, at least initially. Of course, this growth rate cannot be maintained for long, but it does help to explain why bacteria are so numerous. For example, it is estimated that a gram of garden soil contains about 1000 million living bacteria – and an average square centimetre of human skin has a mere 10 million individual bacteria on it.

During growth, the cell contents increase so that after division each daughter cell has sufficient cytoplasm to metabolize and grow. Prior to division, the single circular chromosome, present in the form of a circular strand of DNA helix, divides. The copying process, known as **replication**, starts at a particular sequence of bases. This is the gene that codes for the enzyme which triggers the replication process. After division of the chromosome, a wall is laid down, dividing the cell into two. Daughter cells each have a copy of the chromosome (Figure 1.33).

■ **Figure 1.33**
The steps of the cell cycle and binary fission

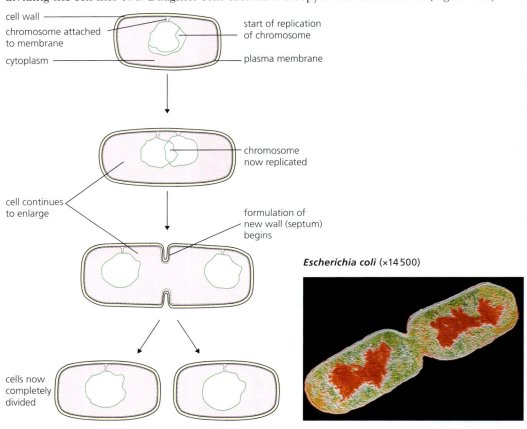

Escherichia coli (×14 500)

14 List the differences between a chromosome of a eukaryotic cell, and of a prokaryotic cell.

15 Distinguish between the following pairs of terms:
   a cell wall and plasma membrane
   b nucleus and nucleoid
   c flagella and pili
   d centriole and chloroplast.

1.3 Membrane structure – *the structure of biological membranes makes them fluid and dynamic*

We have seen that a plasma membrane is a structure common to eukaryotic and prokaryotic cells. The plasma membrane maintains the integrity of the cell (it holds the cell's contents together). Also, it is a barrier across which all substances entering and leaving the cell pass.

■ The structure of the plasma membrane

The plasma membrane is made almost entirely of protein and lipid, together with a small and variable amount of carbohydrate. Figure 1.34 shows how these components are assembled into the plasma membrane. This view of the molecular structure of the plasma membrane is known as the **fluid mosaic model**. The plasma membrane is described as *fluid* because the components (lipids and proteins) are on the move, and *mosaic* because the proteins are scattered about in this pattern.

■ **Figure 1.34**
The fluid mosaic model of the plasma membrane

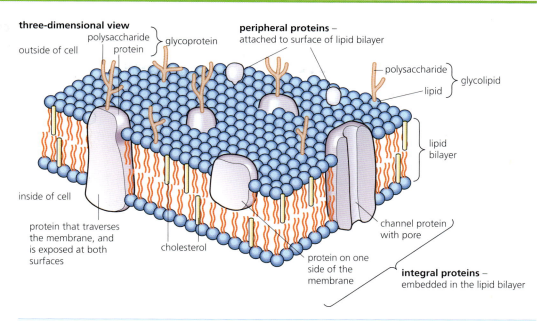

16 **Draw** a diagrammatic cross-section of the fluid mosaic membrane, using Figure 1.34 to help you. **Label** it correctly, using these terms: phospholipid bilayer, cholesterol, glycoprotein, integral protein, peripheral protein.

The phospholipid component

The lipid of membranes is **phospholipid**. The chemical structure of phospholipid is introduced in Figure 1.35 and is covered in detail in Chapter 2, Figure 2.29, page 89.

Take a look at phospholipid structure, now.

You see that phospholipid has a 'head' composed of a glycerol group, to which is attached one ionized phosphate group. This latter part of the molecule has **hydrophilic properties** (meaning 'water-loving'). For example, **hydrogen bonds** readily form between the phosphate head and water molecules (page 69).

The remainder of the phospholipid consists of two long, fatty acid residues consisting of hydrocarbon chains. These 'tails' have **hydrophobic properties** ('water-hating'). So phospholipid is unusual in being partly hydrophilic and partly hydrophobic. This is referred to as '**amphipathic**'.

■ **Figure 1.35**
The phospholipid molecule and its response when added to water (the formation of monolayers and bilayers)

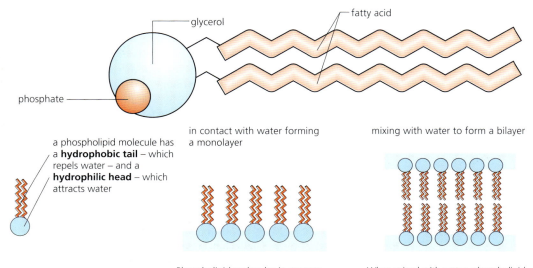

a phospholipid molecule has a **hydrophobic tail** – which repels water – and a **hydrophilic head** – which attracts water

in contact with water forming a monolayer

mixing with water to form a bilayer

Phospholipid molecules in contact with water form a **monolayer**, with heads dissolved in the water and the tails sticking outwards.

When mixed with water, phospholipid molecules arrange themselves into a **bilayer**, in which the hydrophobic tails are attracted to each other.

What are the consequences of this dual nature of phospholipid?

A small quantity of phospholipid in contact with a solid surface (a clean glass plate is suitable) remains as a discrete bubble; the phospholipid molecules do not spread out. However, when a similar tiny drop of phospholipid is added to water it instantly spreads over the entire surface (as a monolayer of phospholipid molecules, in fact)! The molecules float with their hydrophilic 'heads' in contact with the water molecules, and with their hydrocarbon tails exposed above and away from the water, forming a monolayer of phospholipid molecules (Figure 1.35). When more phospholipid is added, the molecules arrange themselves as a bilayer, with the hydrocarbon tails facing together. This is the situation in the plasma membrane.

Additionally, in the lipid bilayer, attractions between the hydrophobic hydrocarbon tails on the inside, and between the hydrophilic glycerol/phosphate heads and the surrounding water on the outside make a stable, strong barrier.

The protein components

The proteins of plasma membranes are **globular proteins** (pages 91–92). Some of these proteins occur partially or fully buried in the lipid bilayer, and are described as integral proteins. Others are superficially attached on either surface of the lipid bilayer and are known as peripheral proteins.

The carbohydrate components

The **carbohydrate** molecules of the membrane are relatively short-chain polysaccharides. They occur only on the outer surface of the plasma membrane. Some of these molecules are attached to the proteins (**glycoproteins**) and some to the lipids (**glycolipids**). Collectively, they are known as the glycocalyx. Its various functions include cell–cell recognition, acting as receptor sites for chemical signals, and the binding of cells into tissues.

Cholesterol

Lipid bilayers have been found to contain molecules of a rather unusual lipid, in addition to phospholipids, present in variable amounts. This lipid is known as **cholesterol** (page 86). Cholesterol has the effect of disturbing the close-packing of the phospholipids, thereby increasing the flexibility of the membrane. We return to this feature shortly.

Nature of Science

Using models as representations of the real world

■ What is the evidence for this model of membrane structure?

What is known about the composition and structure of the plasma membrane has built up from evidence over a period of time. Studies in cell structure (cytology), cell biochemistry and cell behaviour (cell physiology) all contributed. The first ideas about a 'membrane' were based on the observations that:

■ cell contents flow out when the cell surface is ruptured – a membrane barrier is present
■ water-soluble compounds enter cells less readily than compounds that dissolve in lipids; this implies that lipids are a major component of the cell membrane
■ in the presence of water (the environment of life) phospholipid molecules arrange themselves as a bilayer, with the hydrocarbon tails facing together, forming a stable, strong barrier
■ protein is also present in cell membranes as a major component – approximately sufficient to cover both external surfaces of a lipid bilayer.

In response to the evidence, in 1952, James **Danielli** and Hugh **Davson**, two scientists, proposed a membrane structure (which was revised in 1954) in which a lipid bilayer was evenly coated with proteins on both surfaces. Pores were thought to be present in places in the membrane. Early electron micrographs of cell membranes seen in section appeared to support this model (Figure 1.36 A).

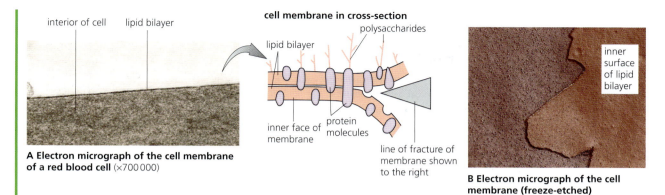

A Electron micrograph of the cell membrane of a red blood cell (×700 000)

B Electron micrograph of the cell membrane (freeze-etched)

■ **Figure 1.36** Plasma membrane structure; evidence from the electron microscope (A = electron micrograph of plasma membrane; B = by freeze-etching)

Later, additional evidence inspired two cytologists (**Singer** and **Nicholson**, in 1972) to propose the fluid mosaic model of membrane structure:

■ attempts to extract the protein from plasma membranes indicated that, while some occurred on the external surfaces and were easily extracted, others were buried within or across the lipid bilayers; these proteins were more difficult to extract
■ freeze-etching studies of plasma membranes confirmed that when a membrane is, by chance, split open along its mid-line, some proteins are seen to occur buried within or across the lipid bilayers (Figure 1.36 B)
■ experiments in which specific components of membranes were 'tagged' by reaction with marker chemicals (typically **fluorescent dyes**) showed component molecules to be continually on the move within membranes – a plasma membrane could be described as strong but 'fluid'
■ lipid bilayers contain molecules of an unusual lipid, cholesterol, the presence of which disturbs the close-packing of the bulk of the phospholipids of the bilayer; the quantity of cholesterol present may vary with the ambient temperatures that cells experience
■ on the outer surface of the plasma membrane, antenna-like carbohydrate molecules form complexes with certain of the membrane proteins (forming **glycoproteins**) and lipids (forming **glycolipids**).

Nature of Science

Falsification of theories

This sequence of discoveries illustrates the part that evidence plays in contradicting earlier conclusions and assumptions. The evidence led to a new hypothesis – here represented by a new model. Scientific knowledge is always tentative, based on the available evidence. It can be **falsified** at any moment by contrary evidence.

TOK Link

The explanation of the structure of the plasma membrane has changed over the years as new evidence and ways of analysis have come to light. Under what circumstances is it important to learn about theories that were later discredited?

17 State the difference between a lipid bilayer and the double membrane of many organelles.

The roles of membrane proteins have been investigated subsequently. Proteins that occur partially or fully buried in the lipid bilayer are described as **integral proteins**. Those that are superficially attached on either surface of the lipid bilayer are known as **peripheral proteins**. Some of these membrane proteins may act as channels for transport of metabolites, or be enzymes and carriers, and some may be receptors or antigens. Those that are involved in transport of molecules across membranes are in the spotlight in the next section.

1.4 Membrane transport – *membranes control the composition of cells by active and passive transport*

Movement of molecules across the plasma membrane of living cells is continual. Into and out of cells pass **water**, **respiratory gases** (oxygen and carbon dioxide), **nutrients** such as glucose, **essential ions** and **excretory products**.

Cells may secrete substances such as **hormones** and **enzymes**, and they may receive **growth substances** and certain hormones. Plants secrete the chemicals that make up their walls through their cell membranes, and assemble and maintain the wall outside the membrane. Certain mammalian cells secrete **structural proteins** such as collagen, in a form that can be assembled outside the cells.

In addition, the plasma membrane is where the cell is identified by surrounding cells and organisms. For example, **protein receptor sites** are recognized by hormones, by neurotransmitter substances from nerve cells, and by other chemicals sent from other cells. Figure 1.37 is a summary of this movement, and Figure 1.38 summarizes the mechanisms of transport across membranes, into which we need to look further.

■ **Figure 1.37**
Movements across the plasma membrane

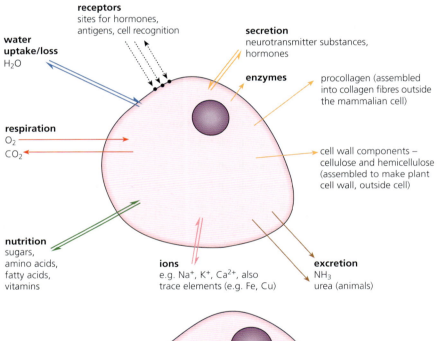

receptors
sites for hormones, antigens, cell recognition

secretion
neurotransmitter substances, hormones

water uptake/loss
H_2O

enzymes

procollagen (assembled into collagen fibres outside the mammalian cell)

respiration
O_2
CO_2

cell wall components – cellulose and hemicellulose (assembled to make plant cell wall, outside cell)

nutrition
sugars, amino acids, fatty acids, vitamins

ions
e.g. Na^+, K^+, Ca^{2+}, also trace elements (e.g. Fe, Cu)

excretion
NH_3
urea (animals)

■ **Figure 1.38**
Mechanisms of movement across membranes

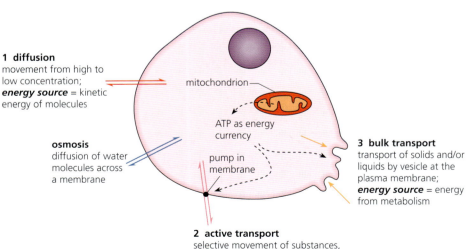

1 diffusion
movement from high to low concentration;
energy source = kinetic energy of molecules

mitochondrion

ATP as energy currency

3 bulk transport
transport of solids and/or liquids by vesicle at the plasma membrane;
energy source = energy from metabolism

osmosis
diffusion of water molecules across a membrane

pump in membrane

2 active transport
selective movement of substances, against a concentration gradient;
energy source = energy from metabolism

■ Movement by diffusion

Atoms, molecules and ions of liquids and gases undergo continuous random movements. These movements result in the even distribution of the components of a gas mixture and of the atoms, molecules and ions in a solution. So, for example, we are able to take a tiny random sample from a solution and analyse it to find the concentration of dissolved substances in the whole solution – because any sample has the same composition as the whole. Similarly, every breath we take has the same amount of oxygen, nitrogen and carbon dioxide as the atmosphere as a whole.

Continuous random movements of all molecules ensures complete mixing and even distribution, given time, in solutions and gases.

Diffusion is the free passage of molecules (and atoms and ions) from a region of their high concentration to a region of low concentration.

Where a difference in concentration has arisen in a gas or liquid, random movements carry molecules from a region of high concentration to a region of low concentration. As a result, the particles become evenly dispersed. The energy for diffusion comes from the **kinetic energy** of molecules. 'Kinetic' means that a particle has energy because it is in continuous motion.

Diffusion in a liquid can be illustrated by adding a crystal of a coloured mineral to distilled water. Even without stirring, the ions become evenly distributed throughout the water (Figure 1.39). The process takes time, especially as the solid has first to dissolve.

■ **Figure 1.39**
Diffusion in a liquid

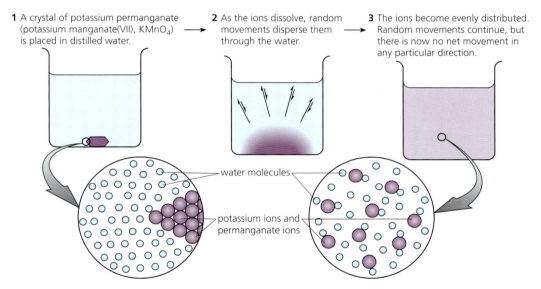

1 A crystal of potassium permanganate (potassium manganate(VII), $KMnO_4$) is placed in distilled water.

2 As the ions dissolve, random movements disperse them through the water.

3 The ions become evenly distributed. Random movements continue, but there is now no net movement in any particular direction.

water molecules

potassium ions and permanganate ions

■ Diffusion in cells

Diffusion across the cell membrane occurs where:

■ The plasma membrane is fully permeable to the solute. The lipid bilayer of the plasma membrane is permeable to non-polar substances, including steroids and glycerol, and also to oxygen and carbon dioxide in solution, all of which diffuse quickly via this route (Figure 1.40).

■ **Figure 1.40**
Diffusion across the
plasma membrane

1 channels for transport of metabolites or water

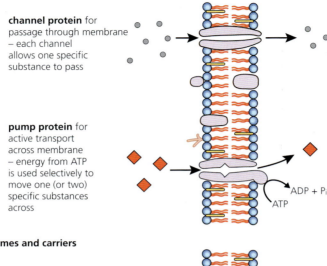

channel protein for
passage through membrane
– each channel
allows one specific
substance to pass

pump protein for
active transport
across membrane
– energy from ATP
is used selectively to
move one (or two)
specific substances
across

ADP + P$_i$

ATP

2 enzymes and carriers

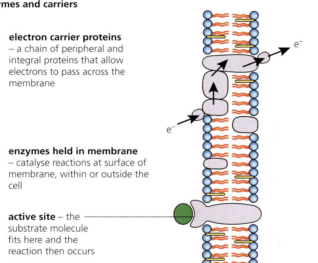

electron carrier proteins
– a chain of peripheral and
integral proteins that allow
electrons to pass across the
membrane

e$^-$

e$^-$

enzymes held in membrane
– catalyse reactions at surface of
membrane, within or outside the
cell

active site – the
substrate molecule
fits here and the
reaction then occurs

3 receptors, antigens, cell–cell recognition and cell binding sites

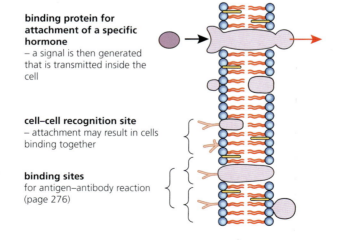

**binding protein for
attachment of a specific
hormone**
– a signal is then generated
that is transmitted inside the
cell

cell–cell recognition site
– attachment may result in cells
binding together

binding sites
for antigen–antibody reaction
(page 276)

The pores in the membrane are large enough for a solute to pass through. Water diffusing across the plasma membrane passes via the protein-lined pores of the membrane, and via tiny spaces between the phospholipid molecules. This latter occurs easily where the fluid mosaic membrane contains phospholipids with unsaturated hydrocarbon tails, for here these hydrocarbon tails are spaced more widely. The membrane is consequently especially 'leaky' to water, for example (Figure 1.41).

■ **Figure 1.41**
How polar water molecules cross the lipid bilayer

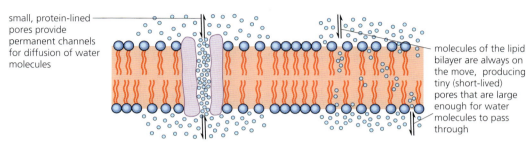

small, protein-lined pores provide permanent channels for diffusion of water molecules

molecules of the lipid bilayer are always on the move, producing tiny (short-lived) pores that are large enough for water molecules to pass through

18 Students were provided with cubes of slightly alkaline gelatine of different dimensions, containing an acid–alkali indicator that is red in alkali but yellow in acid. The cubes were placed in dilute acid solution, and the time taken for the colour in the gelatine to change from red to yellow was measured.

Dimensions/mm	Surface area/mm^2	Volume/mm^3	Time/minutes
10 × 10 × 10	600.0	1000.0	12.0
5 × 5 × 5	150.0	125.0	4.5
4 × 4 × 4	96.0	64.0	4.2
2.5 × 2.5 × 2.5	37.5	15.6	4.0

a For each block, **calculate** the ratio of surface area to volume (SA:V).
b Plot a graph of the time taken for the colour change (vertical or *y* axis) against the SA:V ratio (horizontal or *x* axis).
c **Explain** why the colour changes more quickly in some blocks than others.

Facilitated diffusion

In facilitated diffusion, a substance that otherwise is unable to diffuse across the plasma membrane does so as a result of its effect on particular molecules present in the membrane. In the presence of the substance, these membrane molecules, made of globular protein, form into pores large enough to allow diffusion; they close up again when the substance is no longer present (Figure 1.42). In facilitated diffusion, the energy comes from the kinetic energy of the molecules involved, as is the case in all forms of diffusion. Energy from metabolism is not required. Important examples of facilitated diffusion are the movement of ADP into mitochondria **and** the exit of ATP from mitochondria (page 118).

■ **Figure 1.42**
Facilitated diffusion

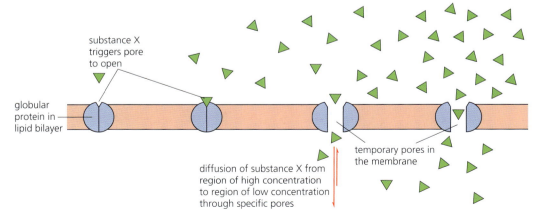

substance X triggers pore to open

globular protein in lipid bilayer

diffusion of substance X from region of high concentration to region of low concentration through specific pores

temporary pores in the membrane

Case study: potassium channels for facilitated diffusion in axons

We will be investigating how nerves work later; you can see the structure of a nerve cell and its axon in Figure 6.42 (page 292). For the moment, we can note that a nerve impulse is transmitted along the axon of a nerve cell by a momentary reversal in electrical potential difference in the axon membrane, brought about by rapid movements of sodium and potassium ions (page 294). These ions pass by facilitated diffusion via pores in the membrane called ion channels. These pores are special channels of globular proteins that span the membrane (Figure 1.43). They have a central channel that can open and close. One type of channel is exclusively permeable to sodium ions, and another to potassium ions.

The potassium channel is voltage-gated. This means that it opens or closes depending on a certain threshold membrane potential being reached. In fact, when the axon has a more positive charge outside than inside, the potassium channels are closed.

Almost immediately that an impulse has passed, there are relatively more positive charges inside the axon and the potassium channels open. Now, potassium ions can exit the axon down an electrochemical gradient, into the tissue fluid outside.

Then, as the interior of the axon starts to become less positive again, the potassium channel is closed, first by action of a 'ball and chain' device. The 'chain' is believed to be a flexible strand of amino acid residues and the 'ball' is globular protein. Finally, when the axon has more positive charge outside than inside, the potassium channel itself returns to the fully closed condition.

Voltage-gated K⁺ channels have positively charged voltage-sensing paddles which are normally attracted to the negatively charged interior surface of the resting axon. In this position the channel is closed mechanically – no K⁺ ions pass.

Once the membrane has depolarized, the paddles are attracted to the outside of the axon membrane and repelled by the positively charged interior. In this position the selective channel gates are opened, and potassium ions diffuse down an electro-chemical gradient.

A 'ball and chain' attached to the interior of the channel protein fits inside the open channel (the flexible chain allows this) and stops diffusion of K⁺ ions while the exterior of the axon is still negatively charged. The ball remains in place until the interior of the axon becomes negatively charged again and the gate itself is closed.

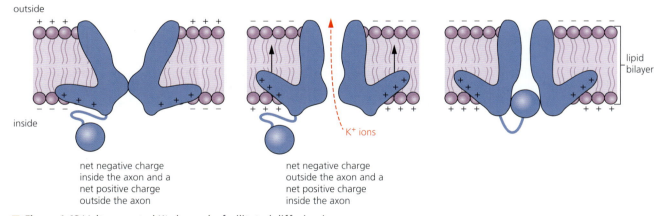

net negative charge inside the axon and a net positive charge outside the axon

net negative charge outside the axon and a net positive charge inside the axon

■ **Figure 1.43** Voltage-gated K⁺ channel – facilitated diffusion in an axon

19 Distinguish between diffusion and facilitated diffusion.

■ Osmosis – a special case of diffusion

Osmosis is a special case of diffusion (Figure 1.44). It is the diffusion of water molecules across a membrane which is permeable to water (**partially permeable**). Since water makes up 70–90% of living cells and cell membranes are partially permeable membranes, osmosis is very important in biology. *Why does osmosis happen?*

Dissolved substances attract a group of polar water molecules around them. The forces that hold the water molecules in this way are weak chemical bonds, including **hydrogen bonds**. Consequently, the tendency for random movement by the dissolved substances and their surrounding water molecules is very much reduced. Organic substances like sugars, amino acids, polypeptides and proteins, and inorganic ions like Na^+, K^+, Cl^- and NO_3^-, have this effect on the water molecules around them.

The stronger the solution (i.e. the more solute dissolved per volume of water), the greater the number of water molecules that are held almost stationary. So, in a very concentrated solution, very many more of the water molecules have restricted movement than in a dilute solution. In pure water, all of the water molecules are free to move about randomly, and do so.

When a solution is separated from water (or a more dilute solution) by a membrane permeable to water molecules (such as the plasma membrane), water molecules that are free to move tend to diffuse, while dissolved molecules and their groups of water molecules hardly move, if at all. So there is a net flow (diffusion) of water, from a more dilute solution into a more concentrated solution, across the membrane. This why the membrane is described as partially permeable.

So we can define **osmosis** as the net movement of water molecules (solvent), from a region of high concentration of water molecules to a region of lower concentration of water molecules, across a selectively permeable membrane. Alternatively, we can state that osmosis is the passive movement of water molecules across a partially permeable membrane, from a region of lower solute concentration to a region of higher solute concentration.

Dissolved solutes generate solute potential

The dissolved solutes present in the cytoplasm and vacuoles of cells generate a force known as the **solute potential**. Solute potential was previously referred to as osmotic pressure or osmotic potential, although these terms have now been abandoned.

Using a simple osmometer (Figure 1.45) we can demonstrate the solute potential of a solution. When the osmometer containing a concentrated solution is lowered into a beaker of water, very many more water molecules stream across the membrane into the solution and very few move in the opposite direction. The solution is diluted and it rises up the attached tube. Osmometers of this sort could be used to compare solute potentials of solutions with different concentrations.

20 When a concentrated solution of glucose is separated from a dilute solution of glucose by a partially permeable membrane, **determine** which solution will show a net gain of water molecules.

21 **Explain** what happens to a fungal spore that germinates after landing on jam made from fruit and its own weight of sucrose.

■ **Figure 1.44** Osmosis

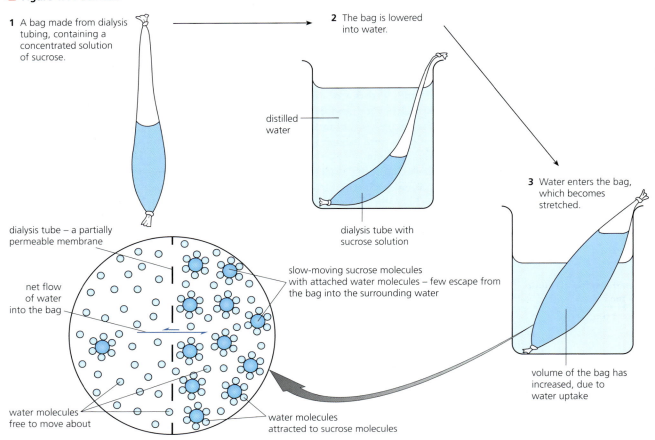

1 A bag made from dialysis tubing, containing a concentrated solution of sucrose.

2 The bag is lowered into water.

distilled water

dialysis tube with sucrose solution

3 Water enters the bag, which becomes stretched.

volume of the bag has increased, due to water uptake

dialysis tube – a partially permeable membrane

net flow of water into the bag

slow-moving sucrose molecules with attached water molecules – few escape from the bag into the surrounding water

water molecules free to move about

water molecules attracted to sucrose molecules

■ **Figure 1.45**
An osmometer to demonstrate solute potential

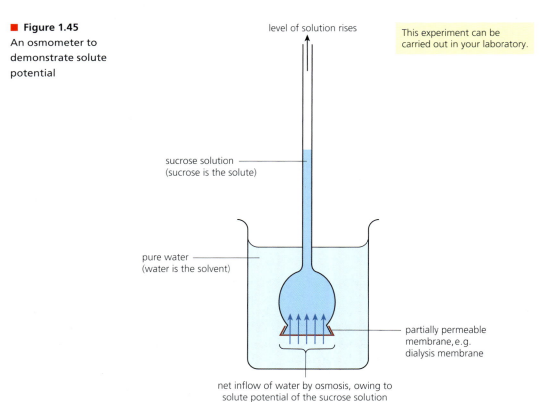

level of solution rises

This experiment can be carried out in your laboratory.

sucrose solution (sucrose is the solute)

pure water (water is the solvent)

partially permeable membrane, e.g. dialysis membrane

net inflow of water by osmosis, owing to solute potential of the sucrose solution

Osmosis in plant and animal cells

The effects of osmosis on plant and animal cells can be quite different.
Can you think why?

First consider osmosis in an individual plant cell, with its cellulose cell wall (Figure 1.46 A). Whether the net direction of water movement is into or out of a plant cell depends on whether the concentration of the cell solution is more or less concentrated than the external solution.

When the external solution is less concentrated (**hypotonic**) than the cell solution there is a net inflow of water into the cell by osmosis, and the cell solution becomes diluted. Then the cell contents become stretched by water uptake, and they press hard against the cell wall. If this happens the cell is described as **turgid**. The pressure that develops (due to the stretching of the wall) eventually becomes so great it prevents further uptake of water. The cell wall has protected the delicate cell contents from damage due to osmosis, but the tissue may be quite rigid due to the internal pressure.

When the external solution is more concentrated (**hypertonic**) than the cell solution there is a net flow of water out of the cell by osmosis, and the cell solution becomes more concentrated. As the volume of cell solution decreases, the cytoplasm pulls away from parts of the cell wall (contact with the cell wall is maintained at points where there are cytoplasmic connections between cells). The cell becomes **flaccid**, and it is said to be **plasmolysed** (from plasmo = cytoplasm, lysis = splitting).

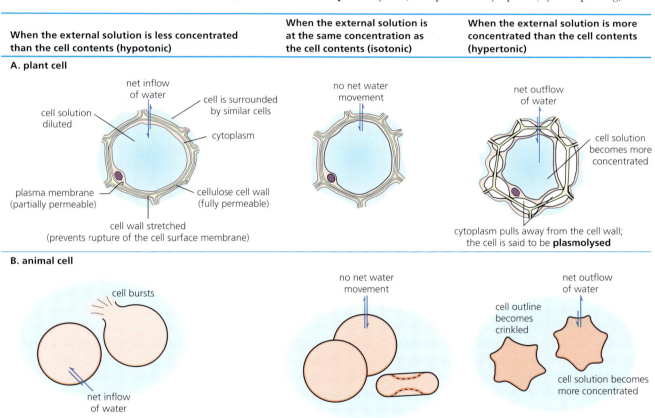

When the external solution is less concentrated than the cell contents (hypotonic)	When the external solution is at the same concentration as the cell contents (isotonic)	When the external solution is more concentrated than the cell contents (hypertonic)

A. plant cell

B. animal cell

◼ **Figure 1.46** Osmosis in plant and animal cells

In the case of an animal cell, the absence of a protective cellulose wall generates a serious problem in terms of water relations. A typical animal cell – a red blood cell is a good example – when placed in pure water or a hypotonic solution will quickly break open from the pressure generated by the entry of an excessive amount of water by osmosis. This is illustrated in Figure 1.46 B. Notice that the same cells, when placed in a hypertonic solution, shrink in size due to net water loss from the cytoplasm.

In mammals and other animals the osmotic concentration of body fluids (blood plasma and tissue fluid) is very carefully regulated, maintaining the same osmotic concentration inside and outside body cells (isotonic conditions), which avoids such problems. This process is an aspect of osmoregulation (Chapter 11).

Osmosis in aquatic unicellular animals

Many unicellular animals survive in fresh water aquatic environments where the external medium is normally hypotonic to their cell solution. These organisms experience a continuous net inflow of water by osmosis and are in danger of disruption of the plasma membrane by high internal pressure.

The protozoan *Amoeba* is one example. In fact, the cytoplasm of amoebae contains a tiny water pump, known as a **contractile vacuole**, which works continuously to pump out excess water. The importance of the contractile vacuole to these organisms is fatally demonstrated if the cytoplasm is temporarily anaesthetized. The animal quickly bursts (Figure 1.47).

■ **Figure 1.47**
Amoeba; the role of the contractile vacuole

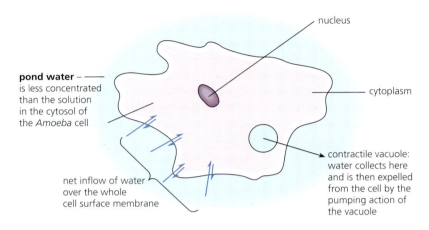

nucleus

pond water –
is less concentrated
than the solution
in the cytosol of
the *Amoeba* cell

cytoplasm

net inflow of water
over the whole
cell surface membrane

contractile vacuole:
water collects here
and is then expelled
from the cell by the
pumping action of
the vacuole

Medical application of osmosis

When human organs are donated for transplant surgery they have to be maintained in a saline solution that is isotonic with the cells of the tissues and organs, in order to prevent damage to cells due to water uptake or loss during transit to the recipient patient.

Medical application of diffusion

In cases of kidney failure, urea and sodium ions may start to accumulate in the blood to harmful levels. In these cases, a treatment known as hemodialysis is prescribed. In hemodialysis, the patient's blood is circulated through an external, partially permeable membrane, arranged so that urea and toxic substances are removed by diffusion. You can see this application of diffusion illustrated in Figure 11.35 (page 479). *Look at this illustration now.*

Nature of Science

Experimental design—accurate quantitative measurement

■ Estimation of osmotic concentration (osmolarity) of plant tissue

When plant cells are bathed in a solution that is isotonic with the cell cytosol there is no net entry or exit of water from the cells. The tissue remains of the same dimensions and mass. Alternatively, if similar plant tissue is placed in a hypertonic solution, the tissue decreases in dimensions and mass. When placed in a hypotonic solution, it increases in dimensions and mass. This observation is the basis of the experiment illustrated in Figure 1.48. Here the aim is to discover the concentration of the bathing solution isotonic with the cells of potato tuber. Study the sequence of steps involved in the experiment. Note the importance of:

- accurate weighing out of the solute (sucrose in this case)
- accurate pipetting of solution from tube to tube in the process of serial dilution
- the use of replicate sample of tissue in each tube
- the accurate measurement of the length of the tissue strips at the end of the experiment.

Examine the graph to identify what molar concentration of sucrose is isotonic with the cytosol of the potato tissue used in this experiment.

■ **Figure 1.48**
The investigation of the osmotic concentration of potato tuber tissue

1 Preparing different concentration of sucrose solution, $0.8\,mol\,dm^{-3} \rightarrow 0.2\,mol\,dm^{-3}$

$100\,cm^3$ of $1\,mol\,dm^{-3}$ solution was taken and the following dilutions carried out:

Volume of distilled water (cm³)	Volume of $1\,mol\,dm^{-3}$ sucrose (cm³)	Concentration of sucrose (mol dm⁻³)
2	8	0.8
4	6	0.6
6	4	0.4
8	2	0.2

2 Preparing the tissue strips and the setting up of the experiment

replicate strip (10 × 1 × 0.5 cm) or cylinders (10 × 1 cm cork borer) cut from a large potato tube.
tissue strips/cylinders washed and 3 placed in each tube

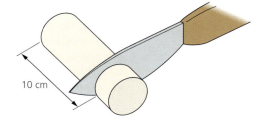

10 cm

3 Measuring the final lengths of the tissue strips

After a period of 30 minutes the tissue strips were retrieved, blotted dry, and their lengths measured accurately. The mean change in length of the 3 strips in each tube was calculated.

4 Graphing the results and estimating the osmotic concentration of potato tuber tissue

A graph of the results was plotted: The relationship between the molar concentration of sucrose solutions and the change in length of tissue strips

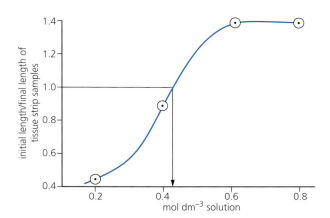

22 **State** the significance of the use of *three* tissue strips in each of the tubes in Figure 1.48.

■ Movement by active transport

We have seen that diffusion is due to random movements of molecules and occurs spontaneously from a high to a low concentration. However, many of the substances required by cells have to be absorbed from a weak external concentration and taken up into cells that contain a higher concentration. Uptake against a concentration gradient cannot occur by diffusion. Instead, it requires a source of energy to drive it. This type of uptake is known as **active transport**.

In active transport, metabolic energy produced by the cell, held as ATP (energy currency – page 115), is used to drive the transport of molecules and ions across cell membranes. Active transport has characteristic features distinctly different from those of movement by diffusion.

1 Active transport occurs against a concentration gradient

Active transport occurs from a region of low concentration to a region of higher concentration. The cytoplasm of a cell normally holds reserves of valuable molecules and ions, like nitrate ions in plant cells or calcium ions in muscle fibres. These useful molecules and ions do not escape; the cell membrane retains them inside the cell. When more useful molecules or ions become available for uptake, they are actively absorbed into the cells. This happens even though the concentration outside the cell is lower than that inside.

2 Active uptake is highly selective

For example, in a situation where potassium chloride (K^+ and Cl^- ions) is available to an animal cell, K^+ ions are more likely to be absorbed, since they are needed by the cell. Where sodium nitrate (Na^+ and NO_3^- ions) is available to a plant cell, it is likely that more of the NO_3^- ions will be absorbed than the Na^+, since this reflects the needs of plant cells.

3 Active transport involves special molecules of the membrane, called pump molecules

The pump molecules pick up particular molecules or ions and transport them to the other side of the membrane, where they are then released. The pump molecules are globular proteins (pages 91–92), sometimes also called carrier proteins. These span the lipid bilayers (Figure 1.34). Movements by the pump molecules require reaction with ATP; this reaction supplies metabolic energy to the process. Most membrane pumps are specific to particular molecules or ions, bringing about selective transport. If the pump molecule for a particular substance is not present, the substance will not be transported.

Active transport is a feature of most living cells. We meet examples of active transport in the gut where absorption occurs (page 252), in the active uptake of ions by plant roots (page 375), in the kidney tubules where urine is formed (page 472) aznd in nerve fibres where an impulse is propagated (page 293).

The protein pumps of plasma membranes are of different types. Some transport a particular molecule or ion in one direction (Figure 1.49), while others transport two substances (like Na^+ and K^+) in opposite directions (Figure 1.50). Occasionally, two substances are transported in the same direction; for example, Na^+ and glucose (Figure 6.42, page 292).

Case study: structure and function of sodium–potassium pumps in axons

A nerve impulse is transmitted along the axon of a nerve cell by a momentary reversal in electrical potential difference in the axon membrane, brought about by rapid movements of sodium and potassium ions. You can see the structure of a nerve cell and its axon in Figure 6.42 (page 290).

Sodium–potassium pumps are globular proteins that span the axon membrane. In the preparation of the axon for the passage of the next nerve impulse, there is **active transport** of potassium (K^+) ions in across the membrane and sodium (Na^+) ions out across the membrane. This activity of the Na^+/K^+ pump involves transfer of energy from **ATP**. The outcome is that potassium and sodium ions gradually concentrate on opposite sides of the membrane.

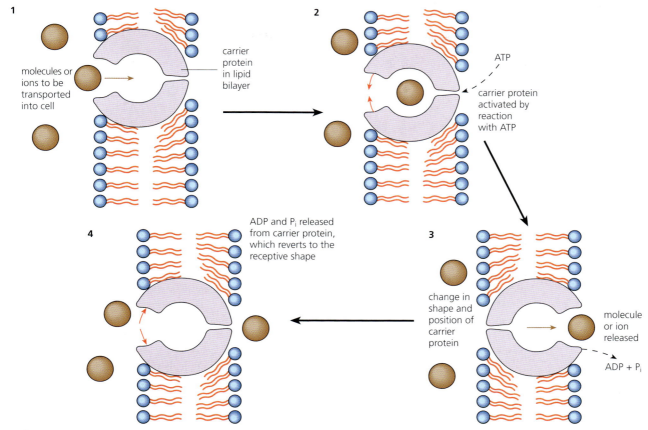

■ **Figure 1.49** Active transport of a single substance

The steps to the cyclic action of these pumps are:

■ with the interior surface of the pump open to the interior of the axon, three sodium ions are loaded by attaching to specific binding sites

■ reaction of the globular protein with ATP now occurs, resulting in the attachment of a phosphate group to the pump protein; this triggers the pump protein to close to the interior of the axon and open to the exterior

■ the three sodium ions are now released and, simultaneously, two potassium ions are loaded by attaching to specific binding sites

■ with the potassium ions loaded, the phosphate group detaches; this triggers a reversal of the shape of the pump protein – it now opens to the interior, again, and the potassium ions are released

■ the cycle is now repeated again.

23 Samples of five plant tissue discs were incubated in dilute sodium chloride solution at different temperatures. After 24 hours, it was found that the uptake of ions from the solutions was as shown in the table (arbitrary units).
Comment on how absorption of sodium chloride occurs, giving your reasons.

	Sodium ions	Chloride ions
Tissue at 5°C	80	40
Tissue at 25°C	160	80

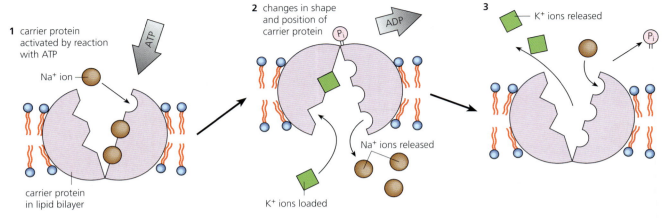

1 carrier protein activated by reaction with ATP

Na+ ion

carrier protein in lipid bilayer

2 changes in shape and position of carrier protein

ATP

ADP

Na+ ions released

K+ ions loaded

3

K+ ions released

■ **Figure 1.50**
The sodium/potassium ion pump

■ Movement by bulk transport

Another mechanism of transport across the plasma membrane is known as **bulk transport**. It occurs by movements of **vesicles** of matter (solids or liquids) across the membrane by processes known generally as **cytosis**. Uptake is called **endocytosis** and export is **exocytosis**.

The strength and flexibility of the fluid mosaic membrane make this activity possible. Energy from metabolism (ATP) is also required. For example, when solid matter is being taken in (**phagocytosis**), part of the plasma membrane at the point where the vesicle forms is pulled inwards and the surrounding plasma membrane and cytoplasm bulge out. The matter, thus, becomes enclosed in a small vesicle.

Vesicles are used to transport materials within cells, for example, between the rough endoplasmic reticulum (RER) and the Golgi apparatus, and on to the plasma membrane (Figures 1.26 and 1.27).

In the human body, there is a huge number of phagocytic cells (phagocytosis means 'cell eating'). These are called the **macrophages**. The macrophages engulf the debris of damaged or dying cells and dispose of it. For example, we break down about 2×10^{11} red blood cells each day. These are ingested and disposed of by macrophages, every 24 hours.

Bulk transport of fluids is referred to as pinocytosis (Figure 1.51).

■ **Figure 1.51**
Transport by cytosis

Movements also take place in the reverse direction, i.e. **exocytosis** of solids and liquids.

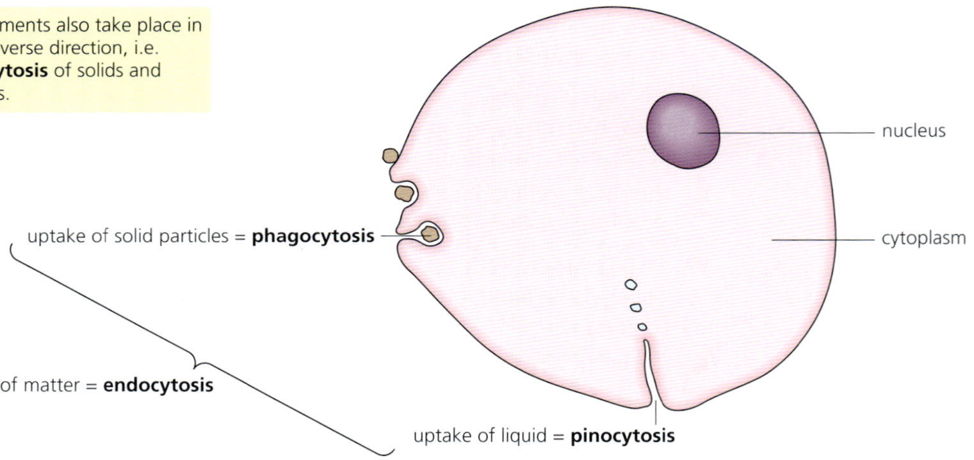

uptake of solid particles = **phagocytosis**

uptake of matter = **endocytosis**

uptake of liquid = **pinocytosis**

nucleus

cytoplasm

24 Distinguish between the following pairs:
 a proteins and lipids in cell membranes
 b active transport and bulk transport
 c endocytosis and exocytosis.

1.5 The origin of cells – *there is an unbroken chain of life from the first cells on Earth to all cells in organisms alive today*

Nature of Science   **Testing the general principles that underlie the natural world**

■ Cells are formed by division of pre-existing cells

At one time it was believed that cells could arise spontaneously – known as '**spontaneous generation**'. This idea was based on the 'mysterious' appearances of living things, such as when:

- the growth of moulds occurred on exposed foods, like cheese
- maggots appeared in exposed meat (certainly by the time it was rotting)
- opened bottles of wine turned cloudy (and the contents turned into vinegar) at favourable temperatures.

At this time, it was not known that 'clean' air carried vast numbers of tiny, viable spores of a range of microorganisms. Pasteur's investigation of this contamination process played an important part in disproving the spontaneous generation theory.

Pasteur's investigation of 'spontaneous generation'

Look at the illustration of Pasteur's experiment in Figure 1.2 (page 3). The results confirmed that the air contains 'invisible' spores of microorganisms. When these spores reach favourable fluids or liquids (such as the nutrient broth that Pasteur used) they 'germinate', giving rise to huge populations of microorganisms by cell division. The result is that nutrient liquids become cloudy, and nutrient solids grow visible colonies and moulds. All these cells have arisen by division of pre-existing cells. Pasteur's experiment may be repeated in laboratories today, as shown in Figure 1.52. *Look at it carefully. Can you say what safety steps are required, and why?*

The culturing of bacteria aseptic (sterile) techniques:
Many species of bacteria are harmless to humans – in fact, very many are indispensable – life as we know it could not proceed without them. However, some species are pathogens, and some strains of certain species are harmful. Consequently, bacteriologists handle all cultures as if they were pathogens, using aseptic techniques:
- nutrients used are first sterilized, as are the culture vessels and other equipment, both before and after us
- a liquid preparation (a broth) may be used for detecting growing microorganisms, as shown here (a solid medium is produced by addition of a gelling agent called agar)

Essential bacteriological procedures are illustrated below.

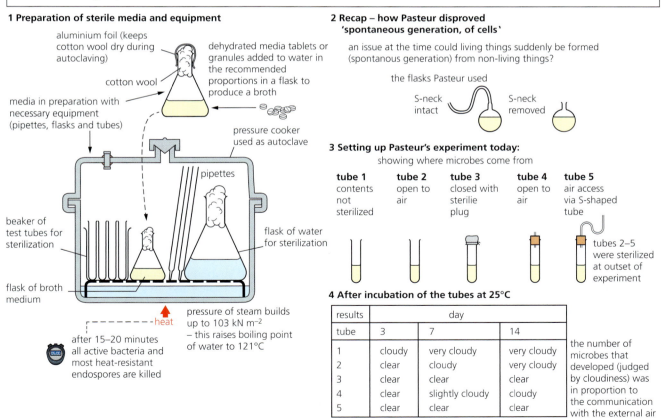

■ **Figure 1.52** Modern demonstrations of Pasteur's experiment

■ The origin of the first cells

In biology, the term 'evolution' specifically means the processes that have transformed life on Earth from its earliest beginnings to the diversity of forms we know about today, living and extinct. It is an organizing principle of modern biology. It helps us make sense of the ways living things are related to each other, for example.

The evolution of life in geological time has involved major steps – none more so than the origin of the first cells. Unless these first cells arrived here from somewhere else in the universe, they must have arisen from non-living materials – starting from the components of the Earth's atmosphere at the time. *About these very first steps we can only speculate.*

The spontaneous origin of life on Earth

The formation of **living cells** from non-living materials would have required the following steps:

■ the synthesis of **simple organic molecules**, such as sugars and amino acids

■ the assembly of these molecules into **polymers** (page 78)

■ the development of **self-replicating molecules**, the nucleic acids

■ the retention of these molecules within **membranous sacs**, so that an internal chemistry developed, different from the surrounding environment.

Experimental evidence for the origin of organic molecules

The molecules that make up living things are built mainly from carbon, hydrogen and oxygen, with some nitrogen, phosphorus, sulfur, and with a range of atoms of relatively few other elements also present. Today, living things make these molecules by the action of enzymes in their cells, but for life to originate from non-living material, the first step was the non-living synthesis of simple organic molecules.

Apparatus like this has been used with various gases to investigate the organic molecules that may be synthesized.

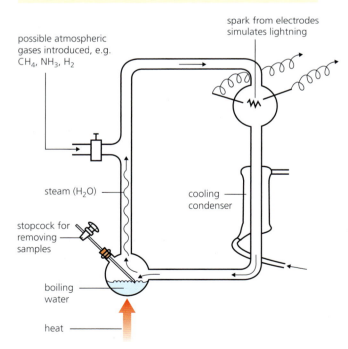

possible atmospheric gases introduced, e.g. CH_4, NH_3, H_2

spark from electrodes simulates lightning

steam (H_2O)

cooling condenser

stopcock for removing samples

boiling water

heat

S.L. Miller and H.C Urey (1953) investigated how simple organic molecules might have arisen from the ingredients present on Earth before there was life. They used a reaction vessel in which particular environmental conditions could be reproduced. For example, strong electric sparks (simulating lightning) were passed through mixtures of methane, ammonia, hydrogen and water vapour for a period of time. They discovered that amino acids (some known components of cell proteins) were formed naturally, as well as other compounds (Figure 1.53).

This approach confirmed that organic molecules can be synthesized outside cells, in the absence of oxygen. The experiment has subsequently been repeated, sometimes using different gaseous mixtures and other sources of energy (UV light, in particular), in similar apparatus. The products have included amino acids, fatty acids and sugars such as glucose. In addition, **nucleotide bases** have been formed and, in some cases, **simple polymers** of all these molecules have been found. So, we can see how it is possible that a wide range of organic compounds could have formed on the pre-biotic Earth, including some of the building blocks of the cells of organisms.

■ **Figure 1.53** Apparatus for simulating early chemical evolution

TOK Link

To what extent can you argue that Miller and Urey's experimental response to a seemingly insoluble issue was (i) an example of a reductionist approach, and (ii) uniquely a scientific response?

Assembly of the polymers of living things

For polymers to be assembled in the absence of cells and enzymes would have required the concentration of biologically important molecules such as monosaccharides (the simple sugars – building blocks for polysaccharides), amino acids (building blocks for proteins) and fatty acids (for lipid synthesis). They would need to come together in 'pockets' where further chemical reactions between them were possible. This might have happened in water close to larval flows of volcanoes or at the vents of sub-marine volcanoes where the environment is hot, the pressure is high, and the gases being vented are often rich in sulfur and other compounds. There is some evidence for the latter.

Origin of self-replicating molecules

For the evolution of life from a mixture of polymers and their monomers, two special situations need to emerge:

■ a 'self-replication' system
■ an ability to catalyse chemical change.

Today, in living cells, these essentials are achieved by **DNA**, the home of the genetic code, and **enzymes**, which are typically of large, globular proteins (page 92). However, neither of these has been synthesized in any experiments that repeat Miller and Urey's demonstration of how biologically important molecules might have been synthesized in the pre-biotic world.

So what may have filled the roles of DNA and enzymes in the origin of life?

A likely answer came as a by-product of a genetic engineering experiment, investigating the enzymes needed to join short lengths of nucleic acid known as **RNA** (page 105). It was discovered that RNA, as well as being information molecules, may also function as enzymes. Perhaps short lengths of RNA combined the roles of 'information molecules' and 'enzymes' in the evolution of life itself.

Universality of the genetic code

Although RNA fragments are fairly inefficient enzymes, they may catalyse the formation of DNA (although sometimes in an error-prone way). We now know that the 64 codons in the genetic code of DNA have the same meaning in nearly all organisms. This supports the idea of a common origin of life on Earth; that the very first DNA has sustained an unbroken chain of life from the first cells on Earth to all cells in organisms alive today. Only the most minor variations in the genetic code have arisen in the evolution and expansion of life since it originated 3500 million years ago.

■ Formation of the first cells

The fossil record tells us the first cells were prokaryotes. *How were these cells assembled?*

25 Suggest likely chemical changes that would have occurred in the first cells, and that would have required a catalyst.

We have seen that a few lipid molecules form a monolayer on the surface of water, and with more lipid present bilayers form – the basis of plasma membranes today. Lengths of these bilayers are likely to have formed **microspheres** (Figure 1.54). Perhaps simple microspheres, surrounding a portion of a pre-biotic 'soup' of polymers and monomers, were the fore-runners of cells. These may have formed membrane systems with distinctive internal chemistry, as they developed a chemical environment different from the surroundings.

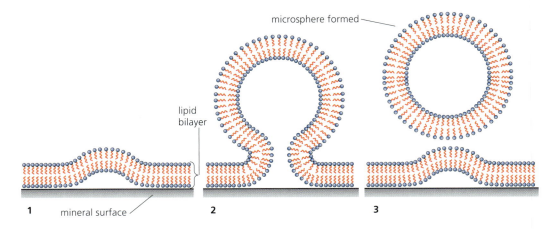

■ Prokaryote to eukaryote

Prokaryote cells differ from microspheres in a number of ways. For example, attached to the plasma membrane in the prokaryote cell is a **circular chromosome** – either of RNA or DNA. Also, a cell wall of complex chemistry is secreted outside the membrane barrier. However, the first prokaryotes could have survived nutritionally on the organic molecules of the pre-biotic soup. In this early environment, with its wealth of organic molecules surrounding simple cells, 'digestion' and 'respiration' would have demanded limited enzymic machinery. These biochemical sophistications would have to evolve with time – if life originated in this manner.

Present-day prokaryotes are similar to fossil prokaryotes, some of which are 3500 million years old. By comparison, the earliest eukaryote cells date back only 1000 million years. *How did eukaryotic cells arise?*

The origin of eukaryotic cells can be explained by the endosymbiotic theory (Figure 1.55). The eukaryotic cell may have formed from large prokaryote cells that came to contain their chromosome (whether of RNA or DNA) in a sac of infolded plasma membrane. If so, a distinct nucleus was now present. But how were the other organelles originated? Remember, membranous organelles are a feature of eukaryotes, additional to their discrete nucleus.

26 **Explain** why we can expect that, of all the fossils found in sedimentary rock, those of the lowest strata may bear the least resemblance to present-day forms.

■ **Figure 1.55**
Origin of the
eukaryotic cell

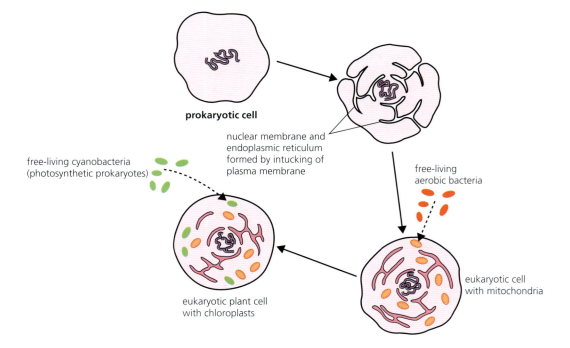

In the evolution of the eukaryotic cell, prokaryotic cells (which had been taken up into food vacuoles for digestion) may have survived as organelles inside the host cell, rather than becoming food items! They would have become integrated into the biochemistry of their 'host' cell over time. This would explain why mitochondria (and chloroplasts) contain a ring of DNA double helix, together with small ribosomes, just like a bacterial cell. It is these features that suggest these organelles are descendants of free-living prokaryotic organisms that came to inhabit larger cells. This concept is known as the **endosymbiotic origin of eukaryotes** (Table 1.7).

■ **Table 1.7**
Evidence for the endosymbiotic theory

Prokaryotes are known to inhabit some eukaryotic cells.
Chloroplasts and mitochondria reproduce by binary fission, just as prokaryotes do.
Chloroplasts and mitochondria contain circular DNA (not associated with histone proteins), like that of prokaryotes.
Chloroplasts and mitochondria contain ribosomes of the size (70S) also found in prokaryotes.
Chloroplasts and mitochondria transcribe mRNA from their DNA, and synthesize specific proteins in their ribosomes, as prokaryotes do.
Chloroplasts and mitochondria are similar in size to prokaryotes.

1.6 Cell division *– cell division is essential but must be controlled*

To recap, multicellular organisms begin life as a single cell which grows and divides. During growth, this cycle is repeated almost endlessly, forming many cells. It is these cells that eventually make up the adult organism. So, new cells arise by division of existing cells, and the cycle of growth and division is called the **cell division cycle**. This cycle has three main stages:

■ interphase

■ division of the nucleus by a process (mitosis) that results in two nuclei, each with an identical set of chromosomes

■ division of the cytoplasm and whole cell (known as cytokinesis).

In fact, in each stage of the cell cycle particular events occur. These events are summarized in Figure 1.56, and they are also discussed below. *Look at the subdivision of interphase now – distinctive features are identified in each stage.*

the cell cycle consists of interphase and mitosis

interphase = $G_1 + S + G_2$

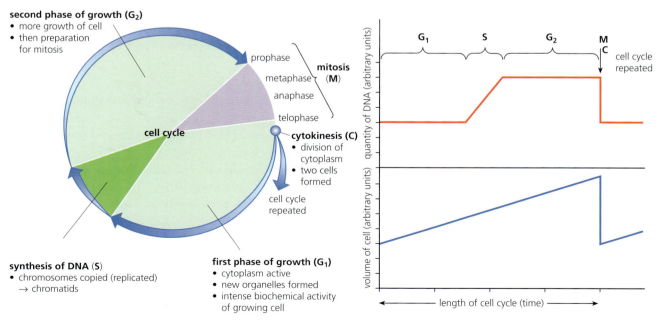

second phase of growth (G_2)
• more growth of cell
• then preparation for mitosis

prophase ⎫
metaphase ⎬ **mitosis** (**M**)
anaphase ⎪
telophase ⎭

cell cycle

cytokinesis (C)
• division of cytoplasm
• two cells formed

cell cycle repeated

synthesis of DNA (**S**)
• chromosomes copied (replicated) → chromatids

first phase of growth (G_1)
• cytoplasm active
• new organelles formed
• intense biochemical activity of growing cell

change in cell volume and quantity of DNA during a cell cycle

quantity of DNA (arbitrary units)

G_1 S G_2 M C   cell cycle repeated

volume of cell (arbitrary units)

← length of cell cycle (time) →

■ **Figure 1.56** The stages of the cell cycle

■ Interphase

Interphase is always the longest part of the cell cycle, but it is of extremely variable length. When growth is fast, as in a developing human embryo and in the growing point of a young stem, interphase may last about 24 hours or less. On the other hand, in mature cells that infrequently divide it lasts a very long period – sometimes indefinitely. For example, some cells, once they have differentiated, rarely or never divide again. Here, the nucleus remains at interphase permanently.

An overview of interphase

When the nucleus of a living cell at interphase is observed by light microscopy, the nucleus appears to be 'resting'. This is not the case. During interphase, the chromosomes are actively involved in protein synthesis. From the chromosomes, copies of the information of particular genes or groups of genes (in the form of mRNA, page 106) are taken for use in the cytoplasm. It is in the ribosomes of the cytoplasm that proteins are assembled from amino acids, combined in sequences dictated by the information from the gene and relayed in the form of mRNA.

The distinctively compact chromosomes, visible during mitosis (Figure 1.58), become dispersed in interphase. They are now referred to as **chromatin**. Amongst the chromatin can be seen one or more dark-staining structures, known as **nucleoli** (singular **nucleolus**). Chemically, the nucleoli consist of protein and RNA, and they are the site of synthesis of the ribosomes. These tiny organelles then migrate out into the cytoplasm.

The steps of interphase

27 **State** what structures of the interphase nucleus can be seen by electron microscopy.

During the first phase of growth (G_1), the synthesis of new organelles takes place in the cytoplasm. This is also a time of intense biochemical activity in the cytoplasm and organelles, and there is an accumulation of energy store before nuclear division occurs again.

Next is a period of synthesis of DNA (**S**), when each chromosome makes a copy of itself. It is said to **replicate**. The two identical structures formed are called **chromatids**. The chromatids remain attached until they divide during mitosis.

Finally, there is a second phase of growth (G_2), which is a continuation of the earlier time of intense biochemical activity and increase in amount of cytoplasm.

■ Control of the cell cycle

Look back at the stages of the cell cycle (Figure 1.56). Note that it consists of distinct phases, represented in shorthand as G_1, S, G_2, M and C.

The cell cycle is regulated by a **molecular control system**. The key points of this system are outlined below, best understood in conjunction with Figure 1.57:

- In the cell cycle there are key **checkpoints** where signals operate. These are stop points which have to be overridden.
- Three checkpoints are recognized – at G_1, G_2 and in M.
- At the G_2 checkpoint, if the 'go-ahead' signal is received here, the cell goes through to M to C, for example.
- The molecular control signal substance in the cytoplasm of cells are proteins known as kinases and cyclins.
- Kinases are enzymes that either activate or inactivate other proteins. Kinases are present in the cytoplasm all the time, though sometimes in an inactive state.
- Kinases are activated by specific cyclins, so they are referred to as cyclin-dependent kinases (**CDKs**).

- Cyclin concentrations in the cytoplasm change constantly. As the concentrations of cyclins increase, they combine with CDK molecules to form a complex which functions as a mitosis-promoting factor (**MPF**).
- As MPF accumulates, it triggers chromosome condensation, fragmentation of the nuclear membrane and, finally, spindle formation – that is, mitosis is switched on.
- By anaphase of mitosis, destruction of cyclins commences (but CDKs persist in the cytoplasm).
- External factors also operate on the cell, either triggering the rise in cyclin concentrations or switching on the destruction of cyclin.

Figure 1.57
The molecular control system of the cell cycle

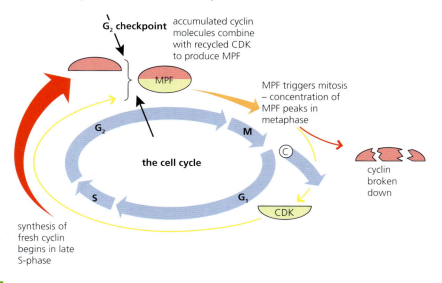

Nature of Science

Serendipity and scientific discoveries

The discovery of cyclins was accidental

The discovery of the proteins that control the cell cycle came partly from work by Tim Hunt, as his team investigated protein synthesis more generally in the eggs of sea urchins. While the synthesis of most new proteins proceeded steadily, as anticipated, a minority of others went through short, abrupt cycles of increasing and decreasing concentration. High threshold levels of these individual proteins were found to correlate with changes in the cell cycle. By this means, and with the contributions of others (including from work on yeasts) the roles of four different proteins in the cell cycle were discovered. They were named cyclins at this stage.

Later, Paul Nurse and Tim Hunt, with Leland Hartwell, were awarded the Nobel Prize in 2001 for their contributions to the discovery of the control of the cell cycle. Tim Hunt makes clear that his discovery of cyclins was accidental in his Nobel Prize lecture, which you can access as a paper at www.nobelprize.org/nobel_prizes/medicine/laureates/2001/hunt-lecture.pdf or in the video made at the ceremony at www.nobelprize.org/mediaplayer/index.php?id=494

The contributing studies of Paul Nurse were made on yeasts. The background of this scientist, currently President of The Royal Society, makes an interesting contrast with that of the others, and helpfully (and encouragingly) establishes how diverse the paths to a distinguished career in research can be: http://en.wikipedia.org/wiki/Paul_Nurse

Mitosis

When cell division occurs, the nucleus divides first. In mitosis, the chromosomes, present as the chromatids formed during interphase, are separated, and accurately and precisely distributed to two daughter nuclei.

Here, mitosis is presented and explained as a process in four phases (Figure 1.58), but remember this is for convenience of description only. Mitosis is a continuous process with no breaks between the phases. *You can follow the events of mitosis in Figure 1.58.*

■ **Figure 1.58** Mitosis in an animal cell

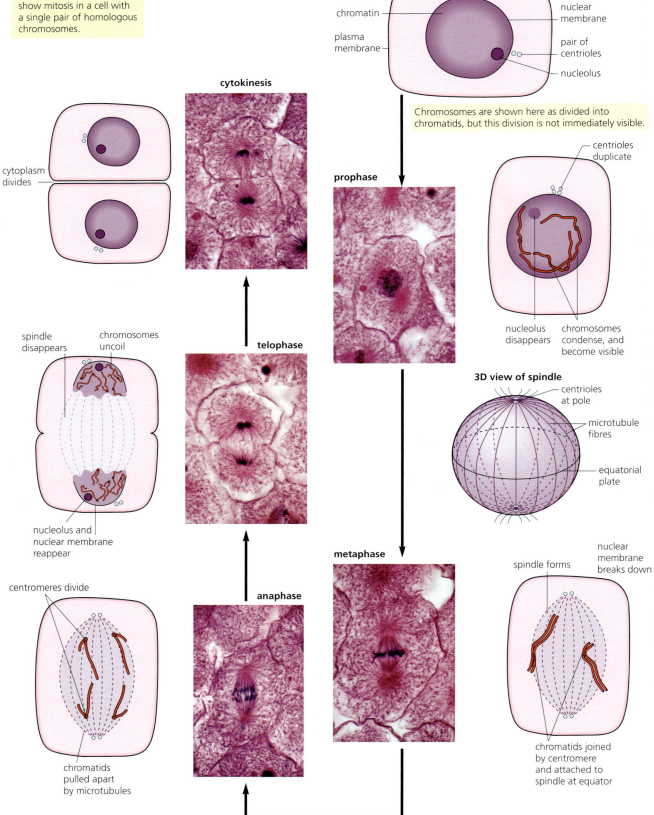

For simplicity, the drawings show mitosis in a cell with a single pair of homologous chromosomes.

interphase

cytoplasm

chromatin

plasma membrane

nuclear membrane

pair of centrioles

nucleolus

Chromosomes are shown here as divided into chromatids, but this division is not immediately visible.

cytokinesis

cytoplasm divides

prophase

centrioles duplicate

nucleolus disappears

chromosomes condense, and become visible

3D view of spindle

centrioles at pole

microtubule fibres

equatorial plate

telophase

spindle disappears

chromosomes uncoil

nucleolus and nuclear membrane reappear

anaphase

centromeres divide

chromatids pulled apart by microtubules

metaphase

spindle forms

nuclear membrane breaks down

chromatids joined by centromere and attached to spindle at equator

- In prophase, the chromosomes become visible as long thin threads. Now, they increasingly shorten and thicken by a process of supercoiling. You can see an electron micrograph of a supercoiled chromosome in Figure 1.60 (page 56). Only at the end of prophase is it possible to see that chromosomes consist of two chromatids held together at the centromere. At the same time, the nucleolus gradually disappears and the nuclear membrane breaks down.

- In **metaphase**, the centrioles move to opposite ends of the cell. Microtubules in the cytoplasm start to form into a spindle, radiating out from the centrioles (Figure 1.58). Microtubules attach to the centromeres of each pair of chromatids, and these are arranged at the equator of the spindle. (Note that, in plant cells, a spindle of exactly the same structure is formed, but without the presence of the centrioles.)

- In **anaphase**, the centromeres divide, the spindle fibres shorten and the chromatids are pulled by their centromeres to opposite poles. Once separated, the chromatids are referred to as chromosomes.

- In **telophase**, a nuclear membrane reforms around both groups of chromosomes at opposite ends of the cell. The chromosomes decondense by uncoiling, becoming chromatin again. The nucleolus reforms in each nucleus. Interphase follows division of the cytoplasm.

Cytokinesis

Division of the cytoplasm, known as **cytokinesis**, follows telophase. During division, cell organelles such as mitochondria and chloroplasts become distributed evenly between the cells. In animal cells, division is by in-tucking of the plasma membrane at the equator of the spindle, 'pinching' the cytoplasm in half (Figure 1.33).

In plant cells, the Golgi apparatus forms vesicles of new cell wall materials, which collect along the line of the equator of the spindle, known as the cell plate. Here the vesicles coalesce to form the new plasma membranes and cell walls between the two cells (Figure 1.59).

■ **Figure 1.59** Cytokinesis in a plant cell

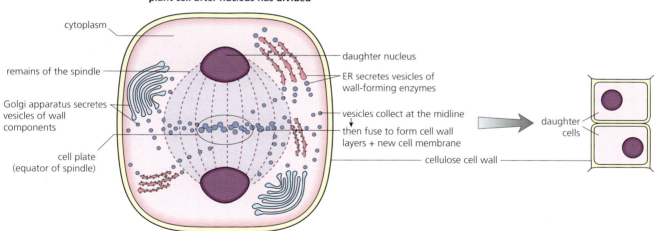

plant cell after nucleus has divided

cytoplasm

remains of the spindle

Golgi apparatus secretes vesicles of wall components

cell plate (equator of spindle)

daughter nucleus

ER secretes vesicles of wall-forming enzymes

vesicles collect at the midline then fuse to form cell wall layers + new cell membrane

cellulose cell wall

daughter cells

The packaging of DNA in the chromosomes

The total length of the DNA of the human chromosomes is over 2 m, shared out between the 46 chromosomes. Each chromosome contains one very long DNA molecule. Of course, chromosomes vary in length but we can estimate that a typical chromosome of 5 μm length contains a DNA molecule that is approximately 5 cm long (that is about 50 000 μm of DNA is packed into 5 μm of chromosome). Today we know that, while some of the proteins of the chromosome are enzymes involved in the copying and repair reactions of DNA, the bulk of chromosome protein has a support and packaging role for DNA.

28 Suggest a main advantage of chromosomes being 'supercoiled' during metaphase of mitosis

One sort of packaging protein is a substance called histone. This is a basic (positively charged) protein containing a high concentration of amino acid molecules with additional base groups (–NH$_2$), such as lysine and arginine. These histones occur clumped together, and provide support to the lengths of the DNA double helix that occur wrapped around them, giving the appearance of beads on a thread. The 'bead thread' is itself coiled up, forming the chromatin fibre. The chromatin fibre is again coiled, and the coils are looped around a 'scaffold' protein fibre, made of a non-histone protein. This whole structure is folded again (super-coiled) into the much-condensed metaphase chromosome (Figure 1.60).

■ **Figure 1.60**
The packaging of DNA in the chromosomes

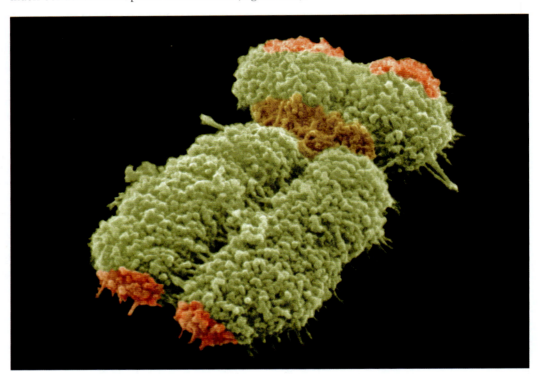

Observing chromosomes during mitosis

Actively dividing cells, such as those at the growing points of the root tips of plants, include many cells undergoing mitosis. This tissue can be isolated, stained with an orcein ethanoic (acetic orcein) stain, squashed, and then examined under the high-power lens of a microscope. Nuclei at interphase appear red–purple with almost colourless cytoplasm, but the chromosomes in cells undergoing mitosis will be visible, rather as they appear in the photomicrographs in Figure 1.58. The procedure is summarized in the flow diagram in Figure 1.61.

From the resulting temporary slides, the proportion of cells with nuclei at interphase or at any stage of mitosis can be calculated from counts of 100 adjacent nuclei. From this data the **mitotic index** – the number of cells undergoing mitosis per thousand cells – can be calculated. For greatest accuracy, the mean of three or more samples of 100 cells should be used. This is important because the mitotic index is used to differentiate benign from malignant tumours – a critical distinction (see below).

■ **Figure 1.61**
The orcein ethanoic stain of an onion root tip squash

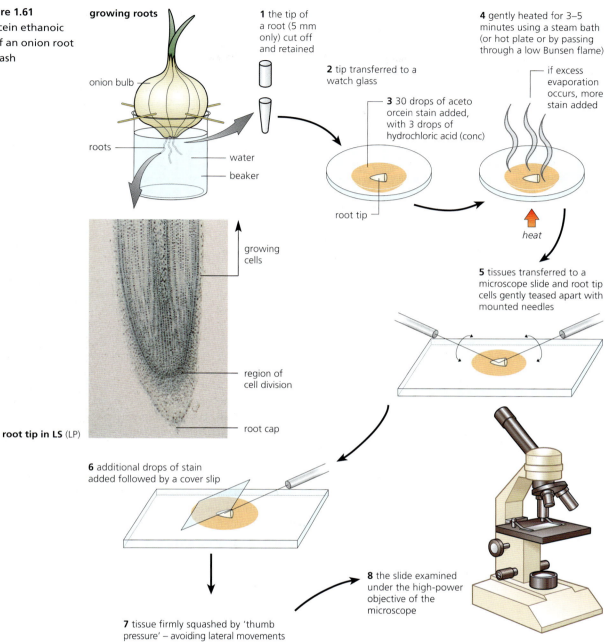

growing roots

onion bulb

roots

water

beaker

1 the tip of a root (5 mm only) cut off and retained

2 tip transferred to a watch glass

3 30 drops of aceto orcein stain added, with 3 drops of hydrochloric acid (conc)

root tip

4 gently heated for 3–5 minutes using a steam bath (or hot plate or by passing through a low Bunsen flame)

if excess evaporation occurs, more stain added

heat

5 tissues transferred to a microscope slide and root tip cells gently teased apart with mounted needles

growing cells

region of cell division

root cap

root tip in LS (LP)

6 additional drops of stain added followed by a cover slip

8 the slide examined under the high-power objective of the microscope

7 tissue firmly squashed by 'thumb pressure' – avoiding lateral movements

Determining the mitotic index

Using a prepared, stained slide of a plant root tip (see Figure 1.61), locate the meristematic region behind the root tip, using the low-power magnification on your microscope. Then focus on this region (with higher-power magnification if necessary), so that you can identify cells where chromosomes are visible (cells undergoing mitosis). Other cells will show their nucleus at interphase. Selecting 100 cells and using a tally chart, record the following data.

Mitotic index of 100 root tip cells	
Number of cells at a stage of mitosis	Number of cells at interphase

Use your data to calculate the mitotic index.

29 Using slides they had prepared to observe chromosomes during mitosis in a plant root tip (Figure 1.61), five students observed and recorded the number of nuclei at each stage in mitosis in 100 cells as shown in the table below.

Stage of mitosis	Number of nuclei counted by				
	student 1	student 2	student 3	student 4	student 5
prophase	64	70	75	68	73
metaphase	13	10	7	11	9
anaphase	5	5	2	8	5
telophase	18	15	16	13	13

a Calculate the mean percentage of dividing cells at each stage of mitosis and present your results as a pie chart.

b Assuming that mitosis takes about 60 minutes to complete in this species of plant, **deduce** what these results imply about the lengths of the four steps.

■ Cancer – diseases of uncontrolled cell division

There are many different forms of cancer, affecting different tissues of the body. Cancer is not thought of as a single disease. Today, in developed countries, one in three people will suffer from cancer at some point in their life and approximately one in four will die from it. In these regions, the commonest cancers are of the lung in males and of the breast in females. However, in many parts of the world, cancer rates are different – often they are significantly lower. Biologists in laboratories throughout the world are researching into the causes and treatment of cancer. You can see the range of common cancers and their incidences worldwide at: http://globocan.iarc.fr/

In all cancers, cells start to divide repeatedly by mitosis, without control or regulation. Where this occurs in the body, the rate of cell multiplication is much faster than the rate of cell death. An irregular mass of cells if formed, called a **tumour** (Figure 1.62).

Sometimes tumour cells break away from this **primary** tumour and are carried to other parts of the body, where they form a **secondary tumour**. This process is known as a **metastasis**. Unchecked, cancerous cells ultimately take over the body at the expense of the surrounding healthy cells, leading to malfunction and death.

So, cancer arises when the cell cycle operates without its normal controls. In a healthy cell, the cell cycle is regulated by a molecular control system in which cyclins are involved (page 52). Cancers are believed to start when changes occur in these genes. These changes are principally caused by damage to the DNA molecules of chromosomes.

A **mutation** is a change in the amount or chemical structure of DNA of a chromosome. Mistakes of different types build up in the DNA of the body cells. The accumulation of mistakes with time explains why the majority of cancers arise in older people.

■ **Figure 1.62**
Steps in the development of a malignant tumour

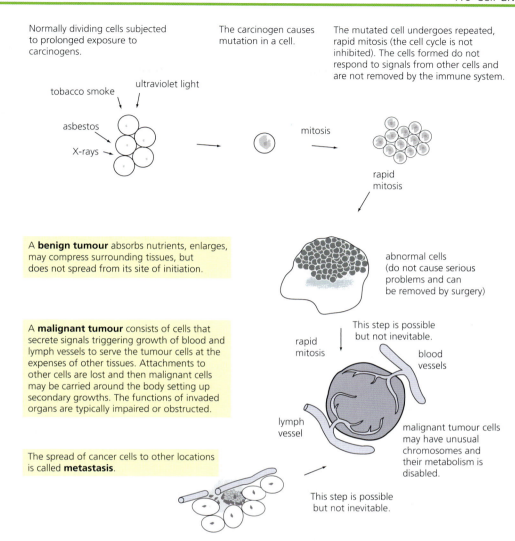

Normally dividing cells subjected to prolonged exposure to carcinogens.

tobacco smoke
ultraviolet light

asbestos

X-rays

The carcinogen causes mutation in a cell.

mitosis

The mutated cell undergoes repeated, rapid mitosis (the cell cycle is not inhibited). The cells formed do not respond to signals from other cells and are not removed by the immune system.

rapid mitosis

A **benign tumour** absorbs nutrients, enlarges, may compress surrounding tissues, but does not spread from its site of initiation.

abnormal cells (do not cause serious problems and can be removed by surgery)

A **malignant tumour** consists of cells that secrete signals triggering growth of blood and lymph vessels to serve the tumour cells at the expenses of other tissues. Attachments to other cells are lost and then malignant cells may be carried around the body setting up secondary growths. The functions of invaded organs are typically impaired or obstructed.

This step is possible but not inevitable.

rapid mitosis

blood vessels

lymph vessel

malignant tumour cells may have unusual chromosomes and their metabolism is disabled.

The spread of cancer cells to other locations is called **metastasis**.

This step is possible but not inevitable.

■ The causes of DNA damage

A factor capable of causing a mutation is called a **mutagen**, and its effects are described as mutagenic. These factors are identified and defined in Table 1.8.

30 **Describe** *three* environmental conditions that may cause normal cells to become cancerous cells.

31 **Describe** how the behaviour of cancerous cells differs from that of normal cells.

■ **Table 1.8**
Principal factors
which can increase
mutation rates and
the likelihood of
cancer

Ionizing radiations

Ionizing radiation includes X-rays and radiation (gamma rays, α particles, β particles) from various radioactive sources. These may trigger the formation of damaging ions inside the nucleus, leading to the break-up of the DNA.

Non-ionizing radiations

Non-ionizing radiations include UV light. This is less penetrating than ionizing radiation but, if it is absorbed by the nitrogenous bases of DNA, may modify them – causing adjacent bases on the DNA strand to bind to each other, instead of binding to their partner on the opposite strand (page 107).

Chemicals

Several chemicals that are carcinogens are present in **tobacco smoke**. Also, prolonged exposure to asbestos fibres may trigger cancer in the linings of the thorax cavity (pleural membranes). The harm usually becomes apparent only many years later.

Virus infection

A specific virus infection, such as with hepatitis B and C viruses, may trigger liver cancer.

Diet

Diet is also linked to both the cause and the prevention of cancer, although the nature of these connections is difficult to establish with any certainty.

■ Oncogenes and cancer

Whatever the cause of a cancer, two types of genes play a part in initiating a cancer if they mutate:

■ **Proto-oncogenes** are genes that code for the proteins that stimulate the cell cycle. Mutations in an oncogene may result in excessive cell division, causing cells to become 'immortal' if their nutrient supply is maintained.

■ **Tumour-suppressing genes** code for proteins that stop a cell cycle if damaged DNA is being copied. Unfortunately, a rare mutation may inactivate the gene which codes for a protein known as p53. When protein p53 is present, copying of faulty DNA is stopped. Then, other enzymes are able to repair the DNA and correct the fault, so cancer is avoided. When p53 is absent, tumour formation is possible.

Cigarette smoking causes lung diseases – the evidence

It was **epidemiology** (the study of the incidence and distribution of diseases and of their control and prevention) that first identified the likely causal links between smoking and disease. More recently, experimental **laboratory-based investigations** have demonstrated *how* cigarette smoke causes disease. Evidence is discussed in Chapter 6.

■ *Examination questions – a selection*

Questions 1–2 are taken from IB Diploma biology papers.

Q1 A red blood cell is 8 μm in diameter. If drawn 100 times larger than its actual size, what diameter will the drawing be in mm?
  A 0.08 mm
  B 0.8 mm
  C 8 mm
  D 80 mm

Standard Level Paper 1, Time Zone 1, May 11, Q3

Q2 The electron micrograph below shows an organelle in a eukaryotic cell. What is the area labelled X and what is the type of reaction occurring there?

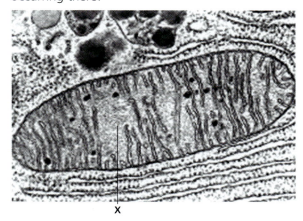

		Reaction
A	matrix	photolysis
B	stroma	Krebs cycle
C	stroma	photolysis
D	matrix	Krebs cycle

Higher Level Paper 1, Time Zone 1, May 11, Q29

Questions 3–9 cover other syllabus issues in this chapter.

Q3 Plasma membranes are fluid due to:
  A the amphipathic properties of phospholipids
  B the water present on the outside of the plasma membrane
  C the integral proteins with polar and nonpolar regions interacting with phospholipids
  D vesicles fusing with the plasma membrane during exocytosis.

Q4 During the cell cycle the concentration of cyclins fluctuates in a cyclical way. A likely explanation for this fluctuation in concentration is:
  A because cyclins are active during the whole cycle due to their enzymatic activity
  B it is the result of cell differentiation as cyclins activate different genes at different times
  C because cyclins are only active at some points of the cycle, inducing changes from phase to phase
  D cyclins were named due to the cyclical activity of cells when duplicating their DNA.

Q5 **a** List three ideas contained within the cell theory. (3)
  b Distinguish between stem cells and cancer cells. (2)

Q6 **a** Describe what happens as cells increase their volume in relationship with their surface area. (2)
  b Cells differentiate during development of embryonic tissue cells. What is the likely cause for this differentiation to occur? (2)

Q7 **a** Draw and label a diagram of a generalised prokaryotic cell. Annotate your diagram with the functions of each named structure. (6)
  b Explain how the size of a prokaryotic cell relates to the size of some organelles typically found in animal cells. (2)
  c Define binary fission. (2)
  d In a table, indicate four major differences between prokaryotes and eukaryotes (4)

Q8 The stages of the cell cycle are interphase, mitosis and cytokinesis. Four distinct phases make up the process of mitosis itself.

  a Outline the events of interphase which establish that this is not a 'resting' stage in the cell cycle. (4)

  b List the major changes that occur to the chromosomes of a nucleus undergoing mitosis during:

    i prophase (the first phase)

    ii anaphase (the third phase). (6)

  c Explain how mitosis produces two genetically identical nuclei. (4)

Q9 Outline the way exocytosis and endocytosis allow cells to export and import substances using vesicles. (8)

2 Molecular biology

ESSENTIAL IDEAS

- Living organisms control their composition by a complex web of chemical reactions.
- Water is the medium of life.
- Compounds of carbon, hydrogen and oxygen are used to supply and store energy.
- Proteins have a very wide range of functions in living organisms.
- Enzymes control the metabolism of the cell.
- The structure of DNA allows efficient storage of genetic information.
- Genetic information in DNA can be accurately copied and can be translated to make the proteins needed by the cell.
- Cell respiration supplies energy for the functions of life.
- Photosynthesis uses the energy in sunlight to produce the chemical energy needed for life.

 Appendix 1: Background chemistry for biologists may be useful when studying this chapter, especially concerning elements, atoms, molecules, ions and compounds, and the ways atoms form molecules.

2.1 Molecules to metabolism – *living organisms control their composition by a complex web of chemical reactions*

The foundations of molecular biology were laid down just 60 years ago by the work of Francis Crick and James Watson, when they established the structure of DNA. When teams or individuals achieve a breakthrough in understanding like this, their discoveries are often named after them. For example, the Krebs cycle in aerobic respiration (page 354) was named after Sir Hans Krebs, and the Calvin cycle in photosynthesis after Melvin Calvin (page 367). Today, **molecular biology** is focused on explaining the range of living processes in terms of the chemical substances involved. This approach has revolutionized biology as a whole, although it is essentially reductionist in nature.

Chemical **elements** are the units of pure substance that make up our world. The Earth is composed of about 92 stable elements in all, present in varying quantities. About 16 elements are required by cells and are, therefore, essential for life. Consequently, the full list of essential elements is a relatively short one. Furthermore, about 99% of living matter consists of just four elements: **carbon**, **hydrogen**, **oxygen** and **nitrogen**.

Why do these four elements predominate in living things?

The elements carbon, hydrogen and oxygen make up the greater part of us because living things contain large quantities of **water**, and also because most other molecules present in cells and organisms are compounds of **carbon** combined with hydrogen and oxygen, including the **carbohydrates**, **lipids** and **nucleic acids**. Compounds containing carbon and hydrogen are known as **organic compounds**.

1 **Distinguish** between the terms 'atom', 'molecule' and 'ion'.

2 **Suggest** where non-organic forms of carbon exist in the biosphere.

The element nitrogen is combined with carbon, hydrogen and oxygen in compounds called amino acids, from which **proteins** are constructed. We will examine the carbon atom and how it forms stable compounds next. Water is the subject of Section 2.2.

■ The carbon atom, how it forms stable compounds, and its significance

Carbon has unique properties – so remarkable we say that they make life possible. These are outlined below.

1 Atoms combine (or 'bond') to form molecules in ways that produce a stable arrangement of electrons in the outer shells of each atom. Atoms are most stable when their outer shell of electrons is complete. The first electron shell of an atom can hold up to two electrons and then it is full. The second shell can hold up to a maximum of eight electrons. Carbon is a relatively small atom. It has four electrons in its second shell, and is able to form four strong, stable bonds. The bonds that carbon atoms form are called **covalent bonds**. In covalent bonding, electrons are shared between atoms – you can see the four covalent bonds in methane in Figure 2.1.

■ **Figure 2.1**
Methane, the simplest organic compound

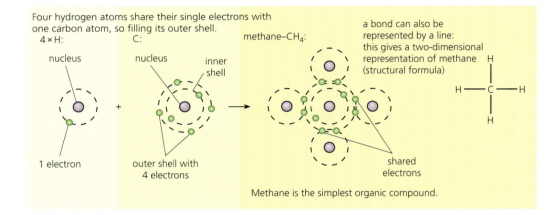

Four hydrogen atoms share their single electrons with one carbon atom, so filling its outer shell.

4 × H: C: methane–CH₄:

a bond can also be represented by a line: this gives a two-dimensional representation of methane (structural formula)

nucleus   nucleus   inner shell

1 electron   outer shell with 4 electrons   shared electrons

Methane is the simplest organic compound.

2 Covalent bonds are the strongest bonds found in biological molecules. This means they need the greatest input of energy to break them. So covalent bonds provide great stability to biological molecules, many of which are very large. Carbon atoms are also able to form covalent bonds with atoms of oxygen, nitrogen and sulfur, forming different groups of organic molecules with distinctive properties.

3 In addition, carbon atoms are able to react with each other to form extended and extremely stable chains. These 'carbon skeletons' may be straight chains, branched chains or rings (Figure 2.2). At least two and a half million organic compounds exist – more than the total of known compounds of all the other elements, in fact.

4 The four covalent bonds of carbon atoms point to the corners of a regular tetrahedron (a pyramid with a triangular base). This is because the four pairs of electrons repel each other and so position themselves away from each other, as far as possible. If there are different groups attached to each of the four bonds around a carbon atom, there are two different ways of arranging the groups (Figure 2.3). This can lead to forms of molecules which are mirror images of each other. Carbon atoms with four different atoms or groups attached are said to be asymmetric. This is another cause of variety among organic molecules.

5 Carbon atoms can form more than one bond between them (Figure 2.4). For example, carbon atoms may share two electrons to form a **double bond**. Carbon compounds that contain double carbon=carbon bonds are known to chemists as 'unsaturated', and carbon compounds containing double bonds are unsaturated compounds. For example, we will meet unsaturated fats of biological importance shortly. In fact, carbon, nitrogen and oxygen all form double bonds. This introduces yet more variety to the range of carbon compounds that make up cells.

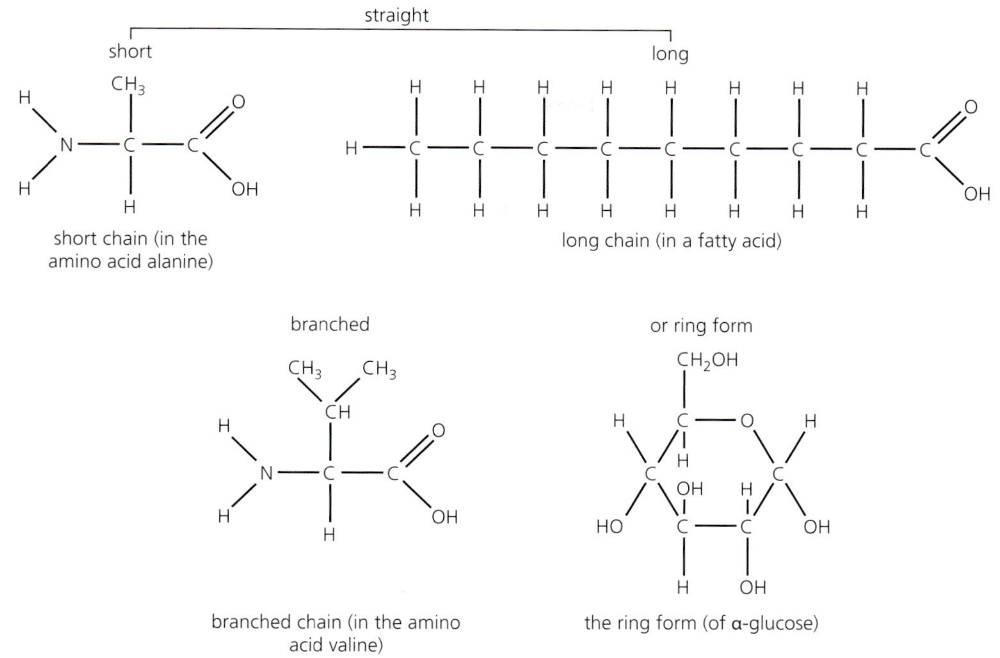

■ **Figure 2.2** Carbon atoms bonded with other carbon atoms to form 'skeletons'

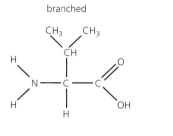

straight

short

long

short chain (in the amino acid alanine)

long chain (in a fatty acid)

branched

or ring form

branched chain (in the amino acid valine)

the ring form (of α-glucose)

■ **Figure 2.3** The tetrahedral carbon atom

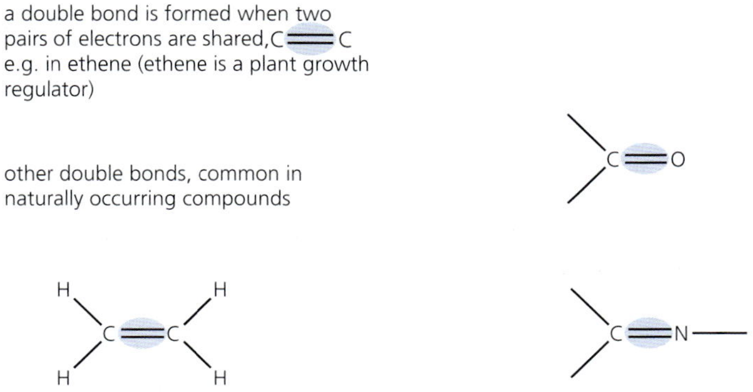

the carbon atom is at the centre of the tetrahedron, a three-dimensional structure, e.g. methane

a)

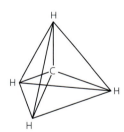

'ball and spring' model

b)

space-filling model

c)

perspective formula

d)

with different groups attached to each of the four bonds, there are two ways of arranging them, each a mirror image of the other

e)

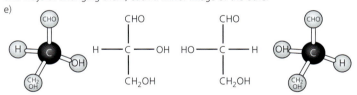

these two forms of glyceraldehyde have very similar chemical properties, but cell enzymes tell them apart – and will only react with one of them

■ **Figure 2.4** Carbon–carbon bonds

a double bond is formed when two pairs of electrons are shared, $C\!=\!=\!C$ e.g. in ethene (ethene is a plant growth regulator)

other double bonds, common in naturally occurring compounds

6 Fortunately, all these carbon compounds fit into a relatively small numbers of 'families' of compounds. Chemical families we often come across in biology include the alcohols, organic acids, ketones and aldehydes. The families are identified by a part of their molecule that is the **functional group**, which gives them their characteristic chemical properties. The chemical structure of these functional groups is shown in Figure 2.5. The remainder of the organic molecule, apart from the functional group, has little or no effect on the chemical properties of the functional group, and is referred to as the **R group**.

■ **Figure 2.5** Some functional groups

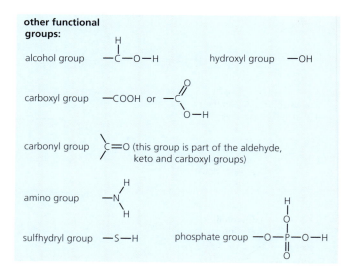

Due to these properties, molecules containing a carbon 'skeleton' exist in vast numbers. We now know that living organisms control their composition and functions by a complex web of chemical reactions. One outcome is the presence in cells of a huge range of organic compounds, to be introduced in following sections.

■ What is metabolism?

There are literally many thousands of chemical reactions taking place within cells and organisms. **Metabolism** is the name we give to these chemical reactions of life. The molecules involved are collectively called **metabolites**. Many metabolites are made in organisms, but others are imported from the environment, such as from food substances, water and the gases carbon dioxide and oxygen.

Metabolism actually consists of chains (linear sequences) and cycles of enzyme-catalysed reactions, such as we see in respiration (page 118), photosynthesis (page 121), protein synthesis (page 110), and in very many other pathways (Figure 2.6). **Enzymes** are biological catalysts (page 92). All these reactions may be classified as one of just two types, according to whether they involve the build-up or breakdown of organic molecules.

■ In **anabolic reactions**, larger molecules are built up from smaller molecules. Examples of anabolism are the synthesis of proteins from amino acids and the synthesis of polysaccharides from simple sugars. The reactions involved include **condensation** reactions (page 78).

■ In **catabolic reactions**, larger molecules are broken down. Examples of catabolism are the digestion of complex foods and the breakdown of sugar in respiration. The reactions involved include **hydrolysis** reactions (page 78).

Overall:

metabolism = anabolism + catabolism

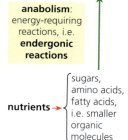

synthesis of complex molecules used in growth and development and in metabolic processes, e.g. proteins, polysaccharides, lipids, hormones, growth factors, hemoglobin, chlorophyll

anabolism: energy-requiring reactions, i.e. **endergonic reactions**

nutrients → sugars, amino acids, fatty acids, i.e. smaller organic molecules

catabolism energy-releasing reactions, i.e. **exergonic reactions**

release of simple substances, e.g. small inorganic molecules, CO_2, H_2O, mineral ions

■ **Figure 2.6** Metabolism, an overview

Metabolism and energy

Chemical energy exists in the structure of molecules. Every molecule contains a quantity of stored energy equal to the quantity of energy needed to synthesize it in the first place. So chemical energy exists in the structural arrangement of the atoms in the molecule, rather than in the chemical bonds between the atoms.

When glucose is oxidized to carbon dioxide and water in aerobic cell respiration energy is transferred. This energy is no longer in store but is on the move; it is active energy. Actually, only part of the stored energy in a molecule is available, known as **free energy**, and can be used to do work. Reactions that release free energy are known as **exergonic reactions** (Figure 2.7). The oxidation of glucose is an example of an exergonic reaction.

On the other hand, reactions that require energy are called **endergonic reactions**. The synthesis of a protein from amino acids is an example of an endergonic reaction.

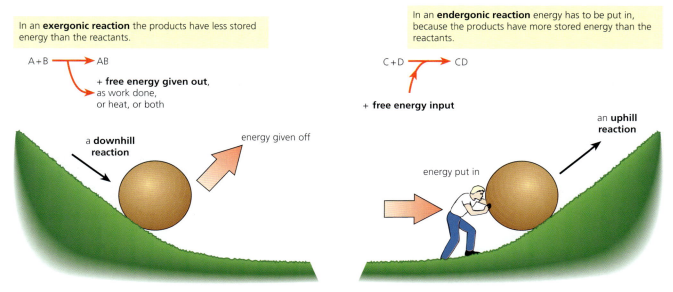

In an **exergonic reaction** the products have less stored energy than the reactants.

A+B ⟶ AB
+ **free energy given out**,
as work done,
or heat, or both

a **downhill reaction**

energy given off

In an **endergonic reaction** energy has to be put in, because the products have more stored energy than the reactants.

C+D ⟶ CD
+ **free energy input**

an **uphill reaction**

energy put in

■ **Figure 2.7** Exergonic and endergonic reactions

The many endergonic reactions that occur in metabolism are made possible by being coupled to exergonic reactions. Coupling occurs through a molecule called **adenosine triphosphate (ATP)**. ATP is referred to as the **energy currency** molecule in biology (Figure 2.62, page 117). Molecules of ATP work in metabolism by acting as common intermediates, linking energy-requiring and energy-yielding reactions. Metabolic processes mostly involve ATP, directly or indirectly.

Now look at the cell metabolites whose chemical structure is shown in Figure 2.8. There are molecules you need to be able to draw, once you are familiar with their structure.

3 The importance of carbon lies in a unique combination of properties. Together, these properties are so remarkable that biologists believe they have made life possible. Establish for yourself that you understand them by closing the book and then recording in **outline** why each is significant.

Molecular diagrams of metabolites	Identification points and further information
glucose – a six carbon sugar ($C_6H_{12}O_6$): the two forms of glucose depend on the positions of the —H and —OH attached to carbon-1 [glucose in pyranose rings] 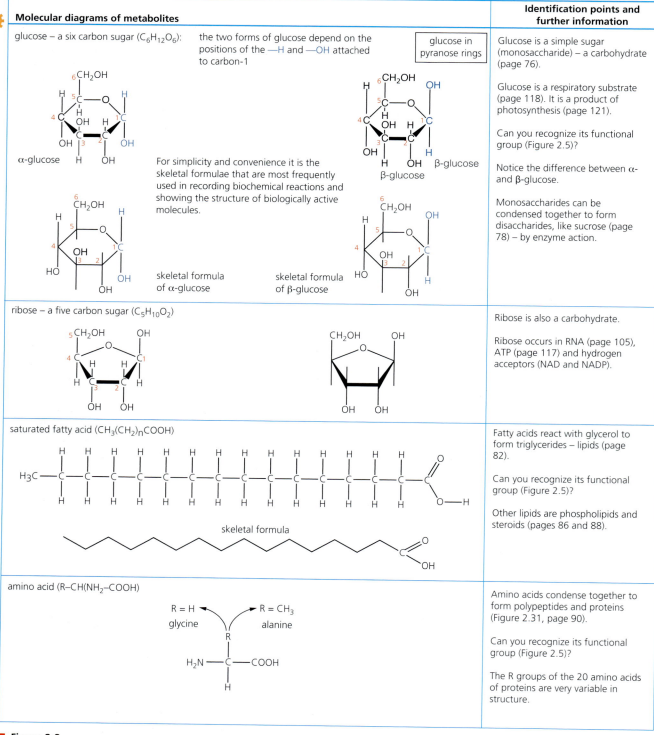 α-glucose β-glucose For simplicity and convenience it is the skeletal formulae that are most frequently used in recording biochemical reactions and showing the structure of biologically active molecules. skeletal formula of α-glucose skeletal formula of β-glucose	Glucose is a simple sugar (monosaccharide) – a carbohydrate (page 76). Glucose is a respiratory substrate (page 118). It is a product of photosynthesis (page 121). Can you recognize its functional group (Figure 2.5)? Notice the difference between α- and β-glucose. Monosaccharides can be condensed together to form disaccharides, like sucrose (page 78) – by enzyme action.
ribose – a five carbon sugar ($C_5H_{10}O_2$)	Ribose is also a carbohydrate. Ribose occurs in RNA (page 105), ATP (page 117) and hydrogen acceptors (NAD and NADP).
saturated fatty acid ($CH_3(CH_2)_nCOOH$) skeletal formula	Fatty acids react with glycerol to form triglycerides – lipids (page 82). Can you recognize its functional group (Figure 2.5)? Other lipids are phospholipids and steroids (pages 86 and 88).
amino acid ($R–CH(NH_2–COOH)$) R = H R = CH_3 glycine alanine	Amino acids condense together to form polypeptides and proteins (Figure 2.31, page 90). Can you recognize its functional group (Figure 2.5)? The R groups of the 20 amino acids of proteins are very variable in structure.

■ **Figure 2.8**

An introduction to sugars, lipids, and amino acids

Drawing molecular diagrams

You should be able to draw molecular diagrams of the molecules listed below. Their structures are shown in Figure 2.8, but you may find them easier to memorize as you become familiar with their main roles in metabolism. Practise drawing them again at these points:

- glucose and ribose – Figure 2.14, page 76
- a saturated fatty acid – Figure 2.21, page 82
- an amino acid – Figure 2.30, page 90.

Nature of Science

Falsification of theories

■ Molecular biology, metabolism and 'vitalism'

It was once believed that organic compounds could be produced only by the chemical processes within living things. The view was that a vital force or 'spark' in life created the molecules of living matter – the chemicals of life could not be reproduced by 'test-tube' reactions. This theory was known as **'vitalism'**.

In 1828 the German chemist Frederick Wöhler heated ammonium cyanate, an inorganic compound, and produced urea. Urea is a typical animal product – produced by the liver (page 468):

$$
\underset{H_2N}{}\overset{\displaystyle \overset{O}{\underset{\|}{C}}}{}\underset{NH_2}{}
$$

Wöhler had synthesized biological material from non-biological substances. The results of his demonstration – once they were widely accepted – ended the idea of vital force. Today, molecular biology explains living processes in terms of the chemical substances involved, without controversy.

2.2 Water – *water is the medium of life*

Living things are typically solid, substantial objects, yet water forms the bulk of their structures – between 65% and 95% by mass of most multicellular plants and animals (about 80% of a human cell consists of water). Despite this, and the fact that water has some unusual properties, water is a substance that is often taken for granted.

The water molecule consists of one atom of oxygen and two atoms of hydrogen combined together by sharing electrons (**covalent bonding**). However, the molecule is triangular rather than linear, and the nucleus of the oxygen atom draws electrons (negatively charged) away from the hydrogen nuclei (positively charged) – with an interesting consequence. Although overall the water molecule is electrically neutral, there is a net negative charge on the oxygen atom and a net positive charge on the hydrogen atoms. In other words, the water molecule carries an unequal distribution of electrical charge within it. This arrangement is known as a **polar molecule** (Figure 2.9).

4 Distinguish between ionic and covalent bonding.

Nature of Science

Using theories to explain natural phenomena

■ Hydrogen bonds

With water molecules, the positively charged hydrogen atoms of one molecule are attracted to negatively charged oxygen atoms of nearby water molecules, causing forces called **hydrogen bonds**. These are weak bonds compared to covalent bonds, yet they are strong enough to hold water molecules together and to attract water molecules to charged particles or to a charged surface. In fact, hydrogen bonds largely account for the unique properties of water. We examine these properties next.

■ The contrasting physical states of water and methane

Water has a relative molecular mass of only 18, yet it is a liquid at room temperature. This is surprising; it contrasts with other small molecules that are **gases**, for example, methane (CH_4) of molecular mass 16. Similarly, neither ammonia (NH_3) of molecular mass 17 nor carbon dioxide (CO_2, 44) are liquids at room temperature. But neither of these molecules contains hydrogen bonds. In gases, molecules are widely spaced and free to move about independently. In liquids, molecules are closer together.

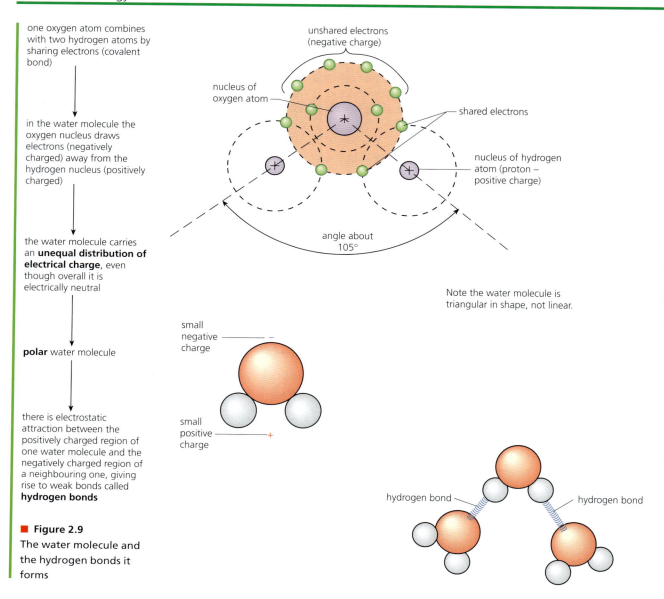

one oxygen atom combines with two hydrogen atoms by sharing electrons (covalent bond)

in the water molecule the oxygen nucleus draws electrons (negatively charged) away from the hydrogen nucleus (positively charged)

the water molecule carries an **unequal distribution of electrical charge**, even though overall it is electrically neutral

polar water molecule

there is electrostatic attraction between the positively charged region of one water molecule and the negatively charged region of a neighbouring one, giving rise to weak bonds called **hydrogen bonds**

unshared electrons (negative charge)

nucleus of oxygen atom

shared electrons

nucleus of hydrogen atom (proton – positive charge)

angle about 105°

Note the water molecule is triangular in shape, not linear.

small negative charge

small positive charge

hydrogen bond

hydrogen bond

■ **Figure 2.9**
The water molecule and the hydrogen bonds it forms

In the case of water, hydrogen bonds pull the molecules very close to each other, which is why water is a liquid at the temperatures and pressure that exist over much of the Earth's surface. As a result, we have a liquid medium with distinctive thermal and solvent properties.

■ Thermal properties of water

Heat energy and the temperature of water

A lot of heat energy is required to raise the temperature of water. This is because much energy is needed to break the hydrogen bonds that restrict the movements of water molecules. This property of water is its **specific heat capacity**. The specific heat capacity of water is the highest of any known substance. Consequently, aquatic environments like streams and rivers, ponds, lakes and seas are very slow to change temperature when the surrounding air temperature changes. Aquatic environments have much more stable temperatures than do terrestrial (land) environments.

Another consequence is that cells and the bodies of organism do not change temperature readily. Bulky organisms, particularly, tend to have a stable body temperature in the face of a fluctuating surrounding temperature, whether in extremes of heat or cold.

Evaporation and heat loss

The hydrogen bonds between water molecules make it difficult for them to be separated and vaporized (to evaporate). This means that much energy is needed to turn liquid water into water vapour (gas). This amount of energy is the **latent heat of vaporization**, and for water it is very high. Consequently, the evaporation of water in sweat on the skin, or in transpiration from green leaves, causes marked cooling. The escaping molecules take a lot of energy with them. You experience this when you stand in a draught after a shower. And since a great deal of heat is lost with the evaporation of a small amount of water, cooling by evaporation of water is economical on water, too.

Heat energy and freezing

The amount of heat energy that must be removed from water to turn it to ice is very great, as is that needed to melt ice. This amount of energy is the **latent heat of fusion** and is very high for water. As a result, both the contents of cells and the water in the environment are always slow to freeze in extreme cold.

■ Cohesive properties of water

Cohesion is the force by which individual molecules stick together. Water molecules stick together as a result of hydrogen bonding. These bonds continually break and reform with other, surrounding water molecules but, at any one moment, a large number are held together by their hydrogen bonds.

Adhesion is the force by which individual molecules cling to surrounding material and surfaces. Materials with an affinity for water are described as **hydrophilic** (page 30). Water adheres strongly to most surfaces and can be drawn up long columns, such as through narrow tubes like the xylem vessels of plant stems, without danger of the water column breaking (Figure 2.11). Compared with other liquids, water has extremely strong adhesive and cohesive properties that prevent it 'breaking' under tension.

Related to the property of cohesion is the property of **surface tension**. The outermost molecules of water form hydrogen bonds with the water molecules below them. This gives water a very high surface tension – higher than any other liquid except mercury. The surface tension of water is exploited by insects that 'surface skate' (Figure 2.10). The insect's waxy cuticle prevents wetting of its body, and the mass of the insect is not great enough to break the surface tension.

Below the surface, water molecules slide past each other very easily. This property is described as low **viscosity**. Consequently, water flows readily through narrow capillaries, and tiny gaps and pores.

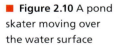
■ **Figure 2.10** A pond skater moving over the water surface

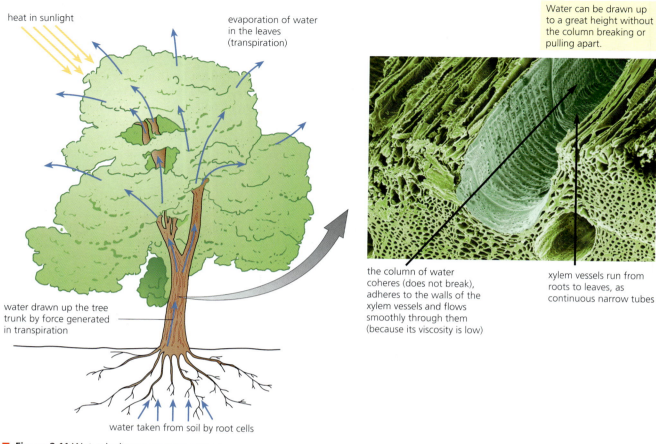

heat in sunlight

evaporation of water in the leaves (transpiration)

Water can be drawn up to a great height without the column breaking or pulling apart.

the column of water coheres (does not break), adheres to the walls of the xylem vessels and flows smoothly through them (because its viscosity is low)

xylem vessels run from roots to leaves, as continuous narrow tubes

water drawn up the tree trunk by force generated in transpiration

water taken from soil by root cells

■ **Figure 2.11** Water is drawn up a tree trunk

■ Solvent properties of water

Water is a powerful solvent for polar substances such as:

■ ionic substances like sodium chloride (Na^+ and Cl^-); all cations (positively charged ions) and anions (negatively charged ions) become surrounded by a shell of orientated water molecules (Figure 2.12)

■ **Figure 2.12** Water as universal solvent

Ionic compounds like NaCl dissolve in water,

$NaCl \rightleftharpoons Na^+ + Cl^-$

with a group of orientated water molecules around each ion:

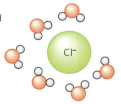

Sugars and alcohols dissolve due to hydrogen bonding between polar groups in their molecules (e.g. — OH) and the polar water molecules:

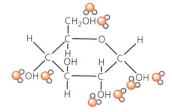

■ carbon-containing (organic) molecules with ionized groups (such as the carboxyl group –COO⁻ and amino group –NH₃⁺); soluble organic molecules like sugars dissolve in water due to the formation of hydrogen bonds with their slightly charged hydroxyl groups (–OH).

5 In an aqueous solution of glucose, **state** which component is the solvent and which is the solute.

Once they have dissolved, molecules (the **solute**) are free to move around in water (the **solvent**) and, as a result, are more chemically reactive than when in the undissolved solid.

On the other hand, non-polar substances are repelled by water, as in the case of oil on the surface of water. Non-polar substances are **hydrophobic** (water-hating).

■ Transport of metabolites in the blood – and their solubilities in water

Blood is the transport medium of the body, many metabolites being carried around the body in the blood plasma (page 254). Yet not all metabolites are soluble in water. In Table 2.1 the ways in which a selection of essential metabolites is carried are compared with their solubility in water.

A summary of the properties of water molecules and the associated benefits to life is given in Table 2.2.

■ **Table 2.1** Transport of metabolites in the blood

Metabolite	Solubility in water	Mechanism of transport in the blood
glucose	highly soluble	dissolved in the blood plasma
amino acids	soluble	dissolved in the blood plasmas
cholesterol	insoluble	in particles called low-density lipoproteins (LDLs), complexes of thousands of cholesterol molecules bound to proteins ('bad cholesterol'), and in high-density lipoprotein particles (HDLs or 'good cholesterol') (page 86)
fats (lipids)	insoluble	absorbed in the gut (into lacteals) as droplets, emulsified by bile salts; transported about the body (from fat store sites to respiring cells) as water-soluble phospholipids (page 88)
oxygen	low solubility in the plasma – about 4.5 ml/l	combined with hemoglobin in the red blood cells (page 254) – about 200 ml per litre of blood can be transported
sodium chloride	highly soluble	as Na⁺ and Cl⁻ ions, dissolved in the plasma

■ **Table 2.2** Water and life – summary

Property	Benefit to life
1 a liquid at room temperature, water dissolves more substances than any other common liquid	liquid medium for living things and for the chemistry of life
2 much heat energy needed to raise the temperature of water	aquatic environment slow to change temperature; bulky organisms have stable temperatures
3 evaporation requires a great deal of heat	evaporation causes marked cooling (much heat is lost by evaporation of a small quantity of water)
4 much heat has to be removed before freezing occurs	cell contents and water in aquatic environments are slow to freeze in cold weather
5 surface water molecules orientate with hydrogen bonds facing inwards	water forms droplets and rolls off surfaces; certain animals exploit surface tension to move over water surface
6 water molecules slide past each other easily (low viscosity)	water flows easily through narrow capillaries, and through tiny spaces (e.g. in soils, spaces in cell walls)
7 water molecules adhere to surfaces	water adheres to walls of xylem vessels as it is drawn up the stem to the leaves, from the roots
8 water column does not break or pull apart under tension	water can be lifted by forces applied at the top, and so can be drawn up xylem vessels of tree trunks by forces generated in the leaves

6 **Explain** the significance to organisms (plants and animals) of water as a coolant and transport medium in terms of its properties.

Water resources – an international issue

The waters of the Earth are constantly circulating. It is solar energy that drives this water cycle and, on average, a molecule of water that evaporates from the sea has returned within about 14 days. The distribution of water resources is shown in Figure 2.13.

An adequate supply of water is a vital requirement of all human communities. A great deal of water is used in industry, including in agriculture (for irrigation) and industrial manufacturing (for cooling and cleaning). Domestic consumption in developed countries demands more than 200 litres per person per day, of which only about 1.5 litres is used for drinking. Much water is wasted. The typical daily water consumption by people in the developing world is about 20 litres. Women and girls in rural Africa spend an average 3 hours per day collecting water.

Global fresh water resources are under threat from rising demand. Shared water resources have generated conflicts between communities in several parts of the world. Growing populations need more water for drinking, hygiene, sanitation and industry. The current sources of water for domestic and industrial uses are surface water (rivers, lakes and reservoirs) and deep groundwater (via wells and springs). The problem of water supply can be overcome in part by better use of available water – much of this is currently wasted. An alternative source is by removal of dissolved salts (desalination) of sea water or brackish water. This is practical where there is a source of cheap energy.

> **7 Design** a pilot plant for the desalination of sea water, given unlimited sunlight.

How much water is there?	Typical use of water in production and manufacturing	Past and future demand
The world's total water amounts to almost 1.5 billion km³ distributed between: salt water 96% fresh water 4% as polar ice/glaciers, underground, in lakes and rivers, as vapour in the air and in the biosphere	The number of litres of water used in the production of: a kilo of beef 16 000 a kilo of microchips 16 000 a kilo of cotton textile 11 000 a hamburger 2400 a kilo of bread 1300 a litre of wine 960 a kilo of crisps 925 a glass of orange juice 850 a litre of beer 300 a kilo of steel 260 *Source: Dr Peter Gleick, President Pacific Institute, California*	The changing pattern of fresh water consumption around the world – figure from the Pacific Institute, California

Water consumption
- Asia
- North America
- Europe
- Africa
- South America
- Australia and Oceania

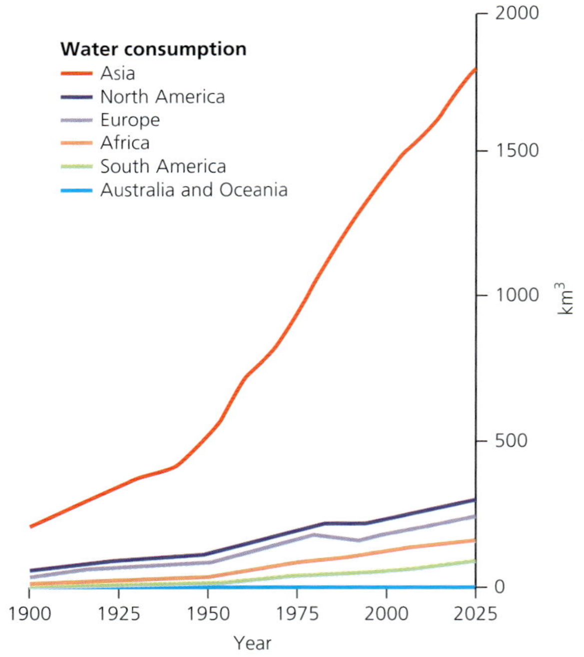

■ **Figure 2.13** Water – the issues

TOK Link

Claims about the memory of water have been categorized as pseudoscientific. What are the criteria that can be used to distinguish scientific claims from pseudoscientific claims?

2.3 Carbohydrates and lipids – *compounds of carbon, hydrogen and oxygen are used to supply and store energy*

■ Carbohydrates

Carbohydrates are the largest group of organic compounds found in living things. They include sugars, cellulose, starch and glycogen. Carbohydrates are compounds that contain only three elements: carbon, hydrogen and oxygen, with hydrogen and oxygen always present in the ratio 2:1 (as they are in water – H_2O), so they can be represented by the general formula $C_x(H_2O)_y$. Table 2.3 is a summary of the three types of carbohydrates commonly found in living things. *Remember, we have already met two monosaccharide sugars (Figure 2.8, page 68) and you are familiar with their structural formulae.*

■ **Table 2.3**
Carbohydrates of cells and organisms

Carbohydrates – general formula $C_x(H_2O)_y$		
Monosaccharides	**Disaccharides**	**Polysaccharides**
simple sugars (e.g. glucose and fructose with six carbon atoms; ribose with five carbon atoms)	two simple sugars condensed together (e.g. sucrose, lactose, maltose)	very many simple sugars condensed together (e.g. starch, glycogen, cellulose)

Monosaccharides – the simple sugars

Monosaccharides are carbohydrates with relatively small molecules. They taste sweet and are soluble in water. In biology, **glucose** is an especially important monosaccharide because:

- all green leaves manufacture glucose using light energy
- our bodies transport glucose in the blood
- all cells use glucose in respiration – we call it one of the respiratory substrates
- in cells and organisms, glucose is the building block for very many larger molecules.

The structure of glucose

Glucose has the chemical or **molecular formula** $C_6H_{12}O_6$. This type of formula tells us what the component atoms are, and the number of each in the molecule. For example, glucose is a 6-carbon sugar, or hexose. But the molecular formula does not tell us the structure of the molecule.

Glucose can be written on paper as a linear molecule but it cannot exist in this form (because each carbon arranges its four bonds into a tetrahedron, so the molecule cannot be 'flat'). Rather, glucose is folded, taking a ring or cyclic form. Figure 2.14 shows the structural formula of glucose.

The carbon atoms of an organic molecule may be numbered. This allows us to identify which atoms are affected when the molecule reacts and changes shape. For example, as the glucose ring forms, the oxygen on carbon-5 attaches itself to carbon-1. Note that the glucose ring contains five carbon atoms and an oxygen atom (Figure 2.14).

Isomers

Compounds that have the same component atoms in their molecules but which differ in the arrangement of the atoms are known as **isomers**. Many organic compounds exist in isomeric forms. For example:

- in the ring structure of glucose the positions of the –H and –OH groups that are attached to carbon atom 1 may interchange, giving rise to two isomers known as **α-glucose** and **β-glucose** (Figure 2.8, page 68)
- an organic compound that has four different chemical groupings attached to a single carbon atom gives rise to a three-dimensional model that can be built in two ways. These are identical except that they are mirror images of each other. The two forms of glyceraldehyde are examples (Figure 2.15). They have the same chemical properties, but in solution they rotate the plane of plane-polarized light in opposite directions, and so they are

optical isomers. One form rotates the plane of polarized light to the right and is said to be dextrorotatory (represented in the name by **D-**), and the other rotates the plane of polarized light to the left and is said to be levorotatory (represented by **L-**).

8 **Suggest** why simple sugars like glucose are not commonly found as a storage form of carbohydrate in cells or tissues.

In the following pages you learn about the roles of α-**D-glucose** and β-**D-glucose** in cell chemistry. The names of organic compounds – and of the linkages between compounds when they combine – can be complex because their names give this information. When we compare the structure of the polysaccharide cellulose with that of starch and glycogen, the significance of this difference will become apparent. Now you know what these names mean.

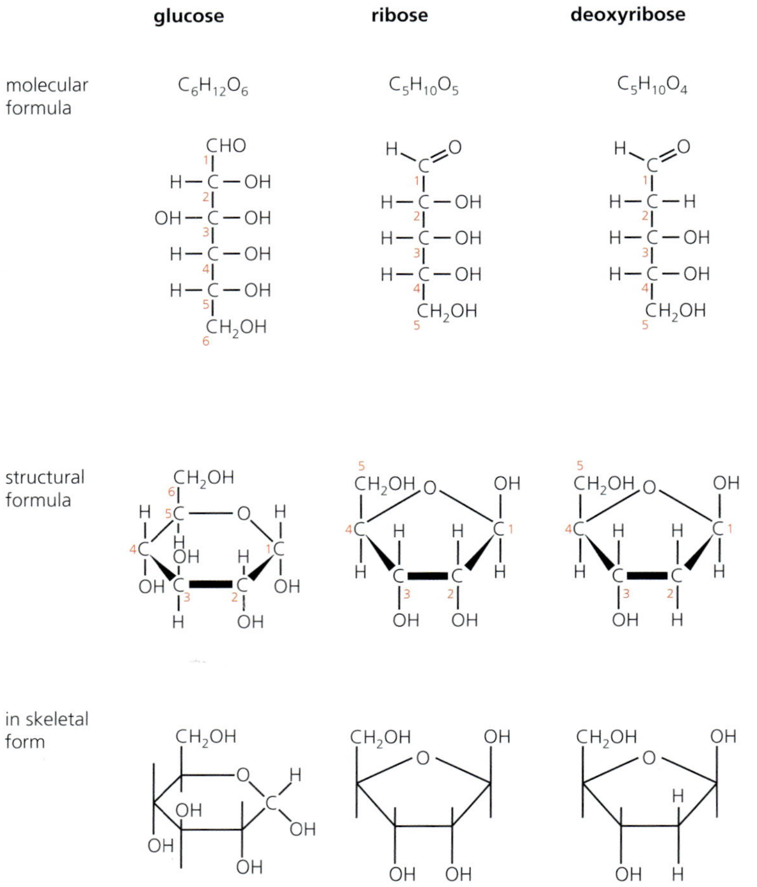

■ **Figure 2.14**
Structural formulae of some monosaccharides

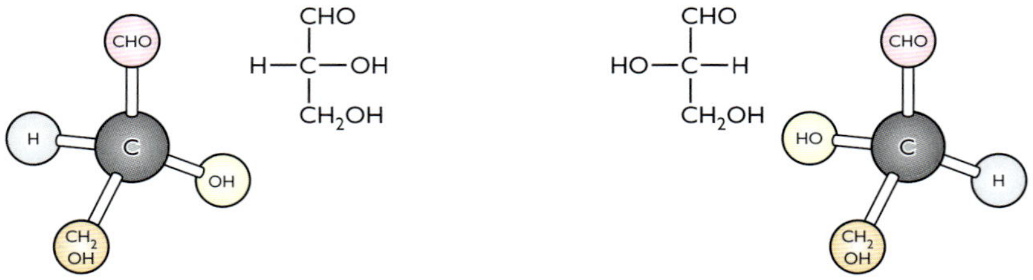

■ **Figure 2.15** Glyceraldehyde:

- the carbon atom is capable of reacting with four other atoms to form covalent bonds
- the four covalent bonds of carbon point to the corners of a regular tetrahedron (a pyramid with a triangular base)
- carbon compounds are three-dimensional molecules. See *Appendix 1, page 5*.

Other monosaccharides of importance in living cells

Glucose, fructose and galactose are examples of hexose sugars that commonly occur in cells and organisms. Other monosaccharide sugars produced by cells and used in metabolism include:

- 3-carbon sugars (**trioses**), early products in photosynthesis (page 367)
- 5-carbon sugars (**pentoses**), namely ribose and deoxyribose.

The pentoses ribose and deoxyribose are components of the nucleic acids (page 105). The structures of both these pentoses are also shown in Figure 2.14.

Test for 'reducing sugars'

Glucose (an aldose) and fructose (a ketose) are reducing sugars (see Figure 2.17). Aldoses and ketoses are '**reducing sugars**'. This means that, when heated with an alkaline solution of copper (II) sulfate (a blue solution, called Benedict's solution), the aldehyde or ketone group reduces Cu^{2+} ions to Cu^+ ions, forming a brick-red precipitate of copper (I) oxide. In the process, the aldehyde or ketone group is oxidized to a carboxyl group (–COOH).

This reaction is used to test for reducing sugar, and is known as **Benedict's test** (Figure 2.16). If no reducing sugar is present, the solution remains blue. The colour change observed depends on the concentration of reducing sugar. The greater the concentration, the more precipitate is formed and the greater the colour change:

blue → green → yellow → red → brown

5 cm³ of Benedict's solution (blue) was added to 10 cm³ of solution to be tested ⟶ test tubes were placed in a boiling water bath for 5 minutes ⟶ tubes were transferred to a rack and the colours compared

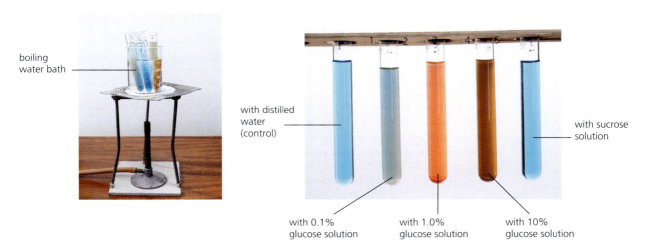

boiling water bath

with distilled water (control)

with sucrose solution

with 0.1% glucose solution

with 1.0% glucose solution

with 10% glucose solution

■ **Figure 2.16** The test for reducing sugar

Disaccharides

Disaccharides are carbohydrates made of two monosaccharides combined together. For example, sucrose is formed from a molecule of glucose and a molecule of fructose chemically combined together.

Condensation and hydrolysis reactions

When two monosaccharide molecules are combined to form a disaccharide, a molecule of water is also formed as a product, so this type of reaction is known as a **condensation reaction**. The linkage between monosaccharide residues, after the removal of H–O–H between them, is called a **glycosidic linkage** (Figure 2.17). This comprises strong, covalent bonds. The condensation reaction is brought about by an enzyme.

■ **Figure 2.17**
Disaccharides and the
monosaccharides that
form them

sucrose   +   water $\overset{\text{hydrolysis}}{\underset{\text{condensation}}{\rightleftharpoons}}$ glucose   +   fructose

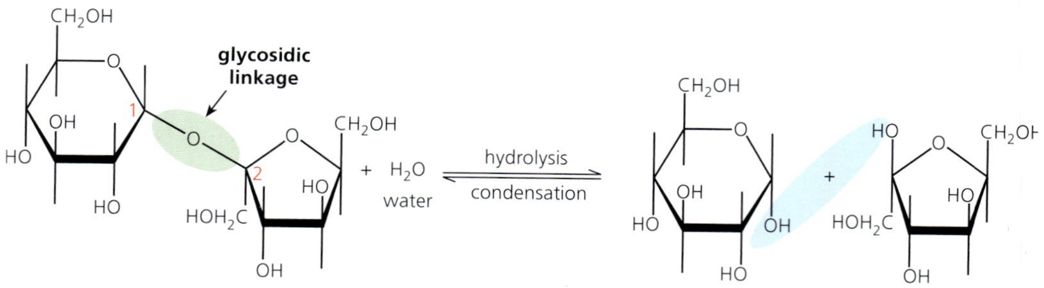

This structural formula shows us how the glycosidic linkage forms/breaks, but the structural formulae of disaccharides should not be memorized.

maltose   +   water $\overset{\text{hydrolysis}}{\underset{\text{condensation}}{\rightleftharpoons}}$ glucose   +   glucose

lactose   +   water $\overset{\text{hydrolysis}}{\underset{\text{condensation}}{\rightleftharpoons}}$ galactose   +   glucose

In the reverse process, disaccharides are 'digested' to their component monosaccharides in a hydrolysis reaction. This reaction involves adding a molecule of water (hydro-) as splitting (-lysis) of the glycosidic linkage occurs. It is catalysed by an enzyme, too, but it is a different enzyme from the one that brings about the condensation reaction.

Apart from sucrose, other disaccharide sugars produced by cells and used in metabolism include:

■ **maltose**, formed by condensation reaction of two molecules of glucose

■ **lactose**, formed by condensation reaction of galactose and glucose.

Polysaccharides

Polysaccharides are built from very many monosaccharide residues condensed together. Each residue is linked by a glycosidic bond. 'Poly' means many and, in fact, thousands of 'saccharide' residues make up a polysaccharide. So a polysaccharide is an example of a giant molecule, a **macromolecule**. Normally, each polysaccharide contains only one type of **monomer**. Cellulose is a good example – built from the monomer glucose.

Cellulose is by far the most abundant carbohydrate – it makes up more than 50% of all organic carbon. (Remember that the gas carbon dioxide, CO_2, and the mineral calcium carbonate, $CaCO_3$, are examples of 'inorganic' carbon.)

The cell walls of green plants and the debris of plants in and on the soil are where most cellulose occurs. It is an extremely strong material – insoluble, tough and durable, and slightly elastic. Cellulose fibres are straight and uncoiled. When they are extracted from plants, cellulose fibres have many industrial uses. We use cellulose fibres as cotton, we manufacture them into paper, rayon fibres for clothes manufacture, nitrocelluose for explosives, cellulose acetate for fibres of multiple uses, and cellophane for packaging.

Cellulose is a polymer of β-**glucose** molecules combined together by glycosidic bonds between carbon-4 of one β-glucose molecule and carbon-1 of the next. Successive glucose units are linked at 180° to each other (Figure 2.18). This structure is stabilized and strengthened by hydrogen bonds between adjacent glucose units in the same strand and, in fibrils of cellulose, by hydrogen bonds between parallel strands, too. In plant cell walls additional strength comes from the cellulose fibres being laid down in layers that run in different directions.

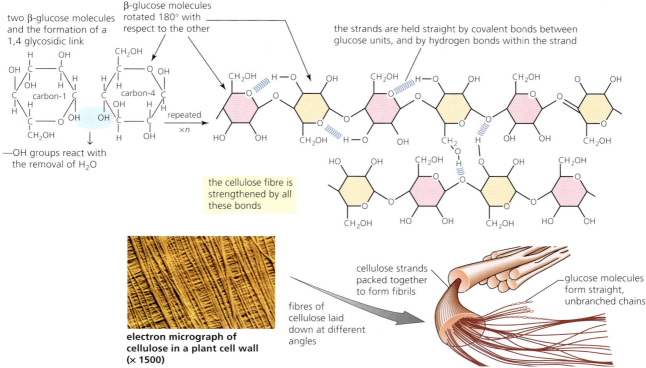

two β-glucose molecules and the formation of a 1,4 glycosidic link

β-glucose molecules rotated 180° with respect to the other

the strands are held straight by covalent bonds between glucose units, and by hydrogen bonds within the strand

—OH groups react with the removal of H_2O

the cellulose fibre is strengthened by all these bonds

electron micrograph of cellulose in a plant cell wall (× 1500)

cellulose strands packed together to form fibrils

fibres of cellulose laid down at different angles

glucose molecules form straight, unbranched chains

■ **Figure 2.18** Cellulose

9 Starch is a powdery material, cellulose is a strong, fibrous substance, yet both are made of glucose. **Identify** the features of the cellulose molecule that account for its strength.

Starch

Starch is a mixture of two polysaccharides:

- **Amylose** is an unbranched chain of several thousand 1,4 linked **α-glucose** units.
- **Amylopectin** has shorter chains of 1,4 linked α-glucose units but, in addition, there are branch points of 1,6 links along its chains (Figure 2.19).

In starch, the bonds between glucose residues bring the molecules together as a **helix**. The whole starch molecule is stabilized by countless hydrogen bonds between parts of the component glucose molecules.

Starch is the major storage carbohydrate of most plants. It is laid down as compact grains in **plastids** called leucoplasts (page 23). Starch is an important energy source in the diet of many animals, too. Its usefulness lies in the compactness and insolubility of its molecule. Also, it is readily hydrolysed to form sugar when required. We sometimes see 'soluble starch' as an ingredient of manufactured foods. Here the starch molecules have been broken down into short lengths, making them more easily dissolved.

We test for starch by adding a solution of iodine in potassium iodide. Iodine molecules fit neatly into the centre of the starch helix, creating a blue–black colour.

Glycogen

Glycogen is a polymer of α-glucose, chemically very similar to amylopectin, although larger and more highly branched. Glycogen is one of our body's energy reserves and is used and respired as needed. Granules of glycogen are seen in liver cells (Figure 2.20), muscle fibres and throughout the tissues of the human body. The one exception is in the cells of the brain. Here, virtually no energy reserves are stored. Instead, brain cells require a constant supply of glucose from the blood circulation.

amylose
(a straight-chain
polymer of α-glucose)

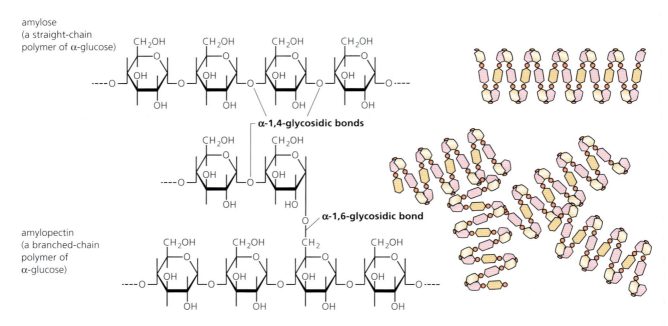

α-1,4-glycosidic bonds

α-1,6-glycosidic bond

amylopectin
(a branched-chain
polymer of
α-glucose)

test for starch with iodine in potassium iodide solution; the blue–black colour comes from a starch/iodine complex

a) Test on a potato tuber cut surface

1% starch solution   0.1% starch solution   0.01% starch solution

b) Test on starch solutions of a range of concentrations

■ **Figure 2.19** Starch

■ **Figure 2.20**
Glycogen granules
in liver cells

electron micrograph of a liver cell (× 7000)

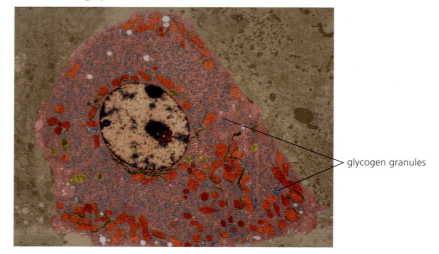

glycogen granules

Examples of some important carbohydrates are given in Table 2.4.

Using *JMol* to compare cellulose, starch and glycogen

You can locate and download the free software '*JMol*' by means of a search engine. Directions on downloading and using *JMol* can be accessed at: http://jmol.sourceforge.net/download/

Open the software, choose 'Molecules to look at' and then the file 'Starch, cellulose and glycogen'. Left click on the molecules to rotate them. Right click to display a menu. Choose 'Zoom' to magnify and make further observations. *Relate what you have seen to the images of these molecules shown in this section of the book* (pages 79–80).

Incidentally, there are other websites that use *JMol*, which you may find easier to use. Use your search engine to locate them if necessary.

■ **Table 2.4**
Key carbohydrates and their roles in animals and plants

10 Define the terms 'monomer' and 'polymer', giving examples from the carbohydrates.

Animals		Plants	
Monosaccharides		**Monosaccharides**	
glucose	• transported to cells in the blood plasma • used as a respiratory substrate for cellular respiration or converted to glycogen (a storage carbohydrate, see below)	glucose	• a first product of photosynthesis
galactose	• used in the production of lactose (milk sugar)	fructose	• produced in cellular respiration as an intermediate of glucose breakdown • used in the production of sucrose
Disaccharides		**Disaccharides**	
lactose	• produced in mammary glands and secreted into the milk as an important component in the diet of very young mammals	sucrose	• produced in green leaves from glucose and fructose • transported in plants in solution, in the vascular bundles
		maltose	• breakdown product in the hydrolysis of starch
Polysaccharides		**Polysaccharides**	
glycogen	• storage carbohydrate formed from glucose in the liver and other cells (but not in brain cells) when glucose is not immediately required for cellular respiration (Figure 2.20)	cellulose	• manufactured in cells and laid down externally, in bundles of fibres, as the main component of the cell walls
		starch	• storage carbohydrate

■ Lipids

Lipids contain the elements carbon, hydrogen and oxygen, as do carbohydrates, but in lipids the proportion of oxygen is much less. Lipids are present as animal **fats** and plant **oils**, and also as the **phospholipids** of cell membranes. Fats and oils seem rather different substances, but their only difference is that at about 20°C (room temperature) oils are liquid and fats are solid. Lipids are insoluble in water. In fact, they generally behave as 'water-hating' molecules, a property described as **hydrophobic**. However, lipids can be dissolved in organic solvents, such as alcohol (e.g. ethanol) and propanone (acetone).

Fats and oils are triglycerides

Fats and oils are compounds called **triglycerides**. They are formed by reactions, between **fatty acids** and an alcohol called **glycerol**, in which water is removed (**condensation** reactions). Three fatty acids combine with one glycerol to form a triglyceride. The structure of a fatty acid commonly found in cells and of glycerol is shown in Figure 2.21, and then the steps to triglyceride formation are shown in Figure 2.22. In cells, enzymes catalyse the formation of triglycerides and also the breakdown of glycerides by **hydrolysis**.

The fatty acids present in fats and oils have long hydrocarbon 'tails'. These are typically of about 16–18 carbon atoms long, but may be anything between 14 and 22. The hydrophobic properties of triglycerides are due to these hydrocarbon tails. A molecule of triglyceride is quite large, but

■ **Figure 2.21** Fatty acids and glycerol, the building blocks

Fatty acid

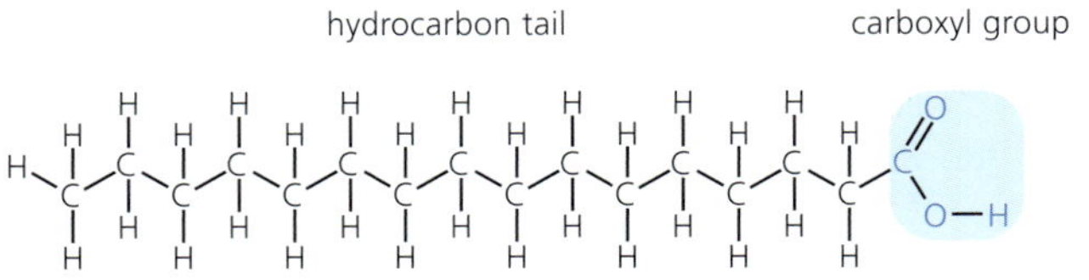

hydrocarbon tail   carboxyl group

this is palmitic acid with 16 carbon atoms

the carboxyl group ionises to form hydrogen ions, i.e. it is a weak acid

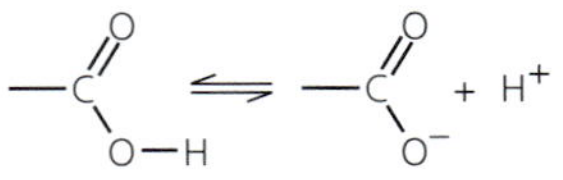

molecular formula of palmitic acid

$CH_3(CH_2)_{14}COOH$

Glycerol

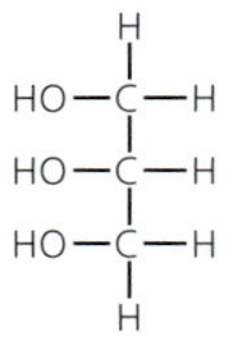

molecular formula of glycerol

$C_3H_5(OH)_3$

11 Distinguish between condensation and hydrolysis reactions. Give an example of each.

relatively small when compared to polymer macromolecules like starch and cellulose. It is only because of their hydrophobic properties that triglyceride molecules clump together (aggregate) into huge globules in the presence of water, giving them the *appearance* of macromolecules.

We describe fatty acid molecules as 'acids' because their functional group (–COOH) tends to ionize (slightly) to produce hydrogen ions – which is the property of an acid:

$$-COOH \rightleftharpoons -COO^- + H^+$$

It is this functional group of each of the three organic acids that reacts with the three –COH functional groups of glycerol to form a triglyceride (Figure 2.22). The bonds formed in this case are known as **ester bonds**.

Saturated and unsaturated lipids

We have seen that the fatty acids combined in a triglyceride may vary in their length. In fact, the fatty acids present in dietary lipids (the lipids we commonly eat) vary in another, more important way, too. To understand this difference we need remember that carbon atoms, combined together in chains, may contain one or more **double bonds**. A double bond is formed when adjacent carbon atoms share *two pairs* of electrons, rather than the single electron pair shared in a single bond (Figure 2.4, page 65).

Carbon compounds that contain double carbon=carbon bonds are known as **unsaturated** compounds. On the other hand, when all the carbon atoms of an organic molecule are combined together by single bonds (the hydrocarbon chain consists of –CH$_2$–CH$_2$– repeated again and again), then the compound is described as **saturated**.

This difference is of special importance in the fatty acids that are components of our dietary lipids (Figure 2.23):

■ Lipids built exclusively from saturated fatty acids are known as **saturated fats**. Saturated fatty acids are major constituents of butter, lard, suet and cocoa butter.

Figure 2.22 Formation of triglyceride

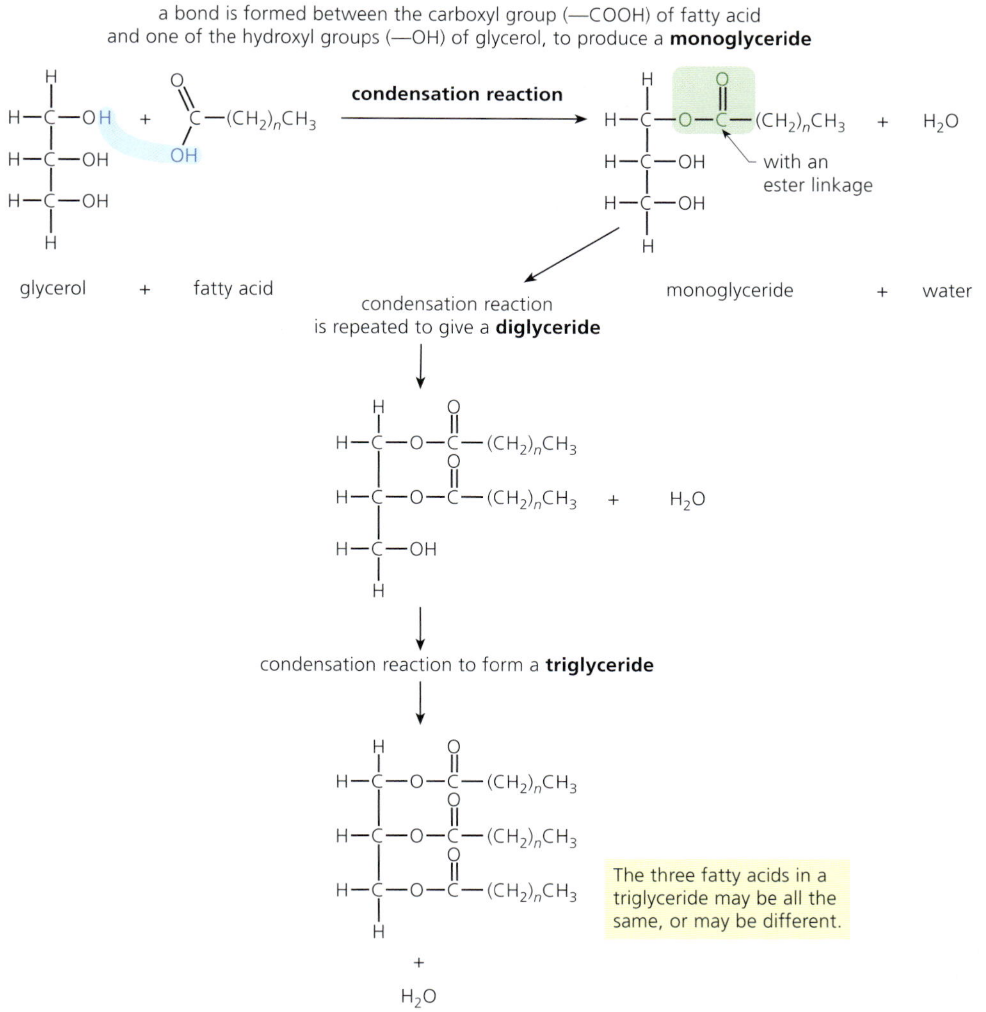
a bond is formed between the carboxyl group (—COOH) of fatty acid and one of the hydroxyl groups (—OH) of glycerol, to produce a **monoglyceride**
condensation reaction
glycerol
+
fatty acid
monoglyceride
+
water
with an ester linkage
condensation reaction is repeated to give a **diglyceride**
condensation reaction to form a **triglyceride**
The three fatty acids in a triglyceride may be all the same, or may be different.

- Lipids built from one or more unsaturated fatty acid are referred to as **unsaturated fats** by dieticians. These occur in significant quantities in many common fats and oils – they make up about 70% of the lipids present in olive oil. Where there is a single double bond in the carbon chain of a fatty acid, the compound is referred to as a **monounsaturated** fatty acid. However, it is possible and common for there to be two or more double bonds in the carbon chain. Lipids with two (and sometimes three) double bonds occur in large amounts in vegetable seed oils, such as maize, soya and sunflower seed oils. These are examples of **polyunsaturated fatty acids**. Fats with unsaturated fatty acids melt at a lower temperature than those with saturated fatty acids, because their unsaturated hydrocarbon tails do not pack so closely together in the way those of saturated fats do. This difference between saturated and polyunsaturated fats is important in the manufacture of margarine and butter-type spreads, since these 'spread better, straight from the fridge'. Polyunsaturated fats are important to the health of our arteries (page 86).

Cis and *trans* fatty acids

In many organic molecules, rotation of one part of the molecule with respect to another part is possible about a single covalent **–C–C–** bond. However, when two carbon atoms are joined by a double bond (–C=C–) there is *no* freedom of rotation at this point in the molecule. We can

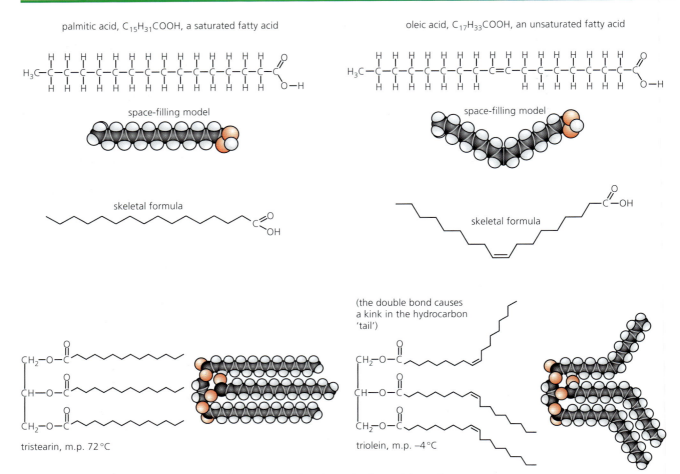

palmitic acid, C₁₅H₃₁COOH, a saturated fatty acid

oleic acid, C₁₇H₃₃COOH, an unsaturated fatty acid

space-filling model

space-filling model

skeletal formula

skeletal formula

(the double bond causes
a kink in the hydrocarbon
'tail')

tristearin, m.p. 72 °C

triolein, m.p. −4 °C

■ **Figure 2.23** Saturated and unsaturated fatty acids and triglycerides formed from them

demonstrate the significance of this in a monounsaturated fatty acid of chemical formula $C_{17}H_{33}COOH$, where the double bond occurs in the mid-point of the hydrocarbon chain, between carbon atoms 9 and 10 (Figure 2.24). In one possible form of this molecule, the two parts of the hydrocarbon chain are on the same side of the double bond. This molecule is the *cis* form of the acid. The alternative form has the two parts of the hydrocarbon chain on opposite sides. This is the *trans* form of the molecule.

■ **Figure 2.24** *Cis* and *trans* fatty acids

$C_{17}H_{33}COOH$, an unsaturated fatty acid

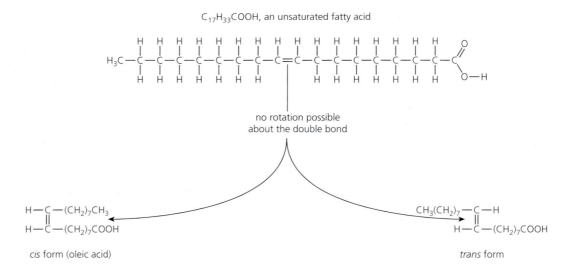

no rotation possible
about the double bond

cis form (oleic acid)

trans form

The *cis* and *trans* fats are significant because:

- our enzymes can 'recognize' the difference between *cis* and *trans* forms of molecules at their active sites (page 94); enzymes of lipid metabolism generally recognize and can trigger metabolism of the *cis* forms, but not the *trans* forms
- diets rich in *trans* fats increase the risk of coronary heart disease by raising the levels of LDL cholesterol in the blood and lowering the levels of HDL cholesterol (page 86).

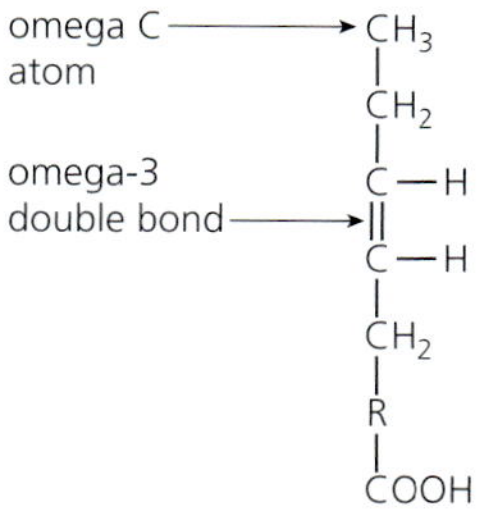

omega C atom → CH₃

omega-3 double bond →

■ **Figure 2.25**
An omega-3 fatty acid

Omega-3 fatty acids

Omega-3 fatty acids are a select group of naturally occurring polyunsaturated fatty acids. They are chemically special in that they have between three and six double bonds in the hydrocarbon tail and, in particular, the first double bond is always positioned between the third and the fourth carbon atom from the opposite (omega) end of the hydrocarbon chain to the carboxyl group, as shown in Figure 2.25. A number of omega-3 fatty acids occur in plant and fish oils, and are thought to be particularly beneficial to health, of which more later.

The roles of fats and oils as energy store and metabolic water source

When triglycerides are oxidized in respiration, a lot of **energy** is transferred (and used to make, for example, ATP – page 115). Mass for mass, fats and oils transfer more than twice as much energy as carbohydrates do, when they are respired (Table 2.5). This is because fats are comparatively low in oxygen atoms (the carbon of lipids is more reduced than that of carbohydrates), so more of the oxygen in the respiration of fats comes from the atmosphere. In the oxidation of carbohydrate, more oxygen is present in the carbohydrate molecule itself. Fat and oils, therefore, form a concentrated energy store.

Because they are **insoluble**, the presence of fat or oil in cells does not cause osmotic water uptake. A fat store is especially typical of animals that endure long unfavourable seasons in which they survive on reserves of food stored in the body. Oils are often a major energy store in plants, their seeds and fruits, and it is common for fruits and seeds – including maize, olives and sunflower seeds – to be used commercially as a source of edible oils for humans.

Complete oxidation of fats and oils produces a large amount of water, far more than when the same mass of carbohydrate is respired. Desert animals like the camel and the desert rat retain much of this **metabolic water** within the body, helping them survive when there is no liquid water for drinking. The development of the embryos of birds and reptiles, while in their shells, also benefits from metabolic water formed by the oxidation of the stored fat in their egg's yolk.

■ **Table 2.5** Lipids and carbohydrates as energy stores – a comparison

Lipids	Role	Carbohydrates
more energy per gram than carbohydrates	**energy store**	less energy per gram than lipids
much metabolic water is produced on oxidation	**metabolic water source**	less metabolic water is produced on oxidation
insoluble, so osmotic water uptake is not caused	**solubility**	sugars are highly soluble in water, causing osmotic water uptake
not quickly 'digested'	**ease of breakdown**	more easily hydrolysed – energy is transferred quickly

Health consequences of the lipid content of diets

1 Diets with an excess of lipids and fatty acids provide more energy-rich items than the body requires. People in such affluent situations are in danger of becoming **overweight** and then **obese** (Figure 2.26), because of the storage of excess fat in the fat cells that make up the adipose tissue stored around the body organs and under the skin.

Body Mass Index
To accurately and consistently quantify body weight in relation to health, the Body Mass Index (BMI) was devised. We calculate our BMI according to the following formula:

$$\frac{\text{body mass in kg}}{(\text{height in m})^2}$$

The 'boundaries' between underweight, normal and overweight are given below.

BMI	Status
below 18.5	underweight
18.5–24.9	normal
25.0–29.8	overweight
30.0 and over	obese

Alternatively, you can use the National Institute for Health website:
www.nhibi.nih.gov/guidelines/obesity/BMI/bmicalc.htm

■ **Figure 2.26**
Defining 'overweight' and 'obese'

People who are chronically overweight have enhanced likelihoods of acquiring **type II diabetes** (page 301) and high blood pressure (**hypertension**, page 266). Also, in an affluent society where people are generally active, there are social and psychological pressures that may cause an obese person to become **depressed**. The popular image of the jolly fat person is often rooted in misunderstanding!

Sometimes, people in affluent communities may suffer from apparently deliberate avoidance of adequate nutrition (anorexia nervosa, Option D). In other communities, health problems can arise from chronic malnutrition due to harvest failure, or due to poor food distribution caused by political instability or warfare.

2   The lipid **cholesterol** is a component of the diet, particularly when animal fats are present. This lipid, a steroid, is of different chemical structure from that of the fatty acids. The 'skeleton' of a steroid is a set of complex rings of carbon atoms; the bulk of the molecule is hydrophobic, but the polar –OH group is hydrophilic (Figure 2.27).

■ **Figure 2.27** The steroid cholesterol

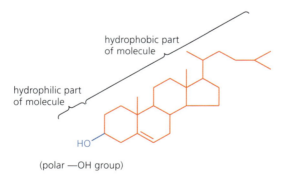

hydrophobic part of molecule

hydrophilic part of molecule

HO

(polar —OH group)

Cholesterol is present in the blood plasma, and has important roles in the body:

■ It is an essential component of the plasma membranes of all cells.

■ The sex hormones progesterone, estrogen and testosterone and also certain growth hormones are produced from it.

■ Bile salts are synthesized from it. These compounds are required for transport of lipids in the blood plasma.

Consequently, cholesterol is essential for normal, healthy metabolism. An adult human on a low-cholesterol diet actually forms about 800 mg of cholesterol in the liver each day. However, diets high in saturated fatty acids (i.e. animal fats) often lead to abnormally high blood-cholesterol levels. Excess blood cholesterol, present as LDL (page 73), may cause **atherosclerosis** – the progressive degeneration of the artery walls. A direct consequence of this is an increased likelihood of the formation and circulation of blood clots, leading to strokes and heart attacks (known as myocardial infarctions, page 266; Figure 2.28).

■ **Figure 2.28**
Correlation studies –
diet and coronary
heart disease (CHD)

Correlation study concerning saturated fat intake and the incidence of CHD

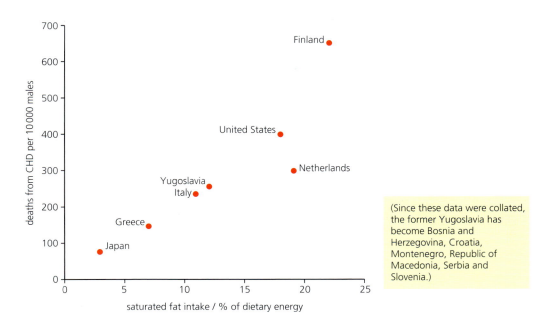

(Since these data were collated, the former Yugoslavia has become Bosnia and Herzegovina, Croatia, Montenegro, Republic of Macedonia, Serbia and Slovenia.)

Understanding correlations – the differences between a positive correlation (left) and a negative correlation (right)

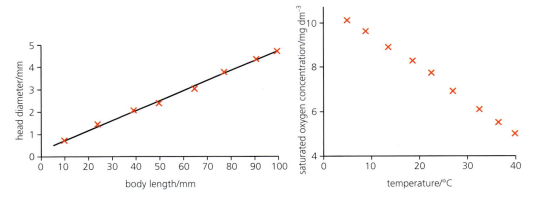

3 **Omega-3 fatty acids**, on the other hand, may help prevent heart disease by reducing the tendency of the blood to form clots and by giving us healthy, well-functioning plasma membranes abound cardiac muscle fibres.

This claim is based on studies of people who have a high daily intake of these fatty acids, such as Greenland Eskimos (from trials known as '**cohort studies**'). Here, diets are exceptionally rich in oily fish meat. Associated with this diet (and lifestyle), a very much reduced likelihood of an acute myocardial infarction (heart attack) than other human groups is observed. It is estimated by dieticians who favour cohort studies that we need about 0.2 g/day (1.5 g/week) of omega-3 fatty acids, which will be taken in if we eat sufficient oily fish, regularly. A plant source rich in omega-3 fatty acids is the walnut.

However, other studies (referred to as '**randomized control trials**') suggest it may be difficult to show the clear benefits of omega-3 fatty acids. This issue may remain a controversial one for some time.

4 Fat deposits are not inert stores, rather they are constantly being broken down and reformed. However, during fat metabolism small amounts of acids are produced, known as **ketone bodies**. In people with an excessively lipid-rich diet, and also those with severe diabetes, these may accumulate and reach toxic levels. Their acidity dangerously lowers the blood pH (from a normal 7.4 to 7.2 or lower), causing a condition known as **acidosis**. This may trigger a coma and can eventually lead to death.

5 Diets that are consistently very **low in energy-rich foods**, including lipids, also cause major health risks. Such diets do not contain sufficient fatty acids, so people typically respire the amino acids derived from protein digestion, rather than using them to build and maintain tissues. On a continuing low-energy diet, muscle proteins are broken down, and the body wastes away. For the nutritionally deprived members of communities in less-developed countries this is a constant danger.

Nature of Science

Evaluating claims

The points above indicate that there are marked variations in the prevalence of different health problems in societies around the world. Claims are based on:

- **epidemiological studies** – these provide circumstantial evidence of health risks; they suggest connections, but they do not establish a cause or biochemical connection
- **clinical studies** of individual patients with health problems attempt to show causal relations between diseases and diets (however, it is not possible to carry out '**controlled' experiments** for ethical reasons).

So, health claims made about lipids need to be assessed in context. The issues continue to present complex challenges for dieticians and politicians in all societies.

> **TOK Link**
>
> It is said there are conflicting views as to the harms and benefits of fats in diets. To what extent is this true in your country? Why?

Other applications of lipids

Subcutaneous fat as a buoyancy aid (and thermal insulation)

Fat is stored in animals as adipose tissue, typically under the skin, where it is known as subcutaneous fat. In aquatic diving mammals, which have a great deal of subcutaneous fat, it is identified as blubber. Blubber undoubtedly gives buoyancy to the body, since fat is not as dense as muscle or bone.

If fat reserves like these have a restricted blood supply (as is normally the case), then the heat of the body is not especially distributed to the fat under the skin. In these circumstances the subcutaneous fat also functions as a heat-insulation layer.

Water-proofing of hair and feathers

Oily secretions of the sebaceous glands, found in the skin of mammals, act as a water repellent, preventing fur and hair from becoming waterlogged when wet. Birds have a preen gland that fulfils the same function for feathers.

Electrical insulation

Myelin lipid in the membranes of Schwann cells, forming the sheaths around the long fibres of nerve cells (page 290), electrically isolates the cell plasma membrane and facilitates the conduction of the nerve impulse there.

■ Phospholipid

A phospholipid has a similar chemical structure to triglyceride; here, one of the fatty acid groups is replaced by phosphate (Figure 2.29). This phosphate is ionized and is, therefore, water soluble. So phospholipids combine the hydrophobic properties of the hydrocarbon tails with hydrophilic properties of the phosphate.

Phospholipid molecules form monolayers and bilayers in water (Figure 1.35, page 31). A phospholipid bilayer is a major component of the plasma membrane of cells.

■ **Figure 2.29**
Phospholipid

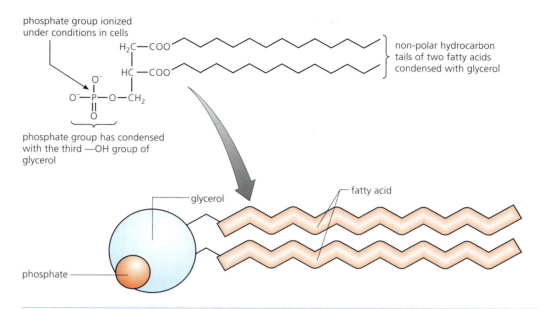

12 **Describe** the property given to a lipid when it combines with a phosphate group.

2.4 Proteins *– proteins have a very wide range of functions in living organisms*

Proteins make up about two-thirds of the total dry mass of a cell. They differ from carbohydrates and lipids in that they contain the element **nitrogen** and, usually, the element sulfur, as well as carbon, hydrogen and oxygen.

Amino acids are the molecules from which peptides and proteins are built – typically several hundred or even thousands of amino acid molecules are combined together to form a protein. Notice that the terms 'polypeptide' and 'protein' can be used interchangeably but, when a polypeptide is about 50 amino acid residues long, it is generally agreed to have become a protein.

■ Amino acids – the building blocks of peptides

Figure 2.30 shows the structure of an amino acid. As their name implies, amino acids carry two groups:

■ an **amino group** (–NH$_2$)
■ an **organic acid group** (carboxyl group –COOH).

These groups are attached to the same carbon atom in the amino acids which get built up into proteins. Also attached here is a side-chain part of the molecule, called an **R group**.

Proteins of living things are built from just **20 different amino acids**, in differing proportions. All we need to note is that the R groups of these amino acids are all very different and consequently amino acids (and proteins containing them) have different chemical characteristics.

Nature of Science   **Looking for patterns, trends and discrepancies**
Most, but not all organisms, assemble proteins from the same amino acids.

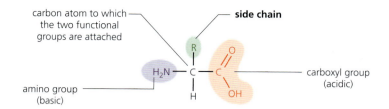

■ **Figure 2.30**
The structure of
amino acids

carbon atom to which
the two functional
groups are attached

side chain

amino group
(basic)

carboxyl group
(acidic)

The 20 different amino acids that make up proteins in cells and organisms differ in their side chains. Below are three illustrations but *details of R groups are not required.*

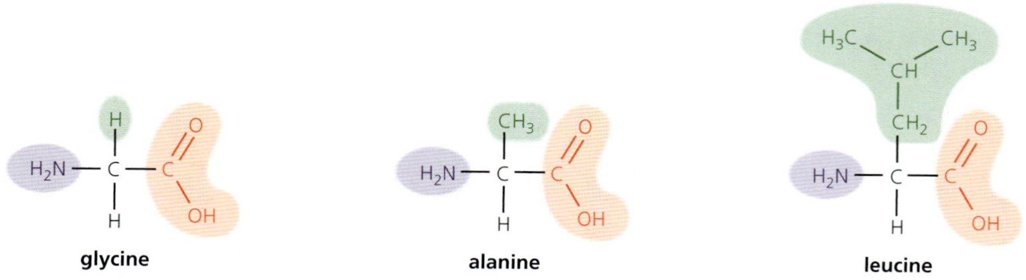

glycine

alanine

leucine

Some amino acids have an additional —COOH group in their side chain (= acidic amino acids).
Some amino acids have an additional —NH_2 group in their side chain (= basic amino acids).

■ Peptide linkages

Two amino acids combine together with the loss of water to form a **dipeptide**. This is one more example of a condensation reaction. The amino group of one amino acid reacts with the carboxyl group of the other, forming a **peptide linkage** (Figure 2.31).

A further condensation reaction between the dipeptide and another amino acid results in a tripeptide. In this way, long strings of **amino acid residues**, joined by peptide linkages, are formed. Thus, peptides or protein chains are assembled, one amino acid at a time, in the presence of a specific enzyme (page 112).

Proteins differ in the variety, number and order of their constituent amino acids. The order of amino acids in the polypeptide chain is controlled by the coded instructions stored in the DNA of the chromosomes in the nucleus (page 110). Changing one amino acid in the sequence of a protein may alter its properties completely.

■ **Figure 2.31** Peptide
linkage formation

amino acids combine together, the amino group of one with the carboxyl group of the other

carboxyl
group

amino
group

**condensation
reaction**

+ H_2O

peptide
linkage

amino acid 1

amino acid 2

dipeptide

water

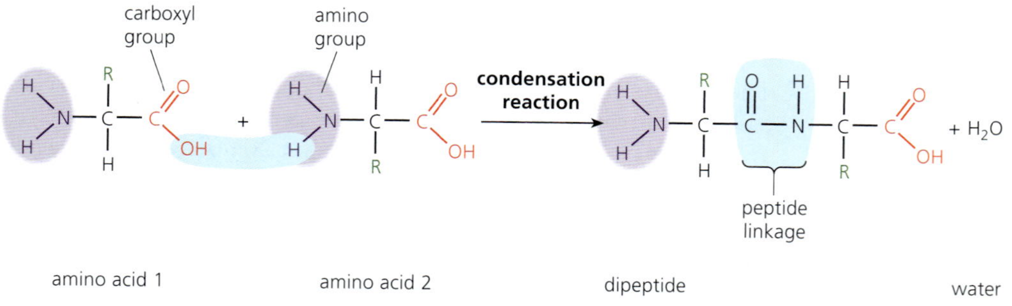

When a further amino acid residue is attached by condensation reaction, a tripeptide is formed.
In this way, long strings of amino acid residues are assembled to form polypeptides and proteins.

■ The structure of proteins

Once the chain is constructed, a protein takes up a specific shape. Shape matters with proteins – their shape is closely related to their function. This is especially the case in proteins that are enzymes, as we shall shortly see. The four levels of structure to a protein are shown in Table 2.6.

When a protein loses its three-dimensional shape, we say it has been **denatured**. Heat or a small deviation in pH from the optimum can have this effect (page 96). Now, when this happens to small proteins, it is found that they generally revert back to their former shape, once the conditions that triggered denaturation are removed. This observation suggested that it is simply the amino acid sequence of a protein that decides its tertiary structure. This may well be true for many polypeptides and small proteins. However, in most proteins within the cell environment, folding is a speedy process in which some accessory proteins, including enzymes are normally involved. These may determine the shape as much or more than the primary structure does.

■ **Table 2.6** The structure of proteins

Primary structure	Secondary structure	Tertiary structure	Quaternary structure
This is the **sequence of the amino acids** in the molecule. Proteins differ in the variety, number and order of their amino acids. This sequence determines the shape and structure of the protein.	This develops when parts of the polypeptide chain take up a particular shape by coiling to produce an α-**helix** or by folding into β-**sheets**. These shapes are permanent, held in place by hydrogen bonds.	This is the precise, compact structure, unique to a particular protein. It arises when the molecule is further folded and held in a particular complex shape, made permanent by four different types of bond that are established between adjacent parts of the chain (Figure 2.32). Some form into long, much-coiled chains (**fibrous proteins**). Others take up a more spherical shape (**globular proteins**).	This arises when two or more proteins are held together, forming a complex, biologically active molecule. An example is the respiratory pigment hemoglobin, found in red blood cells. This molecule consists of four polypeptide chains held around a non-protein heme group, in which an atom of iron occurs.

■ **Figure 2.32** Cross-linking within a polypeptide

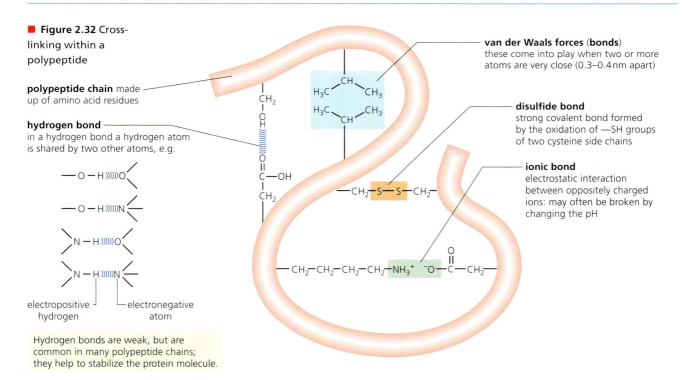

polypeptide chain made up of amino acid residues

hydrogen bond
in a hydrogen bond a hydrogen atom is shared by two other atoms, e.g.

electropositive hydrogen — electronegative atom

Hydrogen bonds are weak, but are common in many polypeptide chains; they help to stabilize the protein molecule.

van der Waals forces (bonds)
these come into play when two or more atoms are very close (0.3–0.4 nm apart)

disulfide bond
strong covalent bond formed by the oxidation of —SH groups of two cysteine side chains

ionic bond
electrostatic interaction between oppositely charged ions: may often be broken by changing the pH

■ Functions of proteins

Living things synthesis many different proteins with a wide range of functions. Table 2.7 illustrates some of these.

RuBisCo – globular protein – enzyme	Insulin – globular protein – hormone
In photosynthesis, CO_2 is combined with an acceptor molecule in the presence of a special enzyme, ribulose bisphosphate carboxylase (RuBisCo for short). The stroma of chloroplasts are packed full of RuBisCo, an enzyme which easily makes up the bulk of all the protein in a green plant. In fact it is the most abundant enzyme present in the living world. The acceptor molecule is a 5-carbon sugar, ribulose bisphosphate (referred to as RuBP), and CO_2 is added in a process known as fixation. Two 3C sugar molecules are formed.	This hormone is a polypeptide produced by the β cells of the islets of Langerhans in the pancreas. It consists of two polypeptide chains linked by disulfide bridges (Figure 2.32). It was the first protein whose amino acid sequence was determined. The effect of insulin in the bloodstream is to promote glucose uptake by cells and to induce the liver to synthesize glycogen. Today, human insulin is obtained from genetically engineered bacteria and used to treat diabetes.
Spider silk – fibrous protein – structural	**Rhodopsin – conjugate protein – pigment**
This is a strong protein fibre produced by spiders and the silk worm. It is composed of a fibrous protein including fibroin. It is extruded as fluid from specialized glands and is used to produce spider's webs and egg and cocoons. The mixture promptly hardens in contact with air.	This pigment, known as 'visual purple', is a light-sensitive conjugated protein found in the rod cells of the retina in mammals. It is a compound of a protein (opsin), a phospholipid and retinal (vitamin A). The effect of light energy on this pigment is to split it into opsin and retinal.
Collagen – fibrous protein – structural	**Immunoglobulins – globular protein – antibody**
This structural protein occurs in skin, tendons, cartilage, bone, teeth, the walls of blood vessels, and the cornea of the eye. The collagen molecule consists of three polypeptide chains, each in the shape of a helix. Chains are wound together as a triple helix forming a stiff cable, strengthened by numerous hydrogen bonds. Many of these triple helixes lie side by side, forming collagen fibres, held together by covalent cross-linkages. The ends of individual collagen molecules are staggered so there are no weak points in collagen fibres, giving the whole structure incredible tensile strength.	These are the antibodies found in the bloodstream – components of our immune response (our main defences once invasion of the body by harmful microorganisms or 'foreign' materials has occurred). An antibody is a glycoprotein secreted by a plasma cell. Antibodies bind to specific antigens that trigger an immune response. This leads to destruction of the antigens (or of any pathogen or other cell to which the antigens are attached). Antibodies have regions that are complementary of the shape of the antigen. Some antibodies are antitoxins and prevent the activity of toxins.

■ **Table 2.7**
The functions of six proteins

13 Deduce which of the proteins listed in Table 2.7 are fibrous and which are globular proteins.

Nature of Science

Looking for patterns, trends and discrepancies

■ Every individual has a unique proteome

Most, but not all, organisms assemble their proteins from the same amino acids, but the proteins of each individual organism are unique. The **proteome** is the entire set of proteins expressed by the genome of the individual organism. It is because the genome, the whole of the genetic information of an organism, is unique to each individual that the proteome it causes to be expressed is also unique.

Proteomics is the study of the structure and function of the entire set of proteins of organisms. The word is derived from 'protein' and 'genome', and the study is made possible, in part, by the capacity of modern computers. Today, the industrial production of proteins by microorganisms (including of genetically modified ones), when cultured in fermenters, is a source of protein for the food, pharmaceutical, and other industries.

14 The possible number of different polypeptides, P, that can be assembled is given by
$$P = A^n$$
where
A = the number of different types of amino acids available
n = the number of amino acid residues in the polypeptide molecule.
Given the naturally occurring pool of 20 different amino acids, **calculate** how many different polypeptides are possible if constructed from 5, 25 and 50 amino acid residues, respectively.

2.5 Enzymes – *enzymes control the metabolism of the cell*

Most chemical reactions do not occur spontaneously (Figure 2.33). In a laboratory or in an industrial process, chemical reactions may be made to occur by applying high temperatures, high pressures, extremes of pH, by maintaining high concentrations of the reacting molecules, or by the use of inorganic catalysts. If these drastic conditions were not applied, very little of the chemical product would be formed. On the other hand, in cells and organisms, many chemical reactions occur simultaneously, at extremely low concentrations, at normal temperatures and under the very mild, almost neutral, aqueous conditions we find in cells.

■ **Figure 2.33**
Can a reaction occur without an enzyme?

1 The reaction: hydrolysis of sucrose to form glucose and fructose

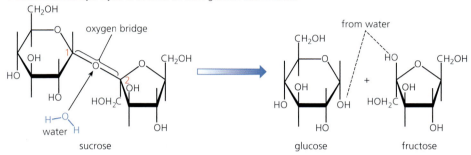

2 Random collision possibilities:
when sucrose and water collide at the wrong angle

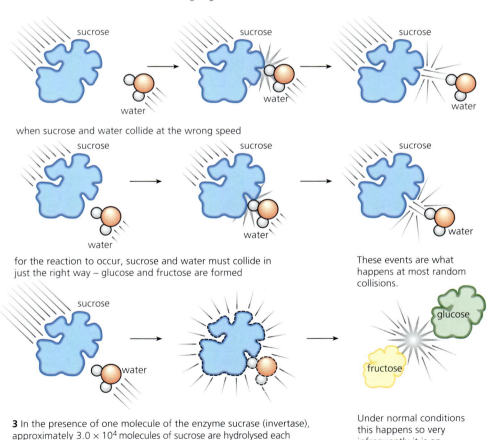

when sucrose and water collide at the wrong speed

for the reaction to occur, sucrose and water must collide in just the right way – glucose and fructose are formed

These events are what happens at most random collisions.

3 In the presence of one molecule of the enzyme sucrase (invertase), approximately 3.0×10^4 molecules of sucrose are hydrolysed each minute!

Under normal conditions this happens so very infrequently it is an insignificant event.

How is this brought about?

It is the presence of **enzymes** in cells and organisms that enables these reactions to occur at incredible speeds, in an orderly manner, yielding products that the organism requires, when they are needed. Sometimes, reactions happen even though the reacting molecules are present in very low concentrations. Enzymes are biological catalysts made of protein. They are truly remarkable molecules. By 'catalyst' we mean a substance that speeds up the rate of a chemical reaction. In general, catalysts:

- are effective in small amounts
- remain unchanged at the end of the reaction.

■ The enzyme active site

In a reaction catalysed by an enzyme, the starting substance is called the **substrate**, and what it is converted to is the **product**. An enzyme works by binding to the substrate molecule at a specially formed pocket in the enzyme. This binding point is called the **active site**. An active site is a region of an enzyme molecule where the substrate molecule binds. Here, as the enzyme and substrate form an **enzyme–substrate complex**, the substrate is raised to a transition state. This complex has the briefest of existences before the substrate molecule is formed into another molecule or broken down into others, by the catalytic properties of the active site. Then the product(s) are released, together with unchanged enzyme (Figure 2.34). The enzyme is available for reuse.

■ **Figure 2.34**
The enzyme–substrate complex and the active site

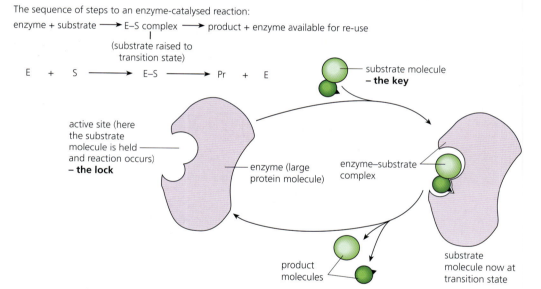

The sequence of steps to an enzyme-catalysed reaction:

enzyme + substrate ⟶ E–S complex ⟶ product + enzyme available for re-use

(substrate raised to transition state)

E + S ⟶ E–S ⟶ Pr + E

substrate molecule – **the key**

active site (here the substrate molecule is held and reaction occurs) – **the lock**

enzyme (large protein molecule)

enzyme–substrate complex

substrate molecule now at transition state

product molecules

Enzymes are typically large, globular protein molecules. Most substrate molecules are quite small molecules by comparison. Even when the substrate molecules are very large, such as certain macromolecules like the polysaccharides, only one bond in the substrate is in contact with the enzyme active site. The active site takes up a relatively small part of the total volume of the enzyme.

Enzyme specificity and the induced-fit model of the active site

Enzymes are highly specific in their action, which makes them different from most inorganic catalysts, such as those used in industry. Enzymes are specific because of the way they bind with their substrate at the active site, which is a pocket or crevice in the protein (Figure 2.35). At the active site, the arrangement of a few amino acid molecules in the protein (enzyme) matches certain groupings on the substrate molecule, enabling the enzyme–substrate complex to form. As it forms, it seems a slight change of shape is induced in the enzyme molecule (**induced-fit hypothesis**, Figure 2.36). It is this change in shape that is important in raising the substrate molecule to the transitional state in which it is able to react.

Meanwhile, other amino acids molecules of the active site bring about the specific catalytic reaction mechanism, perhaps breaking particular bonds in the substrate molecule and forming others. Different enzymes have different arrangements of amino acids in their active sites. Consequently, each enzyme catalyses either a single chemical reaction or a group of closely related reactions.

■ **Figure 2.35**
Enzyme specificity
and the active site

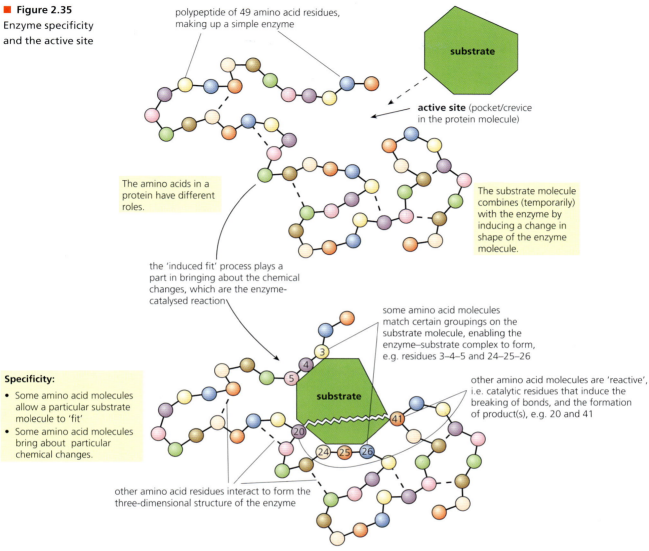

polypeptide of 49 amino acid residues,
making up a simple enzyme

substrate

active site (pocket/crevice
in the protein molecule)

The amino acids in a
protein have different
roles.

The substrate molecule
combines (temporarily)
with the enzyme by
inducing a change in
shape of the enzyme
molecule.

the 'induced fit' process plays a
part in bringing about the chemical
changes, which are the enzyme-
catalysed reaction

some amino acid molecules
match certain groupings on the
substrate molecule, enabling the
enzyme–substrate complex to form,
e.g. residues 3–4–5 and 24–25–26

Specificity:
• Some amino acid molecules
allow a particular substrate
molecule to 'fit'
• Some amino acid molecules
bring about particular
chemical changes.

substrate

other amino acid molecules are 'reactive',
i.e. catalytic residues that induce the
breaking of bonds, and the formation
of product(s), e.g. 20 and 41

other amino acid residues interact to form the
three-dimensional structure of the enzyme

■ **Figure 2.36**
Induced-fit
catalysis by
hexokinase

The enzyme hexokinase catalyses the reaction:

glucose + ATP $\longrightarrow$ glucose-6-phosphate + ADP

Computer-generated image of the induced-fit hypothesis in action:

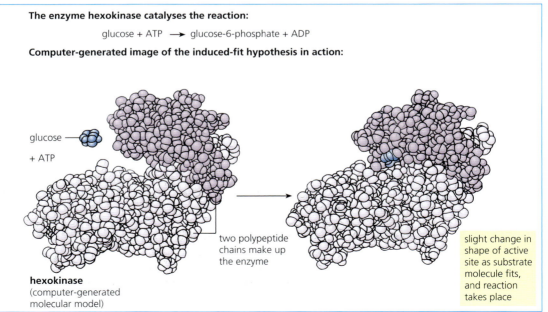

glucose

+ ATP

two polypeptide
chains make up
the enzyme

hexokinase
(computer-generated
molecular model)

slight change in
shape of active
site as substrate
molecule fits,
and reaction
takes place

■ Enzymes control metabolism

There is a huge array of chemical reactions that goes on in cells and organisms. These reactions of metabolism can only occur in the presence of specific enzymes. If an enzyme is not present, the reaction it catalyses cannot occur.

Some enzymes are exported from cells, such as the digestive enzymes (page 251). Enzymes like these are parcelled up and secreted, and they work externally. They are called **extracellular enzymes**. However, very many enzymes remain within cells and work there. These are the **intracellular enzymes**. They are found inside organelles, and in the membranes of organelles, in the fluid medium (cytosol) around the organelles and in the plasma membrane.

Many enzymes are always present in cells and organisms, but some enzymes are produced only under particular conditions, at certain stages or when a particular substrate molecule is present. By making some enzymes and not others, cells can control what chemical reactions happen in the cytoplasm. Later in this chapter we shall see how protein synthesis (and therefore enzyme production) is controlled by the cell nucleus (page 110).

■ Enzymes can be denatured

The rate of an enzyme-catalysed reaction is sensitive to environmental conditions – many factors within cells affect enzymes and, therefore, alter the rate of the reaction being catalysed. In extreme cases, proteins, including enzymes, may become **denatured**. Denaturation is a structural change in a protein that alters its three-dimensional shape. Many of the properties of proteins depend on the three-dimensional shape of the molecule. This is true of enzymes, which are large, globular proteins where a small part of the surface is an active site. Here, the precise chemical structure and physical configuration are critical, but provided the active site is unchanged, substrate molecules can attach and reactions can be catalysed.

Denaturation occurs when the bonds within the globular protein, formed between different amino acid residues, break, changing the shape of the active site.

How denaturation is brought about

Temperature rises and changes in pH of the medium may cause denaturation of the protein of enzymes. Exposure to heat causes atoms to vibrate violently and this disrupts bonds within globular proteins. Protein molecules change chemical characteristics. We see this most dramatically when a hen's egg is cooked. The translucent egg 'white' is a globular protein called albumen, which becomes irreversibly opaque and insoluble. Heat has triggered irreversible denaturation of this globular protein.

Small changes in pH of the medium similarly alter the shape of globular proteins. The structure of an enzyme may spontaneously reform when the optimum pH is restored, but exposure to strong acids or alkalis is usually found to irreversibly denature enzymes.

15 Explain why the shape of globular proteins that are enzymes is important in enzyme action.

■ Factors affecting the rate of enzyme-catalysed reactions

Nature of Science

Experimental design

Accurate, quantitative measurements in enzyme experiments require replicates to ensure reliability.

It was investigation of the effect of change in factors such as temperature, pH and concentration of substrate that particularly helped our understanding of how enzymes work.

The issue of how to design these and similar experiments is the focus of *Appendix 2: 'Investigations, data handling and statistics'* on the accompanying website. It includes an introduction to 'variables' in experiment design.

You should consult it now – the process is illustrated by reference to the next experiment.

Investigating the effects of temperature

Examine the investigation of the effects of temperature on the hydrolysis of starch by the enzyme amylase, shown in Figure 2.37. When starch is hydrolysed by the enzyme amylase, the product is maltose, a disaccharide (page 81). Starch gives a blue–black colour when mixed with iodine solution (iodine in potassium iodide solution), but maltose gives a red colour. The first step in this experiment is to bring samples of the enzyme and the substrate (the starch solution) to the temperature of the water bath before being mixed – a step called pre-incubation.

The progress of the hydrolysis reaction is then followed by taking samples of a drop of the mixture on the end of the glass rod, at half-minute intervals. These are tested with iodine solution on a white tile. Initially, a strong blue–black colour is seen, confirming the presence of starch. Later, as maltose accumulates, a red colour predominates. The endpoint of the reaction is indicated when all the starch colour has disappeared from the test spot. Using fresh reaction mixture each time, the investigation is repeated at a series of different temperatures, say at 10, 20, 30, 40, 50 and 60°C. The time taken for complete hydrolysis at each temperature is recorded and the rate of hydrolysis in unit time is plotted on a graph.

A characteristic curve is the result (Figure 2.38) – although the optimum temperature varies from reaction to reaction and with different enzymes.

How is the graph interpreted?

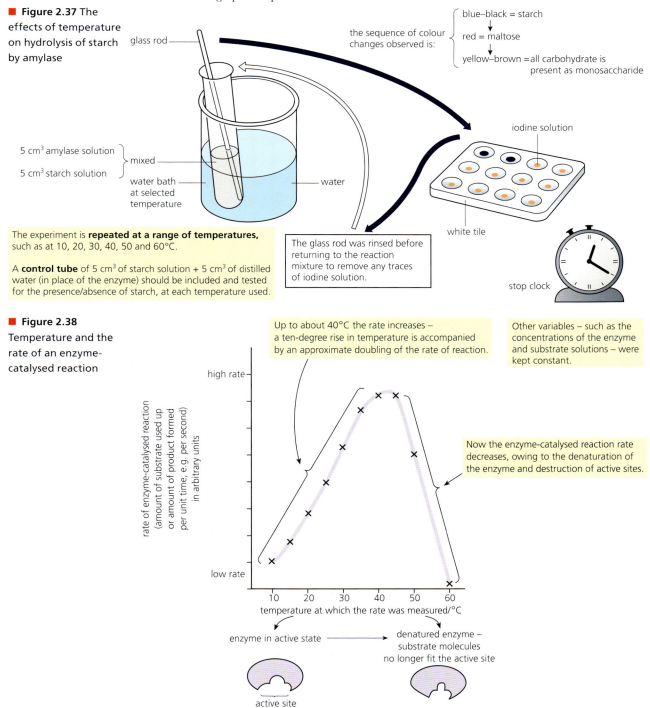

■ **Figure 2.37** The effects of temperature on hydrolysis of starch by amylase

the sequence of colour changes observed is:
blue–black = starch
↓
red = maltose
↓
yellow–brown = all carbohydrate is present as monosaccharide

The experiment is **repeated at a range of temperatures,** such as at 10, 20, 30, 40, 50 and 60°C.

A **control tube** of 5 cm³ of starch solution + 5 cm³ of distilled water (in place of the enzyme) should be included and tested for the presence/absence of starch, at each temperature used.

The glass rod was rinsed before returning to the reaction mixture to remove any traces of iodine solution.

■ **Figure 2.38** Temperature and the rate of an enzyme-catalysed reaction

Up to about 40°C the rate increases – a ten-degree rise in temperature is accompanied by an approximate doubling of the rate of reaction.

Other variables – such as the concentrations of the enzyme and substrate solutions – were kept constant.

Now the enzyme-catalysed reaction rate decreases, owing to the denaturation of the enzyme and destruction of active sites.

enzyme in active state → denatured enzyme – substrate molecules no longer fit the active site

As temperature is increased, molecules have increased active energy and reactions between them go faster. The molecules are moving more rapidly and are more likely to collide and react. In chemical reactions, for every 10°C rise in temperature the rate of the reaction approximately doubles. This property is known as the **temperature coefficient** (Q_{10}) of a chemical reaction.

However, in enzyme-catalysed reactions the effect of temperature is more complex, because proteins are denatured by heat. The rate of denaturation increases at higher temperatures, too. So, as the temperature rises, the amount of active enzyme progressively decreases and the rate is slowed. As a result of these two effects of heat on enzyme-catalysed reactions, there is an apparent optimum temperature for an enzyme.

Not all enzymes have the same optimum temperature. For example, the bacteria in hot thermal springs have enzymes with optima between 80°C and 100°C or higher, whereas seaweeds of northern seas and the plants of the tundra have optima closer to 0°C. Humans have enzymes with optima at or about normal body temperature. This feature of enzymes is often exploited in the commercial and industrial uses of enzymes (Figure 2.43, page 102).

pH

Change in pH can have a dramatic effect on the rate of an enzyme-catalysed reaction (Figure 2.39).

16 In studies of the effect of temperature on enzyme-catalysed reactions, **suggest** why the enzyme and substrate solutions are pre-incubated to a particular temperature before they are mixed.

■ **Figure 2.39**
pH effect on enzyme shape and activity

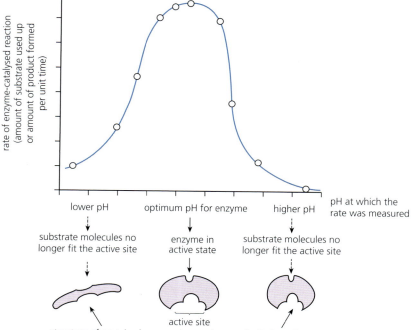

the optimum pH of different human enzymes

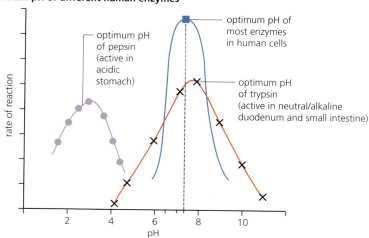

Each enzyme has a limited range of pH at which it functions efficiently. This is often at or close to the neutrality point (pH 7.0). This effect of pH occurs because the structure of a protein (and, therefore, the shape of the active site) is maintained by various bonds within the three-dimensional structure of the protein. A change in pH from the optimum value alters the bonding patterns, progressively changing the shape of the molecule. The active site may be quickly rendered inactive. However, unlike the effects of temperature changes, the effects of pH on the active site are normally reversible – provided, that is, the change in surrounding acidity or alkalinity is not too extreme. As the pH reverts to the optimum for the enzyme, the active site may reappear.

17 Explain what a buffer solution is and why such solutions are often used in enzyme experiments.

Some of the digestive enzymes of the gut have different optimum pH values from the majority of other enzymes. For example, those adapted to operate in the stomach, where there is a high concentration of acid during digestion, have an optimum pH which is close to pH 2.0 (Figure 2.39).

Substrate concentration

The effect of different concentrations of substrate on the rate of an enzyme-catalysed reaction can be shown using an enzyme called catalase. This enzyme catalyses the breakdown of hydrogen peroxide:

$$2H_2O_2 \longrightarrow 2H_2O + O_2$$

Catalase occurs very widely in living things; it functions as a protective mechanism for the delicate biochemical machinery of cells. This is because hydrogen peroxide is a common by-product of reactions of metabolism, but it is also a very toxic substance (since it is a very powerful oxidizing agent, page 348). Catalase inactivates hydrogen peroxide as it forms, before damage can occur.

You can demonstrate the presence of catalase in fresh liver tissue by dropping a small piece into dilute hydrogen peroxide solution (Figure 2.40). *Compare the result obtained with that from a similar piece of liver that has been boiled in water (we have seen that high temperature denatures enzymes).* If you do not wish to use animal tissues, you can use potato or soaked and crushed dried peas.

■ **Figure 2.40**
Liver tissue in dilute hydrogen peroxide solution, a demonstration

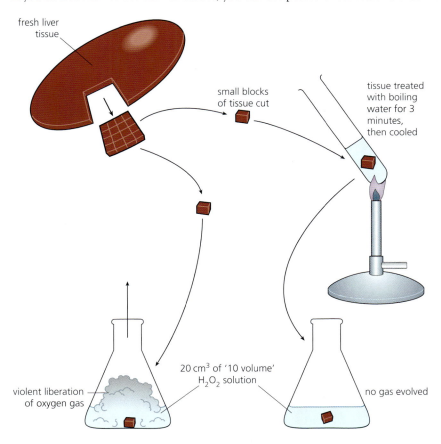

fresh liver tissue

small blocks of tissue cut

tissue treated with boiling water for 3 minutes, then cooled

20 cm³ of '10 volume' H_2O_2 solution

violent liberation of oxygen gas

no gas evolved

When measuring the rate of enzyme-catalysed reactions, we measure the amount of substrate that has disappeared from a reaction mixture or the amount of product that has accumulated in a unit of time. For example, in Figure 2.38 (page 97) it is the rate at which the substrate starch disappears from a reaction mixture that is measured.

Working with catalase, it is convenient to measure the rate at which the product (oxygen) accumulates – the volume of oxygen that has accumulated at 30-second intervals is recorded (Figure 2.41).

Over a period of time, the initial rate of reaction is not maintained, but falls off quite sharply. This is typical of enzyme actions studied outside their location in the cell. The fall-off can be due to a number of reasons, but most commonly it is because the concentration of the substrate in the reaction mixture has fallen. Consequently, it is the initial rate of reaction that is measured. This is the slope of the tangent to the curve in the initial stage of reaction.

■ **Figure 2.41**

Measuring the rate of reaction using catalase

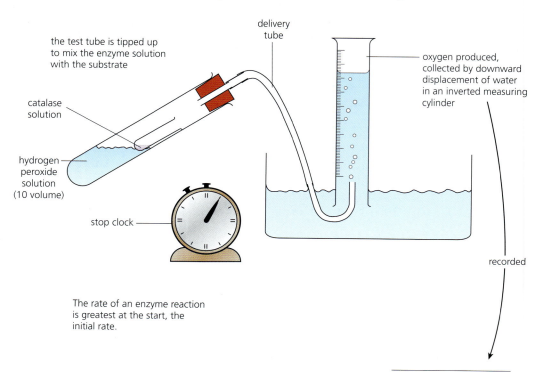

the test tube is tipped up to mix the enzyme solution with the substrate

catalase solution

hydrogen peroxide solution (10 volume)

stop clock

delivery tube

oxygen produced, collected by downward displacement of water in an inverted measuring cylinder

recorded

The rate of an enzyme reaction is greatest at the start, the initial rate.

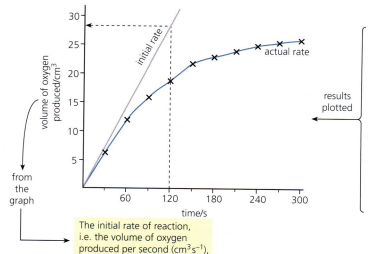

from the graph

Time/s	Gas volume collected/cm³
30	6
60	12
90	16
120	19
150	22
180	23
210	24
240	25
270	25.5
300	26

results plotted

The initial rate of reaction, i.e. the volume of oxygen produced per second ($cm^3 s^{-1}$), can be calculated.

To investigate the effects of substrate concentration on the rate of an enzyme-catalysed reaction, the experiment shown in Figure 2.41 is repeated at different concentrations of substrate, and the initial rate of reaction is plotted in each case. Other variables, such as temperature and enzyme concentration, are kept constant.

When the initial rates of reaction are plotted against the substrate concentration, the curve shows two phases. At lower concentrations, the rate increases in direct proportion to the substrate concentration – but at higher substrate concentrations, the rate of reaction becomes constant, and shows no increase.

Now we can see that the enzyme catalase works by forming a short-lived enzyme–substrate complex. At a low concentration of substrate, all molecules can find an active site without delay. Effectively, there is an excess of enzyme present. Here the rate of reaction is set by how much substrate is present – as more substrate is made available, the rate of reaction increases.

However, at higher substrate concentrations, there comes a point when there is more substrate than enzyme. Now, in effect, substrate molecules have to 'queue up' for access to an active site. Adding more substrate merely increases the number of molecules awaiting contact with an enzyme molecule, so there is now no increase in the rate of reaction (Figure 2.42).

■ **Figure 2.42**
The effect of substrate concentration

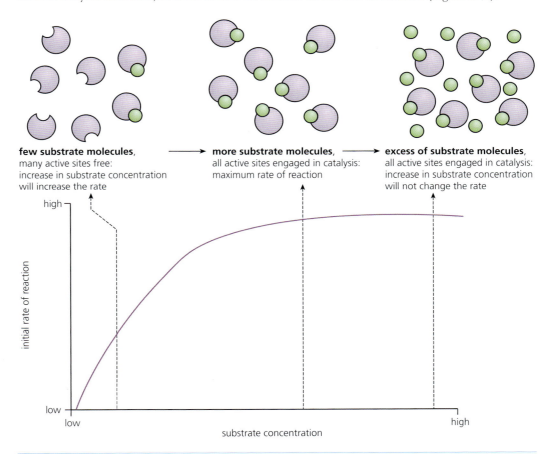

18 In designing the experiment illustrated in Figure 2.41, **identify**:
  a the controlled variables
  b the independent variable
  c the dependent variable
  d one likely potential source of error when the experiment is carried out.
  e **Explain** how such an error is detected and can be avoided, in this experiment.

19 When there is an excess of substrate present in an enzyme-catalysed reaction, **explain** the effect on the rate of reaction of increasing the concentration of:
  a the substrate
  b the enzyme.

Enzymes and a link with homeostasis

The requirements of enzymes for specific conditions – for example, of temperature and pH – have led to the need to control the internal environment of the body, a phenomenon known as **homeostasis** (page 297).

Industrial uses of enzymes

Catalysts are important components of many industrial processes, and today it is often the case that enzymes are used (Figure 2.43). Their use is widespread because they are:

- highly specific, catalysing changes in one particular compound or one type of bond
- efficient, in that a tiny quantity of enzyme catalyses the production of a large quantity of product
- effective at normal temperatures and pressures, and so a limited input of energy (as heat and high pressure) may be required.

Many industrial enzymes are obtained from microorganisms (mainly bacteria and fungi) because these organisms:

- may be grown economically in fermenters throughout the year, rather than being limited to a short growing season
- tend to grow quickly and produce large quantities of enzymes relative to cell mass
- may be bacteria from extreme environments (**extremophiles**, page 227), and therefore have enzymes adapted to function under abnormal conditions of pH or temperature, for example, many enzymes used in biological washing powders
- may be modified genetically with comparative ease, as in the case of the production of human insulin (page 168).

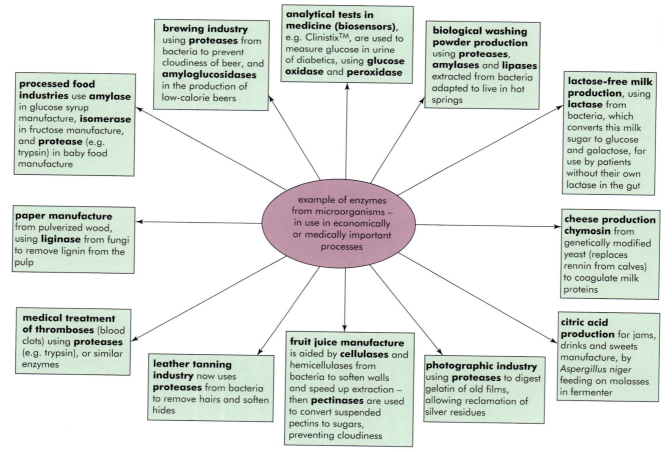

■ **Figure 2.43** Industrial uses of enzymes from microbiological sources

Immobilization of enzymes for industrial use

Enzymes may be used in industrial processes as **cell-free preparations** that are added to a reaction mixture, or they may be **immobilized** and the reactants passed over them. Enzyme immobilization involves the attachment of enzymes to insoluble materials which then provide support. For example, the enzyme may be entrapped between inert fibres or it may be covalently bonded to a matrix. In both cases, the enzyme molecules are prevented from being leached away. The advantages of using an immobilized enzyme in industrial productions are:

- it permits reuse of the enzyme preparation
- the product is enzyme free
- the enzyme may be much more stable and long lasting, due to protection by the inert matrix.

Lactose-free milk – an example of industrial use of enzymes

People who are unable to produce lactase in their pancreatic juice or on the surface of the villi of the small intestine (page 252) fail to digest milk sugar. In these people, the lactose passes on to their large intestine without being hydrolysed to its constituent monosaccharides (page 78). As a result, bacteria in the large intestine feed on the lactose, producing fatty acids and methane, causing diarrhoea and flatulence. Such people are said to be lactose intolerant.

The enzyme lactase is produced in the gut of all human babies while they are dependent on milk. However, it is only found in adults from Northern Europe (and their descendants, wherever they now live) and from a few African peoples. On the other hand, people from the Orient, Arabia and India, most African people and those from the Mediterranean typically produce little or no lactase as adults. Such people may be prescribed **lactose-free milk**; this product can be produced by the application of enzyme technology, using lactase obtained from bacteria. Today, lactose-free milk is obtained by passing milk through a column containing immobilized lactase. The enzyme is obtained from bacteria, purified, and enclosed in capsules (Figure 2.44).

■ **Figure 2.44**
Industrial production of lactose-free milk

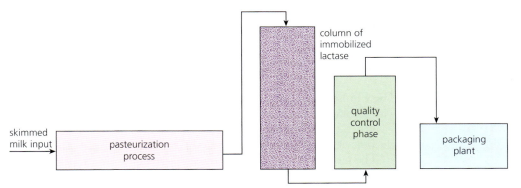

TOK Link

The lactose-free milk available in many developed countries would be especially beneficial if it were available to other communities. Consider this issue by reference to a country where lactose intolerance is more prevalent. What solutions to the underlying problems are practical?

Immobilized enzyme – a laboratory demonstration

Invertase can be immobilized in alginate beads and their effectiveness in hydrolysing sucrose solution investigated (Figure 2.45). The steps to this demonstration are:

1. A solution of sodium alginate (a polysaccharide obtained from the walls of brown algae, capable of holding 200 times its own mass in water) is prepared by dispersing 2 g of the alginate in $100 \, cm^3$ of distilled water at a temperature of 40°C.
2. An alginate–enzyme solution is prepared by stirring $2 \, cm^3$ of invertase concentrate into $40 \, cm^3$ of the cooled alginate solution. (Remember, invertase catalyses the hydrolysis of sucrose [a disaccharide] to two monosaccharide molecules [glucose and fructose] from which it is formed in a condensation reaction.)

3 Immobilized enzyme pellets are produced by filling a syringe with the alginate–invertase solution and arranging for it to drip into a beaker of calcium chloride solution ($100\,cm^3$ of $CaCl_2$, $0.1\,mol\,dm^{-3}$). The insoluble pellets that form may be separated with a nylon or plastic sieve. They should be washed with distilled water at this stage, and allowed to harden.

4 These hardened pellets may be tested by placing them in a tube with a narrow nozzle at the base (the barrel of a large, plastic syringe is suitable). A quantity of sucrose solution ($30\,cm^3$ of 5% sucrose w/v) may be slowly poured down the column of pellets and the effluent collected.

5 The effluent solution is tested for the presence of glucose using a dipstick such as Clinistix™ (Figure 2.46).

The questions to consider when planning and evaluating this experiment include the following:
■ How effective was the immobilized enzyme?
■ How many times can it be used?
■ Is there any way of recovering the enzyme and reusing it, if it were not in pellet form?

> **20 Define** the term 'catalyst'. **List** two differences between inorganic catalysts and enzymes.

a. Preparing immobilized enzymes in alginates: **b. Harvesting alginate-enzyme beads:**   **c. Using immobilized enzyme beads:**

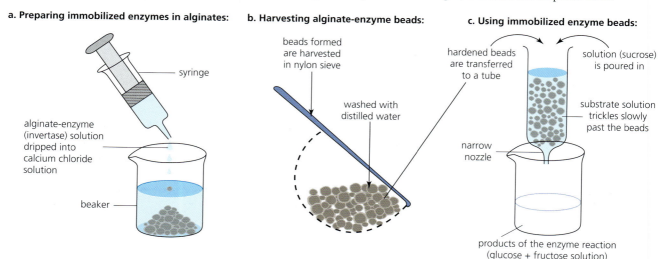

■ **Figure 2.45** Laboratory demonstration of enzyme immobilization

The Clinistix strip contains two immobilized enzymes, glucose oxidase and peroxidase, together with a colourless hydrogen donor compound called chromogen (DH_2).

When the strip is dipped into the sample to be tested, if glucose is present it is oxidized to gluconic acid and hydrogen peroxide.

The second enzyme catalyses the reduction of hydrogen peroxide and the oxidation of chromogen (represented as D in Figure 2.46). The products are water and the oxidized dye, which is coloured.

The more glucose present in the sample, the more coloured dye is formed. The colour of the test strip is then compared to the printed scale to indicate the amount of glucose present.

■ **Figure 2.46** Measurement of glucose using immobilized enzymes in a dipstick (Clinistix)

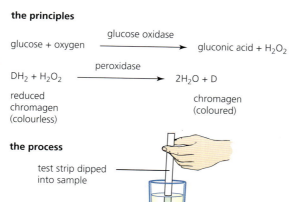

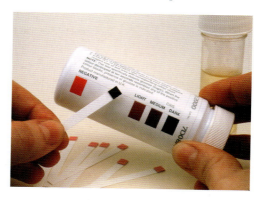

2.6 Structure of DNA and RNA – *the structure of DNA allows efficient storage of genetic information*

Nucleic acids are the **information molecules** of cells found throughout the living world. This is because the code containing the information in nucleic acids, known as the **genetic code**, is a universal one. That means that it makes sense in all organisms. It is not specific to a few organisms or to just one group, such as bacteria.

There are two types of nucleic acid found in living cells: deoxyribonucleic acid (**DNA**) and **ribonucleic acid** (**RNA**). DNA is the genetic material and occurs in the chromosomes of the nucleus. While some RNA also occurs in the nucleus, most is found in the cytoplasm – particularly in the ribosomes. Both DNA and RNA have roles in the day-to-day control of cells and organisms, as we shall shortly see. First, we will look into the structure of nucleotides and the way they are built up to form the unique **DNA double helix**.

■ Nucleotides

A nucleotide consists of three substances combined together. These are:

- a **nitrogenous base** – the four bases of DNA are cytosine (C), guanine (G), adenine (A) and thymine (T)
- a **pentose sugar** – deoxyribose occurs in DNA and ribose in RNA
- **phosphoric acid**.

These components are combined by condensation reaction to form a nucleotide with the formation of two molecules of water. Since any one of the four bases can be incorporated, four different types of nucleotide can be found in DNA. How these components are combined together is shown in Figure 2.47, together with the diagrammatic way the components are represented to illustrate their spatial arrangement. Simple shapes are used rather than complex structural formulae, and these shapes are all that are required here. (*You need to be able to draw simple diagrams, using these symbols.*)

21 Distinguish between a nitrogenous base and a base found in inorganic chemistry.

phosphoric acid

pentose sugars

deoxyribose
ribose

Chemical structures of the components are shown together with the symbols by which they are represented in diagrams. *Note that chemical formulae do not need to be learned.*

nitrogenous bases

adenine

guanine

thymine

cytosine

condensation to form a nucleotide

phosphoric acid

shown diagrammatically as:

base

ribose

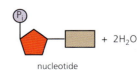

nucleotide

$+ 2H_2O$

■ **Figure 2.47**
The components of nucleotides

■ **Figure 2.48**
How nucleotides make up nucleic acid

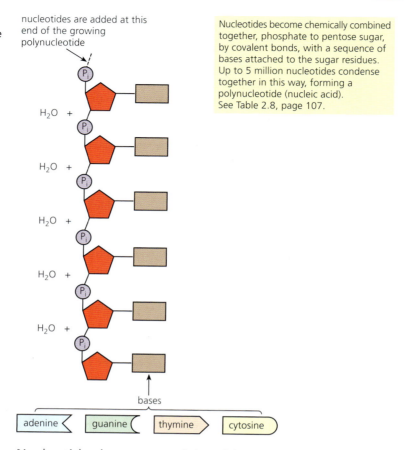

nucleotides are added at this end of the growing polynucleotide

H_2O +

H_2O +

H_2O +

H_2O +

bases

adenine ⟨   guanine (thymine ⟩   cytosine

Nucleotides become chemically combined together, phosphate to pentose sugar, by covalent bonds, with a sequence of bases attached to the sugar residues. Up to 5 million nucleotides condense together in this way, forming a polynucleotide (nucleic acid). See Table 2.8, page 107.

Nucleotides become nucleic acid

Nucleotides may condense together, one nucleotide at a time, to form huge molecules called nucleic acids or **polynucleotides** (Figure 2.48). So, nucleic acids are very long, thread-like macromolecules with alternating sugar and phosphate molecules forming the 'backbone'. This part of the nucleic acid molecule is uniform and unvarying. However, also attached to each of the sugar molecules along the strand is one of the bases, and these project sideways. Since the bases vary, they represent a unique sequence that carries the coded information held by the nucleic acid.

■ The DNA double helix

The DNA molecule consists of two antiparallel polynucleotide strands, paired together, and held by hydrogen bonds. The two strands take the shape of a double helix (Figure 2.49).
The pairing of bases is between **adenine** (**A**) and **thymine** (**T**), and between **cytosine** (**C**) and **guanine** (**G**), simply because these are the only combinations that fit together along the helix. This pairing, known as **complementary base pairing**, also makes possible the very precise way that DNA is copied in a process called **replication**. The existence of the DNA double helix was discovered, and the way DNA holds information was suggested, by Francis Crick and James Watson in 1953 – they received a Nobel Prize for this work (Figure 2.50). They had shown that the structure of DNA allows efficient storage of genetic information.

■ Introducing RNA

RNA molecules are relatively short in length, compared with DNA. In fact, RNAs tend to be from a hundred to thousands of nucleotides long, depending on their particular role. The RNA molecule is a single strand of polynucleotide in which the sugar is ribose. The bases found in RNA are cytosine, guanine, adenine and **uracil** (which replaces thymine of DNA). There are three functional types of RNA: **messenger RNA** (**mRNA**), **transfer RNA** (**tRNA**) and **ribosomal RNA** (Table 2.8). While mRNA is formed in the nucleus and passes out to ribosomes in the cytoplasm, tRNA and ribosomal RNA occur only in the cytoplasm.

■ **Figure 2.49**
The DNA double helix

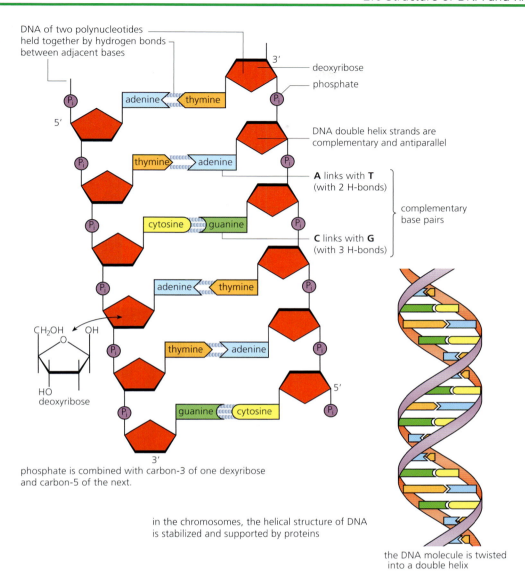

DNA of two polynucleotides held together by hydrogen bonds between adjacent bases

3'

deoxyribose

phosphate

5'

adenine ⟩⟩⟩⟩ thymine

thymine ⟩⟩⟩ adenine

cytosine ⟩⟩⟩ guanine

adenine ⟩⟩⟩ thymine

CH₂OH   OH

O

HO
deoxyribose

thymine ⟩⟩⟩ adenine

3'

guanine ⟩⟩⟩ cytosine

5'

DNA double helix strands are complementary and antiparallel

A links with **T** (with 2 H-bonds)

C links with **G** (with 3 H-bonds)

complementary base pairs

phosphate is combined with carbon-3 of one dexyribose and carbon-5 of the next.

in the chromosomes, the helical structure of DNA is stabilized and supported by proteins

the DNA molecule is twisted into a double helix

■ **Table 2.8**
The differences between DNA and RNA

DNA	Feature	RNA
very long strands, several million nucleotides	**length**	relatively short strands, 100 to several thousand nucleotides
contains deoxyribose	**sugar**	contains ribose
contains bases C, G, A and T (not U)	**bases**	contains C, G, A and U (not T)
consists of two polynucleotide strands of complementary base pairs (C with G and A with T) held by H-bonds in the form of a double helix	**forms**	consists of single strands, and in three functional forms: messenger RNA (mRNA) transfer RNA (tRNA) ribosomal RNA

Using models as representations of the real world

Francis Crick (1916–2004) and **James Watson** (1928–) laid the foundations of a new branch of biology – **cell biology** – and achieved this while still young men. Within two years of their meeting in the Cavendish Laboratory, Cambridge (1951), Crick and Watson had achieved their understanding of the nature of the gene in chemical terms.

Crick and Watson brought together the experimental results of many other workers, and from this evidence they deduced the likely structure of the DNA molecule.

- **Edwin Chergaff** measured the exact amount of the four organic bases in samples of DNA, and found the ratio of A:T and of C:G was always close to 1.

Chergaff's results suggest consistent base pairing in DNA from different organisms.

Organism	Ratio of bases in DNA samples	
	Adenine : Thymine	Guanine : Cytosine
Cow	1.04	1.00
Human	1.00	1.00
Salmon	1.02	1.02
Escherichia coli	1.09	0.99

- **Rosalind Franklin** and **Maurice Wilkins** produced X-ray diffraction patterns by bombarding crystalline DNA with X-rays.

Watson and Crick concluded that DNA is a double helix consisting of:
- two polynucleotide strands with nitrogenous bases stacked on the inside of the helix (like rungs on a twisted ladder)
- parallel strands are held together by hydrogen bonds between the paired bases (A–T, C–G)
- ten base pairs occur per turn of the helix
- the two strands of the double helix are antiparallel.

They built a model. See Figure 2.49 (page 107) for a simplified model of the DNA double helix.

Watson and Crick with their demonstration model of DNA

Rosalind Franklin produced the key X-ray diffraction pattern of DNA at Kings College, London

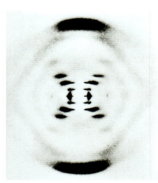

X-ray diffraction pattern of DNA

■ **Figure 2.50** The discovery of the role of DNA

TOK Link

Crick and Watson had a distinctive method of working, including reinterpreting already published data and developing others' studies, leading to the building of models (Figure 2.50). To what extent were their achievements the product of both cooperation and competition? In researching this, these resources are useful:

1 Crick and Watson's original, short 'letter' to *Nature*, published 25 April 1953, if your school or college can make a copy available for you. For a one-off fee, the paper can be obtained from:
   www.nature.com/nature/archive
2 Their own accounts in:
   James D. Watson (1965), *The Double Helix*, Weidenfeld & Nicolson.
   Francis Crick (1988), *What Mad Pursuits*, Penguin.
3 The contribution of Rosalind Franklin (Figure 2.50):
   Brenda Maddox (2003), *The Dark Lady of DNA*, Harper Collins.

2.7 DNA replication, transcription and translation – *genetic information in DNA can be accurately copied and can be translated to make the proteins needed by the cell*

■ Replication – how DNA copies itself

A copy of each chromosome must pass into daughter cells formed by cell division, so the chromosomes must first be copied (**replicated**). *Remember, this process takes place in the interphase nucleus, well before the events of nuclear division (page 52).*

The first step in replication is the '**unzipping**' of the two strands. An enzyme called **helicase** unwinds the DNA double helix at one region, breaks the hydrogen bonds there that hold the strands together and then temporarily keeps the strands of the helix separated. The unpaired nucleotides are now exposed, surrounded by a pool of free-floating nucleotides. In the next step, both strands of DNA act as **templates** in replication. Complementary nucleotides line up opposite each base of the exposed strands – adenine pairs with thymine, cytosine with guanine. Hydrogen bonds then form between the complementary bases, holding them in place.

Finally, a **condensation reaction** links the sugar and phosphate groups of adjacent nucleotides, so forming the new strands. This reaction is catalysed by an enzyme called **DNA polymerase**. The enzyme has a 'proof-reading' role in replication, too – any mistakes that start to happen (such as the wrong bases attempting to pair up) are corrected. The result is that the two strands formed are identical to the original strands. DNA replication is summarized in Figure 2.51.

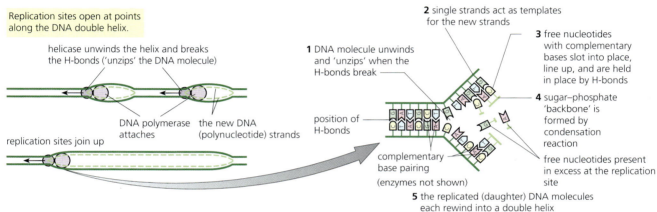

■ **Figure 2.51** DNA replication

Finally, each new pair of strands reforms as a double helix. One strand of each double helix has come from the original chromosome and one is a newly synthesized strand. This arrangement is known as **semi-conservative replication** because half the original molecule stays the same (Figure 2.52).

■ **Figure 2.52**
Semi-conservative versus conservative replication

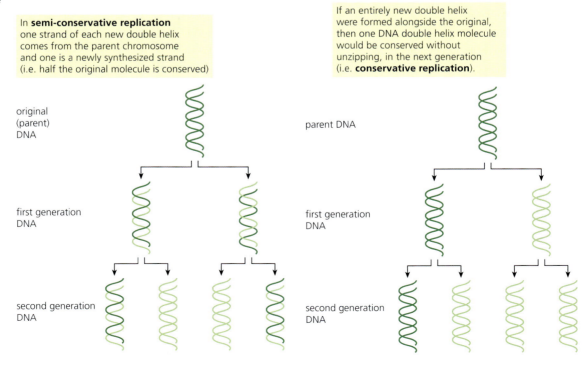

Obtaining evidence for scientific theories

■ The evidence for semi-conservative DNA replication

Experimental evidence that DNA replication is semi-conservative came from an experiment by Meselson and Stahl. In the first step, they grew a culture of the bacterium *E. coli* (page 28) in a medium (food source) where the available nitrogen contained only the heavy nitrogen isotope, ^{15}N. Consequently, the DNA of the bacterium became entirely 'heavy'.

These bacteria were then transferred to a medium of the normal (light) isotope, ^{14}N. New DNA manufactured by the cells was now made of ^{14}N. The change in concentration of ^{15}N and ^{14}N in the DNA of succeeding generations was measured. Interestingly, the bacterial cell divisions in a culture of *E. coli* are naturally synchronized; every 60 minutes they all divide.

The DNA was extracted from samples of the bacteria from each succeeding generation and the DNA in each sample was separated. This was done by placing the sample on top of a salt solution of increasing density, in a centrifuge tube. On being centrifuged, the different DNA molecules were carried down to the level where the salt solution was of the same density. Thus, DNA with 'heavy' nitrogen ended up nearer the base of the tubes, whereas DNA with 'light' nitrogen stayed near the top of the tubes. Figure 2.53 shows the results that were obtained.

> **22 Predict** the experimental results you would expect to see if the Meselson–Stahl experiment (Figure 2.53) was carried on for three generations.

1 **Meselson and Stahl** 'labelled' nucleic acid (i.e. DNA) of the bacterium *Escherichia coli* with 'heavy' nitrogen (^{15}**N**), by culturing in a medium where the only nitrogen available was as ^{15}**NH$_4$**$^+$ ions, for several generations of bacteria.

2 When DNA from labelled cells was extracted and centrifuged in a density gradient (of different salt solutions) all the DNA was found to be 'heavy'.

3 In contrast, the DNA extracted from cells of the original culture (before treatment with ^{15}**N**) was 'light'.

4 Then a labelled culture of *E. coli* was switched back to a medium providing unlabelled nitrogen only, i.e. ^{14}**NH$_4$**$^+$. Division in the cells was synchronized, and:
• after **one generation** all the DNA was of intermediate density (each of the daughter cells contained (i.e. *conserved*) one of the parental DNA strands containing ^{15}**N** alongside a newly synthesized strand containing DNA made from ^{14}N)
• after **two generations** 50% of the DNA was intermediate and 50% was 'light'. This too agreed with semi-conservative DNA replication, given that in only half the cells was labelled DNA present (one strand per cell).

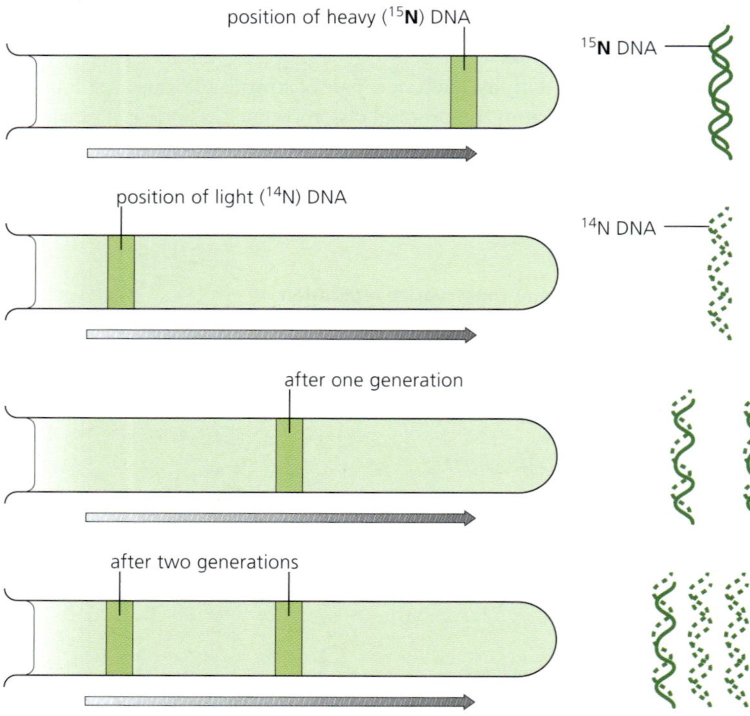

■ **Figure 2.53** DNA replication is semi-conservative

position of heavy (^{15}N) DNA

^{15}N DNA

position of light (^{14}N) DNA

^{14}N DNA

after one generation

after two generations

■ Protein synthesis

We have seen that proteins are linear series of amino acids condensed together. Most proteins contain several hundred amino acid residues, but all are built from only 20 different amino acids.

The unique properties of a protein lie in:

- **which amino acids** are involved in its construction
- **the sequence** in which these amino acids are condensed together. The sequence of bases in DNA dictates the order in which specific amino acids are assembled and combined together.

The DNA molecule in each chromosome is very long, and it codes for a large number of proteins. Within this molecule, the relatively short lengths of DNA that code for single proteins are called **genes**. Proteins are very variable in size and, therefore, so are genes. A very few genes are as short as 75–100 nucleotides. Most are at least 1000 nucleotides long, and some are considerably more.

All proteins are formed in the cytoplasm, some at free-floating ribosomes, and others at the ribosomes on RER. For this to happen, a mobile copy of the information in the genes has to be made, and then transported to these sites of protein synthesis. That copy is made of RNA and is called **messenger RNA (mRNA)**. It is formed by a process called **transcription**. So both DNA and RNA have roles in protein synthesis.

Transcription – the first step in protein synthesis

It is an enzyme called **RNA polymerase** that catalyses the formation of a complementary copy of the genetic code of a gene. This copy takes the form of a molecule of mRNA (Figure 2.54).

In Stage 1 of protein synthesis, the DNA double helix of a particular gene unwinds and hydrogen bonds between the complementary strands of DNA are broken. Inevitably, there is a pool of free nucleotides present.

Then, one strand of the DNA, the **coding strand**, becomes the template for transcription. A single-stranded molecule of RNA is formed by complementary base pairing. Remember, in RNA synthesis, it is uracil which pairs with adenine. Then that mRNA strand leaves the nucleus through pores in the nuclear membrane and passes to ribosomes in the cytoplasm. Here, the information can be 'read' and used in the synthesis of a protein.

The genetic code

Information in the DNA lies in the sequence of the bases, cytosine (C), guanine (G), adenine (A) and thymine (T). This sequence dictates the order in which specific amino acids are assembled and combined together. The code lies in the sequence in *one* of the strands, the **coding strand**. The other strand is complementary to it. The coding strand is always read in the same direction.

The code is a three-letter or **triplet code**, meaning that each sequence of three of the four bases stands for one of the 20 amino acids, and is called a **codon**. With a four-letter alphabet (C, G, A, T) there are 64 possible different triplet combinations (4 × 4 × 4). In other words, the genetic code has many more codons than there are amino acids to be coded. In fact, most amino acids have two or three similar codons that code for them (Figure 2.55). However, some of the codons have different roles. Some represent the 'punctuations' of the code; for example, there are 'start' and 'stop' triplets.

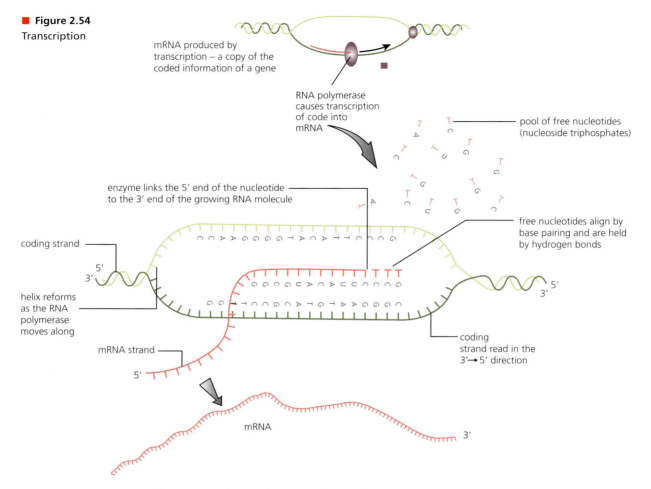

mRNA produced by transcription – a copy of the coded information of a gene

RNA polymerase causes transcription of code into mRNA

pool of free nucleotides (nucleoside triphosphates)

enzyme links the 5′ end of the nucleotide to the 3′ end of the growing RNA molecule

coding strand

free nucleotides align by base pairing and are held by hydrogen bonds

helix reforms as the RNA polymerase moves along

coding strand read in the 3′→5′ direction

mRNA strand

mRNA

The second step in protein synthesis

In Stage 2 of protein synthesis, the amino acids of the pool available for protein synthesis are activated by combining with short lengths of a different sort of RNA, **transfer RNA (tRNA)**. It is this tRNA that translates a three-base sequence into an amino acid sequence. *How does this occur?*

All the tRNAs have a cloverleaf shape, but there is a different tRNA for each of the 20 amino acids involved in protein synthesis. At one end of each tRNA molecule is a site where one particular amino acid of the 20 can be joined. At the other end, there is a sequence of three bases called an **anticodon**. This anticodon is complementary to the codon of mRNA that codes for the specific amino acid (Figure 2.56).

The amino acid becomes attached to its tRNA by an enzyme in a reaction that also requires ATP. These enzymes are specific to the particular amino acids (and types of tRNA) to be used in protein synthesis. The specificity of the enzymes is a way of ensuring the correct amino acids are used in the right sequence.

Translation – the last step in protein synthesis

In Stage 3, the final stage, a protein chain is assembled, one amino acid residue at a time, in tiny organelles called ribosomes. A ribosome moves along the messenger RNA 'reading' the codons from the 'start' codon. In the ribosome, for each mRNA codon, the complementary anticodon of the tRNA–amino acid complex slots into place and is temporarily held in position by hydrogen bonds. While held there, the amino acids of neighbouring tRNA–amino acid complexes are joined by a peptide linkage. This frees the first tRNA which moves back into the cytoplasm for reuse. Once this is done, the ribosome moves on to the next mRNA codon. The process continues until a ribosome meets a 'stop' codon (Figures 2.57 and 2.58).

Amino acid	Abbreviation
alanine	Ala
arginine	Arg
asparagine	Asn
aspartic acid	Asp
cysteine	Cyc
glutamine	Gln
glutamic acid	Glu
glycine	Gly
histidine	His
isoleucine	Ile
leucine	Leu
lysine	Lys
methionine	Met
phenylalanine	Phe
proline	Pro
serine	Ser
threonine	Thr
tryptophan	Trp
tyrosine	Tyr
valine	Val

		Second base			
		A	**G**	**T**	**C**
First base	**A**	AAA Phe AAG Phe AAT Leu AAC Leu	AGA Ser AGG Ser AGT Ser AGC Ser	ATA Tyr ATG Tyr ATT *Stop* ATC *Stop*	ACA Cys ACG Cys ACT *Stop* ACC Trp
	G	GAA Leu GAG Leu GAT Leu GAC Leu	GGA Pro GGG Pro GGT Pro GGC Pro	GTA His GTG His GTT Gln GTC Gln	GCA Arg GCG Arg GCT Arg GCC Arg
	T	TAA Ile TAG Ile TAT Ile TAC Met	TGA Thr TGG Thr TGT Thr TGC Thr	TTA Asn TTG Asn TTT Lys TTC Lys	TCA Ser TCG Ser TCT Arg TCC Arg
	C	CAA Val CAG Val CAT Val CAC Val	CGA Ala CGG Ala CGT Ala CGC Ala	CTA Asp CTG Asp CTT Glu CTC Glu	CCA Gly CCG Gly CCT Gly CCC Gly

■ **Figure 2.55** The 20 amino acids found in proteins, and the genetic code (you will use this figure to deduce which codon corresponds to which amino acid)

23 The sequence of bases in part of a DNA coding strand was found to be: -A-G-A-C-T-G-T-T-C-A-T-T.
Determine the sequence of amino acids this codes for, and where along the length of a gene it occurred.

■ **Figure 2.56**
An amino acid is linked to a specific tRNA

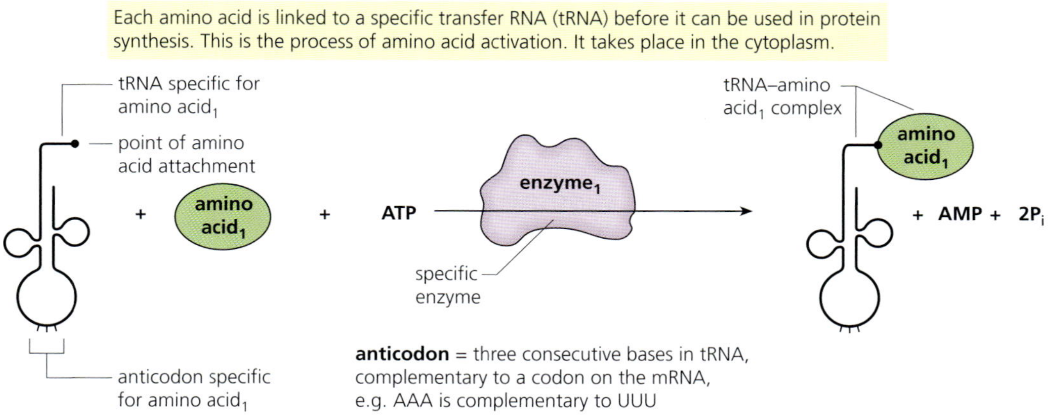

Each amino acid is linked to a specific transfer RNA (tRNA) before it can be used in protein synthesis. This is the process of amino acid activation. It takes place in the cytoplasm.

tRNA specific for amino acid$_1$
point of amino acid attachment
amino acid$_1$
+ ATP
enzyme$_1$
specific enzyme
tRNA–amino acid$_1$ complex
amino acid$_1$
+ AMP + 2P$_i$
anticodon specific for amino acid$_1$

anticodon = three consecutive bases in tRNA, complementary to a codon on the mRNA, e.g. AAA is complementary to UUU

■ Introducing gene technology

Gene technology is an important, practical application of modern genetics – one outcome of the discovery of the structure and role of DNA. For example, the human genes for **insulin** production have been isolated and transferred to a strain of the bacterium *Escherichia coli*. By culturing these genetically modified microorganisms, regular supplies of human insulin have been obtained to treat insulin-dependent diabetes patients (page 167). The **universality** of the genetic code makes this possible. Genes can be transferred between species and result in the same protein being formed.

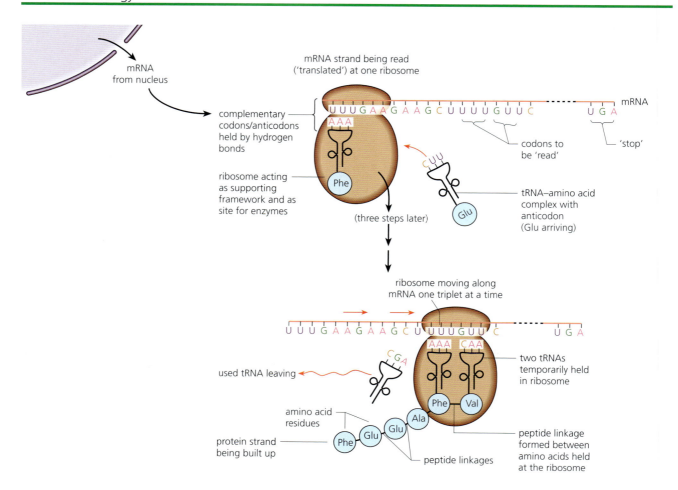

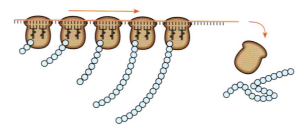

Several ribosomes may move along the mRNA at one time. This structure (mRNA, ribosomes plus growing protein chains) is called a **polysome**.

■ **Figure 2.57** Translation

Sometimes only a very small sample of DNA is available to the genetic engineer. The discovery of the **polymerase chain reaction (PCR)** has permitted fragments of DNA to be copied repeatedly, faithfully and speedily, in a process that has been fully automated. A heat-tolerant DNA polymerase enzyme, obtained from a extremophile bacterium, is used (pages 166–167).

These developments in gene technology are discussed in Chapter 3 (page 164).

24 The sequence of bases in a sample of mRNA was found to be:
   GGU,AAU,CCU,UUU,GUU,ACU,CAU,UGU
   a **Deduce** the sequence of amino acids this codes for.
   b **Determine** the sequence of bases in the coding strand of DNA from which this mRNA was transcribed.
   c Within a cell, **state** where the triplet codes, codons and anticodons are found.

■ **Figure 2.58**
The mRNA genetic dictionary

		2nd base			
		U	**C**	**A**	**G**
1st base	**U**	UUU Phe UUC Phe UUA Leu UUG Leu	UCU Ser UCC Ser UCA Ser UCG Ser	UAU Tyr UAC Tyr UAA *Stop* UAG *Stop*	UGU Cys UGC Cys UGA *Stop* UGG Trp
	C	CUU Leu CUC Leu CUA Leu CUG Leu	CCU Pro CCC Pro CCA Pro CCG Pro	CAU His CAC His CAA Gin CAG Gin	CGU Arg CGC Arg CGA Arg CGG Arg
	A	AUU lie AUC lie AUA lie AUG Met	ACU Thr ACC Thr ACA Thr ACG Thr	AAU Acn AAC Asn AAA Lys AAG Lys	AGU Ser AGC Ser AGA Arg AGG Arg
	G	GUU Val GUC Val GUA Val GUG Val	GCU Aia GCC Aia GCA Aia GCG Aia	GAU Asp GAC Asp GAA Giu GAG Giu	GGU Gly GGC Gly GGA Gly GGG Gly

2.8 Cell respiration – *cell respiration supplies energy for the functions of life*

Living things require **energy** to build and repair body structures and to maintain all the activities of life, such as movement, reproduction, nutrition, excretion and sensitivity. Energy is also required for protein synthesis, and for the active transport of molecules and ions across membranes by membrane pumps. Energy is transferred in cells by the breakdown of nutrients, principally of carbohydrates like glucose. The process by which energy is made available from nutrients in cells is called **cell respiration**. Cell respiration is the controlled release of energy from organic compounds in cells.

■ The calorimeter

We can measure the total amount of energy that can be released from the nutrient glucose by means of a simple **calorimeter** (Figure 2.59). Here, a known amount of glucose is placed in the crucible in a closed environment and burnt in oxygen. The energy, released as heat, is transferred to the surrounding jacket of water and the rise in temperature of the water is measured. Then the energy value of the glucose sample may be calculated, based on the fact that it takes 4.2 joules (J) of heat energy to raise 1 g of water by 1°C. One well-known outcome of this technique is the energy value labelling of manufactured foods on packaging – and the publicity that 'low-calorie' items may receive in slimming diets.

People sometimes talk about 'burning up food' in respiration. In fact, likening respiration to combustion is unhelpful. In combustion, energy in fuel is released in a one-step reaction, as heat (Figure 2.60). Such a violent change would be disastrous for body tissues. In cellular respiration, a large number of small steps occur, each catalysed by a specific enzyme. Energy in respiration is transferred in small quantities – much of the energy is made available to the cells and may be temporarily trapped in the energy currency molecule **adenosine triphosphate** (**ATP**). However, some energy is still lost as heat in each step – we notice how warm we become with strenuous physical activity!

■ ATP – the universal energy currency

Energy that is made available within the cytoplasm is transferred to a molecule called adenosine triphosphate (ATP). ATP is referred to as **energy currency** because, like money, it can be used in different contexts, and it is constantly recycled.

ATP is a **nucleotide** (page 105) with an unusual feature. It carries three phosphate groups linked together in a linear sequence (Figure 2.61). ATP may lose both of the outer phosphate groups, but usually only one at a time is lost. ATP is a relatively small, soluble organic molecule. It occurs in cells at a concentration of 0.5–$2.5\,mg\,cm^{-3}$.

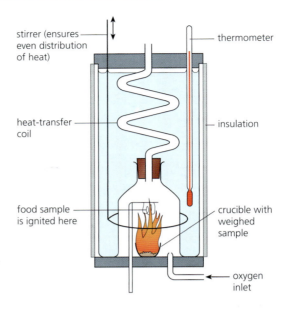

■ **Figure 2.59**
A calorimeter for measuring energy value

Samples are completely oxidized in air.

stirrer (ensures even distribution of heat)

thermometer

heat-transfer coil

insulation

food sample is ignited here

crucible with weighed sample

oxygen inlet

The energy values of foods are published in tables, and those of manufactured and packaged foods may be recorded on the wrapping.

■ **Figure 2.60**
Combustion and respiration compared

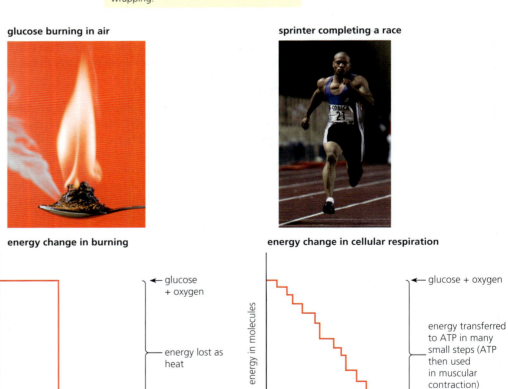

glucose burning in air

sprinter completing a race

energy change in burning

energy change in cellular respiration

energy in molecules

← glucose + oxygen

energy lost as heat

← CO_2 + H_2O

time

energy in molecules

← glucose + oxygen

energy transferred to ATP in many small steps (ATP then used in muscular contraction)

← CO_2 + H_2O

time

Like many organic molecules of its size, ATP contains a good deal of chemical energy locked up in its structure. What makes ATP special as a reservoir of chemical energy is its role as a common intermediate between energy-yielding reactions and energy-requiring reactions and

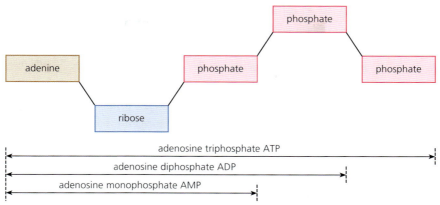

processes. Energy-yielding reactions include many of the individual steps in respiration. Energy-requiring reactions include the synthesis of cellulose from glucose, the synthesis of proteins from amino acids, and the contraction of muscle fibres.

In summary, ATP is a molecule universal to all living things; it is the source of energy for chemical change in cells, tissues and organisms. So ATP is:

■ a substance that moves easily within cells and organisms – by facilitated diffusion

■ a very reactive molecule, able to take part in many steps of cellular respiration and in many reactions of metabolism

■ an immediate source of energy, able to deliver energy in relatively small amounts, sufficient to drive individual reactions.

■ The ATP–ADP cycle and metabolism

In cells, ATP is formed from **adenosine diphosphate** (ADP) and phosphate ion (Pi) using energy from respiration (Figure 2.62). Then, in the presence of enzymes, ATP participates in energy-requiring reactions. The free energy available in ATP is approximately $30–34 \times 10^9\,\text{kJ mol}^{-1}$. Some of this energy is lost as heat in a reaction, but much free energy is made available to do useful work, more than sufficient to drive a typical energy-requiring reaction of metabolism (Figure 2.7, page 67).

Sometimes ATP reacts with water (a hydrolysis reaction) and is converted to ADP and P_i. For example, direct hydrolysis of the terminal phosphate group happens in muscle contraction.

Mostly, ATP reacts with other metabolites and forms phosphorylated intermediates, making them more reactive in the process. The phosphate groups are released later, so both ADP and P_i become available for reuse as metabolism continues.

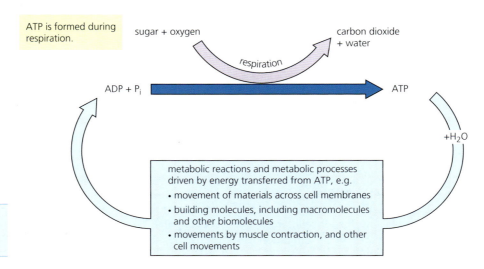

25 Outline why ATP is an efficient energy currency molecule.

The steps of cellular respiration

In the first steps of cellular respiration, the glucose molecule (a 6-carbon sugar) is split into two 3-carbon molecules. Then the products are converted to an organic acid called pyruvic acid (also a 3-carbon compound). Under conditions in the cytoplasm, organic acids are weakly ionized and, therefore, pyruvic acid exists as the **pyruvate** ion.

Obviously, two molecules of pyruvate are formed from each molecule of glucose. In addition, there is a small amount of ATP formed, using a little of the energy that had been locked up in the glucose molecule. No molecular oxygen is required for these first steps of cellular respiration.

Because glucose has been split into smaller molecules, these steps are known as **glycolysis**. The enzymes that catalyse these reactions are found in the cell cytoplasm generally, but not inside an organelle. Throughout cellular respiration, a series of oxidation–reduction reactions occur. (Oxidation and reduction are explained on page 352.)

In summary:

$$\text{glucose} \xrightarrow{\text{enzymes in the cytoplasm}} \text{pyruvate + small amount of ATP}$$

■ Aerobic and anaerobic cellular respiration

While no oxygen is required for the formation of pyruvate by cells in the early steps of cellular respiration, most animals and plants and very many microorganisms do require oxygen for cell respiration, in total. We say that they respire aerobically.

In **aerobic cellular respiration**, sugar is completely oxidized to carbon dioxide and water and much energy is made available. The steps of aerobic respiration can be summarized by a single equation:

$$\text{glucose} \quad + \text{oxygen} \longrightarrow \text{carbon dioxide + water} + \text{ENERGY}$$
$$C_6H_{12}O_6 + 6O_2 \longrightarrow 6CO_2 \quad + 6H_2O + \text{ENERGY}$$

This equation is a balance sheet of the inputs (raw materials) and the outputs (products). It tells us nothing about the separate steps, each catalysed by a specific enzyme, by which cellular respiration occurs. It does not mention pyruvate, for example.

What happens to pyruvate in aerobic respiration?

If oxygen is available to cells and tissues, the pyruvate is completely oxidized to carbon dioxide, water and a large quantity of ATP. Before these reactions take place, the pyruvate first passes into mitochondria by facilitated diffusion. This is because it is only in mitochondria that the required enzymes are found (Figure 2.63).

In summary:

$$\text{pyruvate} \xrightarrow{\text{enzymes in the mitochondria}} \text{carbon dioxide + water + large amount of ATP}$$

In this phase of cellular respiration, the pyruvate is oxidized by:

- removal of hydrogen atoms by hydrogen acceptors (oxidizing agents)
- addition of oxygen to the carbon atoms to form carbon dioxide.

These reactions occur one at a time, each catalysed by a different enzyme.

Also, it is in the mitochondria that the hydrogen (proton) carried by the reduced hydrogen-acceptor molecules reacts with oxygen to form water. The reduced hydrogen acceptor is reoxidized (loses its H) and is available for reuse in the production of more pyruvate. The majority of ATP molecules (the key product of respiration for the cell) are generated in this step, also.

The sites of cellular respiration

The enzymes of cellular respiration occur partly in the cytoplasm (the enzymes of glycolysis) and partly in the mitochondria (the enzymes of pyruvate oxidation and most ATP formation). After formation, ATP passes to all parts of the cell. Both ADP and ATP pass through the mitochondrial membranes by facilitated diffusion. The locations of the different stages of aerobic respiration are shown in Figure 2.63.

■ **Figure 2.63**
The sites of cellular respiration in cells

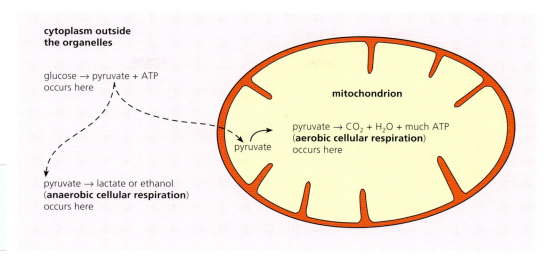

cytoplasm outside the organelles

glucose → pyruvate + ATP
occurs here

pyruvate → lactate or ethanol
(**anaerobic cellular respiration**)
occurs here

mitochondrion

pyruvate → CO_2 + H_2O + much ATP
(**aerobic cellular respiration**)
occurs here

pyruvate

26 State which steps in cellular respiration occur whether or not oxygen is available to cells.

Anaerobic respiration – fermentation

In the absence of oxygen, many organisms (and sometimes tissues in organisms, when these have become deprived of oxygen) will continue to respire pyruvate by different pathways, known as fermentation or **anaerobic respiration**, at least for a short time.

Many species of yeast (*Saccharomyces*) respire anaerobically, even in the presence of oxygen. The products are ethanol and carbon dioxide. You will already be aware that **alcoholic fermentation** of yeast has been exploited by humans for many thousands of years:

■ in bread making – the carbon dioxide causes the bread to 'rise'

■ in wine and beer production.

$$\text{glucose} \longrightarrow \text{ethanol} + \text{carbon dioxide} + \text{ENERGY}$$

Vertebrate muscle tissue can respire anaerobically, too, but in this case it involves the formation of lactic acid rather than ethanol. Once again, under conditions in the cytoplasm, lactic acid is weakly ionized and, therefore, exists as the **lactate** ion.

Lactic-acid fermentation occurs in muscle fibres, but only when the demand for energy for contractions is very great and cannot be met fully by aerobic respiration. In lactic-acid fermentation the sole waste product is lactate.

$$\text{glucose} \longrightarrow \text{lactate} + \text{ENERGY}$$

What happens to pyruvate in anaerobic respiration?

In human skeletal muscle tissue, when oxygen is not available, the pyruvate remains in the cytoplasm and is converted to lactate. In yeast, whether or not oxygen is available, the pyruvate is converted to the alcohol called ethanol. *How is the supply of pyruvate maintained in cells in the absence of oxygen?*

This is a potential problem in the breakdown of pyruvate in the absence of oxygen. Remember, in pyruvate formation, hydrogen-acceptor molecules are reduced (take up hydrogen atoms). Without using oxygen, these must be reoxidized (lose their H) if production of pyruvate is to continue.

The answer is that, in anaerobic cellular respiration, the reduced hydrogen-acceptor molecules donate their hydrogen to form lactate or ethanol from pyruvate. This is how they are reoxidized in the absence of oxygen. In this way, the acceptor molecules are available for further pyruvate synthesis. Pyruvate formation is able to continue.

We can summarize this as:

27 Identify two products of anaerobic respiration in muscle.

$$\text{pyruvate} \xrightarrow[\text{oxidized H acceptor (lost H)}]{\text{reduced H acceptor (carrying H)}} \text{lactate} \qquad \text{pyruvate} \xrightarrow[\text{oxidized H acceptor}]{\text{reduced H acceptor}} \text{ethanol} + CO_2$$

Anaerobic respiration is 'wasteful'

Anaerobic respiration is 'wasteful' of respiratory substrate. This is the case because the total energy yield per molecule of glucose respired, in terms of ATP generated, in both alcoholic and lactic-acid fermentation is limited to the net two molecules of ATP produced in pyruvate formation. No additional energy is transferred in the latter steps and made available in cells.

So we can think of both lactate and ethanol as energy-rich molecules. (This is correct; ethanol is sometimes used as a fuel in cars, for example.) The energy locked up in these molecules may be used later, however. For example, in humans, lactate is transported to the liver and later metabolized aerobically. Energy yields of cellular respiration are compared in Table 2.9.

■ **Table 2.9** Energy yield of aerobic and anaerobic cellular respiration compared

Yield from each molecule of glucose respired		
Aerobic respiration		**Anaerobic respiration**
2 ATPs	**glycolysis**	2 ATPs
up to 36 ATPs	**fates of pyruvate**	nil
38 ATPs	**totals**	2 ATPs

■ Investigating respiration

The rate of respiration of an organism is an indication of its demand for energy. Respiration rate, the uptake of oxygen per unit time, may be measured by means of a **respirometer** (Figure 2.64). The manometer in this apparatus detects change in pressure or volume of a gas. Respiration by tiny organisms (germinating seeds or fly maggots are ideal) that are trapped in the chamber of the respirometer alters the composition of the gas there, once the screw clip has been closed.

Nature of Science

Ethical implications of research

The use of animals in experiments may generate ethical issues. These are discussed in *Appendix 3: Defining ethics and making ethical decisions* on the accompanying website.

■ **Figure 2.64**
A respirometer to measure respiration rate

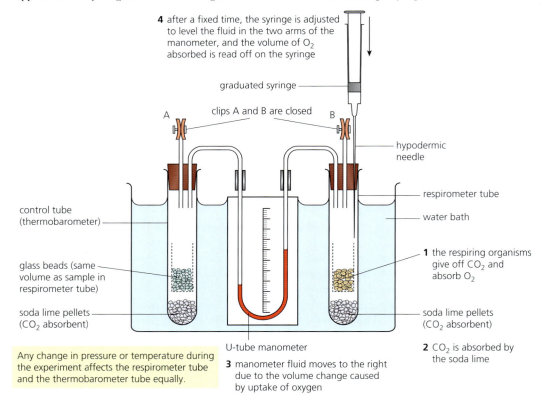

4 after a fixed time, the syringe is adjusted to level the fluid in the two arms of the manometer, and the volume of O_2 absorbed is read off on the syringe

graduated syringe

A clips A and B are closed B

hypodermic needle

respirometer tube

control tube (thermobarometer)

water bath

1 the respiring organisms give off CO_2 and absorb O_2

glass beads (same volume as sample in respirometer tube)

soda lime pellets (CO_2 absorbent)

soda lime pellets (CO_2 absorbent)

2 CO_2 is absorbed by the soda lime

Any change in pressure or temperature during the experiment affects the respirometer tube and the thermobarometer tube equally.

U-tube manometer

3 manometer fluid moves to the right due to the volume change caused by uptake of oxygen

If soda lime is present in the chambers the carbon dioxide gas released by the respiring organism is removed. In this case, only oxygen uptake by the respiring organisms causes a change in volume. As a result, the coloured liquid in the attached capillary tube will move towards the respirometer tube. The resulting reduction in the volume of air in the respirometer tube in a given time period can now be estimated. It is the volume of air from the syringe that must be injected back into the respirometer tube to make the manometric fluid level in the two arms equal again. That volume is equivalent to the volume of oxygen taken up by the respiring organisms.

28 In the respirometer (Figure 2.64), **explain** how changes in temperature or pressure in the external environment are prevented from interfering with measurement of oxygen uptake by respiring organisms in the apparatus.

29 The experiment shown in Figure 2.64 was repeated with maggot fly larvae in tube B, first with the soda lime present and subsequently with water in place of soda lime. The volume change with soda lime was $30\,mm^3\,h^{-1}$, but without soda lime it was $3\,mm^3\,h^{-1}$. **Analyse** these results, explaining the significance of each value.

2.9 Photosynthesis – *photosynthesis uses the energy in sunlight to produce the chemical energy needed for life*

Green plants use the energy of sunlight to produce **sugars** from the inorganic raw materials **carbon dioxide** and **water**, by a process called **photosynthesis**. The waste product is oxygen. Photosynthesis occurs in plant cells that contain the organelles called **chloroplasts** (page 22), including many of the cells of the leaves of green plants. Here, energy of light is trapped by the green pigment **chlorophyll** and becomes the chemical energy in molecules such as **glucose** and in **ATP**. Note that we now say that light energy is *transferred* to organic compounds in photosynthesis, rather than talking of the 'conversion of energy' – a term that was once widely used.

Sugar formed in photosynthesis may temporarily be stored as starch, but sooner or later much is used in metabolism. For example, plants manufacture other carbohydrates, together with the lipids, proteins, growth factors, and all other metabolites they require. For this, they additionally need certain mineral ions, which are absorbed from the soil solution. Figure 2.65 is a summary of photosynthesis and its place in plant metabolism.

30 **Compare** the source of glucose for cellular respiration in mammals and flowering plants.

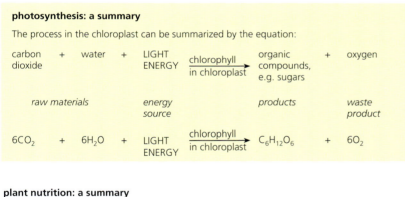

photosynthesis: a summary

The process in the chloroplast can be summarized by the equation:

$$\text{carbon dioxide} + \text{water} + \text{LIGHT ENERGY} \xrightarrow[\text{in chloroplast}]{\text{chlorophyll}} \text{organic compounds, e.g. sugars} + \text{oxygen}$$

raw materials — *energy source* — *products* — *waste product*

$$6CO_2 + 6H_2O + \text{LIGHT ENERGY} \xrightarrow[\text{in chloroplast}]{\text{chlorophyll}} C_6H_{12}O_6 + 6O_2$$

plant nutrition: a summary

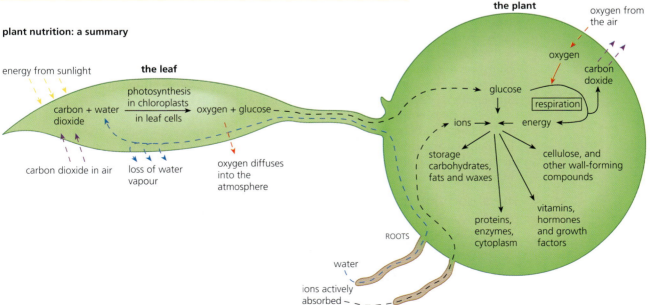

■ **Figure 2.65** Photosynthesis and its place in plant nutrition

■ What is light?

Light is a form of the electromagnetic radiation produced by the Sun. Visible light makes up only a part of the total magnetic radiation reaching the Earth. When this visible 'white' light is projected through a prism, we see a continuous spectrum of light – a rainbow of colours, from red to violet. Different colours have different wavelengths. The wavelengths of electromagnetic radiation and of the components of light are shown in Figure 2.66.

The significance of the spectrum of light in photosynthesis is that not all the colours of the spectrum present in white light are absorbed equally by chlorophyll. Some are even transmitted (or reflected), rather than being absorbed.

■ **Figure 2.66** Electromagnetic radiation and the spectrum of visible light

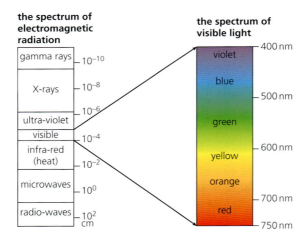

■ Investigating chlorophyll

Some plant pigments are soluble in water but chlorophyll is not. Chlorophyll can be extracted in an organic solvent like propanone (acetone) (Figure 2.67).

> **31** **State** what colour the pigment chlorophyll chiefly reflects or transmits (rather than absorbs).
>
> **32** **Evaluate** the essential safety precautions required when chlorophyll is extracted as shown in Figure 2.67.
>
> **33** **Suggest** why chromatography is a useful technique for the investigation of cell biochemistry.

Chlorophyll as a mixture – introducing chromatography

Plant chlorophyll consists of a mixture of pigments. **Chromatography** is the technique we use to separate components of mixtures, especially when working with small samples. It is an ideal technique for separating biologically active molecules, since biochemists are often able to obtain only very small amounts. Chromatograms are typically run on absorptive paper (**paper** chromatography), powdered solid (**column** chromatography), or on a thin film of dried solid (**thin-layer** chromatography). The technique is illustrated in Figure 2.68.

Look at the chromatogram in Figure 2.68.

The photosynthetic pigments of green leaves are two types of chlorophyll known as **chlorophyll *a*** and **chlorophyll *b***. The other pigments belong to a group of compounds called **carotenoids**. These pigments are, together, involved in the energy transfer processes in the chloroplasts.

The absorption and action spectra of chlorophyll

We have seen that light consists of a roughly equal mixture of all the visible wavelengths, namely indigo, violet, blue, green, yellow, orange and red (Figure 2.66). Now the issue is – how much of each wavelength does chlorophyll absorb? We call this information an absorption spectrum.

The **absorption spectrum** of chlorophyll pigments are obtained by measuring their absorption of indigo, violet, blue, green, yellow, orange and red light, in turn. The results are plotted as a graph showing the amount of light absorbed over the wavelength range of visible light, as shown in Figure 2.69. You can see that chlorophyll absorbs blue and red light most strongly. Other wavelengths are absorbed less or not at all. It is the chemical structure of the chlorophyll molecule that causes absorption of the energy of blue and red light.

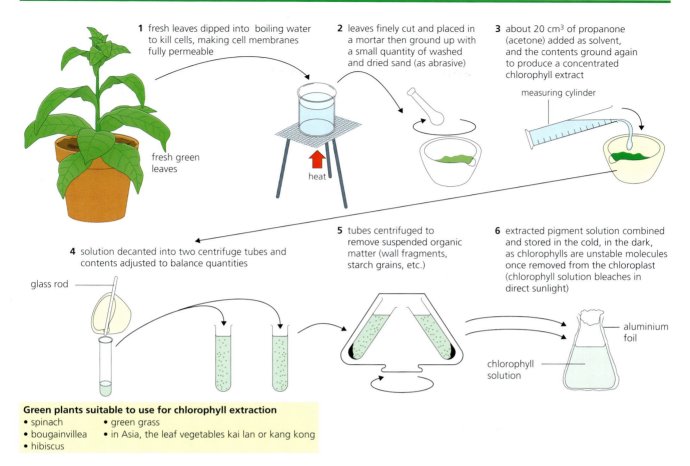

1 fresh leaves dipped into boiling water to kill cells, making cell membranes fully permeable

2 leaves finely cut and placed in a mortar then ground up with a small quantity of washed and dried sand (as abrasive)

3 about 20 cm³ of propanone (acetone) added as solvent, and the contents ground again to produce a concentrated chlorophyll extract

measuring cylinder

fresh green leaves

heat

4 solution decanted into two centrifuge tubes and contents adjusted to balance quantities

5 tubes centrifuged to remove suspended organic matter (wall fragments, starch grains, etc.)

6 extracted pigment solution combined and stored in the cold, in the dark, as chlorophylls are unstable molecules once removed from the chloroplast (chlorophyll solution bleaches in direct sunlight)

glass rod

aluminium foil

chlorophyll solution

Green plants suitable to use for chlorophyll extraction
- spinach
- bougainvillea
- hibiscus
- green grass
- in Asia, the leaf vegetables kai lan or kang kong

■ **Figure 2.67** Steps in the extraction of plant pigments

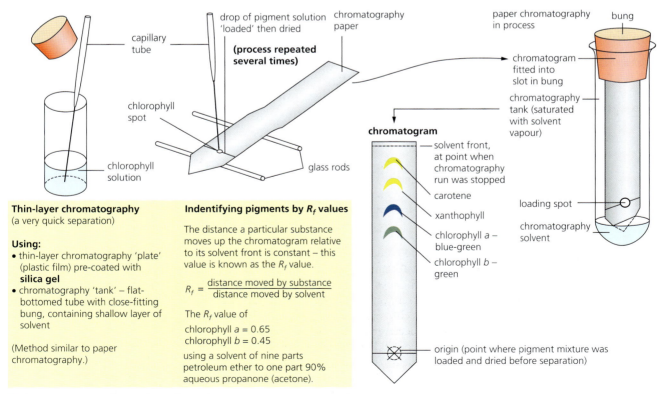

capillary tube

drop of pigment solution 'loaded' then dried

chromatography paper

paper chromatography in process

bung

(process repeated several times)

chlorophyll spot

chlorophyll solution

glass rods

chromatogram

chromatogram fitted into slot in bung

chromatography tank (saturated with solvent vapour)

solvent front, at point when chromatography run was stopped

carotene

xanthophyll

chlorophyll *a* – blue-green

chlorophyll *b* – green

loading spot

chromatography solvent

origin (point where pigment mixture was loaded and dried before separation)

Thin-layer chromatography
(a very quick separation)

Using:
- thin-layer chromatography 'plate' (plastic film) pre-coated with **silica gel**
- chromatography 'tank' – flat-bottomed tube with close-fitting bung, containing shallow layer of solvent

(Method similar to paper chromatography.)

Indentifying pigments by R_f values

The distance a particular substance moves up the chromatogram relative to its solvent front is constant – this value is known as the R_f value.

$$R_f = \frac{\text{distance moved by substance}}{\text{distance moved by solvent}}$$

The R_f value of
chlorophyll *a* = 0.65
chlorophyll *b* = 0.45

using a solvent of nine parts petroleum ether to one part 90% aqueous propanone (acetone).

■ **Figure 2.68** Preparing and running a chromatogram

The **action spectrum** of chlorophyll is the wavelengths of light that bring about photosynthesis. This may be discovered by projecting different wavelengths, in turn and for a unit of time, on aquatic green pond weed. This is carried out in an experimental apparatus in which the rate of photosynthesis can be measured. The gas evolved by a green plant in the light is largely oxygen, and the volume given off in a unit of time is a measure of the rate of photosynthesis. Suitable apparatus is shown in Figure 2.71, of which more later.

The rate of photosynthesis at different wavelengths may then be plotted on a graph on the same scale as the absorption spectrum (Figure 2.69). We have seen that blue and red light are most strongly absorbed by chlorophyll. From the action spectrum, we see that it is these wavelengths that give the highest rates of photosynthesis.

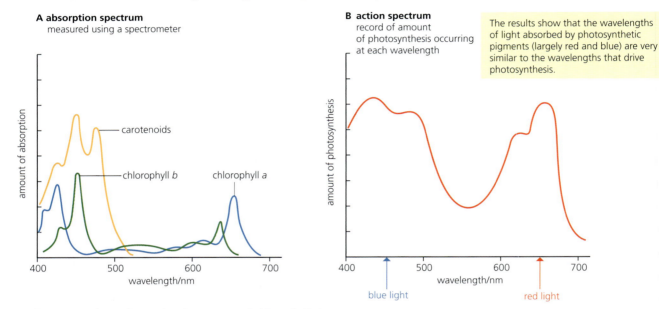

■ **Figure 2.69** Absorption and action spectra of chlorophyll pigments

---

Additional perspectives

Chlorophyll in solution versus chlorophyll in the chloroplast

The chlorophyll that has been extracted from leaves and dissolved in an organic solvent still absorbs light. However, chlorophyll in solution cannot use light energy to make sugar. This is because, in the extraction process, chlorophyll has been separated from the membrane systems and enzymes that surround it in chloroplasts. These are also essential for carrying out the biochemical steps of photosynthesis, as we shall now discover.

■ What happens in photosynthesis?

Photosynthesis is a set of many reactions occurring in chloroplasts in the light. However, these can be conveniently divided into two main steps (Figure 2.70).

1 Light energy is used to split water (photolysis)

This releases the waste product of photosynthesis, oxygen, and allows the hydrogen atoms to be retained on hydrogen-acceptor molecules. The hydrogen is one requirement of Step 2. At the same time, ATP is generated from ADP and phosphate, also using energy from light.

2 Sugars are built up from carbon dioxide

We say that carbon dioxide is fixed to make organic molecules. To do this, both the energy of ATP and of hydrogen atoms from the reduced hydrogen-acceptor molecules is required.

34 State from where a green plant obtains
a carbon and
b hydrogen used to build up glucose molecules.

■ **Figure 2.70**
Two steps of
photosynthesis

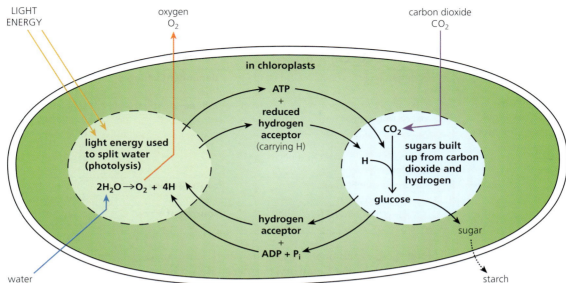

Experimental design
Controlling relevant variables in photosynthesis experiments is essential.

■ Measuring the rate of photosynthesis

An illuminated, freshly cut shoot of a pondweed, when inverted, produces a vigorous stream of
gas bubbles from the base. The bubbles tell us the pondweed is actively photosynthesizing.
At the same time, dissolved carbon dioxide is being removed from the water. Suitable plants
include *Elodea*, *Microphylllum* and *Cabomba*. The rate of photosynthesis can be estimated using:

■ a microburette to measure the volume of oxygen given out in the light (Figure 2.71).
The pondweed is placed in a very dilute solution of sodium hydrogencarbonate, which
supplies the carbon dioxide (as HCO_3^-ions) required by the plant for photosynthesis.
The quantity of gas evolved in a given time, say in 30 minutes, is measured by drawing the
gas bubble that collects into the capillary tube, and measuring its length. This length is
then converted to a volume

■ an oxygen sensor probe connected to a data-logging device

■ a pH meter connected to a data-logging monitor. The uptake of carbon dioxide from the
water will cause the pH to rise.

■ **Figure 2.71**
Measuring the rate of
photosynthesis with a
microburette

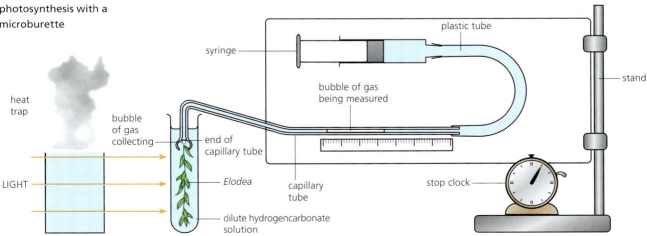

You can use one or more of these techniques to investigate the effects of external conditions on the rate of photosynthesis (Table 2.10).

■ **Table 2.10** Issues in the design of experiments to investigate the effect of external factors on the rate of photosynthesis

Carbon dioxide concentration		Light intensity
O_2 output (bubbles or volume) in unit time	**dependent variable**	O_2 output (bubbles or volume) or pH change in unit time
external CO_2 (i) absence of CO_2 by boiling and cooling water (ii) subsequent stepwise addition of $NaHCO_3$ solution to raise CO_2 by $0.01 \, mol \, dm^{-3}$ – until no further change in O_2 output	**independent variable**	light intensity – systematically positioning the light source (photoflood lamp or 150 W bulb) at 10, 15, 20 and 32 cm from the experimental chamber
temperature, light intensity	**controlled variables**	concentration of $NaHCO_3$, temperature
possibly errors in $NaHCO_3$ solution additions	**sources of error**	possibly the heating effect of the light source

Note: these experiments can be simulated on a computer. Details are given in the section on 'Chapter 8 Chapter Summary, Further study' on the accompanying website.

35 **a** A thermometer is not shown in the apparatus in Figure 2.71. **Predict** why one is required, and **state** where it should be positioned.
  b **Explain** why the cut stem of pondweed was inverted here.

■ External factors and the rate of photosynthesis

The effect of the **concentration of carbon dioxide** on the rate of photosynthesis is shown in Figure 2.72. *Look at this figure now – note the shape of the curve.*

In this experiment, when the concentration of carbon dioxide is at zero there is no photosynthesis, of course. As the concentration is steadily increased, the rate of photosynthesis rises, and the rate of that rise is positively correlated with the increasing carbon dioxide concentration. However, at much higher concentrations of carbon dioxide, the rate of photosynthesis reaches a plateau – now there is no increase in rate with rising carbon dioxide concentration.

■ **Figure 2.72** The effect of carbon dioxide concentration on photosynthesis

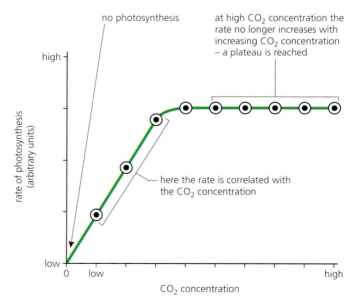

The effect of **light intensity** on the rate is shown in Figure 2.73. Look at this graph now – the shape of the curve is familiar. Then answer question 36.

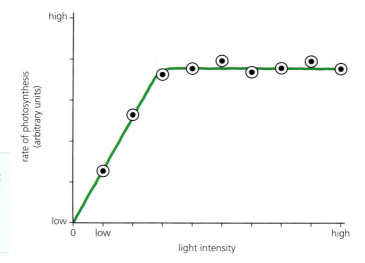

36 Explain what the graph implies about the relationship between light intensity and the rate of photosynthesis.

The effect of **temperature** on the rate of photosynthesis is shown in Figure 2.74. Here the curve of the graph is an entirely different shape. At relatively low temperatures, as the temperature increases, the rate of photosynthesis increases more and more steeply. However, at higher temperatures, the rate of photosynthesis abruptly stops rising and actually falls steeply. The result is a clear optimum temperature for photosynthesis.

■ **Figure 2.74**
The effect of
temperature on
photosynthesis

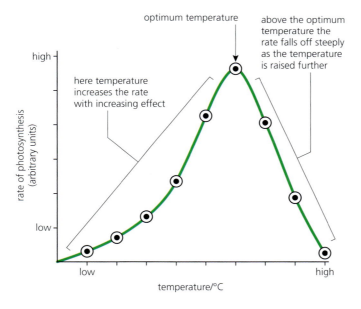

■ The wider significance of photosynthesis

The importance of photosynthesis to the green plant is that it provides the energy-rich sugar molecules from which the plant builds its other organic molecules. But photosynthesis has a wider significance:

37 Suggest the likely fate of starch stored in a green leaf, during periods of darkness.

1   The feeding relationships between organisms are represented in food chains (page 195). Food chains always begin with green plants, or parts of plants. Plants are the **primary producers in the food chain**; other organisms feed on them. Some do so directly (the herbivores); others do so indirectly, by feeding on the herbivores or on organisms that do.

So, virtually all life is dependent on green plant nutrition, including human life. It was an environmentalist that summarized our dependence on green plants in the phrase: *'Have you thanked a green plant today?'*!

2   Changes in the Earth's oceans and rock depositions were the first significant consequences of the oxygen liberated by the earliest photosynthetic organism (prokaryotes), about 3 500 million years ago. It was this output from aquatic organisms that caused the oxidation of huge quantities of dissolved iron in the oceans, triggering precipitation of sedimentary rocks rich in iron ores. Later, there was some release of oxygen into the atmosphere, and then terrestrial rock deposits also started to be oxidized. All these iron ores are the basis of our modern steel industry. The atmospheric oxygen concentration only started to rise significantly about 750 million years ago (mya) – long after the bulk of this iron oxidation was underway.

3   Some of the oxygen originating from green plants is converted into **ozone** in the upper atmosphere, due to the action of ultra-violet (UV) radiation from the Sun. (Ozone is a form of oxygen that contains three atoms of oxygen combined together.) As a result, ozone occurs naturally in the Earth's atmosphere as a layer found in the region called the stratosphere. The effect of this ozone layer is to protect terrestrial life from UV radiation by significantly reducing the quantity of UV that reaches the Earth's surface. UV light is very harmful to living things because it is absorbed by the organic bases (adenine, guanine, thymine, cytosine and uracil) of nucleic acids (DNA and RNA), causing them to be modified (mutation, page 132). Consequently, the conversion of oxygen to ozone and the maintenance of the high-level ozone layer have been important in the evolution of terrestrial life (UV light does not penetrate water and so cannot reach aquatic organisms). The ozone in our upper atmosphere is important to the survival of life today.

4   Green plants also maintain the **composition of the atmosphere** today. For example, the quantity of **carbon dioxide** removed by plants in photosynthesis each day is almost equal to that added to the air from respiration (page 121) and from the burning of fossil fuels. This is illustrated by the carbon cycle (page 201). Photosynthesis is also the only natural process that releases **oxygen** into the atmosphere. All the oxygen present in the air (about 21%) is a waste product of photosynthesis. This is the source of oxygen for aerobic respiration.

Examination questions – a selection

Questions 1–2 are taken from IB Diploma biology papers.

Q1 What sequence of processes is carried out by the structure labelled X during translation?

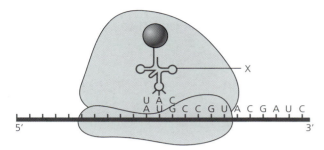

A Combining with an amino acid and then binding to an anticodon
B Binding to an anticodon and then combining with an amino acid
C Binding to a codon and then combining with an amino acid
D Combining with an amino acid and then binding to a codon

Standard Level Paper 1, Time Zone 1, May 10, Q9

Q2 Discuss the relationship between **one** gene and **one** polypeptide. (7)

Standard Level Paper 2, Time Zone 2, May 12, Q6c

Questions 3–10 cover other syllabus issues in this chapter.

Q3 When organic compounds started to be created under artificial conditions, it showed that a vital force is not needed to produce them. This affirmation indicates that
A organic compounds can be the result of a combination of molecules based on the nature of its components
B organic compounds are made of any component found in nature
C the reactions leading to the synthesis of organic compounds only occur inside living organisms
D combining inorganic compounds will give rise to organic ones.

Q4 What is the difference between anabolic and catabolic reactions?
A Anabolism is the synthesis of organic compounds and catabolism is the synthesis of inorganic compounds.
B Anabolic and catabolic reactions can be explained by condensation reactions only.
C Anabolism is the formation of polymers by condensation, and catabolism is the breakdown of macro molecules by hydrolysis.
D Hydrolysis and condensation reactions, both are part of anabolism and catabolism.

Q5 What determines the three-dimensional conformation of a protein?
A The interactions between the amino groups in the amino acids
B The amino acid sequence
C The peptide bonds between the amino acids
D More than one polypeptide is acting together

Q6 Carbohydrates, lipids and proteins make up the bulk of the organic compounds found in organisms.
a Identify a monosaccharide commonly found in both animal and plant cells. (1)
b State the polysaccharide with a role as energy store found in skeletal muscle, and what monomer it is constructed from. (2)
c Draw a generalised structural formula for:
  i an amino acid
  ii a fatty acid. (4)
d Define the type of chemical reaction involved in the formation of a peptide linkage between two amino acids, and state the other product, in addition to a dipeptide. (2)
e List three advantageous features of lipids as energy stores for terrestrial animals. (3)

Q7 **a** Describe three factors that affect enzymes. Start with the one that affects enzymes the most.
b Lactase is an example of an enzyme used in modern food processing. Outline the way lactase is used when making sugar-free milk. (6)

Q8 Cell respiration can be defined as the controlled release of energy from organic compounds such as glucose. Identify at least three processes that create ATP, organizing them by starting with the one that produces most ATP per molecule of glucose. (3)

Q9 State the role of both light energy and water in photosynthesis. (2)

Q10 a State at which period in a cell cycle the replication process takes place. (1)

b For the process of DNA replication, draw a diagrammatic representation of the structure of a 'replication fork' in which two important enzymes essential to replication are shown *in situ*. Annotate your diagram to make clear the roles of these enzymes. (6)

c Explain the part played by complementary base pairing in replication. (2)

d Explain why the replication process is described as semi-conservative. (2)

3 Genetics

ESSENTIAL IDEAS

- Every living organism inherits a blueprint for life from its parents.
- Chromosomes carry genes in a linear sequence that is shared by members of a species.
- Alleles segregate during meiosis allowing new combinations to be formed by the fusion of gametes.
- The inheritance of genes follows patterns.
- Biologists have developed techniques for artificial manipulation of DNA, cells and organisms.

3.1 Genes – *every living organism inherits a blueprint for life from its parents*

A **gene** is a heritable factor that influences a specific character. By 'character' we mean some feature of an organism like 'height' in the garden pea plant or 'blood group' in humans. 'Heritable' means genes are factors that pass from parent to offspring during reproduction.

Species vary in the number of genes they have – some have many more than others. Table 3.1 lists the numbers of genes present in a range of common organisms. Notice that the list includes one bacterium, as well as certain plants and animals, and that the water flea has more genes than a human but the fruit fly has less.

■ **Table 3.1** Estimated approximate numbers of protein-coding genes

Homo sapiens (human)	20 000	*Oryza sativa* (rice)	41 500
Canis familiaris (domestic dog)	19 000	*Vitis vinifera* (grape)	30 450
Drosophila melanogaster (fruit fly)	14 000	*Arabidopsis thaliana* (rockcress)	27 000
Plasmodium (malarial parasite)	5 000	*Saccharomyces cerevisiae* (yeast)	6 000
Daphnia (water flea)	31 000	*Escherichia coli* (bacterium)	4 300

Genes are located on **chromosomes**. Each gene occupies a specific position on a chromosome; therefore each eukaryotic chromosome is a linear series of genes. Furthermore, the gene for a particular characteristic is *always* found at the same position or **locus** on a particular chromosome. (The plural of locus is **loci**.) For example, the gene controlling height in the garden pea plant is always present in the exact same position on one particular chromosome of that plant. However, that gene for height may code for 'tall' or it may code for 'dwarf', as we shall shortly see. In other words, there are different forms of genes. In fact, each gene has two or more forms and these are called **alleles**. The word 'allele' just means 'alternative form'.

Now, the chromosomes of eukaryotic cells occur in pairs called **homologous pairs**. ('Homologous' means 'similar in structure'.) One of each pair came originally from one parent, and the other from the other parent. So, for example, the human has 46 chromosomes, 23 coming originally from each parent in the process of sexual reproduction. Homologous chromosomes resemble each other in structure and they contain the same sequence of genes. In Figure 3.1 we see some of the genes and their alleles in place on a homologous pair of chromosomes.

■ The chemical nature of the gene

A eukaryotic chromosome contains a single, extremely long molecule of DNA (Figure 2.49, page 107). So we can also define the gene as a specific length of that DNA molecule. In fact, it consists of hundreds or (more typically) thousands of **base pairs** long. We have also seen that, in the process of protein synthesis, the sequence of bases in nucleic acid dictates the order in which specific amino acids are assembled. This information, called the **genetic code**, is a three-letter or triplet code, meaning that each sequence of three bases stands for one of the 20 amino acids, and is known as a **codon**. A gene is a long sequence of codons.

■ **Figure 3.1**
Genes and alleles of
a homologous pair
of chromosomes

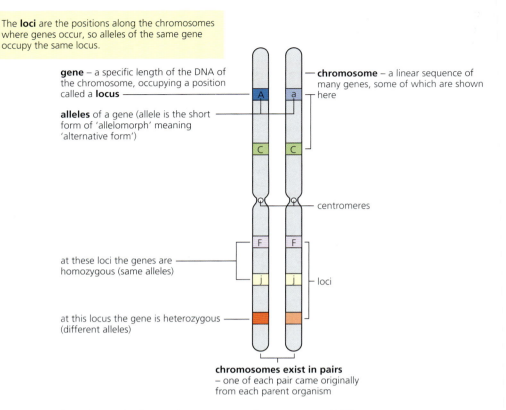

The **loci** are the positions along the chromosomes where genes occur, so alleles of the same gene occupy the same locus.

gene – a specific length of the DNA of the chromosome, occupying a position called a **locus**

alleles of a gene (allele is the short form of 'allelomorph' meaning 'alternative form')

chromosome – a linear sequence of many genes, some of which are shown here

centromeres

at these loci the genes are homozygous (same alleles)

loci

at this locus the gene is heterozygous (different alleles)

chromosomes exist in pairs – one of each pair came originally from each parent organism

■ How do the alleles of a particular gene differ?

Alleles differ from each other in the bases they contain but the differences are typically very small, limited to one or only a few bases in most cases. The example of the allele for sickle-cell hemoglobin and the allele for normal hemoglobin is discussed below.

■ The genome – the genetic information of an organism

The total of all the genetic information in an organism is called the **genome** of the organism. The genome of a prokaryotic organism, such as the bacterium *E. coli* (Figure 1.32, page 29), consists of the DNA that makes up the single circular chromosome found in the bacterium, together with the DNA present in the small, circular plasmids that also occur in the cytoplasm.

The genome of a eukaryotic organism such as a human consists of the DNA present in the 46 chromosomes in the nucleus, together with the DNA of the plastids found in mitochondria. Of course, in a green plant plasmids are also present in the chloroplasts.

■ New genes are formed by mutation

DNA can change. Normally, the sequence of nucleotides in DNA is maintained without change, but very occasionally alterations do happen. If a change occurs we say a **mutation** has occurred.

A **gene mutation** involves a change in the sequence of bases of a particular gene. At certain times in the cell cycle mutations are more likely to occur than at other times. One such occasion is when the DNA molecule is replicating. We have noted that the enzyme DNA polymerase, which brings about the building of a complementary DNA strand, also 'proof-reads' and corrects most errors (page 109). However, gene mutations can and do occur spontaneously during this step.

Also, certain conditions or chemicals may cause change to the DNA sequence of bases. These can include ionizing radiation (page 163), UV light and various chemicals.

Base substitution mutation and sickle-cell condition

An extremely common gene mutation occurs in hemoglobin. This oxygen-transporting pigment of red blood cells is made of four polypeptide molecules – two known as α **hemoglobin** and two as β **hemoglobin**. These interlock to form a compact molecule (Figure 3.2).

■ **Figure 3.2**
Sickle-cell anemia,
an example of a
gene mutation

Anemia is a disease typically due to a deficiency in healthy red cells in the blood.

Hemoglobin occurs in red cells – each contains about
280 million molecules of hemoglobin. A molecule consists
of two α hemoglobin and two β hemoglobin subunits,
interlocked to form a compact molecule.

The **mutation** that produces sickle cell hemoglobin (**Hg^S**) is
in the gene for β hemoglobin. It results from the substitution
of a single base in the sequence of bases that make up all the
codons for β hemoglobin.

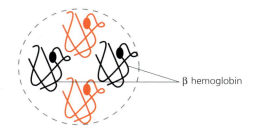

β hemoglobin

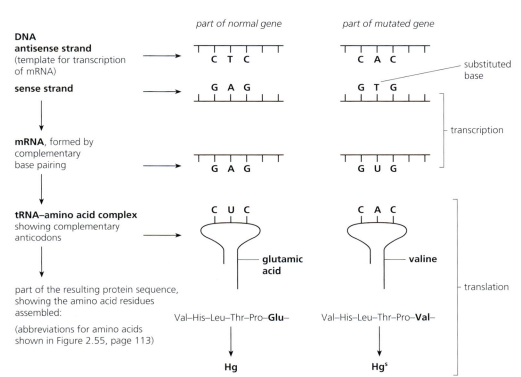

**DNA
antisense strand**
(template for transcription
of mRNA)

part of normal gene

C T C

part of mutated gene

C A C

substituted
base

sense strand

G A G

G T G

transcription

mRNA, formed by
complementary
base pairing

G A G

G U G

tRNA–amino acid complex
showing complementary
anticodons

C U C

glutamic
acid

C A C

valine

translation

part of the resulting protein sequence,
showing the amino acid residues
assembled:

(abbreviations for amino acids
shown in Figure 2.55, page 113)

Val–His–Leu–Thr–Pro–**Glu**–

Hg

Val–His–Leu–Thr–Pro–**Val**–

Hg^s

drawing based on a photomicrograph of a blood smear,
showing blood of a patient with sickle cells present among
healthy red cells

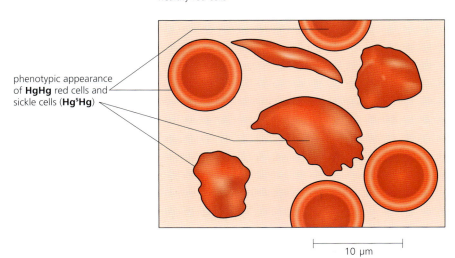

phenotypic appearance
of **HgHg** red cells and
sickle cells (**Hg^{s}Hg**)

10 μm

The gene that codes for the amino acid sequence of β hemoglobin occurs on chromosome 11 and is prone to a substitution of the base A to T in a codon for the amino acid glutamic acid – the sixth amino acid in this polypeptide. As a consequence of this base substitution, the amino acid valine appears at that point.

The presence of a non-polar valine in the β hemoglobin creates a hydrophobic spot in the otherwise hydrophilic outer section of the protein. This tends to attract other hemoglobin molecules, which bind to it. In tissues with low partial pressures of oxygen (such as a tissue with a high rate of aerobic respiration) the sickle-cell hemoglobin molecules in the capillaries readily clump together into long fibres. These fibres distort the red blood cells into sickle shapes. In this condition, the red blood cells cannot transport oxygen. Also, sickle cells may get stuck together, blocking smaller capillaries and preventing the circulation of normal red blood cells. The result is that people with sickle cells suffer from anemia – a condition of inadequate delivery of oxygen to cells. This unusual hemoglobin molecule is known as **hemoglobin S (Hg^s)**.

People with a single allele for hemoglobin S (**Hg Hg^s**) have less than 50% hemoglobin S. Such a person is said to have **sickle-cell trait** and they are only mildly anemic. However, those with both alleles for hemoglobin S (**Hg^s Hg^s**) are described as having **sickle-cell anemia** – a serious condition that may trigger heart and kidney problems, too.

An advantage in having sickle-cell trait?

Malaria is the most important of all insect-borne diseases. It is in Africa south of the Sahara that about 80% of the world's malaria cases are found, and here that most fatalities due to the disease occur. Here, millions of people are infected at any one time. It is estimated that 1.5 million sufferers, mostly children under 5 years, die each year.

Malaria is caused by *Plasmodium*, a protozoan, which is transmitted from an infected person to another person by blood-sucking mosquitoes of the genus *Anopheles*. Only the female mosquito is the vector (the male mosquito feeds on plant juices).

Plasmodium completes its lifecycle in red blood cells, but it seems that people with sickle-cell trait are protected to a significant extent (Figure 3.3). Where malaria is endemic in Africa, possession of one mutant gene (having sickle-cell trait, but not full anemia) seems to be advantageous. Natural selection or 'survival of the fittest' ensures this allele persists and fewer of the alleles for normal hemoglobin are carried into the next generation.

distribution of hemoglobin S is virtually the same as that of malaria

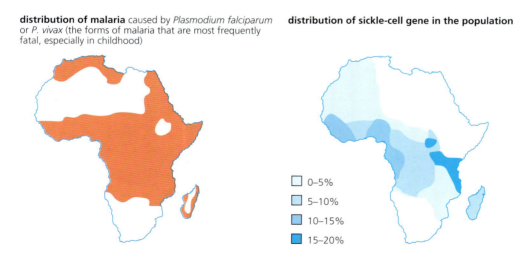

distribution of malaria caused by *Plasmodium falciparum* or *P. vivax* (the forms of malaria that are most frequently fatal, especially in childhood)

distribution of sickle-cell gene in the population

- ☐ 0–5%
- ☐ 5–10%
- ☐ 10–15%
- ■ 15–20%

Sickle-cell anemia is a genetically transmitted disease of the blood caused by an abnormal form of hemoglobin. Red cells with 'sickle' hemoglobin do not carry oxygen and can cause blockage of arterioles. Normally, sickle-cell anemia confers a disadvantage, but the malarial parasite is unable to complete its lifecycle in red cells with hemoglobin S, so the gene for sickle-cell is selected for in regions of the world where malaria occurs.

■ **Figure 3.3** How sickle-cell trait may confer an advantage

TOK Link

The distribution of sickle-cell alleles in Africa appears to confer an advantage for the people in regions where the malarial parasite is endemic (Figure 3.3). However, this type of evidence is **circumstantial**. How could a causal link be established?

1 **Explain** how knowledge of the cause of sickle-cell trait and sickle-cell anemia supports the concept of a gene as a linear sequence of bases.

■ The Human Genome Project

The Human Genome Project (HGP), an initiative to map the entire human genome, was a publicly funded project that was launched in 1990. The ultimate objective of the HGP was to discover the base sequence of the entire human genome. The work was shared among more than 200 laboratories around the world, avoiding duplication of effort. However, in 1998 the task became a race when a commercially funded company set out to achieve the same outcome in only three years, by different techniques. Because of a fear that a private company might succeed, patent the genome, and then sell access to it (rather than making the information freely available to all), the HGP teams accelerated their work.

In fact, both teams were successful well ahead of the projected completion dates, because of rapid improvements in base sequencing techniques. On the 26th June 2000, a joint announcement established that the sequencing of the human genome had been achieved. At the same time as the HGP was underway, teams of scientists set about the sequencing of the DNA of other organisms. Initially, this included the common human gut bacterium, *E. coli,* the fruit fly and the mouse. Since then, more than 30 non-human genomes have been sequenced to date.

It is important to realize that the existence of these sorts of investigations is dependent on the use of computers to store vast quantities of data, and to analyse and compare data sets.

A spin-off of the HGP has been an international study of human variation (HapMap Project: http://hapmap.ncbi.nlm.nih.gov/abouthapmap.html). The sequencing of the human genome has shown that all humans share the vast majority of their base sequences, but also that there are many single nucleotide polymorphisms which contribute to human diversity. Also important is the realization that ethical, legal and social issues are generated by the project.

Nature of Science

Developments in scientific research follow improvements in technology

■ Gene sequencing – the process

To sequence a genome, the entire genome is broken up into manageable pieces and then the fragments are separated so that they can be sequenced individually. Copies of these DNA fragments are made in a way such that the sequential positions of the four nucleotides in the fragment can be identified.

Single-stranded copies are made in the presence of non-standard nucleotides and an excess of standard components. The non-standard nucleotides occasionally block the copying process, so that many copies of the sequence are made, all of different lengths. The four non-standard nucleotides are each tagged with a differently coloured fluorescent marker. This process is discussed in more detail on page 322.

Samples are separated according to length, by capillary electrophoresis machine. This procedure distinguishes DNA fragments that differ in size by only a single nucleotide. After separation, a laser beam makes the fluorescent markers fluoresce. Then an optical detector linked to a computer deduces the base sequence from the sequence of colours detected. This automated procedure is entirely dependent on improvements in technology. An animated sequence of this process can be accessed via your web browser, appropriately interrogated.

■ **Figure 3.4** Here the order of the nucleotide bases is being determined in a molecule of DNA by a largely automated process

Sequencing the rice genome

Rice is the staple food for more than 60% of the world's population (more wheat is produced than rice, but wheat is also used extensively in animal feeds). The plant was domesticated in south-west Asia, probably 12 000 years ago – much earlier than the first recorded cultivation of wheat (in the Fertile Crescent). If this is so, rice has certainly fed more people over a longer period than any other crop.

In 1997, scientists from many nations agreed to collaborate on the sequencing of the rice genome – one of the smallest genomes of all the cereal crops. The completed sequence was published in 2002. This information has assisted plant breeders' efforts to assimilate useful genes from wild rice species into cultivated varieties to improved resistance to disease, for example.

Outcomes of the HGP

Ultimately, most, if not all, aspects of biological investigation will benefit from the results of this project. At this stage, we can illustrate the impact of the HGP with three examples.

1 How many individual genes do we have and how do they work?

Genes may be located in base-sequence data by the detection of sequences that are uninterrupted by 'start' and 'stop' codons. Such regions are more likely to code for a protein. The three billion bases that make up the human genome represent about 20 000 genes, far fewer than was expected. *Drosophila* (the fruit fly) has almost half our number of genes, and a rice plant has many more than 40 000 genes.

How is the structural, physiological and behavioural complexity of humans delivered by so relatively few genes?

The secret of our complexity may partly lie in the issue of so-called 'junk DNA' (sections of chromosomes that do not code for proteins). Also, it is not the number of our genes that is the issue, but how we use them. Promoters and enhancers are associated with many genes and may determine which cells express which genes, when they are expressed, and at what level.

Another issue is the effect of 'experience'. The genetic code of DNA is not altered by the environment (except in cases of mutation) – information flows out of genes and not back into them. However, while our genes are largely immune to direct outside influence, our experience may regulate the expression of particular genes. For example, genes are switched on and off by promoters, and this may occur in response to external factors. If so, our genes may be responding to our actions, as well as causing them.

2 Locating the cause of genetic disorders

Another outcome of the HGP is the ability to locate genes that are responsible for human genetic disorders. More than half of all genetic disorders are due to a **mutation of a single gene** that is commonly recessive. To prove a gene is associated with a disease, it must be shown that patients have a mutation in that gene, but that unaffected individuals do not. The outcome of the mutation in afflicted people is the loss of the ability to form the normal product of the gene. Common genetic diseases include cystic fibrosis (page 161), sickle-cell disease (page 132) and hemophilia (page 160).

3 Development of the new discipline of bioinformatics

Bioinformatics is the storage, manipulation and analysis of biological information via computer science. At the centre of this development are the creation and maintenance of databases concerning nucleic acid sequences and the proteins derived from them. Already, the genomes of many prokaryotes and eukaryotes have been sequenced, as well as that of humans. This huge volume of data requires organization, storage and indexing to make practical use of the subsequent analyses. These tasks involve applied mathematics, informatics, statistics and computer science.

Use of the Gene Bank database to determine differences in base sequence of a gene in two species

These are the steps to a comparison of the nucleotide sequences of the cytochrome c oxidase gene of humans with that of the Sumatran orang-utan, using the National Center for Biotechnology Information web services which provide access to biomedical and genomic information:

1   You will need two websites. Get them ready in two tabs.
    A http://www.ncbi.nlm.nih.gov/gene
    B http://blast.ncbi.nlm.nih.gov/Blast.cgi
2   Go to website A.
    a   Type in 'human cytochrome c oxidase' in the search bar and press search.
    b   Click on number 2, the blue link for COX6B1, in the search results.
3   This page gives more information about the gene.
    a   Scroll down until you find NCBI Reference Sequences (RefSeq).
    b   Click on the first mRNA and protein(s) link NM_001863.4. This will give the nucleotide sequence for the gene.
4   Click on FASTA. This will give you the sequence of the nucleotides in the DNA.
    a   Now you need to copy the whole sequence to your clipboard.
5   Go to website B on your second tab.
    a   Click on 'nucleotide blast'.
    b   Paste your nucleotide sequence into the box.
    c   Select 'others' next to 'Database' (this will include comparisons with other species).
    d   Scroll down and click on BLAST. This will compare the human cytochrome c oxidase sequence with others.
6   Scroll down to the descriptions and find *Pongo abelii* (Sumatran orang-utan).
    a   What % match to the human sequence does it have? (Use the ident column – 96%.)
    b   Click on the species link or scroll down to the alignments and compare the gene sequences base by base.
        i   What type of mutations can you see? (Substitutions and deletions?)
7   Scroll back to the top.
    a   Click on 'distance tree of results'. An evolutionary tree will appear in a new tab. You can change this to 'slanted' to get a more familiar diagram. This will show the evolutionary relationship of various species based on the cytochrome c oxidase gene.

You can make comparisons with other species listed in the descriptions using this procedure. Also, other genes, proteins and evolutionary trees may be compared using this procedure.

■ **Figure 3.5**

Screen image of cytochrome oxidase base sequence comparisons of human and Sumatran orang-utan DNA

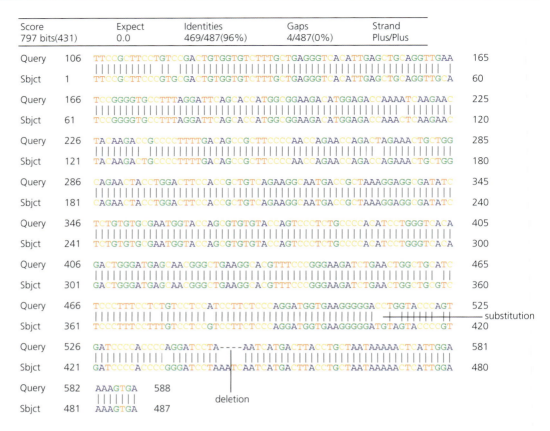

Score 797 bits(431)	Expect 0.0	Identities 469/487(96%)	Gaps 4/487(0%)	Strand Plus/Plus

```
Query  106  TTCCGCTTCCTGTCCGACTGTGGTGTCTTTGCTGAGGGTCACATTGAGCTGCAGGTTGAA  165
            |||||||||| || ||||||||||||||||||||||||||||||||||||||||||||| |
Sbjct    1  TTCCGCTTCCCGTGCGACTGTGGTGTCTTTGCTGAGGGTCACATTGAGCTGCAGGTTGCA   60

Query  166  TCCGGGGTGCCTTTAGGATTCAGCACCATGGCGGAAGACATGGAGACCAAAATCAAGAAC  225
            |||||||||||||||||||||||||||||||||||||||||||||||||||| |||||||
Sbjct   61  TCCGGGGTGCCTTTAGGATTCAGCACCATGGCGGAAGACATGGAGACCAAACTCAAGAAC  120

Query  226  TACAAGACCGCCCTTTTGACAGCCGCTTCCCCAACCAGAACCAGACTAGAAACTGCTGG  285
            ||||||||  ||||||||||||||||||||||||||||||||||||||| |||||||||
Sbjct  121  TACAAGACTGCCCCTTTTGACAGCCGCTTCCCCAACCAGAACCAGACCAGAAACTGCTGG  180

Query  286  CAGAACTACCTGGACTTCCACCGCTGTCAGAAGGCAATGACCGCTAAAGGAGGCGATATC  345
            |||||||||||||||||||||||||||||||||||||||||||||||||||||||||||
Sbjct  181  CAGAACTACCTGGACTTCCACCGCTGTCAGAAGGCAATGACCGCTAAAGGAGGCGATATC  240

Query  346  TCTGTGTGCGAATGGTACCAGCGTGTGTACCAGTCCCTCTGCCCCACATCCTGGGTCACA  405
            ||||||||||||||||||||||||||||||||||||||||||||||| |||||||||| |
Sbjct  241  TCTGTGTGCGAATGGTACCAGCGTGTGTACCAGTCCCTCTGCCCCACATCCTGGGTCACA  300

Query  406  GACTGGGATGAGCAACGGGCTGAAGGCACGTTTCCCGGGAAGATCTGAACTGGCTGCATC  465
            |||||||||||||||||||||||||||||||||||||||||||||||||||||||| ||
Sbjct  301  GACTGGGATGAGCAACGGGCTGAAGGCACGTTTCCCGGGAAGATCTGAACTGGCTGCGTC  360

Query  466  TCCCTTTCCTCTGTCCTCCATCCTTCTCCCAGGATGGTGAAGGGGGACCTGGTACCCAGT  525
            ||||||||| ||||| ||| ||||||| ||||||||||||||||||||  +|+||||| ||               substitution
Sbjct  361  TCCCTTTCCTTTGTCCTCCGTCCTTCTCCCAGGATGGTGAAGGGGGATGTAGTACCCCGT  420

Query  526  GATCCCCACCCCAGGATCCTA----AATCATGACTTACCTGCTAATAAAAACTCATTGGA  581
            |||||||||||||||||||||    |||||||||||||||||||||||||||||||||||
Sbjct  421  GATCCCCACCCCGGGATCCTAAATCAATCATGACTTACCTGCTAATAAAAACTCATTGGA  480
                                  deletion

Query  582  AAAGTGA  588
            |||||||
Sbjct  481  AAAGTGA  487
```

3.2 Chromosomes *– chromosomes carry genes in a linear sequence that is shared by members of the species*

■ The single chromosome of prokaryotes

The structure of a bacterial cell was introduced in Figure 1.31, and contrasted with the structure of a eukaryotic cell in Table 1.6 (page 29).

Remind yourself of these key differences now.

The single chromosome of a prokaryote consists of a circular DNA molecule, not associated with protein. Prokaryotes typically also contain small extra DNA molecules in their cytoplasm, known as **plasmids** (Figure 3.34, page 169). Plasmids:

■ are small and circular, and also without protein attached to their DNA

■ carry additional genes, and may occur singly or exist as multiple copies within the cell

■ may or may not be passed on to daughter cells at the time of cell division

■ are absent from eukaryotic cells, with the exception of yeast cells

■ are exploited by genetic engineers in the transfer of genes from one organism to another (page 169).

■ The chromosomes of eukaryotes

Each eukaryotic chromosome consists of a single, extremely long DNA molecule associated with proteins. About 50% of the chromosome is protein, in fact. Some of these proteins are enzymes involved in copying and repair reactions of DNA, but the bulk are histones. Histones are large, globular proteins. The DNA molecule is wound around them, giving the impression of a string of beads (Figure 7.9, page 328) when observed in a stained preparation of a nucleus at interphase. So, the histones have a support and packaging role.

The DNA molecule runs the full length of the chromosome. It consists of two anti-parallel polynucleotide strands held together by hydrogen bonds, and taking the shape of a double helix (Figure 2.49, page 107). In effect, the DNA molecule is a long sequence of genes.

Nature of Science

Developments in research follow improvements in techniques

■ The length of DNA in a cell

The development of the technique of autoradiography enabled advancement in biological research in several fields. Examples illustrated elsewhere in this book include in the discovery of the carbon path in photosynthesis (Figure 8.21, page 366) and the pathway of sugar translocation in the plant stem (Figure 9.19, page 390). Using a technique that included autoradiography, a research biologist, John Cairns, produced images of DNA molecules from *E. coli* (page 29). The process involved:

- incubating cultures of *E. coli* with radioactive thymine so that, after two generations, the DNA of the bacteria was radioactive
- digesting the cell walls with lysozyme, so that the DNA present was released onto the surface of a membrane
- applying a film of photographic emulsion to the surface of the membrane, and holding it there in the dark for many weeks
- finally, examining the developed film microscopically to locate where radioactive atoms had decayed and caused darkening of the photographic negative. In this way, the length (and shape) of the bacterial DNA was disclosed.

Cairns established the DNA of *E. coli* was a single circular DNA molecule of length 1100 μm. It occurs packed into a bacterial cell of only 2 μm diameter!

Subsequently, the same technique was applied to study eukaryotic chromosomes. The length of DNA in cells is phenomenal. In a human cell, the DNA held in the nucleus measures about 2 m in total length. The integrity of this huge molecule has to be maintained during the process of cell division. This involves it being packaged and folded away, yet parts also have to be accessed periodically for transcription (page 108).

■ **Table 3.2** A comparison of genome size

Total number of base pairs (bp) in haploid chromosomes		
T2 phage, a virus specific to a bacterium	3569	(3.5 kb)
Escherichia coli, a gut bacterium	4 600 000	(4.6 Mb)
Drosophila melanogaster, fruit fly	123 000 000	(123 Mb)
Oryza sativa, rice	430 000 000	(430 Mb)
Homo sapiens	3 200 000 000	(3.2 Gb)
Paris japonica, 'canopy plant'	150 000 000 000	(150 Gb)

■ The features of chromosomes

There are five features of the chromosomes of eukaryotic organisms which it is helpful to revise now.

1 The number of chromosomes per species is fixed

The number of chromosomes in the cells of different species varies, but in any one species the number of chromosomes per cell is normally constant (Table 3.3). For example, the mouse has 40 chromosomes per cell, the onion has 16, humans have 46, and the sunflower has 34. Each species has a characteristic chromosome number. Note, these are all *even* numbers.

■ **Table 3.3** Diploid chromosome numbers compared

Homo sapiens (human)	46	*Pan troglodytes* (chimpanzee)	48
Canis familiaris (dog)	78	*Drosophila melanogaster* (fruit fly)	8
Mus musculus (mouse)	40	*Helianthus annuus* (sunflower)	34
Parascaris equorum (roundworm)	2	*Oryza sativa* (rice)	24

2 The shape of a chromosome is characteristic

Chromosomes are long thin structures of a fixed length. Somewhere along the length of the chromosome is a narrow region called the **centromere**. Centromeres may occur anywhere along the chromosome, but they are always in the same position on any given chromosome. The position of the centromere and the length of chromosome on each side enable scientists to identify particular chromosomes in photomicrographs.

3 The chromosomes of a cell occur in pairs called homologous pairs

We have seen that the chromosomes of a cell occur in pairs, called **homologous pairs** (Figure 3.1). One of each pair came originally from the male parent and one from the female parent. This is why chromosomes occur in homologous pairs.

Cells in which the chromosomes are in homologous pairs are described as having a **diploid** nucleus. We describe this as **2n** where the symbol '**n**' represents one set of chromosomes. A cell which has one chromosome of each pair has a **haploid** nucleus. We represent this as **n**. A sex cell has a haploid nucleus – formed as a result of the nuclear division known as meiosis (page 142).

Karyotype and karyograms

The number and type of chromosomes in the nucleus is known as the **karyotype**. In Figure 3.6 on the left-hand side, the karyotype of a diploid human male cell is shown, much enlarged. These chromosomes are seen at an early stage of the nuclear division called mitosis (mitotic metaphase, page 54). You can see that at this stage each chromosome is present as two chromatids, held together by its centromere.

For the image on the right, the individual chromosomes were cut out from a copy of the original photograph. These were then arranged in homologous pairs in descending order of size, and numbered. A photograph of this type is called a **karyogram**.

Sex chromosomes

In the karyogram you can see that the final pair of chromosomes is not numbered. Rather, they are labelled **X** and **Y**. These are known as the **sex chromosomes**; they decide the sex of the individual – a male in this case. We will return to this issue later. All the other chromosomes (pairs numbered 1 to 22) are called **autosomes**. Karyograms are used by genetic counsellors to detect the presence of (rare) chromosomal abnormalities, such as Down's syndrome (pages 145–147).

human chromosomes of a male (karyotype)
(seen at the equator of the spindle during nuclear division)

chromosomes arranged as homologous pairs in descending order of size

homologous chromosomes

each chromosome has been replicated (copied) and exists as two chromatids held together at their centromeres

images of chromosomes cut from a copy of this photomicrograph can be arranged and pasted to produce a **karyogram**

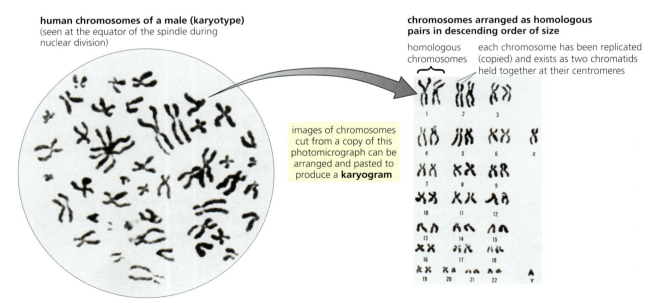

■ **Figure 3.6** Chromosomes as homologous pairs, seen during nuclear division

4 Chromosomes, their genes and loci

We have seen that chromosomes carry genes in a linear sequence, that the position of a gene is called a **locus** (plural, **loci**), and that each gene has two or more forms, called **alleles** (Figure 3.1).

The two alleles may carry exactly the *same* 'message' – the same sequence of bases. A diploid organism that has the same allele of a gene at the gene's locus on both copies of the homologous chromosomes in its cells is described as **homozygous**.

Alternatively, the two alleles may be *different*. A diploid organism that has different alleles of a gene at the gene's locus on both copies of the homologous pair is **heterozygous**.

Use of databases to identify the locus of a human gene

You can use a search engine to locate online databases to identify the locus of a particular gene. One such site is: www.ncbi.nlm.nih.gov/gene

Once you have accessed the site, you may enter the name of a gene to obtain information on its locus, starting with which chromosome the gene is located on. Genes you might search for include:

Gene role	gene name
Testis determining factor – switches fetal development to 'male' (page 159)	TDF
Chloride channel protein – a mutant allele causes cystic fibrosis (page 162)	CFTR

5 Chromosomes are copied precisely

Between nuclear divisions, while the chromosomes are uncoiled and cannot be seen, each chromosome is copied. It is said to **replicate**.

Replication occurs in the cell cycle, during interphase (page 52). The two identical structures formed are called **chromatids** (Figure 3.7). The chromatids remain attached by their centromeres until they are separated during nuclear division. After division of the centromeres, the chromatids are recognized as chromosomes again.

Of course, when chromosomes are copied, the critical event is the copying of the DNA double helix that runs the length of the chromosome. Replication occurs in a very precise way, brought about by specific enzymes, as we have already discussed (page 109).

■ **Figure 3.7** One chromosome as two chromatids

sister chromatids attached at the centromere, making up one chromosome

the centromere, a small constriction on the chromatids, is not a gene and does not code for a protein, as genes do

each chromatid is a copy of the other, with its linear series of genes (individual genes are too small to be seen)

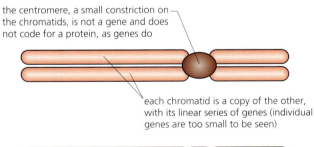

2 **Construct** and **annotate** a flow chart to illustrate the cell cycle (page 51).

chromatids separate during nuclear division

centromere divides

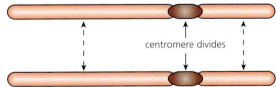

3.3 Meiosis – *alleles segregate during meiosis allowing new combinations to be formed by the fusion of gametes*

Divisions of the nucleus occur by a very precise process, ensuring the correct distribution of chromosomes between the new cells (daughter cells). There are two types of nuclear division, known as **mitosis** and **meiosis**.

In mitosis, the daughter cells produced have the same number of chromosomes as the parent cell, typically two of each type, known as the **diploid (2*n*)** state. Mitosis is the nuclear division that occurs when an organism grows, when old cells are replaced, and when an organism reproduces asexually. Mitosis is explained in Chapter 1, page 53.

In meiosis, the daughter cells contain half the number of chromosomes of the parent cell. That is, one chromosome of each type is present in the nuclei formed; this is known as the **haploid (*n*)** state. Meiosis is the nuclear division that occurs when sexual reproduction occurs, normally during the formation of the gametes.

The differences between mitosis and meiosis are summarized in Figure 3.8.

■ **Figure 3.8** Mitosis and meiosis, the significant differences

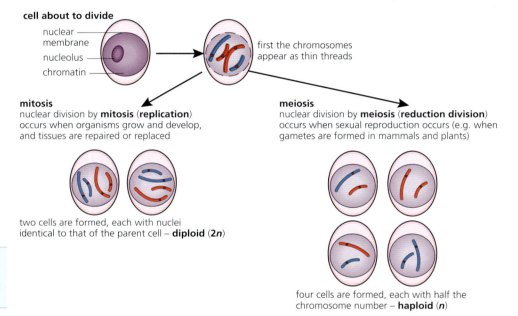

cell about to divide

nuclear membrane
nucleolus
chromatin

first the chromosomes appear as thin threads

mitosis
nuclear division by **mitosis (replication)** occurs when organisms grow and develop, and tissues are repaired or replaced

two cells are formed, each with nuclei identical to that of the parent cell – **diploid (2*n*)**

meiosis
nuclear division by **meiosis (reduction division)** occurs when sexual reproduction occurs (e.g. when gametes are formed in mammals and plants)

four cells are formed, each with half the chromosome number – **haploid (*n*)**

3 Suggest why it is essential that nuclear division is a precise process.

■ Meiosis, the reduction division

Nature of Science

Making careful observations

■ The discovery of meiosis

Meiosis was discovered by careful microscope examination of dividing germ-line cells. This was possible after the discovery of dyes that, when applied to tissues, specifically stained the contents of the nucleus. First, chromosomes were observed, described and named. Further careful studies then revealed the steps of mitosis and meiosis. The unravelling of the complexities of meiosis followed the observation of the doubling of the chromosome number at fertilization. Appreciation of a need for a reductive division preceded its discovery.

Germ-line cells are found in the gonads (testes and ovaries). Meiosis is part of the life cycle of every organism that reproduces sexually. In meiosis, four daughter cells are produced – each having half the number of chromosomes of the parent cell. Halving of the chromosome number of gametes is essential because at fertilization the number is doubled.

How does meiosis work?

Meiosis involves **two divisions** of the nucleus, known as **meiosis I** and **meiosis II**. As in mitosis, chromosomes replicate to form chromatids during interphase, before meiosis occurs. Then, early in meiosis I, **homologous chromosomes pair up**. By the end of meiosis I, homologous chromosomes have separated again, but the chromatids they consist of do not separate until meiosis II. Thus, meiosis consists of two nuclear divisions but only **one replication of the chromosomes**.

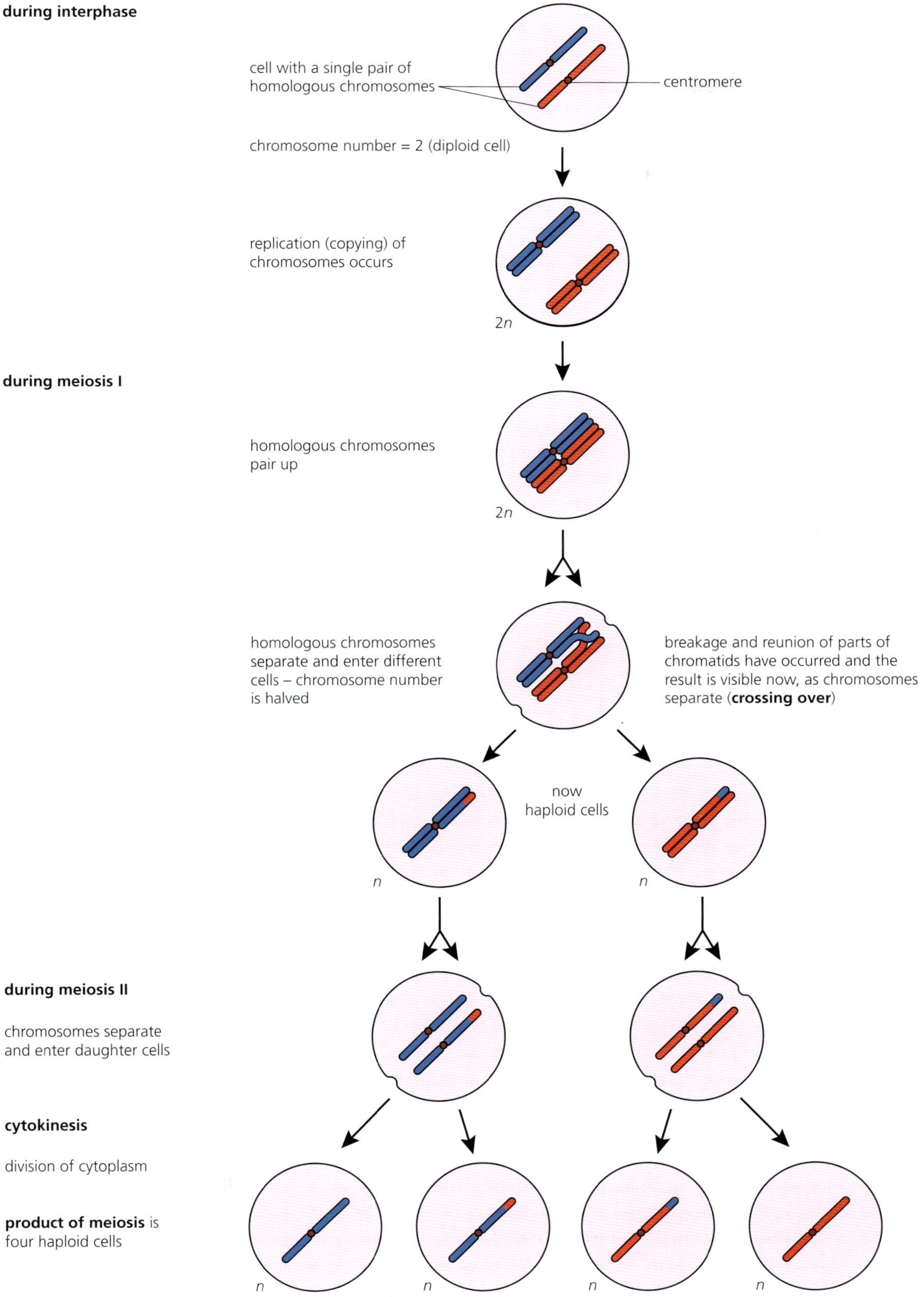

during interphase

cell with a single pair of homologous chromosomes

centromere

chromosome number = 2 (diploid cell)

replication (copying) of chromosomes occurs

2*n*

during meiosis I

homologous chromosomes pair up

2*n*

homologous chromosomes separate and enter different cells – chromosome number is halved

breakage and reunion of parts of chromatids have occurred and the result is visible now, as chromosomes separate (**crossing over**)

now haploid cells

n

n

during meiosis II

chromosomes separate and enter daughter cells

cytokinesis

division of cytoplasm

product of meiosis is four haploid cells

n   *n*   *n*   *n*

■ **Figure 3.9** What happens to chromosomes in meiosis?

You will probably find difficult the drawing of diagrams to show the stages of meiosis from direct observation of prepared microscope slides. For one thing, in books, the stages are typically illustrated in a cell with very few chromosomes. This is so events can be shown clearly. In reality, most species have many more chromosomes. Your diagrams may need to show the events in just one homologous pair of chromosomes, but you should make this simplification clear in your annotations.

The process of meiosis

The key events of meiosis are summarized in Figure 3.9.

In the **interphase** (page 52) that *precedes* meiosis, the chromosomes are replicated as **chromatids**, but between meiosis I and II there is no further interphase, so no replication of the chromosomes occurs *during* meiosis.

As meiosis begins, the chromosomes become visible. At the same time, homologous chromosomes pair up. (Remember, in a diploid cell each chromosome has a partner that is the same length and shape and with the same linear sequence of genes. It is these partner chromosomes that pair.)

When the homologous chromosomes have paired up closely, each pair is called a **bivalent**. Members of the bivalent continue to shorten – a process known as condensation.

During the coiling and shortening process within the bivalent, the chromatids frequently break. Broken ends rejoin more or less immediately. When non-sister chromatids from homologous chromosomes break and rejoin they do so at exactly corresponding sites, so that a cross-shaped structure called a chiasma is formed at one or more places along a bivalent. The event is known as a **crossing over** because lengths of genes have been exchanged between chromatids.

Then, when members of the bivalents start to repel each other and separate, the bivalents are (initially) held together by one or more chiasmata. This temporarily gives the bivalent an unusual shape. So, crossing over is an important mechanical event (as well as genetic event).

Next, the spindle forms. Members of the bivalents become attached by their centromeres to the fibres of the spindle at the equatorial plate of the cell. Spindle fibres pull the homologous chromosomes apart to opposite poles, but the **individual chromatids remain attached** by their centromeres.

Meiosis I ends with two cells each containing a single set of chromosomes made of two chromatids. These cells do not go into interphase, but rather continue smoothly into meiosis II. This takes place at right angles to meiosis I, and is exactly like mitosis. Centromeres of the chromosomes divide and **individual chromatids now move to opposite poles**. Following division of the cytoplasm, there are four cells – each with half the chromosome number of the original parent cell. (The four cells are said to be haploid.)

Meiosis and genetic variation

The four haploid cells produced by meiosis differ genetically from each other for two reasons:

■ There is **crossing over** of segments of individual maternal and paternal homologous chromosomes. These events result in new combinations of genes on the chromosomes of the haploid cells produced.

■ There is **independent assortment** (random orientation) of maternal and paternal homologous chromosomes. This happens because the way the bivalents line up at the equator of the spindle in meiosis I is entirely random. Which chromosome of a given pair goes to which pole is unaffected by (independent of) the behaviour of the chromosomes in other pairs.

Independent assortment is illustrated in Figure 3.10 in a parent cell with a diploid number of 4 chromosomes. In human cells, the number of pairs of chromosomes is 23; the number of possible combinations of chromosomes that can be formed by random orientation during meiosis is 2^{23}, which is over 8 million. We see that independent assortment alone generates a huge amount of variation in the coded information carried by different gametes into the fertilization stage.

Finally, in fertilization the fusion of gametes from different parents promotes genetic variation too.

■ **Figure 3.10**
Genetic variation
due to independent
assortment

Independent assortment is illustrated in a parent cell with two pairs of homologous chromosomes (four bivalents). The more bivalents there are, the more variation is possible. In humans, for example, there are 23 pairs of chromosomes giving over 8 million combinations.

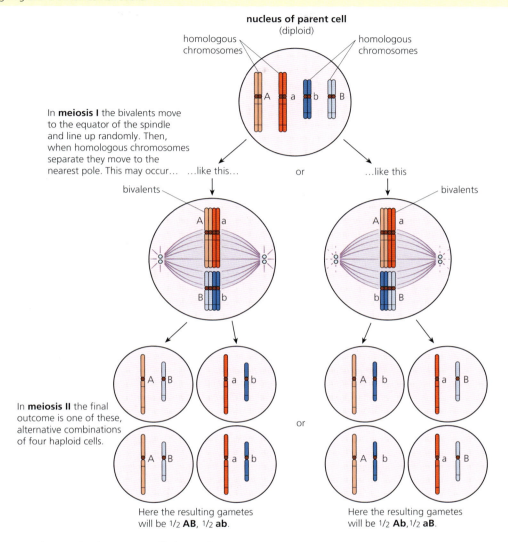

Errors in meiosis – non-disjunction

Very rarely, errors occur in the precisely controlled movements of the chromosomes during meiosis. The outcome is an alteration to part of the chromosome set. For example, chromosomes that *should* separate and move to opposite poles during the nuclear division of gamete formation *fail* to do so. Instead, a pair of chromosomes can move to the same pole. This malfunction event is referred to as **non-disjunction**. It results in gametes with more than and less than the haploid number of chromosomes.

For example, people with **Down's syndrome** have an extra chromosome 21, giving them a total of 47 chromosomes. How this non-disjunction arises is illustrated in Figure 3.11.

The symptoms of Down's syndrome are variable but, when severe, they include congenital heart and eye defects. The incidence of all forms of chromosomal abnormalities increases significantly with age (Figure 3.12). Women over the age of 40 who become pregnant are advised to have the chromosomes of the fetus assessed by screening (page 148). Other examples of non-disjunction are identified in Table 3.4.

4 **Evaluate** the extent to which the data in Figure 3.12 supports the suggestion that human birth defects are a function of maternal age.

■ **Figure 3.11**
Down's syndrome,
an example of non-disjunction

An extra chromosome causes Down's syndrome. The extra one comes from a meiosis error. The two chromatids of chromosome 21 fail to separate, and both go into the daughter cell that forms the secondary oocyte.

karyotype of a person with Down's syndrome

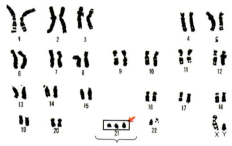

an extra chromosome 21

Steps of non-disjunction in meiosis
(illustrated in nucleus with only two pairs of homologous chromosomes – for clarity)

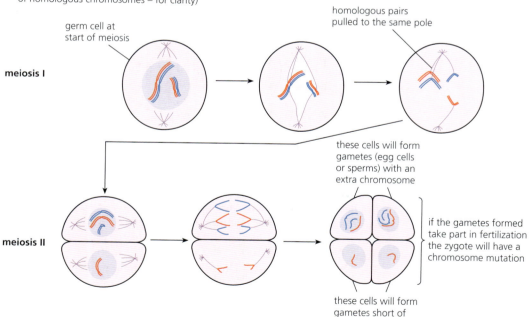

■ **Figure 3.12** Effect of maternal age on chromosome abnormalities

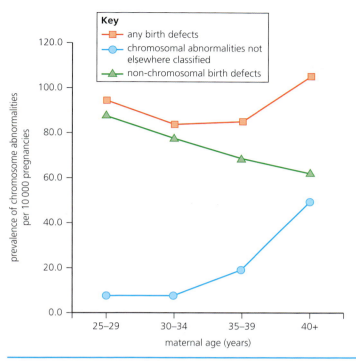

■ **Table 3.4** Other examples of non-disjunction

Changes in the number of chromosomes involving only part of the chromosome set are commonest in the sex chromosomes in humans
Klinefelter's syndrome – XXY (male) The extra X chromosome results when a male parent's X and Y chromosomes fail to segregate, producing XY sperm; one of these then fertilizes a normal (X) ovum. The effects of this defect differ widely, but an XXY male is likely to be sterile and have limited development of secondary sexual characteristics.
Turner's syndrome – presence of one X chromosome and no Y chromosome (female) The affected individuals are sexually underdeveloped and are sterile.

How non-disjunctions (chromosome mutations) are detected

Chromosome mutations are detected by karyotyping the chromosome set of a fetus. The two methods for obtaining fetal cells are known as **amniocentesis** and **chorionic villus sampling**. When and how these procedures may be carried out is described in Figure 3.13. Both techniques carry a slight risk of miscarriages of the fetus, so they are only recommended in cases where there is significant likelihood of genetic defect. Parents must consent to the procedure. Look at the outcome of the genetic screening of a fetus shown in Figure 3.14 and then answer question 5.

5   **Analyse** the human karyogram in Figure 3.14 to determine the gender of the patient and whether non-disjunction has occurred.

TOK Link

In 1922 the number of chromosomes counted in a human cell was 48. This remained the established number for 30 years, even though a review of photographic evidence from the time clearly showed that there were 46. Why are we slow to change our beliefs?

■ **Figure 3.13**
Screening of a fetus in the uterus

chorionic villus sampling – withdrawal of a sample of the fetal tissue part-buried in the wall of the uterus in the period 8–10 weeks into the pregnancy; the tiny sample is of cells that are actively dividing and can be analysed quickly.

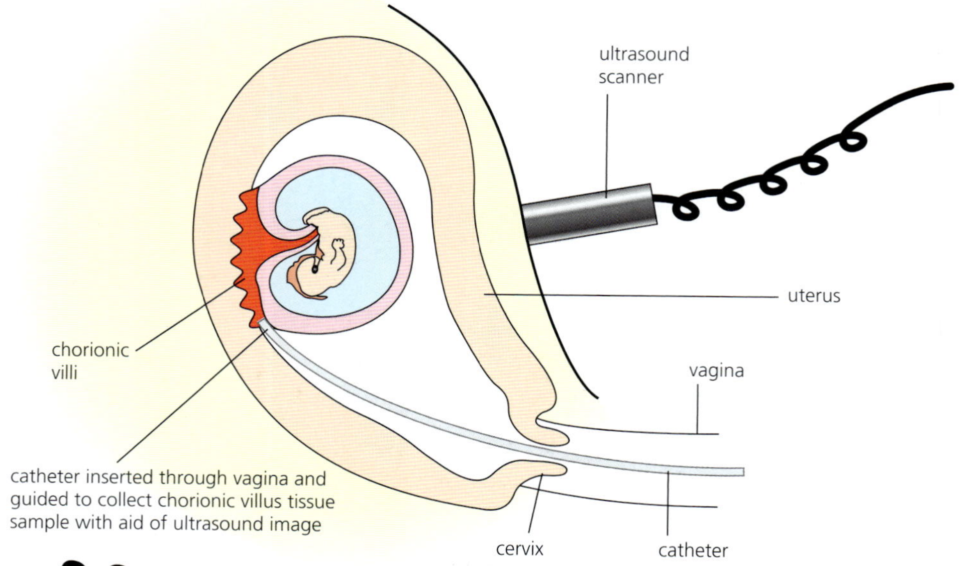

ultrasound scanner
uterus
chorionic villi
vagina
catheter inserted through vagina and guided to collect chorionic villus tissue sample with aid of ultrasound image
cervix
catheter

amniocentesis – withdrawal of a sample of amniotic fluid in the period 16–30 weeks of gestation; the fluid contains cells from the surface of the embryo

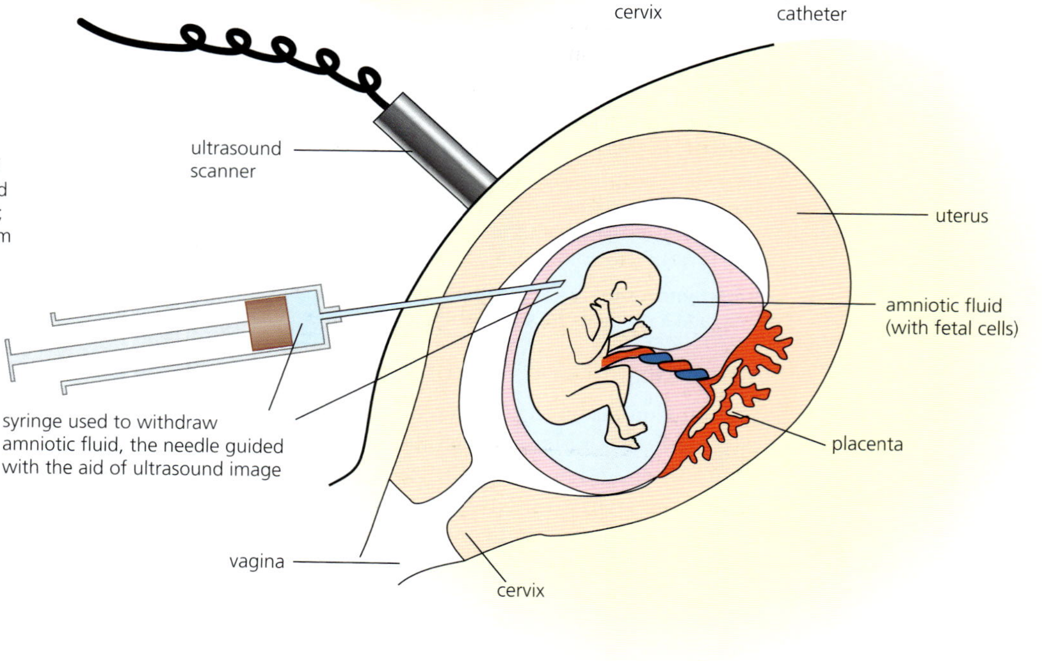

ultrasound scanner
uterus
amniotic fluid (with fetal cells)
syringe used to withdraw amniotic fluid, the needle guided with the aid of ultrasound image
placenta
vagina
cervix

■ **Figure 3.14**
Human karyogram used in a genetic screening for counselling

1
2
3
4
5
6
7
8
9
10
11
12
13
14
15
16
17
18
19
20
21
22
X Y

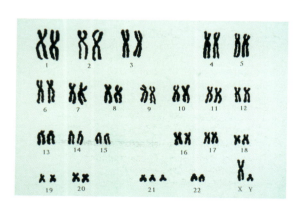

3.4 Inheritance – *the inheritance of genes follows patterns*

The mechanism of inheritance was successfully investigated before chromosomes had been observed or genes were detected. It was **Gregor Mendel** who made the first discovery of the fundamental laws of heredity (Figure 3.15).

Nature of Science

■ **Figure 3.15** Gregor Mendel – founder of modern genetics

Making quantitative measurements with replicates to ensure reliability

Gregor Mendel was born in Moravia in 1822, the son of a peasant farmer. As a young boy, he worked to support himself through schooling, but at the age of 21 he was offered a place in the monastery at Bruno (now in the Czech Republic). The monastery was a centre of research in natural sciences and agriculture, as well as in the humanities. Mendel was successful there. Later, he became Abbot.

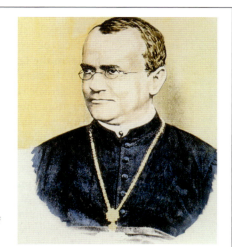

Mendel discovered the principles of heredity by studying **the inheritance of seven contrasting characteristics of the garden pea plant**. These did not 'blend' on crossing, but retained their identities, and were inherited in fixed mathematical ratios.

He concluded that hereditary factors (we now call them genes) determine these characteristics, that these factors occur in duplicate in parents, and that the two copies of the factors segregate from each other in the formation of gametes.

Today, we often refer to Mendel's laws of heredity, but Mendel's results were not **presented as laws** – which may help to explain the difficulty others had in seeing the significance of his work at the time.

Mendel was successful because:

- his experiments were carefully planned, and used large samples
- he carefully recorded the numbers of plants of each type but expressed his results as ratios
- in the pea, contrasting characteristics are easily recognized
- by chance, each of these characteristics was controlled by a single factor (gene)* rather than by many genes, as most human characteristics are
- pairs of contrasting characters that he worked on were controlled by factors (genes) on separate chromosomes*
- in interpreting results, Mendel made use of the mathematics he had learnt.

* Genes and chromosomes were not known then.

Features of the garden pea

round v. wrinkled seeds

green v. yellow cotyledons (seed leaves)

dwarf v. tall plants

Taking repeated measurements and large numbers of readings can improve reliability in data collection.

Initially, Mendel investigated the inheritance of a single contrasting characteristic, known as a **monohybrid cross**. Mendel had noticed that the garden pea plant was either tall or dwarf.

How was this contrasting characteristic controlled?

Mendel's investigation of the inheritance of height in pea plants began with plants that always 'bred true' (Figure 3.16). This means that the tall plants produced progeny that were all tall and the dwarf plants produced progeny that were all dwarf, when each was allowed to self-fertilize. Self-fertilization is the normal condition in the garden pea plant.

Mendel crossed true-breeding tall and dwarf plants and found the progeny (the F_1 generation) were all tall. The offspring were allowed to self-pollinate (and so self-fertilize) to produce the second (F_2) generation. The progeny of this cross consisted of both tall and dwarf plants in the ratio of 3 tall:1 dwarf.

In Mendel's interpretation of the monohybrid cross he argued that, because the dwarf characteristic had apparently disappeared in the F_1 generation and reappeared in the F_2 generation, there must be a factor controlling dwarfness that remained intact from one generation to another. However, this factor for 'dwarf' did not express itself in the presence of a similar factor for tallness. In other words, as characteristics, tallness is dominant and dwarfness is recessive; the dominant allele totally masks the effects of a recessive (non-dominant) allele. Logically, there must be two independent factors for height, one received from each parent. A sex cell (gamete) must contain only one of these factors.

Mendel saw that a 3:1 ratio could be the product of randomly combining two pairs of unlike factors (**T** and **t**, for example). This can be shown using a grid, now known as a Punnett grid, after the mathematician who first used it (Figure 3.17).

Mendel's conclusions from the monohybrid cross were that:

- within an organism there are breeding factors controlling characteristics such as 'tall' and 'dwarf'
- there are two factors in each cell
- one factor comes from each parent
- the factors separate in reproduction and either one can be passed on to an offspring
- the factor for 'tall' is an alternative form of the factor for 'dwarf'
- the factor for 'tall' is dominant over the factor for 'dwarf'.

■ **Figure 3.16** The steps of Mendel's monohybrid cross

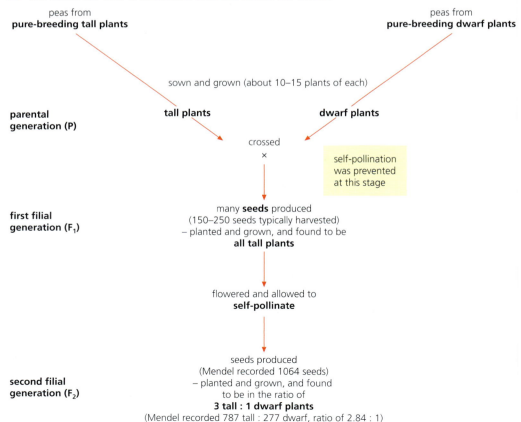

■ **Figure 3.17** Genetic diagram showing the behaviour of alleles in Mendel's monohybrid cross

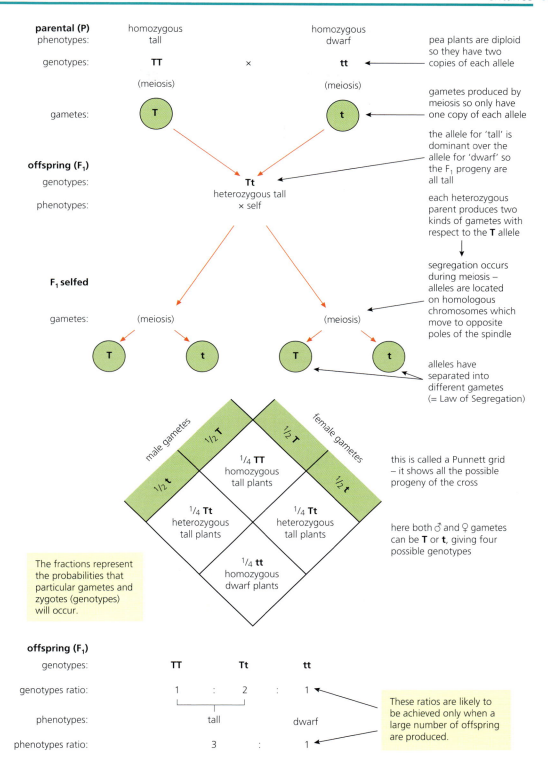

Notice the use of 'P', 'F₁' and 'F₂' for the three generations involved. Look carefully at the use of a Punnett grid to represent the process and products of the F₁ cross. This device shows clearly the combining process of the gametes. The fractions represent the probabilities that particular gametes and zygotes will occur.

Mendel never stated his discoveries as laws, but it would have been helpful if he had done so. For example, he might have said, 'Each characteristic of an organism is determined by a pair of factors of which only one can be present in each gamete.'

Today we call a similar statement Mendel's First Law, the **Law of Segregation**:

'The characteristics of an organism are controlled by pairs of alleles which separate in equal numbers into different gametes as a result of meiosis.'

■ Comparison of predicted and actual outcomes of genetic crosses

The actual results that Mendel obtained in three of his later investigations of inheritance in the garden pea plant were as shown in Table 3.5.

■ **Table 3.5** Mendel's later experimental results

Character	Cross	Number of F_2 counted	Number showing dominant character	Number showing recessive character
Position of flowers	axil × terminal	585	651 axial	207 terminal
Colour of seed coat	grey × white	929	705 grey	224 white
Colour of cotyledons	yellow × green	8023	6022 yellow	2001 green

Now answer the question below.

6 a These experiments followed on from Mendel's cross shown in Figure 3.16 and developed from the conclusions he made as a result of the first monohybrid cross. **State** what ratio of offspring he would have predicted from these crosses.
  b **Calculate** the actual ratios obtained in each of these crosses.
  c **Suggest** what chance events may influence the actual ratios of offspring obtained in breeding experiments like these. (We will return to this issue at a later point. For example, the statistical test known as the chi-squared test can be applied to the results of genetic crosses to test whether the actual results are close enough to the predicted results to be significant.)

■ Genotype, phenotype and the test cross

The alleles that an organism carries (present in every cell) make up the genotype of that organism. A genotype in which the two alleles of a gene are the same is said to 'breed true' or, more scientifically, to be **homozygous** for that gene. In Figure 3.17, the parent pea plants (P generation) were either homozygous tall or homozygous dwarf.

If the alleles are different, the organism is **heterozygous** for that gene. In Figure 3.17, the progeny (F_1 generation) were heterozygous tall.

7 **State** whether a person with sickle-cell trait (page 134) is homozygous or heterozygous for sickle-cell hemoglobin.

So the **genotype** is the genetic constitution of an organism. Alleles interact in various ways and with environmental factors. The outcome is the phenotype. The **phenotype** is the way in which the genotype of the organism is expressed – including the appearance of the organism. In Mendel's monohybrid cross (Figure 3.17) the heights of the plants were their phenotypes.

If an organism shows a recessive characteristic in its phenotype (like the dwarf pea) it must have a homozygous genotype (**tt**). But if it shows the dominant characteristic (like the tall pea) then it may be either homozygous for a dominant allele (**TT**) or heterozygous for the dominant allele (**Tt**). In other words, **TT and Tt look alike**; they have the same phenotype but different genotypes.

Plants like these can only be distinguished by the offspring they produce in a particular cross. When the tall heterozygous plants (**Tt**) are crossed with the homozygous recessive plants (**tt**), the cross yields 50% tall and 50% dwarf plants (Figure 3.18). This type of cross has become known as a **test cross**. If the offspring are all tall, then we know that the tall plants under test are homozygous plants (**TT**). Of course, sufficient plants have to be used to obtain these distinctive ratios.

8 **Construct** a table to explain the relationship between Mendel's Law of Segregation and meiosis.

■ **Figure 3.18** Genetic diagram of Mendel's test cross

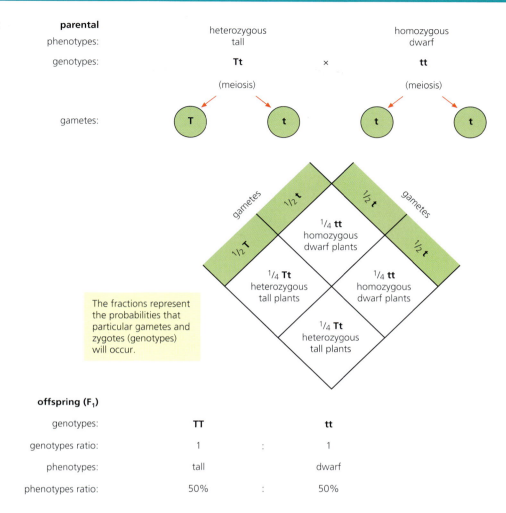

parental
phenotypes: heterozygous   homozygous
                  tall   dwarf

genotypes: **Tt**   ×   **tt**

                  (meiosis)   (meiosis)

gametes: T   t   t   t

½ T   ½ t   ½ t   ½ t

¼ **tt**
homozygous
dwarf plants

¼ **Tt**   ¼ **tt**
heterozygous   homozygous
tall plants   dwarf plants

¼ **Tt**
heterozygous
tall plants

The fractions represent
the probabilities that
particular gametes and
zygotes (genotypes)
will occur.

offspring (F₁)

genotypes:	**TT**		**tt**
genotypes ratio:	1	:	1
phenotypes:	tall		dwarf
phenotypes ratio:	50%	:	50%

Modification of the 3:1 monohybrid ratio

In certain types of monohybrid cross the 3:1 ratio is not obtained. Two of these situations are illustrated next.

Codominance – when both alleles are expressed

In the case of some genes, both alleles may be expressed simultaneously, rather than one being dominant and the other recessive. For example, in the common garden flower *Antirrhinum*, when red-flowered plants are crossed with white-flowered plants, the F₁ plants have pink flowers. When pink-flowered *Antirrhinum* plants are crossed, the F₂ offspring are found to be red, pink and white in the ratio 1:2:1 respectively.

Pink colouration of the petals occurs because both alleles are expressed in the heterozygote – both a red and a white pigment system are present. Red and white are **codominant alleles**. In genetic diagrams, codominant alleles are represented by a capital letter for the gene, and different superscript capital letters for the two alleles, in recognition of their equal influence (Figure 3.19).

■ **Figure 3.19**
Codominance in
the garden flower,
Antirrhinum

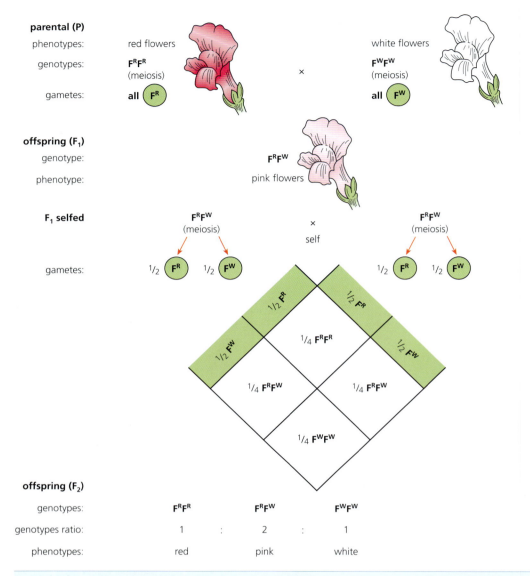

parental (P)

phenotypes: red flowers — white flowers

genotypes: F^RF^R (meiosis) × F^WF^W (meiosis)

gametes: all F^R — all F^W

offspring (F$_1$)

genotype: F^RF^W

phenotype: pink flowers

F$_1$ selfed F^RF^W (meiosis) × self F^RF^W (meiosis)

gametes: ½ F^R ½ F^W — ½ F^R ½ F^W

½ F^R ½ F^R

½ F^W ½ F^W

¼ F^RF^R

¼ F^RF^W ¼ F^RF^W

¼ F^WF^W

offspring (F$_2$)

genotypes:	F^RF^R		F^RF^W		F^WF^W
genotypes ratio:	1	:	2	:	1
phenotypes:	red		pink		white

9 **Construct** for yourself (using pencil and paper) a monohybrid cross between cattle of a variety that has
a gene for coat colour with red and white codominant alleles. Homozygous red and homozygous white
parents cross to produce roan offspring (red and white hairs together). **Predict** what offspring you would
expect and in what proportions, when a sibling cross (equivalent to selfing in plants) occurs between roan
offspring.

More than two alleles exist for a particular locus

The genes introduced so far exist in two forms (two alleles), for example the height gene of the
garden pea exists as tall and dwarf alleles. This means that in genetic diagrams we can represent
alleles with a single letter (here, **T** or **t**) according to whether they are dominant or recessive.

For simplicity we began by considering inheritance of a gene for which there are just two
alleles. However, we now know that not all genes are like this. In fact, most genes have more
than two alleles, and these are cases of genes with **multiple alleles**.

With multiple alleles, we choose a single capital letter to represent the locus at which
the alleles may occur, and the individual alleles are then represented by an additional single
letter (usually capital) in a superscript position – as with codominant alleles. An excellent
example of multiple alleles is found in the genetic control of the ABO blood group system in
humans. Human blood belongs to one of four groups: A, B, AB or O. Table 3.6 lists the possible
phenotypes and the genotypes that may be responsible for each blood group.

■ **Table 3.6**
The ABO blood groups – phenotypes and genotypes

Phenotype	Genotypes
A	I^AI^A or I^AI^i
B	I^BI^B or I^BI^i
AB	I^AI^B
O	I^iI^i

■ **Figure 3.20**
Inheritance of blood groupings A, B, AB and O

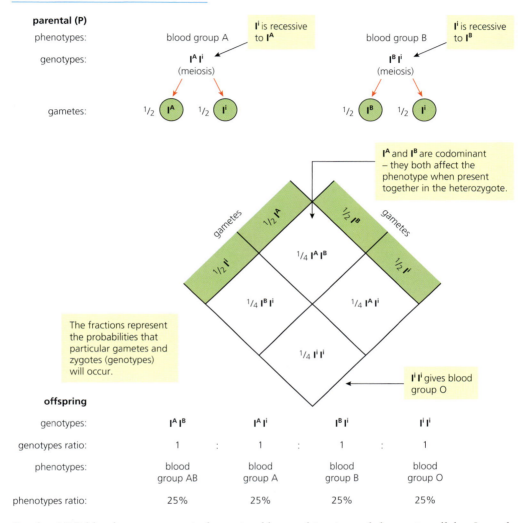

parental (P)

phenotypes: blood group A I^i is recessive to I^A blood group B I^i is recessive to I^B

genotypes: $I^A I^i$ (meiosis) $I^B I^i$ (meiosis)

gametes: $1/2$ I^A $1/2$ I^i $1/2$ I^B $1/2$ I^i

I^A and I^B are codominant – they both affect the phenotype when present together in the heterozygote.

gametes $1/2$ I^A $1/2$ I^B gametes

$1/2$ I^i $1/4$ $I^A I^B$ $1/2$ I^i

$1/4$ $I^B I^i$ $1/4$ $I^A I^i$

The fractions represent the probabilities that particular gametes and zygotes (genotypes) will occur.

$1/4$ $I^i I^i$

$I^i I^i$ gives blood group O

offspring

genotypes:	$I^A I^B$		$I^A I^i$		$I^B I^i$		$I^i I^i$
genotypes ratio:	1	:	1	:	1	:	1
phenotypes:	blood group AB		blood group A		blood group B		blood group O
phenotypes ratio:	25%		25%		25%		25%

So, the ABO blood group system is determined by combinations of alternative alleles. In each individual, only two of the three alleles exist, but they are inherited as if they were alternative alleles of a pair. However, I^A and I^B are codominant alleles and both I^A and I^B are dominant to the recessive I^i. In Figure 3.20 the way in which the alternative blood groups may be inherited is shown.

10 One busy night in an understaffed maternity unit, four children were born at about the same time. The babies were muddled up by mistake; it was not certain which child belonged to which family. Fortunately, the children had different blood groups: A, B, AB and O.

The parents' blood groups were also known:
- Mr and Mrs Jones   A × B
- Mr and Mrs Lee   B × O
- Mr and Mrs Gerber   O × O
- Mr and Mrs Santiago   AB × O

The nurses were able to decide which child belonged to which family. **Deduce** how this was done.

■ Human inheritance investigated by pedigree chart

Studying human inheritance by experimental crosses (with selected parents, sibling crosses, and the production of large numbers of progeny) is out of the question. Instead, we may investigate the pattern of inheritance by researching a family pedigree, where appropriate records of the ancestors exist. A human pedigree chart uses a set of rules. These are identified in Figure 3.21.

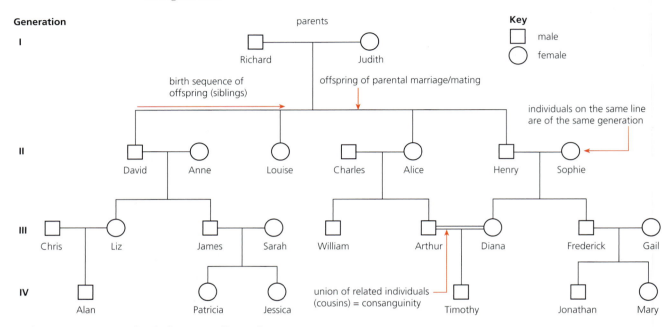

■ **Figure 3.21** An example of a human pedigree chart

11 In the human pedigree chart in Figure 3.21 **state**:
  a who are the female grandchildren of Richard and Judith
  b who are Alan's (i) grandparents and (ii) uncles
  c how many people in the chart have parents unknown to us
  d the names of two offspring who are cousins.

Analysis of pedigree charts to deduce the pattern of inheritance

We can use a pedigree chart to detect conditions that are due to dominant and recessive alleles. In the case of a characteristic due to a dominant allele, the characteristic tends to occur in one or more members of the family in every generation (Figure 3.23). On the other hand, a recessive characteristic is seen infrequently, often skipping many generations.

For example, albinism is a rare inherited condition of humans (and other mammals) in which the individual has a block in the biochemical pathway by which the pigment melanin is formed. Albinos have white hair, very light-coloured skin and pink eyes. Albinism shows a pattern of recessive monohybrid inheritance in humans. In the chart shown of a family with albino members, albinism occurs infrequently, skipping two generations altogether (Figure 3.22).

Brachydactyly is a rare condition of humans in which the fingers are very short. Brachydactyly is due to a mutation in the gene for finger length. Unusually, the mutant allele is dominant, so the condition shows a pattern of dominant monohybrid inheritance; that is, it tends to occur in every generation in a family (Figure 3.23).

■ **Figure 3.22** Pedigree chart of a family with albino members

Albino people must be homozygous for the recessive albino allele (**pp**). People with normal skin pigmentation may be homozygous normal (**PP**) or carriers (**Pp**).

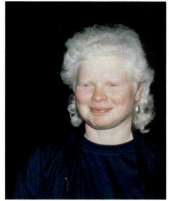

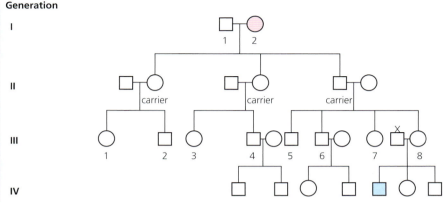

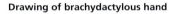

This is a typical family tree for inheritance of a characteristic controlled by a recessive allele. In generation I, individual 2 must be **pp** (albino). In generation II, all offspring must be carriers (**Pp**) because they have received a recessive allele (**p**) from their mother. In generation III, individual 8 must be a carrier and her partner **X** must also be a carrier, since their offspring include a son that is albino.

■ **Figure 3.23**
Brachydactyly, and pedigree chart of a family with brachydactylous genes

X-ray of bones of hand of normal length

Drawing of brachydactylous hand

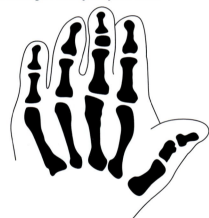

Pedigree chart of family with brachydactylous alleles

12 If a homozygous normal-handed parent (**nn**) had a child with a heterozygous brachydactylous parent (**Nn**), **calculate**, using a Punnett grid, the probability of an offspring with brachydactylous hands. **Construct** a genetic diagram to show your workings.

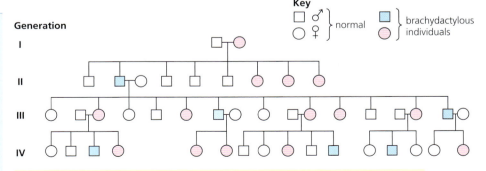

In **generation I** the parents are assumed to be normal male (**nn**) and brachydactylous female (heterozygous **Nn**), as the ratio of offspring is similar to that of a test cross.

In **each subsequent generation** about half the offspring are brachydactylous (i.e. **Nn** or **NN**) and half are normal (**nn**).

■ Sex chromosomes and gender

In humans, gender is determined by specific chromosomes known as the **sex chromosomes**. Each of us has one pair of sex chromosomes (either XX or XY chromosomes), along with the 22 other pairs (known as **autosomal chromosomes**).

Egg cells produced by meiosis all carry an X chromosome, but 50% of sperms carry an X chromosome and 50% carry a Y chromosome. At fertilization, an egg cell may fuse with a sperm carrying an X chromosome, leading to a female offspring. Alternatively, the egg cell may fuse with a sperm carrying a Y chromosome, leading to a male offspring. So, the gender of offspring in humans (and all mammals) is determined by the male partner. Also, we would expect equal numbers of male and female offspring to be produced over time by a breeding population (Figure 3.24).

■ **Figure 3.24** X and Y chromosomes and the determination of sex

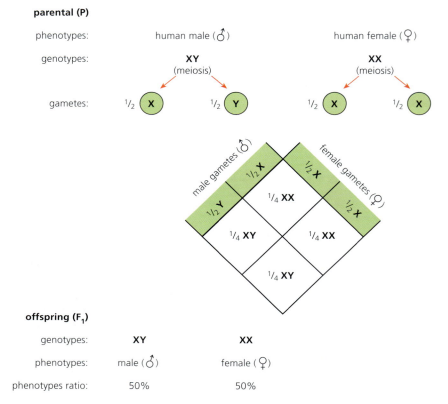

Human X and Y chromosomes and the control of gender

Initially, male and female embryos develop identically in the uterus. At the seventh week of pregnancy, however, a cascade of developmental events is triggered, leading to the growth of male genitalia if a Y chromosome is present in the embryonic cells.

On the Y chromosome is the **prime male-determining gene**. This gene codes for a protein called **testis-determining factor** (**TDF**). TDF functions as a molecular switch; on reaching the embryonic gonad tissues, TDF initiates the production of a relatively low level of testosterone. The effect of this hormone at this stage is to inhibit the development of female genitalia, and to cause the embryonic genital tissues to form testes, scrotum and penis.

In the absence of a Y chromosome, the embryonic gonad tissue forms an ovary. Then, partly under the influence of hormones from the ovary, the female reproductive structures develop.

Pairing of X and Y chromosomes in meiosis

We know that homologous chromosomes pair up early in meiosis (Figure 3.10, page 145). Pairing is an essential step in the mechanism of meiosis. However, only a very small part of the X and Y chromosomes of humans have complementary alleles and can pair up during meiosis. The bulk of the X and Y chromosomes contain genes that have no corresponding alleles on the other (Figure 3.25).

The short Y chromosome carries genes specific for male sex determination and sperm production, including the gene that codes for TDF, which, as we have seen, switches development of embryonic gonad tissue to testes early in embryonic development.

The X chromosome carries an assortment of genes, very few of which are concerned with sex determination.

We shall examine the effects of some of the alleles of these genes next.

■ **Figure 3.25**
The pairing of
X and Y chromosomes
in meiosis

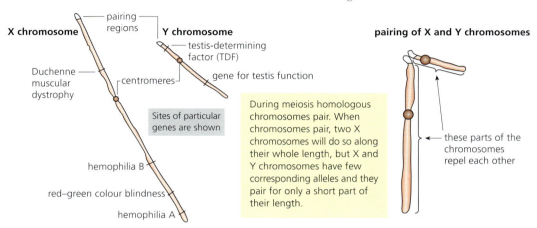

During meiosis homologous chromosomes pair. When chromosomes pair, two X chromosomes will do so along their whole length, but X and Y chromosomes have few corresponding alleles and they pair for only a short part of their length.

Sex linkage

Genes present on the sex chromosomes are inherited with the sex of the individual. They are said to be **sex-linked characteristics**. Sex linkage is a special case of linkage occurring when a gene is located on a sex chromosome (usually the X chromosome).

The inheritance of these sex-linked genes is different from the inheritance of genes on the autosomal chromosomes. This is because the X chromosome is much longer than the Y chromosome (since many of the genes on the X chromosome are absent from the Y chromosome). In a male (XY), most alleles on the X chromosome lack a corresponding allele on the Y and will be apparent in the phenotype **even if they are recessive**.

Meanwhile, in a female (XX), a single recessive gene on one X chromosome may be masked by a dominant allele on the other X and would not be expressed. A human female can be homozygous or heterozygous with respect to sex-linked characteristics, whereas males have only one allele.

A heterozygous individual with a recessive allele of a gene that does not have an effect on their phenotype is known as a **carrier**; they carry the allele but it is not expressed. So, female carriers are heterozygous for sex-linked recessive characteristics. Of course, the unpaired alleles of the Y chromosome are all expressed in the male. However, the alleles on the (short) Y chromosome are mostly concerned with male structures and male functions. Examples of recessive conditions controlled by genes on the X chromosome are red–green colour blindness and hemophilia.

Red–green colour blindness

A red–green colour-blind person sees green, yellow, orange and red as the same colour. The condition afflicts about 8% of males, but only 0.4% of females in the human population. This is because a female with normal colour vision may by homozygous for the normal colour-vision allele ($X^B X^B$) or she may be heterozygous for normal colour vision ($X^B X^b$). For a female to be red–green colour blind, she must be homozygous recessive for this allele ($X^b X^b$) – this occurs extremely rarely. On the other hand a male with a single recessive allele for red–green colour vision ($X^b Y$) will be affected.

The inheritance of red–green colour blindness is illustrated in Figure 3.26. It is helpful for those who are red–green colour blind to recognize their inherited condition. Red–green colour blindness is detected by the use of multicoloured test cards.

13 State how the genetic constitution of a female who is red–green colour blind is represented. **Explain** why it is impossible to have a 'carrier' male.

■ **Figure 3.26**
Detection and
inheritance of
red–green colour
blindness

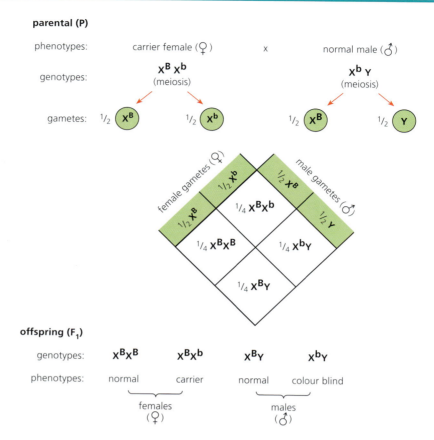

Colour blindness is detected
by multicoloured test cards.
A mosaic of dots is arranged
on the cards so that those
with normal vision see a
pattern that is not visible to
those with colour blindness.

Hemophilia

If a break occurs in the circulatory system of a mammal, there is a risk of uncontrolled bleeding. Usually, this risk is averted by the blood clotting mechanism (page 272). Hemophilia is a rare, genetically determined condition in which the blood does not clot normally. The result is frequent and excessive bleeding.

There are two forms of hemophilia, known as hemophilia A and hemophilia B. They are due to a failure to produce adequate amounts of particular blood proteins that are essential to the complex blood clotting mechanism. Today, hemophilia is effectively treated by the administration of the clotting factor that the patient lacks.

Hemophilia is a sex-linked condition because the genes controlling production of these blood proteins are located on the X chromosome. Hemophilia is caused by a recessive allele. As a result, hemophilia is largely a disease of the male – since a single X chromosome carrying the defective allele (X^hY) will result in disease. For a female to have the disease, she must be homozygous for the recessive gene (X^hX^h), but this condition is usually fatal in the uterus, typically resulting in a natural abortion.

A female with only one X chromosome with the recessive allele (X^HX^h) is described as a carrier. She has a normal blood-clotting mechanism. When a carrier is partnered by a normal male, there is a 50% chance of the daughters being carriers and a 50% chance of the sons having hemophilia (Figure 3.27).

14 Hemophilia results from a sex-linked gene. The disease is most common in males, but the hemophilia allele is on the X chromosome. **Explain** this apparent anomaly.

■ **Figure 3.27**
The inheritance of hemophilia

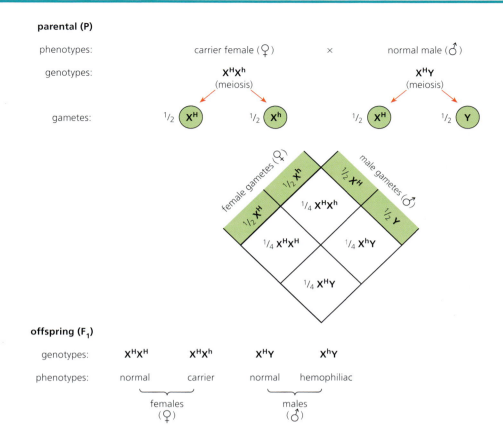

parental (P)

phenotypes: carrier female (♀)   ×   normal male (♂)

genotypes: $X^H X^h$ $X^H Y$
(meiosis)   (meiosis)

gametes: ½ X^H ½ X^h ½ X^H ½ Y

female gametes (♀)   male gametes (♂)

½ X^h ½ X^H

½ X^H ¼ $X^H X^h$ ½ Y

¼ $X^H X^H$ ¼ $X^h Y$

¼ $X^H Y$

offspring (F₁)

genotypes: $X^H X^H$ $X^H X^h$ $X^H Y$ $X^h Y$

phenotypes: normal   carrier   normal   hemophiliac

females (♀)   males (♂)

■ Genetic disease and its inheritance

Genetic diseases are caused by inheritance of a gene or genes. Most arise from a mutation involving a single gene. About 4000 disorders of humans have a genetic basis, and they affect between 1 and 2% of the human population. More than half of the known genetic diseases are due to a mutant allele that is recessive and in these cases a person must be homozygous for the mutant gene for the condition to be expressed. However, people with a single mutant allele are 'carriers' of that genetic disease. Quite surprising numbers of us are carriers of one or more conditions.

Common genetic diseases include sickle-cell disease (pages 133–134), Duchene muscular dystrophy, severe combined immunodeficiency (SCID), familial hypercholesterolemia, hemophilia (page 160), thalassemia and cystic fibrosis.

Cystic fibrosis is the most common genetic disease among Caucasians (4% of people of European descent carry one defective allele for cystic fibrosis). The disease is due to a mutation of a single gene on chromosome 7 and it affects the epithelial cells of the body. The normal CF gene codes for a protein which functions as an ion pump. The pump transports chloride ions across membranes and water follows the ions, so epithelia are kept smooth and moist. The mutated gene codes for no protein or for a faulty protein. The outcomes are that epithelia remain dry and there is a build-up of thick, sticky mucus. The effects are felt:

■ in the pancreas – here, secretion of digestive juices by the gland cells in the pancreas is interrupted by blocked ducts

■ the sweat glands, where salty sweat is formed – a feature exploited in diagnosis

■ in the lungs which become blocked by mucus and are prone to infection – this effect can be life-threatening

■ by adult patients in their reproductive organs –the epithelial membranes here are affected too.

■ **Figure 3.28**
The inheritance
of cystic fibrosis

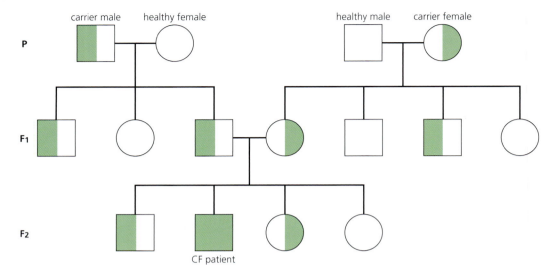

■ **Figure 3.28**
The inheritance
of cystic fibrosis

Cystic fibrosis patients have two copies of the mutated CF gene (one from each parent).
Carriers have one healthy gene and one mutated gene – so they may pass on a mutated CF gene to an offspring.

In Figure 3.28 we see the pedigree chart typical of a recessive characteristic – the condition is seen infrequently, often skipping many generations.

A small percentage of genetic diseases are caused by a dominant allele. **Huntington's disease** is due to an autosomal dominant allele on chromosome 4. The disease is extremely rare (1 case per 20 000 live births). Appearance of the symptoms is delayed until the age of 40–50 years, by which time the affected person – unaware of the presence of the disease – may have passed on the dominant allele to his or her children.

The disease takes the form of progressive mental deterioration, accompanied by involuntary muscle movements (twisting, grimacing and staring in 'fear').

In a pedigree chart of a family with Huntington's disease, the condition ends to occur in one or more members of the family in every generation – a characteristic due to a dominant allele.

What causes generate genetic disease and cancers?

An abrupt change in the structure, arrangement or amount of DNA of chromosomes may change the characteristics of an organism or of an individual cell in which they occur. This is because the change, a **mutation**, results in the alteration or non-production of a cell protein (via the **mRNA** which codes for it). While many mutations are neutral in outcome, some are harmful and some are lethal.

Mutations that occur in body cells of multicellular organisms are called **somatic mutations**. They are only passed on to the immediate descendants of that cell. However, we have already seen that somatic mutations can also be the cause of a cancer (page 58). Somatic mutations do occur with increasing frequency as an organism ages; cancers are more common in elderly people than in the young. These mutations are eliminated when the cells die.

On the other hand, mutations occurring in the cells of the gonads (**germ-line mutations**) give rise to gametes with an altered genome. These can be inherited by the offspring and so be the cause of genetic changes in future generations – including genetic diseases.

Changes in the sequence of bases in the DNA of a gene may occur spontaneously as a result of errors in DNA replication, although these are extremely rare since DNA polymerase has a built-in checking mechanism that works as it operates (page 109).

Alternatively, mutations can be caused by environmental agents that we call **mutagens**. These can include ionizing radiation in the form of X-rays, cosmic rays and radiation from radioactive isotopes (α, β and γ rays). Any of these can cause the break-up of the DNA molecule. Non-ionizing mutagens include UV light and various chemicals, including carcinogens in tobacco smoke (tar compounds). These act by modifying the chemistry of the base pairs of DNA.

Catastrophic world events involving ionizing radiation

World events where huge populations have been suddenly exposed to catastrophic levels of ionizing radiation are listed in Table 3.7.

■ Table 3.7 Violent exposure of large populations to radiation

6 August 1945 at Hiroshima, Japan	Explosion of an atomic bomb consisting of uranium[235], equivalent to 18 000 tonnes of TNT.
9 August 1945 at Nagasaki, Japan	Explosion of an atomic bomb consisting of plutonium[239], equivalent to 20 000 tonnes of TNT.
26 April 1986 at Chernobyl, Ukraine	Explosion and meltdown of an atomic reactor. The radiation released is estimated at least 200 times that released by the atomic bomb explosions in Japan in 1945.

Deaths at these horrific events result from physiological traumas due to blast, heat, fire and radiation effects. These forces claim many lives in very distressing circumstances, both immediately and in the weeks that follow. If genetic changes are induced by exposure to radiation (both from the initial exposure and subsequently, because some of the radioisotopes released have very long half-lives), these become detectable much later. To be confident of cause and effect of symptoms that develop, comparisons with data from the study of control groups is essential.

For example, assessments of the effects of radiation exposure in survivors of the atomic bombs in Japan over subsequent decades were conducted on 93 000 atomic-bomb survivors set against a control group of 27 000 people. The incidence of cancer in survivors aged 70 years who were exposed at age 30 years was significantly higher for both men and women than in the control groups.

However, detection of human germ-cell mutations has been difficult. While very high doses of radiation applied to experimental animals have caused major disorders among offspring, little evidence of clinical effects has yet been seen in children of A-bomb survivors. Given the relatively low average dose to survivors, the results are not surprising. However, it is unlikely that humans are entirely free from induction of germ cell mutations following earlier irradiation.

Because of the unprecedented scale of the more recent nuclear power-station accident at Chernobyl, no one can predict what the future holds for those involved and who continue to be exposed to raised background levels of radiation. Several million people living there received the highest known exposure to radiation in this 'atomic age'. Huge increases in cancers, particularly breast cancer and thyroid cancer occurred. The epidemic of thyroid cancer has yet to abate.

■ Figure 3.29 The city of Pripyat and the Chernobyl power plant are surrounded by forest

However, the actual number of deaths directly attributable to Chernobyl has been lower than expected. A United Nations report in the past decade concluded:

■ among the 200 000 workers exposed in the first year after the accident, 2200 radiation deaths can be expected

■ the total of deaths attributable to radiation may reach 4000, based on the estimated doses received.

In a zone of about 18 miles diameter around the abandoned nuclear power site at Chernobyl radiation levels remain far too high for humans to be permitted to return either to the abandoned town or to surrounding agricultural land. The place is not deserted, however. With the passage of 20 years since the meltdown, the area has become a natural woodland with a rich flora and fauna. It is now home to more than 66 different species of mammals and 280 species of birds. The current radiation levels are apparently not shortening these animals' natural life expectancies. When, if ever, humans can safely return there to live may be a different matter.

3.5 Genetic modification and biotechnology
– biologists have developed techniques for artificial manipulation of DNA, cells and organisms

Today, biologists have developed techniques for artificial manipulation of DNA, cells and organisms. These gene technologies have followed on from the discovery of the **structure and role of DNA as the universal code**, the application of **electron microscopy** to cell biology, and the ability to **isolate viable cell organelles** for *in vitro* studies in biochemistry. These skills, together with advances in **enzymology**, are the basis of gene technology.

Today, this new technology includes the following processes:

■ **Genetic engineering** – the transfer of a gene from one organism (the donor) to another (the recipient). For example, genes coding for human insulin, for human growth hormone, and for blood clotting factors (Factor VIII) have all been transferred to bacterial cells. Human genes have also be transferred to eukaryotic cells, such as yeasts and potato.

■ **DNA sequencing** – the creation of genomic libraries of the precise sequence of nucleotides in the DNA of individual organisms. The nucleotide sequence in the whole human genome was the product of the Human Genome Project (HGP, page 135).

■ **Genetic fingerprinting** – in which DNA is analysed in order to identify the individual from which the DNA was taken. It is now commonly used in forensic science (for example, to identify someone from a blood or other body fluid sample and to identify botanical evidence from samples of pollen and plant fragments) and to establish the genetic relatedness of individuals. It is also used to determine whether individuals of endangered species have been bred in captivity or captured in the wild.

The gene technologist's tool-kit

It was the discovery and isolation of four naturally occurring enzymes that now makes possible the manipulation of individual genes. We meet these enzymes in other contexts, but here it is significant that some of these enzymes are now obtained from bacteria we call extremophiles (page 227). In these organisms, enzymes are adapted to function at optimum rates under extreme conditions, including, for example, very high or low temperatures. We shall discuss how these individual enzymes work, shortly. They are introduced in Table 3.8.

■ **Table 3.8**
The genetic
technologist's tool-kit
of enzymes

Enzyme	Natural source	Application in genetic engineering
restriction enzymes (restriction endonucleases)	cytoplasm of bacteria (naturally used to combat viral infection by breaking up viral nucleic acid)	break DNA molecules into shorter lengths, each acting at a specific nucleotide sequence
DNA ligase	found with nucleic acid in the nucleus of *all* organisms	joins together DNA molecules during replication
DNA polymerase	found with nucleic acid in the nucleus of *all* organisms	synthesizes nucleic acid strands, guided by the template strand of nucleic acid
reverse transcriptase	retroviruses only	synthesis of a DNA strand (cDNA) complementary to an existing RNA strand

These enzymes are extracted mainly from microorganisms or viruses, and are used to manipulate nucleic acids in very precise ways.

15 The virus called HIV is able to infect a human cell (page 278). Investigate and **outline** the role of reverse transcriptase in this process.

Gene technologies – their significance for science and society

Gene technologies have applications in biotechnology, medicinal drug production (in the pharmaceuticals industry), the treatment of genetic diseases, in agriculture and in forensic science. Consequently, genetic technologies generate many potential benefits for humans, but there are potential hazards, too. Sometimes, the economic advantages may be out-weighed by environmental drawbacks or dangers. Thus, gene technologies raise **ethical issues** that require balanced and informed judgments to be made by communities.

■ Two processes central to genetic engineering

1 Electrophoresis

Gel electrophoresis is a process that is used to separate molecules such as proteins and fragments of nucleic acids. It is widely applied in many studies of DNA. For example, it is central to investigation of the sequence of bases in particular lengths of DNA, known as **DNA sequencing**. It is also used in the identification of individual organisms and species, known as **genetic profiling**. We will review these applications shortly.

The principle and practice of electrophoresis

In electrophoresis, proteins or nucleic acid fragments (either DNA or RNA) are separated on the basis of their net charge and mass. Separation is due to the following factors:

- Differential migration of these molecules through a supporting medium – typically this is either agarose gel (a very pure form of agar) or polyacrylamide gel (PAG). In these media, the tiny pores in the gel act as a **molecular sieve**; small particles can move quite quickly, whereas larger molecules move much more slowly
- The **electrical charge** that molecules carry – it is the phosphate groups in DNA fragments that give them a net negative charge. Consequently, when these molecules are placed in an electric field, they migrate towards the positive pole (anode).

There is a double principle of electrophoretic separation: separation occurs on the basis of **size** *and* **charge**.

DNA fragments are produced by the actions of one or more restriction enzymes. Different restriction enzymes cut at particular base sequences, as and where these occur along the length of the DNA. Consequently, fragments of different lengths are produced.

A series of groves or wells are cut close to one end of the gel, which is then submersed in a salt solution that conducts electricity. Then a small quantity of a mixture to be separated is placed in a well. Several different mixtures can be separated in a single gel, at one time (Figure 3.30). Animations of electrophoresis are available at:

- www.dnalc.org/resources/animations/gelelectrophoresis.html
- http://learn.genetics.utah.edu/content/labs/gel/

■ **Figure 3.30**
Electrophoretic
separation of DNA
fragments

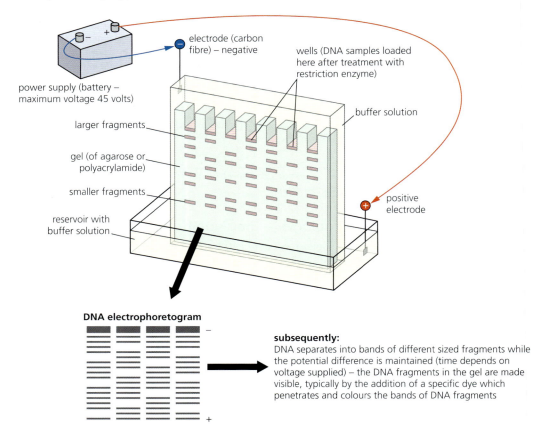

electrophoresis in progress

electrode (carbon fibre) – negative

power supply (battery – maximum voltage 45 volts)

wells (DNA samples loaded here after treatment with restriction enzyme)

buffer solution

larger fragments

gel (of agarose or polyacrylamide)

smaller fragments

positive electrode

reservoir with buffer solution

DNA electrophoretogram

subsequently:
DNA separates into bands of different sized fragments while the potential difference is maintained (time depends on voltage supplied) – the DNA fragments in the gel are made visible, typically by the addition of a specific dye which penetrates and colours the bands of DNA fragments

After separation, the fragments are not immediately visible – they are tiny and transparent. They can be identified by gene probes and DNA stains:

■ **Gene probes** – these consist of single-stranded DNA with a base sequence that is complementary to that of a particular fragment or gene whose position or presence is sought. The probe must be made radioactive so that, when the treated gel is exposed to X-ray film, the presence of that particular probe and complementary fragment will be disclosed. Alternatively, the probe can have a fluorescent stain attached. It will then fluoresce distinctively in UV light, thereby indicating the presence of the particular fragment or gene that is sought.

16 Explain how electrophoresis works. **Outline** the role of the buffer and power supply in electrophoresis.

■ **Stains** – these immediately locate the position of all DNA fragments once applied. Stains include:

☐ ethidium bromide – DNA fragments fluoresce in short-wave UV radiation

☐ methylene blue – stains gel and DNA, but colour fades quickly.

2 The polymerase chain reaction

In the **polymerase chain reaction (PCR)**, double-stranded DNA is **amplified**, meaning many copies are made (Figure 3.31). Often, it is only possible to produce or recover a very small amount of DNA (such as in genetic engineering or at a crime scene). Minute DNA samples can be amplified by PCR. In PCR, DNA is replicated in an entirely automated process, *in vitro*, to produce a large amount of the sequence. A single molecule is sufficient as the starting material, should this be all that is available. The products are exact copies.

The heat-resistant polymerase enzyme used in PCR is obtained from a bacterium found in hot springs. The steps in PCR are of special interest, as they show us the importance to genetic engineering of the discovery of the extremophiles, with their uniquely adapted enzymes.

■ **Figure 3.31**
The polymerase
chain reaction

The polymerase chain reaction involves a series of steps, each taking a matter of minutes.
The process involves a heating and cooling cycle and is automated.
Each time it is repeated in the presence of excess nucleotides, the number of copies of the
original DNA strand is doubled.

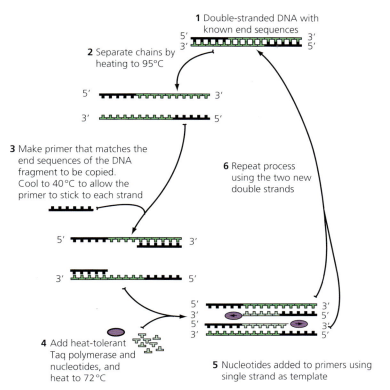

1 Double-stranded DNA with
known end sequences

2 Separate chains by
heating to 95°C

3 Make primer that matches the
end sequences of the DNA
fragment to be copied.
Cool to 40°C to allow the
primer to stick to each strand

6 Repeat process
using the two new
double strands

4 Add heat-tolerant
Taq polymerase and
nucleotides, and
heat to 72°C

5 Nucleotides added to primers using
single strand as template

Note: 'Primers' are short sequences of single-stranded DNA made synthetically with base
sequences complementary to one end (the 3' end) of DNA.
Remember: DNA polymerase synthesizes a DNA strand in the 3' to 5' direction.

■ Gene technology applied to human insulin production

One of the earliest successful applications of these techniques was when the human genes for
insulin production were transferred to a strain of the bacterium *Escherichia coli*. Insulin consists
of two short polypeptides linked together by sulfide bridges. Once the hormone is assembled
from its component polypeptides, it enables body cells to regulate the blood sugar level
(page 302). Regular supplies of insulin are required to treat insulin-dependent diabetes. Cultures
of *E. coli* have been 'engineered' to manufacture and secrete human insulin, when cultured in
a bulk fermenter with appropriate nutrients. The insulin is extracted and made available for
clinical use. We look into this technique now (Figure 3.32).

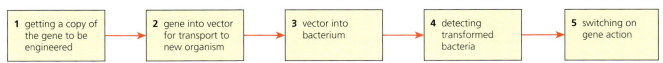

| **1** getting a copy of the gene to be engineered | **2** gene into vector for transport to new organism | **3** vector into bacterium | **4** detecting transformed bacteria | **5** switching on gene action |

■ **Figure 3.32** The steps to genetic engineering of *E. coli* for insulin production

Obtaining a copy of the human insulin genes by isolating mRNA

One way to obtain a copy of the genes for insulin is to start with messenger RNA (mRNA),
rather than searching for the gene itself among the chromosomes. The human pancreas contains
patches of cells (islets of Langerhans, Figure 6.53, page 302) where insulin is produced. Here the

insulin genes in the nuclei of the cells are transcribed to produce mRNA. This passes out of the nucleus to the ribosomes in the cytoplasm where the sequence of bases of the mRNA is translated into the linear sequence of amino acids of the insulin proteins.

There is a particular advantage in making a copy of a gene from the mRNA it codes for. This is explained later in this chapter (page 173). First, let's consider the process. It involves:

- mRNA for insulin being isolated from a sample of human pancreas tissue
- the use of the enzyme **reverse transcriptase** (obtained from a retrovirus other than HIV) alongside the isolated mRNA to form a single strand of DNA
- the conversion of this DNA into double-stranded DNA, using **DNA polymerase**. In this way the gene is manufactured (Figure 3.33). This form of an isolated gene is known as **complementary DNA (cDNA)**.

■ **Figure 3.33** Using reverse transcriptase to build the gene for human insulin

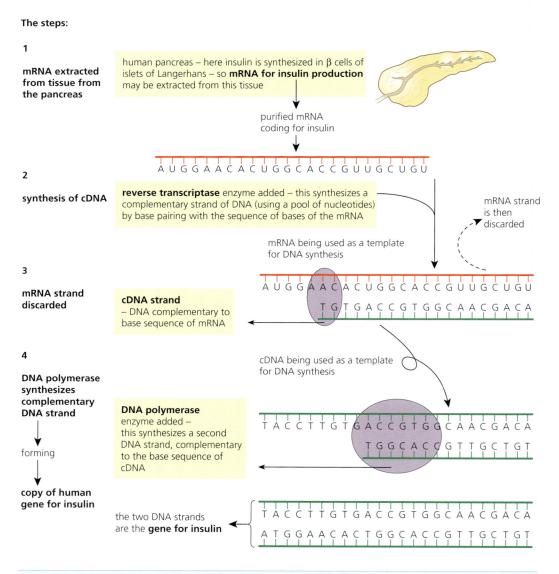

The steps:

1

mRNA extracted from tissue from the pancreas

human pancreas – here insulin is synthesized in β cells of islets of Langerhans – so **mRNA for insulin production** may be extracted from this tissue

purified mRNA coding for insulin

A U G G A A C A C U G G C A C C G U U G C U G U

2

synthesis of cDNA

reverse transcriptase enzyme added – this synthesizes a complementary strand of DNA (using a pool of nucleotides) by base pairing with the sequence of bases of the mRNA

mRNA strand is then discarded

mRNA being used as a template for DNA synthesis

3

mRNA strand discarded

cDNA strand – DNA complementary to base sequence of mRNA

A U G G A A C A C U G G C A C C G U U G C U G U
T G T G A C C G T G G C A A C G A C A

4

DNA polymerase synthesizes complementary DNA strand

↓

forming

↓

copy of human gene for insulin

DNA polymerase enzyme added – this synthesizes a second DNA strand, complementary to the base sequence of cDNA

cDNA being used as a template for DNA synthesis

T A C C T T G T G A C C G T G G C A A C G A C A
T G G C A C C G T T G C T G T

the two DNA strands are the **gene for insulin**

T A C C T T G T G A C C G T G G C A A C G A C A
A T G G A A C A C T G G C A C C G T T G C T G T

17 **Construct** a table to show the differences between RNA and DNA.

Inserting the DNA into a plasmid vector

In genetic engineering it is the **plasmids** of bacteria that are commonly used as the vector for the transference of amplified genes. Many bacterial cells have plasmids in addition to their chromosome. They are small, circular, DNA molecules that are passed on to daughter cells when the bacterium divides. They were seen in Figure 1.32 (page 29), and are shown in Figure 3.34. Plasmids have first to be isolated from the cytoplasm of a sample of the strain of bacteria being used for the amplification process.

Then the DNA of the plasmid is cut open using a **restriction enzyme** (restriction endonuclease). The restriction enzyme chosen cleaves the DNA at a specific sequence of bases, known as the 'restriction site'. The restriction enzyme selected leaves exposed specific DNA sequences, referred to as 'sticky ends'. These are short lengths of unpaired bases, and are formed at each cut end (Figure 3.34).

■ **Figure 3.34** Gene 'splicing' – the role of the restriction enzyme and ligase

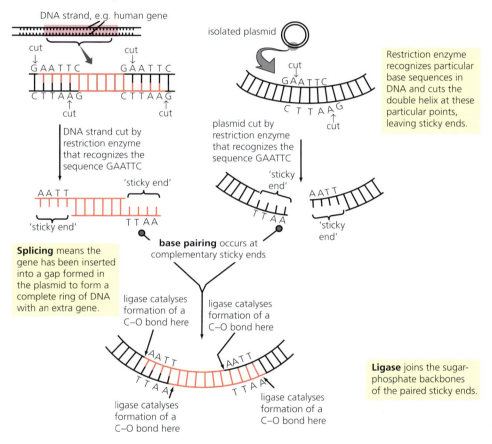

18 Explain
a what is meant by a 'sticky end'
b how one sticky end attaches to a complementary sticky end.

Now, copies of the same sticky-end sequence must be created on the insulin cDNA. This is done by addition of short lengths of DNA that are then 'trimmed' with the same restriction enzyme. In this way, complementary 'sticky ends' now exist at the free ends of the cut plasmids and the cDNA for insulin, making it possible for them to be 'spliced' together into one continuous ring of DNA. The enzyme **ligase** catalyses the formation of a new C–O bond between the ribose and phosphate of the DNA 'backbones' being joined together. Energy in the form of ATP is needed to bring the reaction about. The steps are summarized in Figure 3.35.

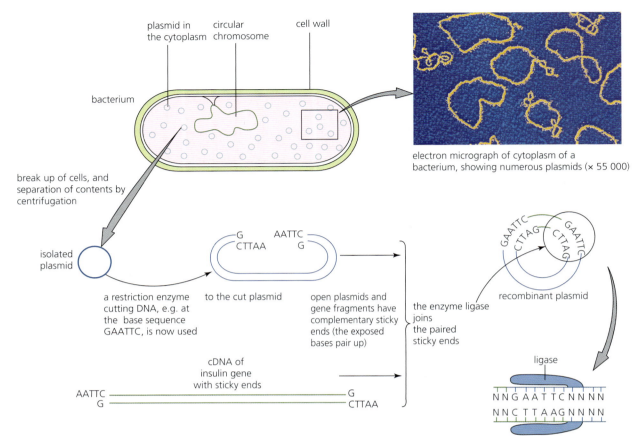

■ **Figure 3.35** Using a plasmid as a vector for the 'insulin' gene

A note on restriction enzymes

Restriction enzymes occur naturally in bacteria. They are believed to have evolved as a defence mechanism against invading viruses, for inside the bacterial cell they selectively cut up foreign nucleic acids. Many different restriction enzymes have been identified as useful by genetic engineers, and are named after the bacterium from which they are isolated (genus, species and strain, Table 3.9).

■ **Table 3.9** Some commonly employed restriction enzymes

Name	Source organism	Recognition sequence
EcoRI	*Escherichia coli*	GAATTC
BamH1	*Bacillus amyloliquefaciens*	GGATCC
HindIII	*Hemophilus influenza*	AAGCTT

Inserting the plasmid vector into the host bacterium

In the next step, recombinant plasmids are returned to bacterial cells. This is a challenge, since the cell wall is a barrier to entry. The ways in which this may be done are shown in Figure 3.36. Once this has been brought about, the bacteria are described as '**transformed**'.

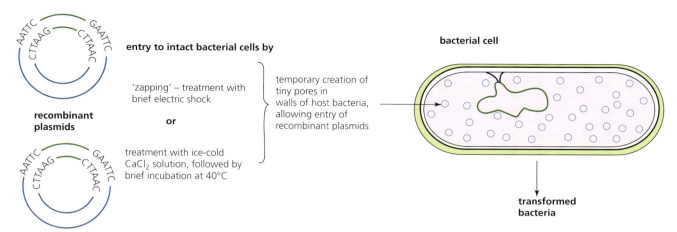

■ **Figure 3.36** The return of recombinant plasmids to the bacterium

Identifying transformed bacteria prior to cloning

Unfortunately, only a limited number of the cells of a culture of bacteria, treated as shown in Figure 3.36, successfully take up recombinant plasmids; the proportion of transformed bacteria may be as low as 1%. Only these bacteria will be able to synthesize insulin. It is these bacteria that must be cloned and then cultured in a fermenter, if a significant amount of product (insulin) is to be produced.

Successfully transformed bacteria may be selected by the use of a plasmid with **antibiotic resistance genes (R plasmids)**. The process is summarized in Figure 3.37. Here the transformed bacteria are cloned and then cultured in a fermenter. Note that, while the antibiotic resistance genes of R-plasmids were the first used 'markers', because of the long-term safety issues that their use raises, other markers have been developed. We return to this issue later.

A new development in insulin production

19 Suggest why the genes of prokaryotes are generally easier to modify than those of eukaryotes.

Insulin consists of two polypeptide chains, A and B, joined together by covalent disulfide bonds ($-CH_2-S-S-CH_2-$) (Figure 2.32, page 91). Two separate genes code for each of the polypeptides that make up insulin. As a consequence, separate plasmids had to be engineered, one for polypeptide A and one for B. Genetically engineered *E. coli* containing *both* types of plasmid were then selected. Finally, the two polypeptide chains were further treated to promote disulfide bond formation and development of the quaternary structure of insulin.

These complications are now avoided by the use of genetically engineered yeast cells for insulin production, rather than *E. coli*. (Remember, yeast is a eukaryotic cell with a range of organelles, but it also contains plasmids.) In a eukaryotic cell, proteins for secretion from the cell are manufactured by ribosomes of the RER, and transferred to the Golgi apparatus (Figure 1.27, page 25). Here, insulin is converted into its normal quaternary structure *prior* to discharge from the transformed yeast cells.

Alternative markers for genetic engineering

A marker is a gene which is transferred along with the required gene during the process of genetic engineering for recognition purposes; it is used to identify those cells to which the gene has been successfully transferred. In the production of insulin in *E. coli* the original markers were antibiotic resistance genes (Figure 3.37).

Now alternatives are preferred. This is because of the potential risk that the antibiotic resistance gene could be accidentally transferred to other bacteria, including pathogenic strains of *E. coli* or even other pathogens. The outcome would be an 'engineered' disease-causing microorganism that was resistant to one or more antibiotics.

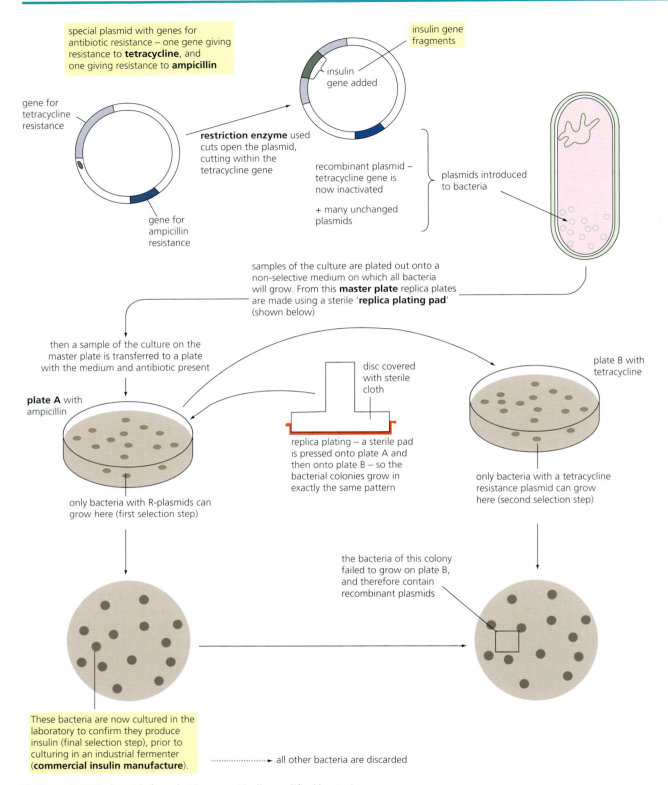

special plasmid with genes for antibiotic resistance – one gene giving resistance to **tetracycline**, and one giving resistance to **ampicillin**

gene for tetracycline resistance

gene for ampicillin resistance

insulin gene added

insulin gene fragments

restriction enzyme used cuts open the plasmid, cutting within the tetracycline gene

recombinant plasmid – tetracycline gene is now inactivated

+ many unchanged plasmids

plasmids introduced to bacteria

samples of the culture are plated out onto a non-selective medium on which all bacteria will grow. From this **master plate** replica plates are made using a sterile '**replica plating pad**' (shown below)

then a sample of the culture on the master plate is transferred to a plate with the medium and antibiotic present

disc covered with sterile cloth

plate B with tetracycline

plate A with ampicillin

replica plating – a sterile pad is pressed onto plate A and then onto plate B – so the bacterial colonies grow in exactly the same pattern

only bacteria with R-plasmids can grow here (first selection step)

only bacteria with a tetracycline resistance plasmid can grow here (second selection step)

the bacteria of this colony failed to grow on plate B, and therefore contain recombinant plasmids

These bacteria are now cultured in the laboratory to confirm they produce insulin (final selection step), prior to culturing in an industrial fermenter (**commercial insulin manufacture**).

all other bacteria are discarded

■ **Figure 3.37** R plasmids for selecting genetically modified bacteria

Genes that produce fluorescent substances are now used as markers (Figure 3.38) instead of antibiotic resistance genes, alongside the cDNA of the desired gene. Both genes are linked to a special promoter. The marker gene is expressed only when the desired gene has been successfully inserted into plasmids and these are present in transformed bacteria. Transformed bacterial colonies glow under UV light!

■ **Figure 3.38**
The use of fluorescent markers

The crystal jellyfish (*Aequorea victoria*) occurs in the plankton off the west coast of North America. It shows bright green fluorescence when exposed to light in the blue to ultraviolet range.

Fluorescence is due to the presence of a green fluorescent protein (GFP). This protein and the gene that codes for it have been extracted.

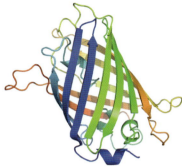

The tertiary structure of GFP protein

The GFP gene is now used as a marker gene in genetic engineering. For example, it may be inserted into plasmids alongside a cloned gene and its promoter. When these plasmids are returned to bacteria and the promoter is activated, the transformed bacteria will be detected by exposure to blue light.

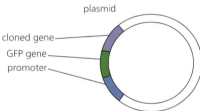

plasmid

cloned gene
GFP gene
promoter

The tertiary structure of GFP protein

The GFP gene has been introduced and maintained in the genome, and then expressed in species of bacteria, yeast, fish, plants, insects and mammals. For example, these GM mice glow green under blue light because the GFP gene has been introduced into their DNA.

Another method of isolating genes for genetic engineering

Converting mRNA coding for the insulin genes into cDNA, as described above, has the advantage of yielding copies of the genes *without* their short lengths of 'nonsense' DNA (known as introns). Within the cell, after transcription of the coding strand in the nucleus, editing out of the introns is a natural and essential stage. However, it is not easy for a genetic engineer to carry out this process *in vitro*.

20 **Distinguish** between these pairs:
   a genotype and genome
   b restriction endonuclease and ligase
   c a bacterial chromosome and a plasmid.

An alternative method is also used to generate genes. It is possible to synthesize a copy of a gene in a laboratory where the linear sequence of amino acids of the polypeptide or protein is known. The data in the genetic code give the sequence of nucleotides from which to construct a copy of the gene.

■ Producing transgenic eukaryotes

Transgenic organisms have genetic material introduced artificially from another organism. Manipulating genes in eukaryotes is a more difficult process than in prokaryotes. There are several reasons for this, including the following.

- Plasmids, the most useful vehicle for moving genes, do not occur in eukaryotes (except in yeasts) and, if introduced, do not always survive there to be replicated.
- Eukaryotes are diploid organisms, so two alleles for every gene are required to be engineered into the nucleus. By comparison, prokaryotes have a single, circular 'chromosome', so only one copy of the gene is required.

Producing transgenic animals

Despite the difficulties of engineering eukaryotic cells, several varieties of transgenic animal have been produced. For example, sheep have been successfully genetically modified to produce a special human blood protein, known as **AAT**. This protein enables us to **maintain lung elasticity** which is essential in breathing movements. Patients with a rare genetic disease are unable to manufacture AAT protein, and they develop emphysema (destruction of the walls of the air sacs, so that the lungs remain full of air during expiration).

21 **Explain** why a gene that is inserted into the nucleus of a fertilized egg cell is also passed to the progeny of the animal that forms from the zygote.

The chemical industry is unable to manufacture AAT protein in the laboratory on a practical scale. However, the human gene for AAT protein production has been identified and isolated, and it has been cloned into sheep, together with a promoter gene (a sheep's milk protein promoter) attached to it. Consequently the sheep's mammary glands produce the human protein and secrete it in their milk, during natural lactations. Thus, human AAT protein is made available for use with patients.

Producing transgenic plants

Many commercially valuable plant species have been genetically engineered and field trials undertaken. Transgenic flowering plants may be formed using tumour-forming *Agrobacterium*. This soil-inhabiting bacterium sometimes invades broad-leaved plants at the junction of stem and root, and there forms a huge growth called a tumour or crown gall. The gene for tumour formation occurs naturally in a plasmid in the bacterium, known as a Ti plasmid.

22 **Explain** why plants formed from the gall tissue will contain the recombinant gene that has been added to the Ti plasmid.

Useful genes may be added to the Ti plasmid, using a restriction enzyme and ligase, and the recombinant plasmid placed back into *Agrobacterium*. A host crop plant is then infected by the modified bacterium. The gall tissue that results may be cultured into independent plants, all of which also carry the useful gene. Commercially valuable plant species that have been genetically engineered (and field trials undertaken) include cotton, tobacco, oilseed rape, maize, potatoes, soya and tomatoes.

Resistance to insect attack

The bacterium *Bacillus thuringiensis* carries a gene which codes for a protein (*Bt* toxin) that is toxic to insects. This gene has been isolated and introduced into certain crop plants, including commercial maize. The toxin is present in the cells of these genetically engineered plants (*Bt* plants) in a harmless form. It requires the alkaline conditions of an insect's gut to be activated. (The highly acidic conditions in the stomach of vertebrates, for whom the crops are intended, destroys the protein toxin before it can be activated.) There is a question as to whether harmless insects, like the monarch butterfly, are at risk from *Bt* plants (Table 3.10).

The monarch butterfly under threat?

This beautiful butterfly hibernates along the coast of southern California and in fir forest in Mexico, but in spring it undertakes a spectacular migration (of about 3000 miles) in huge numbers, to its breeding ground in the vast American corn belt. Here it feeds on milkweed plants. The observed populations of Monarchs have been severely reduced in many years.

The case against *Bt* crops	*Bt*-transgenic crops are not the reason for population decline
Bt corn plants express the *Bt* gene in their pollen, as well as in the cells of their leaves. In laboratory conditions, Monarch caterpillars that are enclosed with milkweed plants heavily dusted with *Bt* corn pollen are seven times more likely to die than the controls.	To be at risk in the wild, the Monarch larvae would need to be present at the time corn pollen is released, and to be exposed to huge amounts of corn pollen on their food plants.
	Exceptional weather events in recent years have severely reduced the butterfly populations (an unseasonal snowstorm in 1995 killed 5–7 million Monarchs).
	Natural predators kill many – only 5% of Monarch caterpillars survive, even on standard corn crops.
	Human activities are the greatest threat – the butterfly's Mexican habitats are being destroyed by logging and their Californian habitats by urban development.

■ **Table 3.10**
The monarch butterfly, and the case for and against *Bt* toxin

Other beneficial improvements to crop plants

Consider the following benefits.

■ Tolerance to the herbicide glyphosate – the herbicide affects only the weeds and not the crop plant.

■ Rice plants that produce β-carotene (a precursor of vitamin A) – this vitamin is essential for vision in dim light, and a deficiency leads to blindness. More recently, it has also been shown that vitamin A is crucial to the functioning of the human immune system; a deficiency of vitamin A is the cause of many childhood deaths from such common infections as diarrhoeal disease and measles. Unmodified rice is deficient in β-carotene. For vast numbers of humans, rice is the staple crop.

23 Suggest what advantage would result from the eventual transfer of genes for nitrogen fixation from nodules in leguminous plants to cereals such as wheat.

Nature of Science | Assessing risks associated with scientific research

■ Benefits and hazards of gene technology

Genetically modified organisms can be produced for specific purposes, relatively quickly in some cases. They offer enormous benefits. However, geneticists are actually producing *new* organisms when genes are transferred, so there may be hazards as well as benefits (Table 3.11).

24 Occasional articles in the press make criticisms of genetic modification of food organisms, sometimes with alarmist headlines. Search the web for articles that give opposing views. **Construct** the following to use in a discussion with your peers:
a a concise summary of the criticisms that are frequently made, avoiding extremist language or unnecessary exaggeration
b a list of balancing arguments in favour of genetic modification of food organisms
c a concise statement of your own view on this issue.

TOK Link

The use of DNA for securing convictions in legal cases is well established, yet even universally accepted theories are overturned in the light of new evidence. What criteria are necessary for assessing the reliability of evidence?

The benefits so far	Hazards that have been anticipated
In addition to human insulin production by modified *E. coli*, human growth hormone and blood clotting Factor VIII have been produced by genetically modified bacteria.	Could a usually harmless organism, such as the human gut bacterium *E. coli*, be transformed into a harmful pathogen that escapes the laboratory and enters the population at large?
Sugar beet plants have been genetically modified to be 'tolerant' of glyphosate herbicide (these plants are able to inactivate the herbicide when it is sprayed over them, but the surrounding weeds are killed (Option B).	The accidental transfer of herbicide-resistance from crop plants, such as sugar beet, to a related wild plant is a possibility. It may then transmit to other plants, resulting in 'super weeds' whose spread would be hard to prevent.
Natural resistance to attack by chewing insects has resulted from the engineering of the genes for *Bt* toxin into crop plants (potato, cotton and maize), reducing the need for extensive aerial spraying of expensive insecticides, themselves harmful to wildlife.	Natural insecticidal toxin, engineered into crop plants, whilst an effective protection from browsing insects, might also harm pollinating species – bees and butterflies. Also, the presence of the *Bt* gene in the environment may lead to insects with resistance.
Bioremediation is the removal of toxic compounds, carcinogens and pollutants, such as industrial solvents, by microorganisms that are genetically modified to degrade these substances into safer molecules. The process is slow.	Genes move between bacterial populations with time. Could the antibiotic-resistance genes in plasmid vectors (page 172) become accidentally transferred into a pathogenic organism?
There are on-going attempts to use gene technology to treat genetic diseases, such as cystic fibrosis (page 161) and severe combined immunodeficiency (SCID). Some cancers may also yield to this approach, in the future.	There are well-publicized anxieties about food products that are made from genetically engineered species – might the food become toxic or trigger some allergic reactions? So far there has been no evidence to support these concerns.

DNA profiling (genetic fingerprinting)

The bulk of our DNA is not composed of 'genes' and so does not code for proteins. These extensive 'non-gene' regions include short sequences of bases, repeated many times. While some of these sequences are scattered throughout the length of the DNA molecule, many are joined together in major clusters. It is these major lengths of the non-coding, 'nonsense' DNA that are used in genetic profiling. They may be referred to as 'satellite' and 'microsatellite' DNA, but, alternatively, they are known as '**variable number tandems repeats**' (**VNTRs**).

We inherit a distinctive combination of these apparently non-functional VNTRs, half from our mother and half from our father. Consequently, each of us has a unique sequence of nucleotides in our DNA (except for identical twins, who share the same pattern).

To produce a genetic 'fingerprint', a sample of DNA is cut with a restriction enzyme which acts close to the VNTR regions. Electrophoresis is then used to separate pieces according to length and size, and the result is a pattern of bands (Figure 3.39). The steps to DNA fingerprinting are listed in Table 3.12.

	To produce a DNA 'fingerprint' or profile
1	A sample of cells is obtained from blood, semen, hair root or body tissues, and the DNA is extracted. (Where a tiny quantity of DNA is all that can be recovered, this is precisely copied by means of the **polymerase chain reaction**, page 167.)
2	The amplified DNA is cut into small, double-stranded fragments using a particular **restriction enzyme**, chosen because it 'cuts' close to but not within the satellite DNA.
3	The resulting DNA fragments are of varying lengths and are separated by **gel electrophoresis** into bands (invisible at this stage, Figure 3.39).
	Subsequent steps include 'Southern blotting' and are illustrated in Figure 3.39
4	The gel is treated to split the DNA into single strands and then a copy is transferred to a membrane.
5	Selected, radioactively labelled DNA probes are added to the membrane to bind to particular bands of DNA, and then the excess probes are washed away. Alternatively, the probes may be labelled with a fluorescent stain which shows up under UV light.
6	The membrane is now overlaid with X-ray film which becomes selectively 'fogged' by emission from the retained labelled probes.
7	The X-ray film is developed, showing up the positions of the bands (fragments) to which probes have attached. The result is a profile with the appearance of a 'bar code'.

Southern blotting (named after the scientist who devised the routine):
• extracted DNA is cut into fragments with restriction enzyme
• the fragments are separated by electrophoresis
• fragments are made single-stranded by treatment of the gel with alkali.

1 Then a copy of the distributed DNA fragments is produced on nylon membrane:

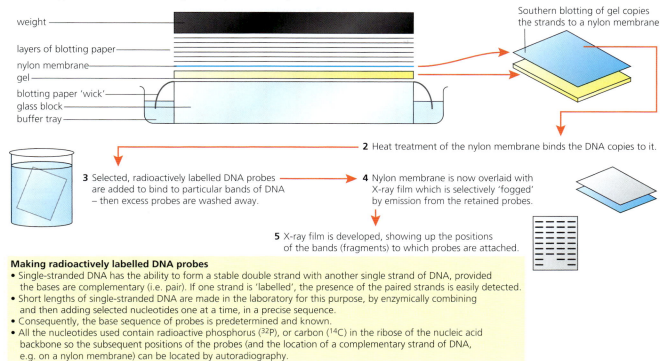

weight

layers of blotting paper

nylon membrane

gel

blotting paper 'wick'

glass block

buffer tray

Southern blotting of gel copies the strands to a nylon membrane

2 Heat treatment of the nylon membrane binds the DNA copies to it.

3 Selected, radioactively labelled DNA probes are added to bind to particular bands of DNA – then excess probes are washed away.

4 Nylon membrane is now overlaid with X-ray film which is selectively 'fogged' by emission from the retained probes.

5 X-ray film is developed, showing up the positions of the bands (fragments) to which probes are attached.

Making radioactively labelled DNA probes
• Single-stranded DNA has the ability to form a stable double strand with another single strand of DNA, provided the bases are complementary (i.e. pair). If one strand is 'labelled', the presence of the paired strands is easily detected.
• Short lengths of single-stranded DNA are made in the laboratory for this purpose, by enzymically combining and then adding selected nucleotides one at a time, in a precise sequence.
• Consequently, the base sequence of probes is predetermined and known.
• All the nucleotides used contain radioactive phosphorus (^{32}P), or carbon (^{14}C) in the ribose of the nucleic acid backbone so the subsequent positions of the probes (and the location of a complementary strand of DNA, e.g. on a nylon membrane) can be located by autoradiography.

What a probe is and how it works

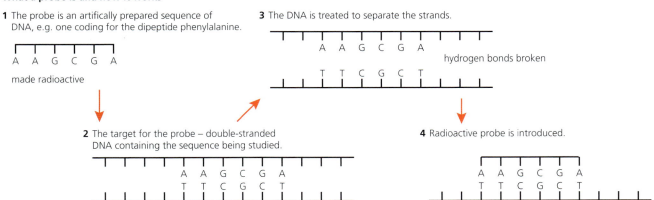

1 The probe is an artifically prepared sequence of DNA, e.g. one coding for the dipeptide phenylalanine.

A   A   G   C   G   A

made radioactive

2 The target for the probe – double-stranded DNA containing the sequence being studied.

A   A   G   C   G   A
T   T   C   G   C   T

3 The DNA is treated to separate the strands.

A   A   G   C   G   A

hydrogen bonds broken

T   T   C   G   C   T

4 Radioactive probe is introduced.

A   A   G   C   G   A
T   T   C   G   C   T

■ **Figure 3.39** Steps to genetic profiling

Applications of genetic fingerprinting

Applications in forensic science include the following.

■ **Identification of suspects** – samples are taken from the scenes of serious and violent crimes, such as rape attacks. DNA from both victims and suspects, as well as from others who have certainly not been involved in the crime (used as control samples), is profiled. The greatest care is required; there must be no possibility of cross-contamination if the outcome of testing is to be meaningful.

■ **Identification of corpses** – bodies that are otherwise too decomposed for recognition may be identified, as might parts of the body remaining after bomb blasts or other violent incidents, including natural disasters.

Other applications include establishing paternity and conservation:

> **25 Explain** why the composition of the DNA of identical twins challenges an underlying assumption of DNA fingerprinting, but that of non-identical twins does not.

- **Determining paternity** – a range of samples of DNA are analysed side by side, of the people who are possibly related. The banding patterns are then compared (Figure 3.40). Because a child inherits half its DNA from its mother and half from its father, the bands in a child's DNA fingerprint that do not match its mother's must come from the child's father.
- **Species conservation** – it is possible to determine whether individuals of endangered species in captivity have been bred there or were captured in the wild. The technique has wide applications in studies of wild animals, for example, concerning breeding behaviour, and in the identification of unrelated animals as mates in captive breeding programmes of animals in danger of extinction (Option C).

DNA profiling in forensic investigation

Identification of criminals

At the scene of a crime (such as a murder), hairs – with hair root cells attached – or blood may be recovered. If so, the resulting DNA profiles may be compared with those of DNA obtained from suspects.

Examine the DNA profiles shown to the right, and suggest which suspect should be interviewed further.

Identification in a rape crime involves the taking of vaginal swabs. Here, DNA will be present from the victim and also from the rapist. The result of DNA analysis is a complex profile that requires careful comparison with the DNA profiles of the victim and of any suspects. A rapist can be identified with a high degree of certainty, and the innocence of others established.

Identification of a corpse which is otherwise unidentifiable is achieved by taking DNA samples from body tissues and comparing their profile with those of close relatives or with DNA obtained from cells recovered from personal effects, where these are available.

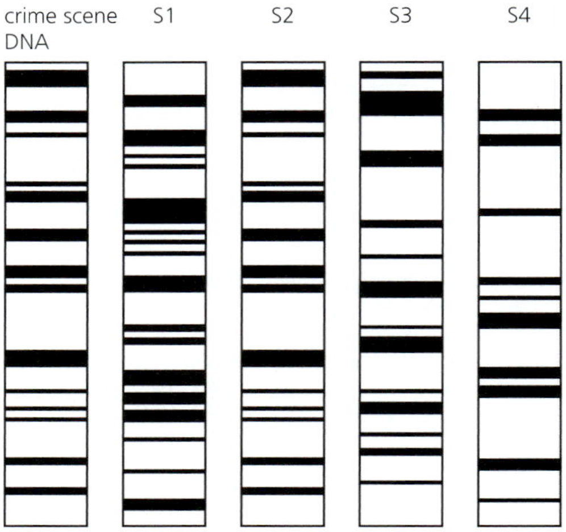

DNA profiles used to establish family relationships

Is the male (M) the parent of both children?

Examine the DNA profiles shown to the right.

Look at the children's bands (C).

Discount all those bands that correspond to bands in the mother's profile (F).

The remaining bands match those of the biological father.

DNA fingerprinting has also been widely applied in biology. In ornithology, for example, DNA profiling of nestlings has established a degree of 'promiscuity' in breeding pairs, the male of which was assumed to be the father of the whole brood. In birds, the production of a clutch of eggs is extended over a period of days, with copulation and fertilization preceding the laying of each egg. This provides the opportunity for different males to fertilize the female.

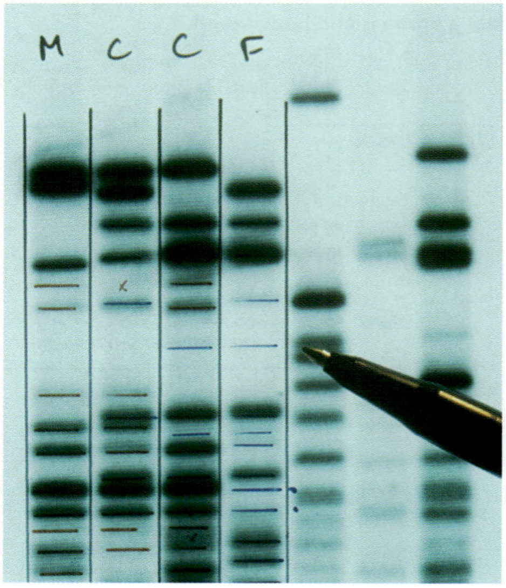

■ **Figure 3.40** DNA profiles used to investigate relatedness

■ Clones

Clones are identical things – a group of genetically identical organisms are clones. So too are a group of cells derived from a single parent cell.

Clones occur naturally. For example, they are formed by asexual reproduction in plants. Cloning has been used widely in commercial plant propagation for many years. You will be carrying out cloning when you investigate the rooting of stem cuttings (Figure 3.41).

What will you investigate?
Formulate your hypothesis, for example:
- plants need oxygen for root growth.
- hormone rooting powder promotes root growth.

Select a plant
Most readily available plants are suitable, such as:
- willow (*Salix* sp)
- honeysuckle (*Lonicera* sp.)
- blackberry (*Rubis* sp).

Selecting and preparing the shoots
How many will you need for treatment and control?

shoots selected of similar, standard length and number of leaves

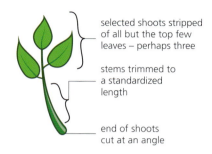

selected shoots stripped of all but the top few leaves – perhaps three

stems trimmed to a standardized length

end of shoots cut at an angle

Possible experimental apparatus

Rooting in a solid medium	Rooting in water
The effect of rooting powder:	**The requirement for oxygen:**

inert medium (vermiculite or perlite) moistened with water or dilute culture solution added

cut stem treated with hormone rooting powder

cut stem untreated

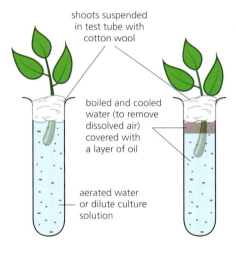

shoots suspended in test tube with cotton wool

boiled and cooled water (to remove dissolved air) covered with a layer of oil

aerated water or dilute culture solution

Conduct and management issues:
What environment and control will the experimental apparatus be held in?
What external conditions can and must be controlled?
Time frame – how long will the experimental run for?
When and how will observations and measurements be taken and recorded?

Finally – comclusion and points for discussion:
What changes in method would you recommend if the experimental were to be repeated?
What and where in the stem are the tissue(s) that have generated roots?
What are the roles of roots in the growth of the plants?
In what sense can the products of your experiment be described as 'clones'?

■ **Figure 3.41**
Investigating the rooting of stem cuttings

- Humans that are identical twins are clones. About 1 in every 270 pregnancies in the UK produces identical twins. This results when the first-formed cells of the blastula (Figure 11.48, page 491) divide into separate groups prior to implantation.

- Among vertebrates asexual reproduction does not occur, but cloning is common among some species of non-vertebrates. Examples are the individuals produced by budding in hydra.

- Animals can be cloned at the early embryo stage by breaking up the embryo into individual cells. These stem cells are still capable of developing into all the tissue types of the adult organism (a condition known as **pluripotency**).

Steps in the production of Dolly the clone

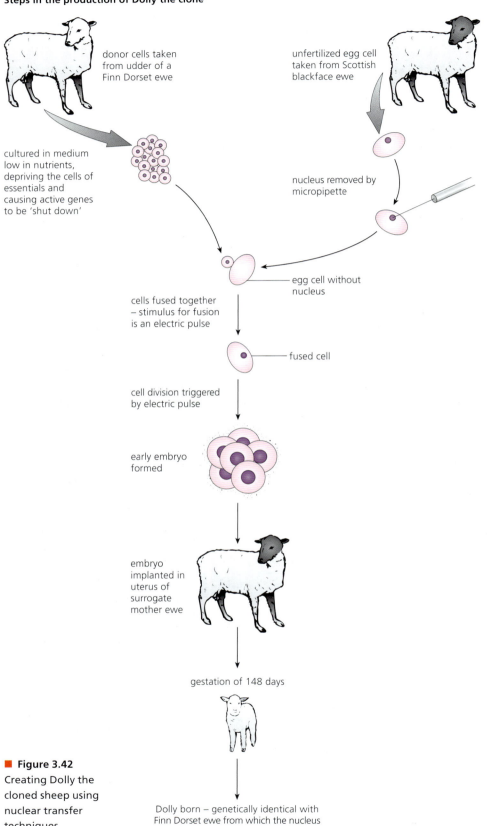

donor cells taken from udder of a Finn Dorset ewe

unfertilized egg cell taken from Scottish blackface ewe

cultured in medium low in nutrients, depriving the cells of essentials and causing active genes to be 'shut down'

nucleus removed by micropipette

egg cell without nucleus

cells fused together – stimulus for fusion is an electric pulse

fused cell

cell division triggered by electric pulse

early embryo formed

embryo implanted in uterus of surrogate mother ewe

gestation of 148 days

■ **Figure 3.42**
Creating Dolly the cloned sheep using nuclear transfer techniques

Dolly born – genetically identical with Finn Dorset ewe from which the nucleus was taken (see genetic profile)

other ewes

cells from Finn Dorset (udder cells used for cloning)

cells from Dolly

other ewes

1 2 U C D 3 4

Genetic profile of Finn Dorset ewe cells, Dolly's cells and other ewes

■ Genetic engineers clone genes when a single gene is introduced into a plasmid and then bacteria are induced to take up the recombinant plasmid (Figures 3.34 and 3.35). Once this has occurred, the plasmid is replicated many times in the cytoplasm. In this way, very many identical copies of the gene are produced. This **molecular cloning** is an important step in recombinant DNA technology.

■ In recent years, animal clones have been produced by nuclear transfer techniques. Dolly the sheep was the first mammal to be cloned from non-embryonic cells. This was achieved at the Roslin Institute, Edinburgh in 1996. Dolly was produced from a fully differentiated udder cell taken from a six-year-old ewe. The isolated cell was induced to become 'dormant' or genetically quiescent, and then converted to the embryonic state by fusion with an egg cell from which the nucleus had been removed, taken from a different ewe. The process, known as **somatic-cell transfer**, is summarized in Figure 3.42.

Human cloning?

Preliminary experiments have shown that it may be possible to clone humans, too. However, the ethical issues of such a move have stimulated opposition and comment from many people and organizations, and human cloning is currently banned in many countries. In 2004, the UK's then Human Fertilization and Embryology Authority (HFEA) gave a Newcastle biomedical team permission to create human embryos that were clones of patients. They also used the somatic cell transfer technique.

The embryos created in these Newcastle experiments could not be implanted in a uterus because reproductive cloning is illegal in the UK. Instead, the team work on embryonic **stem cells** derived from the embryos.

We have seen that stem cells have the ability to produce a variety of cells (page 15). It is thought that these cells may have great potential for treating disease by replacing damaged tissue. The Newcastle team was investigating possible cures for degenerative diseases such as multiple sclerosis, and Alzheimer's and Parkinson's diseases. Unlike an organ transplant, stem cells are not expected to be rejected by the patient's body, as the process ensures the engineered cells have the same DNA as the patient.

Since human clones can form naturally, and because experiments in human cloning are being planned, permitted and funded in several countries around the world, this controversial issue needs our careful, informed consideration. Arguments for and against human cloning are listed in Table 3.13.

What views do you hold?

Points against human cloning	Points in favour of human cloning
Human beings might be planned and produced with the sole intention of supplying 'spare parts' for a related person with a health problem.	Parents at high risk of producing offspring with genetic disease would have the opportunity of healthy children.
	Infertile couples could have children of their own.
Cloning could facilitate the process of 'improving' humans, by designing and delivering a race of 'superior' people (in the same way as attempts were made to 'improve' humans by the arbitrary standards adopted by now discredited eugenics movements).	Cloning technologies are being developed to deliver organs for transplants that are entirely compatible and are not rejected by the recipient's immune system.
	New treatments for genetic diseases are planned and may shortly be achieved.
Cloning techniques are experimental and unreliable, possibly resulting in the death of many embryos and newborns – there are still so many unknown factors operating.	Cloning techniques are as safe and reliable as other comparable medical procedures, given this early stage in their development and our limited experiences.
The traditional concept of 'family' is of a group of people with individuality, and with a clear sense of personal worth.	Clones are not true duplicates because environmental factors and personal experiences influence development and who we are.
Clones might have diminished rights and a lessened sense of individuality.	Identical twins have a strong sense of individuality and personal worth.
Some aspects of human life should exist above the values and standards of the laboratory.	It takes time for all new developments to be accepted.
Many believers consider that cloning is against the will of their god.	Most improvements in medical technologies have received trenchant opposition which has receded with familiarity, and as the advantages are perceived.
Human inventiveness must not extend to tampering with 'nature' in respect of human life issues.	

■ **Table 3.13** The cases for and against human cloning

■ *Examination questions – a selection*

Questions 1–4 are taken from IB Diploma biology papers.

Q1 What is chorionic villus sampling?

 A Sampling cells from the placenta
 B Sampling cells from the fetal digestive system
 C Sampling fetal cells from the amniotic fluid
 D Sampling stem cells from the umbilical cord

 Standard level Paper 1, Time Zone 0, Nov 10, Q15

Q2 What would be the expected result if a woman carrier for colour blindness and a colour-blind man had many children?

 A All offspring will be colour blind.
 B All male offspring will be colour blind and all females normal.
 C All males will be normal and all females will be colour blind.
 D All females will be carriers of colour blindness or colour blind.

 Standard Level Paper 1, Time Zone 0, Nov 12, Q14

Q3 What information can be concluded from the karyogram?

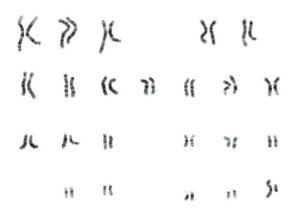

 A The person is a normal male.
 B The person is a normal female.
 C The person is a male with Down syndrome.
 D The person is a female with Down syndrome.

 Standard Level Paper 1, Time Zone 1, May 13, Q16

Q4 Clouded leopards live in tropical rainforests of South-East Asia. The normal spots (brown with a black outline) are dominant and black spots are recessive. The trait is sex-linked. A male with black spots was crossed with a female with normal spots. She had four cubs, two males and two females. For each sex, one cub had normal spots and the other cub had black spots. Deduce the genotype of the mother. Show your work in a Punnett grid. (3)

 Standard Level Paper 2, Time Zone 2, May 13, Question 2b

Questions 5–10 cover other syllabus issues in this chapter.

Q5 What was achieved by the Human Genome Project?

 A The sequencing of only the genes and their loci in humans.
 B The sequencing of chromosomes that make up somatic homologous pairs in meiosis.
 C The sequencing of the entire base sequence for human genes.
 D All the genes presented in an individual which are different to others.

Q6 **a** List the chromosomes features used to create a karyogram. (2)
 b Name two techniques used to obtain sample cells to obtain the karyogram. (2)
 c Identify the name of the process that leads to Down's syndrome during meiosis. (1)
 d Outline two processes that occur in meiosis and are not present in mitosis. (4)

Q7 Mendel carried out a breeding experiment with garden pea plants in which pure-breeding pea plants grown from seeds with a smooth coat were crossed with plants grown from seeds with a wrinkled coat. All the seeds produced (called the F1 generation) were found to have a smooth coat. When plants were grown from these seeds and allowed to self-pollinate, the second generation of seeds (the F2 generation) included both smooth and wrinkled seeds in the ratio of 3:1.

a In Mendel's explanation of his results he used the term 'hereditary factor'. Identify our term for 'factors' and state where in the cell they occur. (2)

b The parent plants have diploid cells, while the gametes (sex cells) are haploid. Define what we mean by *haploid* and *diploid*. (4)

c By reference to the above experiment, define the following terms and give an example of each:
  i   *homozygous* and *heterozygous* (4)
  ii *dominant* and *recessive* (4)
  iii *genotype* and *phenotype*. (4)

d Using appropriate symbols for the alleles for smooth and wrinkled coat, construct a genetic diagram (including a Punnett grid) to show the behaviour of the alleles in this experiment. (6)

Q8 a Define the term *mutation*. (2)

  b By means of specific examples, explain the difference between chromosome mutation and gene mutation. (4)

Q9 The following are sequences of amino acids from the same section belonging to a transmembrane protein found in *Pongo abelii* (orangutan), *Homo sapiens* (humans) and *Canis lupus familiaris* (dog).

Pongo abelii	NYF————————SLIFNVILFFKMKFLFWS FKSVATKI
Homo sapiens	MFLQWVLCAAI WLVALVVNLILH CPK–––FWPFAMLGGCI
Canis lupus familiaris	MFLQWVLCAAI WSIALVVNLILH CPK–––FWPFAMVGGCI

  a Describe how scientists use the differences in the sequences of amino acids to find out evolutionary relationship among species. (2)

  b Orangutans are more closely related to humans than dogs. Explain why this section of the protein shows a bigger difference between orangutans and humans than the differences between dogs and humans. (4)

Q10 a Identify what is achieved by gel electrophoresis and PCR techniques. (2)

  b Define: *genetic modification*, *clone* and *DNA profiling*. (3)

  c Outline the potential risks and benefits of using a named genetic modified crop. (8)

4 Ecology

ESSENTIAL IDEAS

- The continued survival of living organisms including humans depends on sustainable communities.
- Ecosystems require a continuous supply of energy to fuel life processes and to replace energy lost as heat.
- Continued availability of carbon in ecosystems depends on carbon cycling.
- Concentrations of gases in the atmosphere affect the climates experienced at the Earth's surface.

Ecology is the study of living things in their environment. It is an essential component of modern biology. Understanding the relationships between organisms and their environment is just as important as knowing about the structure and physiology of animals and plants. **Environment** is a term we commonly use to mean 'surroundings'. In biology, we talk about the environment of cells in an organism, or the environment in which the whole organism lives. So 'environment' is a rather general, unspecific term – but useful, nonetheless.

4.1 Species, communities and ecosystems – *the continued survival of living organisms including humans depends on sustainable communities*

■ Species

There are vast numbers of different types of living organism in the world – almost unlimited diversity, in fact. Up to now, about 2 million species have been described and named in total. *But what we mean by 'species'?*

By the term 'species' we refer scientifically to a particular type of living thing. A species is a group of individuals of common ancestry that closely resemble each other and are normally capable of interbreeding to produce fertile offspring.

There are three issues to bear in mind about this definition.

- Some (very successful) species reproduce asexually, without any interbreeding at all. Organisms that reproduce asexually are very similar in structure, showing little variation between individuals.

- Occasionally, members of different species breed together. However, where such cross-breeding occurs, the offspring are almost always infertile.

- Species change with time; new species evolve from other species. The fact that species do change means they are not constant always easy to define. However, evolutionary change takes place over a long period of time. On a day-to-day basis, the term 'species' is satisfactory and useful.

So, a species is a group of organisms that is reproductively isolated, interbreeding to produce fertile offspring. Organisms belonging to a species have morphological (structural) similarities, which are often used to define the species.

■ Populations, communities and ecosystems

Members of a species may be reproductively isolated in separate populations. A **population** consists of all the individuals of the same species in a habitat at any one time. The members of a population have the chance to interbreed, assuming the species concerned reproduces sexually. The boundaries of populations are often hard to define, but those of aquatic organisms living in a small pond are clearly limited by the boundary of the pond (Figure 4.1).

A **community** consists of all the living things in a habitat – the total of all the populations, in fact. So, for example, the community of a well-stocked pond would include the populations of rooted, floating and submerged plants, the populations of bottom-living animals, the populations of fish and non-vertebrates of the open water, and the populations of surface-living organisms – a very large number of organisms, in fact.

Finally, a community forms an **ecosystem** by its interactions with the non-living (abiotic) environment. An ecosystem is defined as a community of organisms and their surroundings, the environment in which they live and with which they interact. An ecosystem is a basic functional unit of ecology, since the organisms that make up a community cannot realistically be considered apart from their physical environment.

Examples such as a woodland or a lake illustrate two important features of an ecosystem, namely that it is:

- a largely **self-contained** unit, since most organisms of the ecosystem spend their entire lives there and their essential nutrients are endlessly recycled around and through it

- an **interactive** system, in that the kinds of organism that live there are largely decided by the physical environment and in that the physical environment is constantly altered by the organisms.

The organisms of an ecosystem are called the **biotic** component, and the physical environment is known as the **abiotic** component.

Within any ecosystem, organisms are normally found in a particular part or habitat. The **habitat** is the locality in which an organism occurs. So, for example, within a woodland, the tree canopy is the habitat of some species of insects and birds, while other organisms occur in the soil. Within a lake, habitats might include a reed swap and open water. Incidentally, if the occupied area is extremely small, we call it a **microhabitat**. The insects that inhabit the crevices in the bark of a tree are in their own microhabitat. Conditions in a microhabitat are likely to be very different from conditions in the surrounding habitat.

■ The pond or lake as an ecosystem

Figure 4.1 represents a **transect** through a fresh water ecosystem – a pond or lake. Notice that a range of habitats within the ecosystem are identified, and that the feeding relationships of the community of different organisms are highlighted. We will consider feeding relationships between organisms next.

■ **Figure 4.1**
A pond or lake as an ecosystem

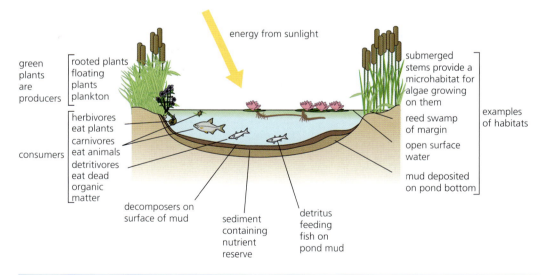

1 **Apply** one or more of the terms shown below to describe each of the listed features of a fresh water lake.

population   ecosystem   habitat   abiotic factor   community   biomass

a the whole lake
b all the frogs of the lake
c the flow of water through the lake
d all the plants and animals present
e the total mass of vegetation growing in the lake
f the mud of the lake
g the temperature variations in the lake

■ Feeding relationships – producers, consumers and decomposers

Think of an ecosystem with which you are very familiar. Perhaps it is one near your home, school or college. It might be savannah, a forest, a lake, woodland or meadow. Whatever you have in mind, it will certainly contain a community of plants, animals and microorganisms, all engaged in their characteristic activities. Some of these organisms will be much easier to observe than others, possibly because of their size, or the times of day (or night) at which they feed, for example.

The essence of survival of organisms is their activity. To carry out their activities organisms need **energy**. We have already seen that the immediate source of energy in cells is the molecule **ATP** (page 115), which is produced by respiration. The energy of ATP has been transferred from sugar and other organic molecules – the respiratory substrates. These organic molecules are obtained from nutrients as a result of the organism's mode of nutrition.

We know that green plants make their own organic nutrients from an external supply of inorganic molecules, using energy from sunlight in photosynthesis (page 121). The nutrition of a typical green plant is described as **autotrophic** (meaning 'self-feeding') and, in ecology, green plants are known as **producers**. There are a very few exceptions to this (Figure 4.2).

An autotroph is an organism that synthesizes its organic molecules from simple inorganic substances

Nature of Science

Looking for patterns, trends and discrepancies

Broomrape (*Orobanche* sp.) is a 'root parasite', attaching to the root systems of its various host plants, below ground. Above ground, the shoots are virtually colourless (chlorophyll-free), and the leaves reduced to small bracts. *Why*?
Once established, the plant is seen to concentrate on reproduction, seed production and seed dispersal. This suggests that the task of reaching fresh hosts is a major challenge in the life-cycle of a parasite. *Which of these features are evident in the plant shown here*?

■ **Figure 4.2** Not quite all green plants are autotrophic

■ Classifying a species from a knowledge of its mode of nutrition

1 Autotrophs versus consumers

So, the great majority of green plants are entirely autotrophic in their nutrition. In this they play a key part in food chains, as we shall shortly see. In contrast to green plants, animals and most other types of organism use only existing nutrients, which they obtain by digestion and then absorption into their cells and tissues for use. Consequently, animal nutrition is dependent on plant nutrition, either directly or indirectly. In ecology, animals are known as **consumers** and animal nutrition is described as **heterotrophic** (meaning 'other nutrition').

■ A heterotroph is an organism that obtains organic molecules from other organisms.

■ A consumer is an organism that ingests other organic matter that is living or recently killed.

■ **Figure 4.3**
False-colour micrograph of *Euglena*, a species that is both autotrophic and heterotrophic

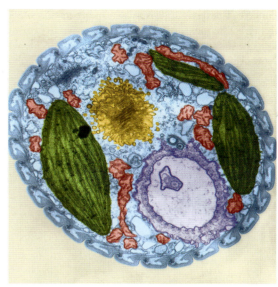

In **heterotrophic nutrition**, bacteria are taken into food vacuoles by phagocytosis and the contents digested by hydrolytic enzymes from lysosomes. *Find the Golgi apparatus and lysosomes in the cytosol.*

In **autotrophic nutrition**, photosynthesis occurs in the chloroplasts. There is a light-sensitive 'eyespot' present which enables *Euglena* to detect the light source.

Notice the plasma membrane has a ridged appearance here – this arrangement is supported by a system of microtubules below.

Note that some of the consumers, known as **herbivores**, feed directly and exclusively on plants. Herbivores are **primary consumers**. Animals that feed exclusively on other animals are **carnivores**. Carnivores that feed on primary consumers are known as **secondary consumers**. Carnivores that feed on secondary consumers are called **tertiary consumers**, and so on.

2 Detritivores and saprotrophs

Eventually, all producers and consumers die and decay. Organisms that feed on dead plants and animals, and on the waste matter of animals, are described as **saprotrophs** (meaning 'putrid feeding') and, in ecology, these feeders are known as **detritivores** or **decomposers**.

- ■ **A saprotroph is an organism that lives on or in dead organic matter, secreting digestive enzymes into it and absorbing the products of digestion.**

- ■ **A detritivore or a decomposer is an organism that ingests dead organic matter.**

2 **Construct** a dichotomous key in the form of a flow chart, classifying species on the basis of their alternative modes of nutrition.

Feeding by saprotrophs releases inorganic nutrients from the dead organic matter, including carbon dioxide, water, ammonia, and ions such as nitrates and phosphates. Sooner or later, these inorganic nutrients are absorbed by green plants and reused. We will look in more detail at the cycling of nutrients in the **biosphere** later in this chapter.

■ Practical ecology: Testing for associations between species

The distribution of two or more species in a habitat may be entirely random. Alternatively, factors such as specific abiotic conditions may bring about close association of some species – plant A may tend to grow close to plant B. For example, soils rich in calcium ions typically support distinctively different populations from those found on dry acid soils. If we want to discover whether there is a particular association between two species in a habitat, we need reliable data on their distribution; this is obtained by **random sampling**. In this way, every individual in the community has an equal chance of being selected and so a representative sample is assured.

Quadrats are commonly used to study populations and communities. A quadrat is a square frame which outlines a known area for the purpose of sampling. The choice of size of quadrat varies depending on the size of the individuals of the population being analysed. For example, a 10 cm² quadrat is ideal for assessing epiphytic *Pleurococcus*, a single-celled alga, commonly found growing on damp walls and tree trunks. Alternatively, a 1 m² quadrat is far more useful for analysing the size of two herbaceous plant populations observed in grassland, or of the earthworms and the slugs that can be extracted from between the plants or from the soil below.

Quadrats are placed according to **random numbers**, after the area has been divided into a grid of numbered sampling squares (Figure 4.4). The presence or absence in each quadrat of the two species under investigations is then recorded. The data is then subjected to statistical test. The chi-squared (χ^2) test is used to examine data that falls into discrete categories – as it does in this case. It tests the significance of the deviations between numbers observed (**O**) in an investigation and the number expected (**E**). The measure of deviation, known as chi-squared, is converted into a probability value using a chi-squared table. In this way, we can decide whether the differences observed between our sets of data are likely to be real or, alternatively, obtained just by chance.

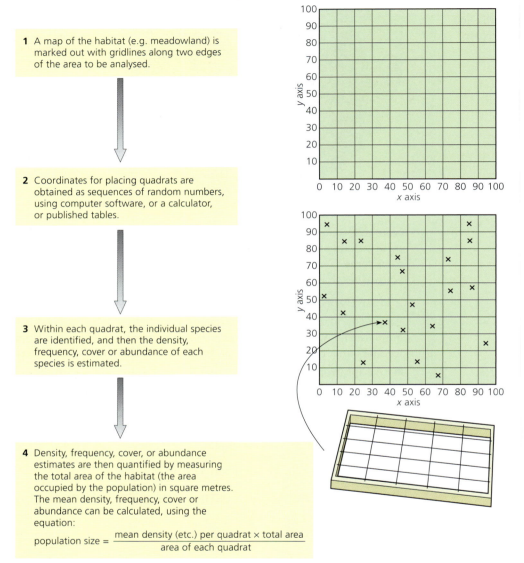

■ **Figure 4.4** Random locating of quadrats

1 A map of the habitat (e.g. meadowland) is marked out with gridlines along two edges of the area to be analysed.

2 Coordinates for placing quadrats are obtained as sequences of random numbers, using computer software, or a calculator, or published tables.

3 Within each quadrat, the individual species are identified, and then the density, frequency, cover or abundance of each species is estimated.

4 Density, frequency, cover, or abundance estimates are then quantified by measuring the total area of the habitat (the area occupied by the population) in square metres. The mean density, frequency, cover or abundance can be calculated, using the equation:

$$\text{population size} = \frac{\text{mean density (etc.) per quadrat} \times \text{total area}}{\text{area of each quadrat}}$$

■ Recognizing and interpreting statistical significance

Two moorland species and the chi-squared test

This example examines whether the moorland species bell heather (*Erica onerea*) and common heather, also known as ling (*Calluna vulgaris*), tend to occur together. Moorlands are upland areas with acidic and low-nutrient soils, where heather plants dominate. Heathers have long woody stems and grow in dense clumps. They have colourful, bright flowers. The question here is whether there is a statistically significant association between ling and bell heather on an area of moorland. As scientists we would carry out a statistical test to work out the probability of getting results that indicate there is *no* association between the two species – indicating the **null hypothesis** is true. The null hypothesis in this example would be that there is no statistically significant association between bell heather and ling in an area of moorland; that is, their distributions are independent of each other. If our results do not support the null hypothesis, then there is an association.

Ling *Calluna vulgaris*

Bell heather *Erica tetralix*

■ **Figure 4.5** A moorland ecosystem and two common plants found there

1 The measurements and results

In order to sample the two species, the presence or absence of each species was recorded in each of 200 quadrats. The quadrats were located at random on a 100 m by 100 m area of moorland (Table 4.1).

■ **Table 4.1** Observed results – the distribution of ling and bell heather

	Bell heather present	Bell heather absent	Total
Ling present	89	45	134
Ling absent	31	35	66
	120	80	200

2 The calculations

a **Expected results**: assuming that the two species are randomly distributed with respect to each other, the probability of ling being present in a quadrat is:

column total/total number of quadrats

$$= \frac{134}{200}$$
$$= 0.67$$

Similarly, the probability of bell heather being present in a quadrat is:

$$\frac{120}{200}$$
$$= 0.60$$

The probability of both species occurring together, assuming random distribution between each species, is: $0.60 \times 0.67 = 0.40$. The number of quadrats in which both species can be expected is therefore $0.40 \times 200 = \mathbf{80}$.

Having worked out the number of expected quadrats where the species are found together, other expected values can be calculated by subtracting from the totals. For example, the expected number of quadrats with bell heather but no ling is $120 - 80 = 40$. Expected values follow the assumption that totals for each row and column do not change, because the relationship shown by the data is assumed to represent the true relative frequency of each species (Table 4.2).

■ **Table 4.2** The full expected results

	Bell heather present	Bell heather absent	Total
Ling present	80	54	134
Ling absent	40	26	66
	120	80	200

Now the calculated values can be checked by using the ratios represented in the table of observed results (Table 4.1).

For example, the expected number of quadrats where there is no ling and no bell heather can be calculated as follows:

Probability of no ling in a quadrat $= \dfrac{66}{200} = 0.33$

Probability of no bell heather in a quadrat $= \dfrac{80}{200} = 0.40$

Probability of neither species in a quadrat $= 0.33 \times 0.40 = 0.13$

Number of expected quadrats with neither species present $= 0.13 \times 200 = 26$

(Note that this figure agrees with the estimated value in Table 4.2).

b **Statistical test:** the observed and expected results are recorded in Table 4.3.

■ **Table 4.3**
Observed (O) and expected (E) distribution of ling and bell heather

		Bell heather present	Bell heather absent	Total
Ling present	O	89	45	134
	E	80	54	
Ling absent	O	31	35	66
	E	40	26	
		120	80	200

Then, chi squared is calculated from the formula:

$$\chi^2 = \Sigma \dfrac{(O - E)^2}{E}$$

So, chi squared in this example

$$= \dfrac{(89 - 80)^2}{80} + \dfrac{(45 - 54)^2}{54} + \dfrac{(31 - 40)^2}{40} + \dfrac{(35 - 26)^2}{26}$$
$$= 1.01 + 1.50 + 2.03 + 3.11$$
$$= 7.65$$

To find whether this result is statistically significant or not, the value must be compared to a **critical value** (Table 4.4). To locate the critical value, the appropriate degrees of freedom need to be calculated.

Degrees of freedom = (number of columns – 1) × (number of rows – 1)

In this case, degrees of freedom $= (2 - 1) \times (2 - 1) = 1$

■ **Table 4.4**
Critical values for the χ^2 test

Degrees of freedom	0.05 level of significance
1	3.84
2	5.99
3	7.81
4	9.49

The chi-squared value of 7.65 is larger than the critical value of 3.84 for 1 degree of freedom at the probability level of $p = 0.05$ (the 5% probability level). The null hypothesis is therefore rejected; there is a statistically significant association between bell heather and ling in this area of moorland. So, the distributions of the two species are not independent of each other – the distribution of the two species is associated.

c The value of chi squared may also be obtained using a programmed pocket calculator or a computer program such as:

www.socscistatistics.com/tests/chisquare/Default2.aspx

3   Carry out a χ^2 test to see if there is an association between bell heather and bilberry from the observed results shown in the table.

From your calculations, **deduce** whether the two species are associated or whether they tend to occupy different microhabitats on this moorland.

	Bell heather present	Bell heather absent	Total
Bilberry present	12	55	67
Bilberry absent	88	45	133
	100	100	200

■ The need for sustainability in human activities

We have noted that ecosystems are largely self-contained and self-sustaining units. They have the potential to maintain sustainability over long periods of time. Most organisms of an ecosystem spend their entire lives there. Here, their essential nutrients will be endlessly recycled around and through them. This illustrates a key feature of environments – that they naturally self-regulate. The basis of sustainability is the flow of energy through ecosystems and the endless recycling of nutrients. This is summarized in Figure 4.6, and is the focus of Section 4.2.

Unfortunately, we humans often destabilize ecosystems. This is a result of our presence in large numbers over much of the globe, and our profligate use of space and resources. Our demands for food for expanding populations, and for materials and minerals for homes and industries tend to destroy ecosystems.

Today, the impact of humans on the environment is very great indeed. Conservation attempts to manage the environment so that, despite human activities, a balance is maintained. The aims are to preserve and promote habitats and wildlife, and to ensure natural resources are used in a way that provides a sustainable yield. Conservation is an active process, not simply a case of preservation, and there are many different approaches to it. More effective family planning and population control in human communities could be a highly significant factor in some areas of the world.

4   For an ecosystem near your home, school or college, **list** the ways in which you feel the human community has adversely changed the environment.
Can you **suggest** a practical way that this harm could be reduced or reversed?

Investigating the self-sustainability of ecosystems – using mesocosms

The sustainability of an ecosystem may change when an external 'disturbing' factor that disrupts the natural balance is applied. Investigation of this may be attempted in natural habitats or in experimental, enclosed systems. Both approaches have advantages and drawbacks (Table 4.5).

■ **Table 4.5** Alternative approaches to investigating ecosystem sustainability

	A natural ecosystem, for example an entire pond or lake	A small-scale laboratory model aquatic system (a mesocosm)
Advantages	realistic – actual environmental conditions are experienced	able to control variables – opportunity to measure the degree of stability or extent of change in a community, and to investigate the precise impact of a disturbing factor
Disadvantages	variable conditions – minimum or non-existent control over 'controlled variables'	unrealistic – possibly of disputed relevance and applicability to natural ecosystems

Case study: An investigation of eutrophication

In water enriched with inorganic ions (such as from raw sewage or fertilizer 'run-off' from surrounding land), plant growth is typically luxuriant. The increase in concentration of ammonium, nitrate and phosphate ions particularly increases plant growth. When seasonal water temperature rises, the aquatic algae undergo a population explosion – causing an algal bloom, for example. This process is known as **eutrophication**.

Later, when the algal bloom has died back, the organic remains of the plants are decayed by saprotrophic aerobic bacteria. The water becomes deoxygenated, and anaerobic decay occurs with hydrogen sulfide. A few organisms can survive in these conditions and prosper, but the death of many aquatic organisms results.

Can mesocosms be set up to investigate eutrophication, so avoiding the destruction of a natural ecosystem?

■ **Figure 4.6**
An experimental mesocosm apparatus

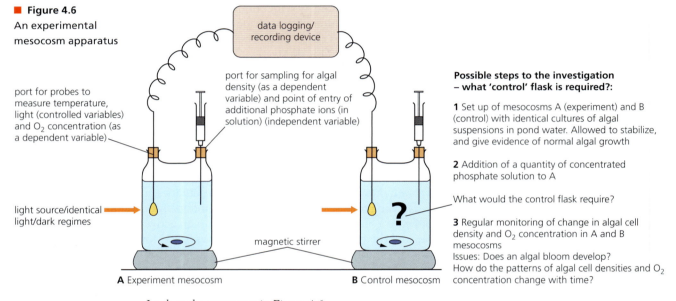

Possible steps to the investigation
– what 'control' flask is required?:

1 Set up of mesocosms A (experiment) and B (control) with identical cultures of algal suspensions in pond water. Allowed to stabilize, and give evidence of normal algal growth

2 Addition of a quantity of concentrated phosphate solution to A

What would the control flask require?

3 Regular monitoring of change in algal cell density and O_2 concentration in A and B mesocosms
Issues: Does an algal bloom develop? How do the patterns of algal cell densities and O_2 concentration change with time?

Look at the apparatus in Figure 4.6.

- Does the figure show an appropriate practical investigation of eutrophication under controlled laboratory conditions? What changes might be made?
- Here, two dependent variables have been proposed. Why?
- If the additional phosphate ions added to mesocosm (A) resulted in an algal bloom, how could the control (B) be arranged to establish that influx of phosphate ions caused it?
- How would you expect the oxygen concentrations to change over an extended period in both mesocosms (A) and (B)?

■ Cycling of nutrients

Nutrients provide the chemical elements that make up the molecules of cells and organisms. We recognize that all organisms are made of carbon, hydrogen and oxygen, together with mineral elements nitrogen, calcium, phosphorus, sulfur and potassium, and several others, in increasingly small amounts. Plants obtain their essential nutrients as carbon dioxide and water, from which they manufacture sugar. With the addition of mineral elements, absorbed as ions from the soil solution, they build up the complex organic molecules they require (Figure 2.65, page 121). Animals, on the other hand, obtain nutrients as complex organic molecules of food which they digest, absorb and assimilate into their own cells and tissues.

Recycling of nutrients is essential for the survival of living things, because the available resources of many elements are limited. When organisms die, their bodies are broken down to simpler substances (for example, CO_2, H_2O, NH_3 and various ions), as illustrated in Figure 4.7. Nutrients are released.

Scavenging actions of **detritivores** often begin the process of breakdown and decay, but saprotrophic **bacteria and fungi** always complete the breakdown. Elements that are released may become part of the soil solution, and some may react with chemicals of soil or rock particles, before becoming part of living things again through reabsorption by plants. Ultimately, both plants and animals depend on the activities of saprotrophic microorganisms to release matter from dead organisms for reuse.

The complete range of recycling processes by which essential elements are released and reused involve both living things (the biota) and the non-living (abiotic) environment. The latter consists of the atmosphere, hydrosphere (oceans, rivers and lakes) and the lithosphere (rocks and soil). All the essential elements take part in such cycles. One example is the carbon cycle (page 201).

In summary, the supply of nutrients in an ecosystem is finite and limited. By contrast, there is a continuous, but variable, supply of energy in the form of sunlight. This we focus on next.

5 Explain how it is that animal life is dependent on the actions of saprotrophs.

dead animal

1 break up
of animal body by scavengers and detritivores, e.g. carrion crow, magpie, fox

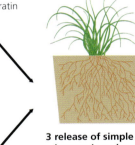

2 succession of microorganisms
– mainly bacteria, feeding:
- firstly on simple nutrients such as sugars, amino acids, fatty acids
- secondly on polysaccharides, proteins, lipids
- thirdly on resistant molecules of the body, such as keratin and collagen

3 release of simple inorganic molecules
such as CO_2, H_2O, NH_3, ions such as Na^+, K^+, Ca^{2+}, NO_3^-, PO_4^-, all available to be reabsorbed by plant roots for reuse

2 succession of microorganisms
– mainly fungi, feeding:
- firstly on simple nutrients such as sugars, amino acids, fatty acids
- secondly on polysaccharides, proteins, lipids
- thirdly on resistant molecules such as cellulose and lignin

1 break up
of plant body by detritivores, e.g. slugs and snails, earthworms, wood-boring insects

dead plant

■ **Figure 4.7** The sequence of organisms involved in decay

■ **Figure 4.8**
Cycling of nutrients and the flow of energy within an ecosystem – a summary

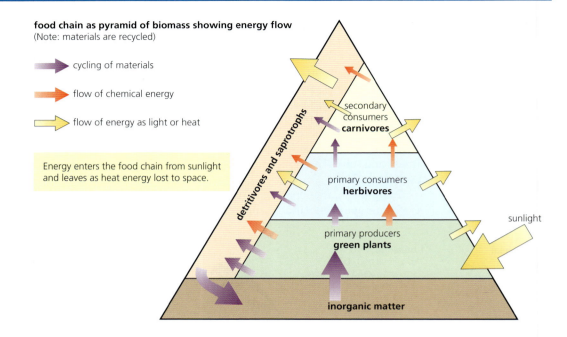

food chain as pyramid of biomass showing energy flow
(Note: materials are recycled)

➡ cycling of materials

➡ flow of chemical energy

➡ flow of energy as light or heat

Energy enters the food chain from sunlight and leaves as heat energy lost to space.

detritivores and saprotrophs

secondary consumers **carnivores**

primary consumers **herbivores**

primary producers **green plants**

sunlight

inorganic matter

4.2 Energy flow – *ecosystems require a continuous supply of energy to fuel life processes and to replace energy lost as heat*

■ Most ecosystems rely on a supply of energy from sunlight

We can demonstrate the dependence of ecosystems on sunlight by drawing up food chains

Drawing up a food chain

Look at Figure 4.9. A feeding relationship in which a carnivore eats a herbivore, which itself has eaten plant matter, is called a **food chain**. In Figure 4.9, light is the initial energy source, as it is in most other food chains. Note that, in a food chain, the arrows point *to* the consumers and so indicate the direction of energy flow. Food chains from contrasting ecosystems are shown in Figure 4.10.

■ **Figure 4.9**
A food chain

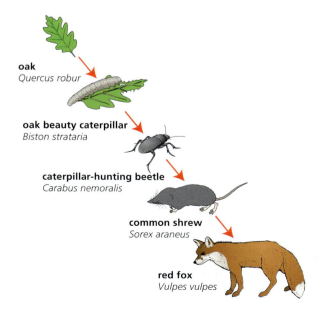

oak
Quercus robur

oak beauty caterpillar
Biston stataria

caterpillar-hunting beetle
Carabus nemoralis

common shrew
Sorex araneus

red fox
Vulpes vulpes

■ **Figure 4.10**
Food chains
from contrasting
ecosystems

A in a tropical rainforest

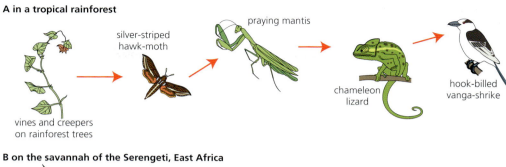

vines and creepers
on rainforest trees

silver-striped
hawk-moth

praying mantis

chameleon
lizard

hook-billed
vanga-shrike

B on the savannah of the Serengeti, East Africa

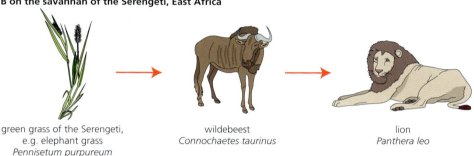

green grass of the Serengeti,
e.g. elephant grass
Pennisetum purpureum

wildebeest
Connochaetes taurinus

lion
Panthera leo

6 Using the information in the food web in Figure 4.11 B, **construct** two individual food chains from the marine ecosystem, each with at least three linkages (four organisms). **Identify** each organism with its common name, and **state** whether each is a producer, primary consumer, secondary consumer, etc.

Food webs

In an ecosystem, the food chains are not isolated. Rather, they interconnect with other chains. This is because most prey species have to escape the attentions of more than one predator. Predators, as well as having preferences, need to exploit alternative food sources when any one source becomes scarce. They also take full advantage of gluts of food as particular prey populations become temporarily abundant.

Consequently, individual food chains may be temporary and are interconnected so that they may form a **food web**. Two examples of food webs are shown in Figure 4.11, one from a woodland ecosystem and one from a marine ecosystem.

Examine these now.

Note that food chains tell us about the feeding relationships of organisms in an ecosystem, but they are entirely **qualitative** relationships (we know which organisms are present as prey and as predators), rather than providing **quantitative** data (we do not know the numbers of organisms at each level).

Trophic levels

Food chains show a sequence of organisms in which each is the food of the next organism in the chain, so each organism represents a feeding or **trophic level**. Chains typically start with a producer and end with a consumer – perhaps a secondary or tertiary consumer. The trophic levels of organisms in some of the food chains and food webs in this book are classified in Table 4.6.

■ **Table 4.6**
An analysis of trophic levels

Trophic level	Woodland	Rainforest	Savannah
producer	oak	vines and creepers on rainforest trees	grass
primary or first consumer	caterpillar	silver-striped hawk-moth	wildebeest
secondary consumer	beetle	praying mantis	lion
tertiary consumer	shrew	chameleon lizard	
quaternary consumer	fox	hook-billed vanga-shrike	

7 **Identify** the initial energy source for almost all communities.

There is not a fixed number of trophic levels in a food chain, but typically there are three, four or five levels only. There is an important reason why stable food chains remain quite short. We will come back to this point, shortly.

A a woodland food web

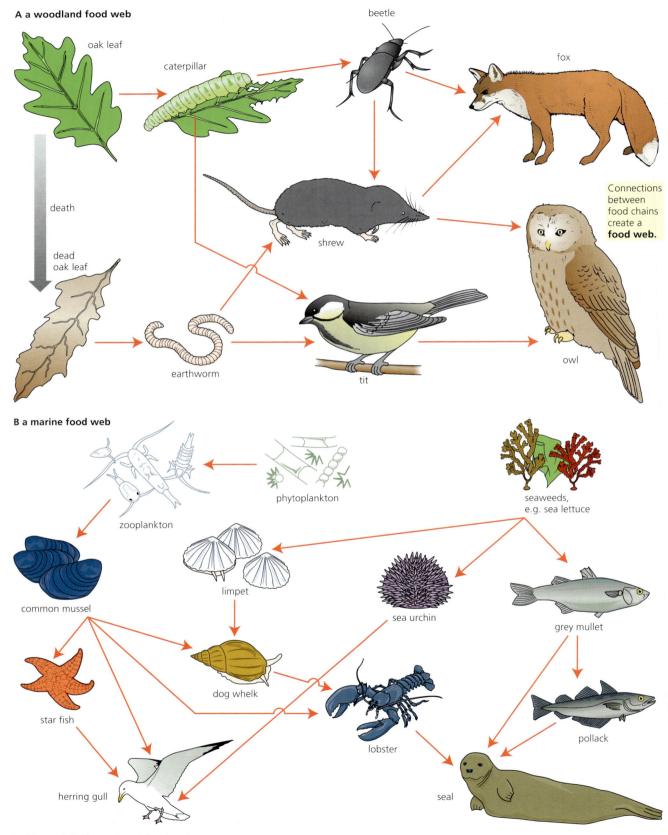

Connections between food chains create a **food web.**

B a marine food web

■ **Figure 4.11** Examples of food webs

8 Suggest what trophic levels humans occupy. Give examples.

Sometimes it can be difficult to decide at which trophic level to place an organism. For example, an **omnivore** feeds on both plant matter (primary consumer) and on other consumers (secondary consumer, or higher). In the woodland food chain of Figure 4.9, the fox more commonly feeds on beetles than shrews, because there are many more beetles about and they are easier to catch.

Energy flow

Between each trophic level there is an energy transfer. At the base of the food chain, green plants, the producers, transfer light energy into the chemical energy of sugars in photosynthesis. However, much of the light energy reaching the green leaf is not retained in the green leaf. Some is reflected away, some transmitted, and some lost as heat energy. Meanwhile, sugars are converted into lipids, amino acids and other metabolites within the cells and tissues of the plant. Some of these metabolites are used in the growth and development of the plant and, through these reactions, energy is locked up in the organic molecules of the plant body.

The reactions of respiration and of the rest of the plant's metabolism produce heat, another form of energy, as a waste product. Inevitably, chemical energy is transferred every time the tissues of a green plant are eaten by herbivores. Finally, on the death of the plant, the remaining energy passes to detritivores and saprotrophs when dead plant matter is broken down and decayed. The diverse routes of energy flow through a primary producer are summarized in Figure 4.12.

Consumers eat producers or primary or secondary consumers, according to where they occur in a food chain. In this way, energy is transferred to the consumer. The consumer, in turn, transfers energy in the muscular movements by which it hunts and feeds, and as it seeks to escape from predators. Some of the food it has eaten remains undigested, passing through the consumer unchanged, and is lost as waste products in faeces and as urea. Also, heat energy, a waste product of the reactions of respiration and of the animal's metabolism, is continuously lost as the consumer grows, develops and forms body tissues. If the consumer is itself caught and consumed by another, larger consumer, energy is again transferred. Finally, on the death of the consumer, the remaining energy passes to detritivores and saprotrophs when dead matter is broken down and decayed. Ultimately, the energy in matter that is decayed is lost as heat.

In summary, the energy conversions that occur in organisms are:

■ light energy to chemical energy in photosynthesis
■ chemical energy to heat as a waste product of the anabolic reactions of metabolism and in respiration
■ chemical energy to electrical energy in nerve impulses and kinetic energy in muscle contraction
■ chemical energy in dead matter is lost as heat as a waste product of decay.

Since living organisms cannot convert heat energy to other forms of energy, all energy reaching the Earth from the Sun is ultimately lost from ecosystems as heat energy into space. Energy flow through consumers is shown in Figures 4.12 and 4.13.

■ The concept of energy flow explains the limited length of food chains

Energy is transferred from one organism to another in a food chain, but only some of the energy transferred becomes available to the next organism in the food chain. In fact, only about 10% of what is eaten by a consumer is built into that organism's body, and so is potentially available to be transferred on through predation or browsing. There are two consequences of this.

Nature of Science   **Using theories to explain natural phenomena**

■ The energy loss at transfer between trophic levels is the reason why food chains are short. Few transfers can be sustained when so little of what is eaten by one consumer is potentially available to those organisms in the next step in the food chain. Consequently, it is very uncommon for food chains to have more than four or five links between producer (green plant) and top carnivore.

■ Feeding relationships of a food chain may be structured like a pyramid. At the start of the chain is a very large amount of living matter (biomass) of green plants. This supports a smaller biomass of primary consumers, which, in turn, supports an even smaller biomass of secondary consumers. A generalized ecosystem pyramid diagram, representing the structure of an ecosystem in terms of the biomass of the organisms at each trophic level, is shown in Figure 4.14.

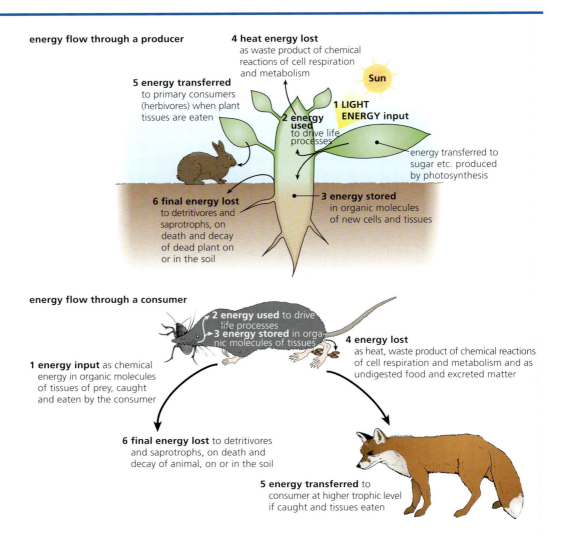

energy flow through a producer

4 heat energy lost as waste product of chemical reactions of cell respiration and metabolism

5 energy transferred to primary consumers (herbivores) when plant tissues are eaten

2 energy used to drive life processes

Sun

1 LIGHT ENERGY input

energy transferred to sugar etc. produced by photosynthesis

6 final energy lost to detritivores and saprotrophs, on death and decay of dead plant on or in the soil

3 energy stored in organic molecules of new cells and tissues

energy flow through a consumer

2 energy used to drive life processes
3 energy stored in organic molecules of tissues

4 energy lost as heat, waste product of chemical reactions of cell respiration and metabolism and as undigested food and excreted matter

1 energy input as chemical energy in organic molecules of tissues of prey, caught and eaten by the consumer

6 final energy lost to detritivores and saprotrophs, on death and decay of animal, on or in the soil

5 energy transferred to consumer at higher trophic level if caught and tissues eaten

A linear food chain – energy flow

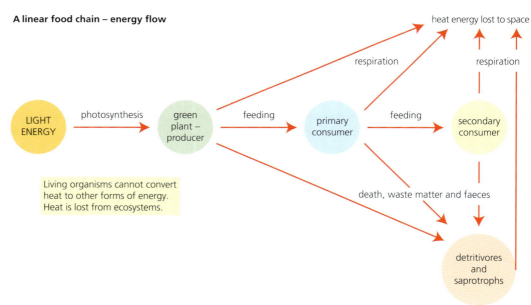

heat energy lost to space

respiration

respiration

LIGHT ENERGY → photosynthesis → green plant – producer → feeding → primary consumer → feeding → secondary consumer

Living organisms cannot convert heat to other forms of energy. Heat is lost from ecosystems.

death, waste matter and faeces

detritivores and saprotrophs

Only energy taken in at one trophic level and then built in as chemical energy in the molecules making up the cells and tissues is available to the next trophic level. This is about 10% of the energy.

The reasons are as follows.

- Much energy is used for cell respiration to provide energy for growth, movement, feeding, and all other essential life processes.
- Not all food eaten can be digested. Some passes out with the faeces. Indigestible matter includes bones, hair, feathers, and lignified fibres in plants.
- Not all organisms at each trophic level are eaten. Some escape predation.

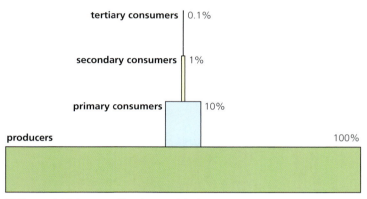

■ **Figure 4.14** A generalized pyramid of energy

9 **Explain** why, in a food chain, a large amount of plant material supports a smaller mass of herbivores and an even smaller mass of carnivores.

Representing energy flow using pyramids of energy

An early attempt to quantify energy transfer in a natural community was carried out on a river system in Florida, USA, over 50 years ago. Here an American ecologist recorded the energy held at each of four trophic levels. His results are recorded in Figure 4.15 A. Using this data, we can calculate the percentage energy transfer from producers (green aquatic plants) to the primary consumers:

$$\frac{14\,000}{87\,000} \times 100\% = 16.1\%$$

Now, this figure is significantly more than 10%. Are we to assume that a figure of about 10% of energy transferred is an inaccurate, sweeping generalization? Probably not. In this river significantly more of energy of these particular primary producers was transferred to primary consumers because the plants concerned were almost entirely of highly digestible matter. They lacked any woody tissues, such as those common to most terrestrial plants.

Now complete the calculation in question 10.

Another investigation of energy transfer involved a farm animal, as shown Figure 4.15 B. You can see that, of the energy intake by the cow, approximately half passes through – lost in faeces and urine. However, in this illustration, the energy locked up in the new biomass in the cow has not been calculated and recorded.

Now complete the calculations in question 11.

 ## ■ Food chains and world hunger

Today, the number of humans in the world population is huge. This frequently places excessive demands on the food supply and on the resources that are used in its production. So much so that, around the world, there are local populations of people with too little to eat. World hunger is a major problem of which we can't fail to be aware.

There is a potential ethical challenge for well-nourished people living alongside other humans who may be starving and who may die prematurely due to malnourishment. In the light of this, some humans opt for a vegetarian diet. Vegetarians do not eat meat and most do not eat fish, but the majority consume animal products, such as milk, cheese and eggs.

■ **Figure 4.15**
Energy transfer studies in a river ecosystem and in an agricultural context

A pyramid of energy from a river system in Florida

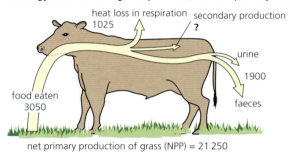

tertiary consumers   90   figures are in kJ m^{-2} yr^{-1}

secondary consumers ☐1600

primary consumers   14 000

producers   87 000

B energy transfer from grass (producer) to cow (primary consumer)

heat loss in respiration   secondary production
1025   ?

urine
1900

food eaten
3050   faeces

net primary production of grass (NPP) = 21 250

10 **Calculate** the percentage energy transfer between primary and secondary consumers in the data from the river system in Florida (Figure 4.15 A).

11 From the data in Figure 4.15 B, **calculate**
  a the energy value of the new biomass of the cow, and then express this as a percentage of the energy consumed by the cow
  b the percentage energy transfer between primary and secondary consumer.

The rationale for this type of vegetarianism is that, when we eat meat rather than food of plant origin, we **extend food chains by a least one trophic level**. The effect of this is all too evident from your calculations in questions 10 and 11 above, for example. In effect, vegetarians observe that we waste energy by choosing to eat animal products rather than matter of plant origin, thereby leaving less food for others. This is an important issue when considering the eating of animal products.

4.3 Carbon cycle – *continued availability of carbon in ecosystems depends on carbon cycling*

As we have seen, nutrients of the ecosystem are recycled and reused. All the chemical elements of nutrients circulate between living things and the environment. The process of exchange or 'flux' (back and forth movements) of materials is a continuous one. These movements take more or less circular paths, and have a biological component and a geochemical component. Consequently, these movements are known as **biogeochemical cycles**.

We are going to examine the **carbon cycle** as an example (Figure 4.16). *Look at it now. Locate the element carbon in both an abiotic and a biotic component.* Use this comprehensive representation to construct a specific carbon cycle for the terrestrial ecosystem in which you live.

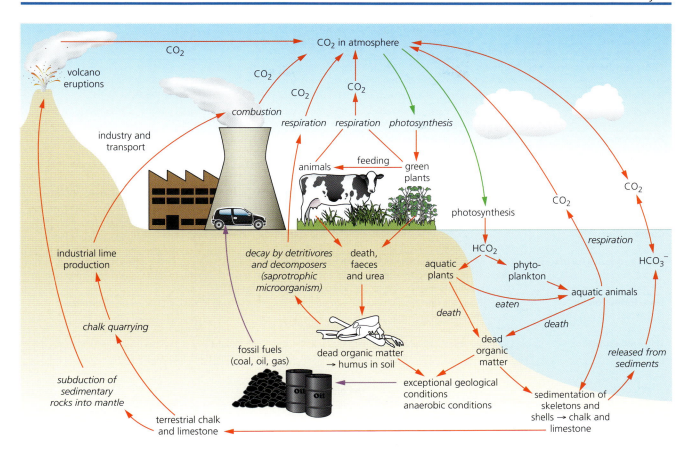

■ **Figure 4.16** The carbon cycle

The carbon cycle illustrates the fact that biogeochemical cycles have two 'pools' of nutrients.

■ The **reservoir pool** is large, non-biological and slow moving. In the carbon cycle, this is chiefly the carbonates locked up in the chalk and limestone deposits of the Earth.

■ The **exchange pool** is much smaller, more active, and sited at the points of exchange between the living and non-living parts of the cycle.

Have you identified the reservoir and the exchange pool in the diagram of the carbon cycle?

The processes by which carbon moves between these pools, and between living things and the environment are summarized in Table 4.7. Read through the table and locate each process in the carbon cycle.

■ Carbon fluxes

Notice that the size of the pools and the fluxing of carbon between them is summarized in Figure 4.17. Examine this data carefully, too. For example, the data on rates of flux will help confirm your identifications of the reservoir pool and exchange pool of carbon.

■ **Table 4.7** How carbon is circulated between living things and the environment

Processes by which carbon is circulated	
Photosynthesis in terrestrial and aquatic plants	Autotrophs convert CO_2 from the atmosphere into carbohydrates and other organic compounds. Aquatic plants use dissolved CO_2 and hydrogencarbonate ions (as HCO_3^-) from the water in the same way.
Respiration in all living things	CO_2 is produced as a waste product and diffuses out into the atmosphere or water.
Decay by saprotrophic microorganisms	Dead organic matter is decomposed to CO_2, water, ammonia and mineral ions by microorganisms. The CO_2 produced diffuses out into the atmosphere or water (as HCO_3^- ions).
Peat formation	In acidic and anaerobic conditions, dead organic matter is not fully decomposed but accumulates as peat. Peat decays slowly when exposed to oxygen, releasing CO_2 into the atmosphere. Peat in past geological areas was converted to coal, oil or gas and these fossil fuels accumulated. They are now progressively exploited.
Methane formation from organic matter under anaerobic conditions	Organic matter held under anaerobic conditions (such as in waterlogged soil or in the mud of deep ponds) is decayed by methane-producing bacteria. Methane accumulates in the ground in porous rocks or under water, but may progressively escape into the atmosphere. In air and light, methane (CH_4) is oxidized to CO_2 and water.
Combustion of fossil fuels	Releases CO_2 into the atmosphere. Since the start of the Industrial Revolution in Europe, CO_2 has been released at an increasing rate.
Shell and bone formation by organisms	Many organisms combine HCO_3^- with calcium ions and other minerals to form carbonate shells and bones. These may accumulate as sediments and come to form sedimentary rocks (chalk and limestone) in geological time. Reef-building coral are particular examples of this.
Lime formation for agriculture and building materials	Terrestrial chalk and limestone are quarried and converted to lime in kilns. CO_2 is released into the atmosphere in the process.
Volcanic eruptions	On a geological timescale, sedimentary rocks (including chalk and limestone) become subducted into the Earth's mantle. When volcanoes erupt, molten rocks and CO_2 are released into the atmosphere.

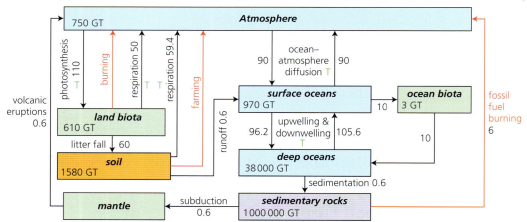

The global carbon cycle
Red arrows are flows that are related to human activities
Green T = flows that are sensitive to temperature

units are gigatonnes of carbon – one gigatonne = one billion metric tonnes = 10^{15} g

The global carbon cycle showing the size of the carbon reservoirs (in gigatonnes [GT]) and the exchange between reservoirs (GT/year). Figures are estimates.

numbers next to flows are the approximate annual flows in GT/year

■ **Figure 4.17** Carbon fluxes due to processes in the carbon cycle

12 **Outline** the difference in the rates of 'flux' between the atmosphere and land biota and between the deep ocean and sedimentary rocks.

13 **Estimate** what percentage of the Earth's carbon is locked up in the reserve pool.

Nature of Science

Making accurate, quantitative measurements

■ **The monitoring of the level of atmospheric carbon dioxide**

The concentration of atmospheric carbon dioxide has become critically important today. It is important to obtain reliable data on the concentration of carbon dioxide and methane in the atmosphere. Appropriately, accurate measuring devises were established at the Mauna Loa monitoring station on Hawaii, beginning in 1957 as part of the International Geophysical Year initiative, to monitor the global environment. Measurements of carbon dioxide and methane are now the responsibility of two scientific institutions – the Scripps Institution of Oceanography, and the National Oceanic and Atmospheric Administration. Monthly data is posted.

The most recent data for carbon dioxide levels available from the on-going investigation are shown in Figure 4.18. Notice the annual rhythm shown in the atmospheric carbon dioxide concentration (lower in the summer months and higher in the winter months). This is due to photosynthesis on land in the Northern Hemisphere, which impacts on the composition of the global atmosphere. The unmistakable trend in this data is discussed in Section 4.4 below.

■ **Figure 4.18**
Atmospheric carbon dioxide measured at the Mauna Loa Observatory, Hawaii

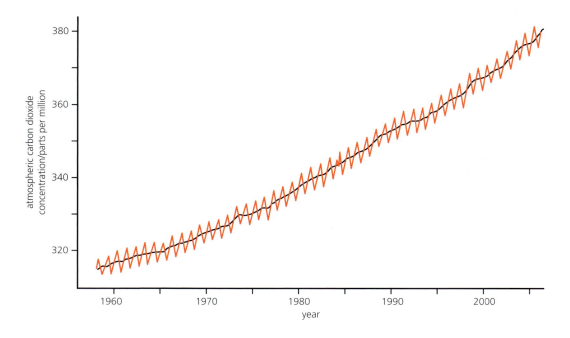

14 In the cycling of carbon in nature, **state** in what forms inorganic carbon can exist in:
  a the atmosphere
  b the hydrosphere
  c the lithosphere.

4.4 Climate change – *concentrations of gases in the atmosphere affect climates experienced at the Earth's surface*

■ The greenhouse effect

The radiant energy reaching the Earth from the Sun includes visible light (short wave radiation) and infra-red radiation (longer wave radiation – heat). It is the infra-red that principally warms up the sea and the land. As it is warmed, the Earth itself radiates infra-red radiation back towards space. However, much of this heat does not escape from our atmosphere. Some is reflected back by clouds and much is absorbed by gases in the atmosphere, which are warmed. In this respect, the atmosphere is working like the glass in a greenhouse, which is why this phenomenon is called the **greenhouse effect** (Figure 4.19). We must recognize that the greenhouse effect is very important to life on the Earth – without it, surface temperatures would be altogether too cold for life.

■ **Figure 4.19**
The greenhouse effect

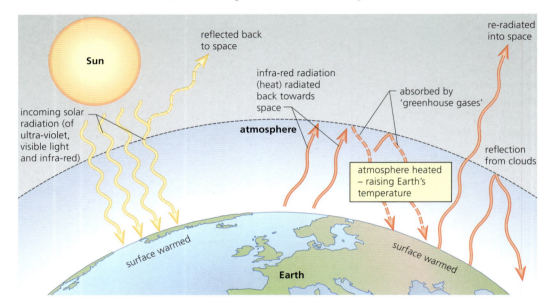

The gases in the atmosphere that absorb infra-red radiation are referred to as **greenhouse gases**. **Carbon dioxide** and **water vapour** are the most significant greenhouse gases. Other gases, including **methane** and **nitrogen oxides**, have less impact. Note that ozone is another atmospheric pollutant, but it is not the cause of or a contributor to the enhanced greenhouse effect.

The contribution of each gas to the greenhouse effect is largely a product of the properties of the gas and of how abundant it is at any time. For example, methane is considerably more powerful as a greenhouse gas than the same mass of carbon dioxide. However, methane is present in lower concentrations than carbon dioxide. Also, it is a relatively short-lived component of the atmosphere – in the light, methane molecules are steadily oxidized to carbon dioxide and water (Table 4.7). In Table 4.8 figures for the typical contributions of the chief greenhouse gases are listed.

■ **Table 4.8**
The direct consequence of greenhouse gases to the 'greenhouse effect'

Compound	Approximate contribution/%
water vapour (+ clouds)	36–72
carbon dioxide	9–26
methane + nitrogen oxides	4–9

An enhanced greenhouse effect leading to global warming?

Today, carbon dioxide is present in the atmosphere at a concentration of about 380–400 parts per million (ppm). We might expect the amount of atmospheric carbon dioxide to be maintained by a balance between the fixation of this gas during photosynthesis and release of carbon dioxide into the atmosphere by respiration, combustion and decay by microorganisms – an interrelationship illustrated in the carbon cycle (Figure 4.16). In fact, photosynthesis does withdraw almost as much carbon dioxide during daylight hours as is released into the air by all the other processes, day and night – but not quite as much. As a result, the level of atmospheric carbon dioxide is now rising (Figure 4.18).

How current and historic levels of atmospheric carbon dioxide are known

The composition of the atmosphere has changed over time – this is beyond dispute.

The **long-term records** of changing levels of greenhouse gases (and associated climate change) are based on evidence obtained from ice cores drilled in the Antarctic and Greenland ice sheets. The ice there has formed from accumulation of layer upon layer of frozen snow, deposited and compacted there over thousands of years. Gases from the surrounding atmosphere were trapped as the layers built up. Data on the composition of the bubbles of gas obtained from different layers of these cores from the Vostok ice in East Antarctic is a record of how the carbon dioxide and methane concentrations have varied over a period of 400 000 years of Earth's history. Similarly, variations in the concentration of oxygen isotopes from the same source indicate how temperature has changed during the same period (Figure 4.20).

■ **Figure 4.20**
Three types of data recovered from the Vostok ice cores over 400 000 years of Earth history

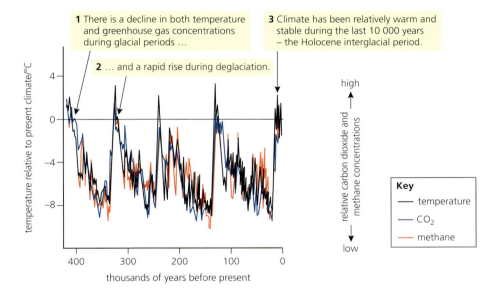

1 There is a decline in both temperature and greenhouse gas concentrations during glacial periods …

2 … and a rapid rise during deglaciation.

3 Climate has been relatively warm and stable during the last 10 000 years – the Holocene interglacial period.

Key
— temperature
— CO_2
— methane

15 From the data in Figure 4.20 **explain** why it is that the atmospheric carbon dioxide varies between high and low values within each 12-month period of the graph.

The causes of historic and current levels of atmospheric carbon dioxide

From the graph in Figure 4.20, we can see that the level of atmospheric carbon dioxide has varied quite markedly. Note that there have been periods in Earth's history when it was especially raised. We assume these rises were triggered by volcanic eruptions and the weathering of chalk and limestone at these times. Today, we estimate that carbon dioxide from volcanoes contributes only about 1% of the amounts released by human activities.

However, since the beginning of the Industrial Revolution in the developed countries of the world (the past 200 years or so), there has been a sharp and accelerating rise in the level of this greenhouse gas. This is attributed to the burning of coal and oil. These 'fossil fuels' were mostly laid down in the Carboniferous Period. As a result, we are now adding to our atmosphere carbon that had been locked away for about 350 million years. This is an entirely new development in geological history. There is a correlation between rising atmospheric concentrations of carbon dioxide and rising average global temperatures. Many climate scientists argue this development poses a major environmental threat to life as we know it.

■ **Table 4.9**
Changing levels of atmospheric CO_2 – recorded and predicted

	CO₂/ppm
pre-Industrial Revolution level	280 (±10)
by mid 1970s	330
by 1990	360
by 2007	380
by 2013	400
by 2050 (if current rate is maintained)	**500**

Evidence of environmental change triggered by 'greenhouse gases'

A report of the Intergovernmental Panel on Climate Change (IPCC) has predicted highly significant environmental impacts as a consequence of global warming, triggered by current and rising levels of carbon dioxide. Some of these are already evident (Table 4.10).

■ **Table 4.10**
Theatres of environmental change attributed to global warming

Polar ice melt

The Arctic is a highly sensitive region – ice cover varies naturally according to the season. However, since 1979 the size of the summer polar ice cap has shrunk by more than 20%. In this period, the decline in the ice has been, on average, more than 8% per decade. At this rate, there may be no ice in the summer of 2060. The associated Greenland ice is similarly in decline.

In Antarctica the picture is less clear. The Antarctic Peninsula has warmed, and 13 000 km² of ice has been lost in the past 50 years. Also, major sections of the Antarctic ice shelf have broken off. Meanwhile, at times, the interior ice has become cooler and thicker, as circular winds around the land mass have prevented warmer air reaching the interior. Warmer seas may be eroding the ice from underneath, but the IPCC predicts that the Antarctic's contribution to rising sea levels will be small.

Glacier retreats

Since 1980, retreat by glaciers has been happening worldwide and rapidly. Mid-latitude mountain ranges such as the Himalayas, the Tibetan plateau, Alps, Rockies and the southern Andes, plus the tropical summit of Kilimanjaro show the greatest losses of glacier ice. Rivers below these mountain ranges are glacier-fed, and so the melting of glaciers will have an increasing impact on the water supplies for a great many people.

Rising sea levels

The impact of global warming on sea levels is due to the thermal expansion of sea water and the widespread melting of ice. The global average sea level rose at an average rate of 1.8 mm per year in the period 1960–2003, but during the latter part of that period the rate was far higher than at the beginning. If this acceleration continues at the current rate, sea levels could rise by at least 30 cm in this century. This phenomenon will threaten low-lying islands and countries (including Bangladesh), and major city communities such as London, Shanghai, New York and Tokyo.

Changing weather and ocean current patterns

At the poles, cold, salty water sinks and is replaced by surface water that is warmed in the tropics. Now, melting ice decreases ocean salinity which then slows the great ocean currents – these convey heat energy from warmer to colder regions through their pattern of convection. So, for example, as the Gulf stream (which, to date, has kept temperatures in Europe relatively warmer than in Canada) slows down, more heat is retained in the Gulf of Mexico. Here, hurricanes get their energy from hot water, and are becoming more frequent and more severe.

Also, alteration in the patterns of heat and rainfall distribution over continental land masses are predicted to cause Russia and Canada to experience the largest mean temperature rises, followed by several Asian countries and already drought-ridden countries in West Africa. Least warming is anticipated in Ireland and Britain in the Northern Hemisphere, and New Zealand, Chile, Uruguay and Argentina in the south. The most immediately vulnerable populations are already impoverished communities in parts of Africa; the least vulnerable is the wealthy population of Luxembourg.

Coral bleaching

Corals are colonies of small animals embedded in a calcium carbonate shell that they secrete. They form under-water structures, known as coral reefs, in warm, shallow water where sunlight penetrates. Microscopic (photosynthetic) algae live sheltered and protected in the cells of corals. The relationship is one of mutual advantage (a form of symbiosis called mutualism), for the coral gets up to 90% of its organic nutrients from these organisms. Coral reefs are the 'rainforests of the oceans' – the most diverse of ecosystems known. Although they cover less than 0.1% of the surface of the oceans, these reefs are home to about 25% of all marine species.

When under environmental stress due to high water temperature, the algae are expelled (causing loss of colour) and the coral starts to die. Mass bleaching events occurred in the Great Barrier Reef in 1999 and 2002.

Another cause of stress is increasing acidity. As carbon dioxide dissolved in the oceans increases, the pH decreases and the water becomes more acidic. This is known as ocean acidification. This prevents the corals from building and maintaining their calcium carbonate skeletons. The reefs are dissolving.

Today, coral reefs are dying all around the world. The effects of thermal and acidic stress are likely to be exacerbated under future climate scenarios.

Global warming – international responses

Release of greenhouse gases occurs locally but has a global impact. International cooperation to reduce emissions is essential. What actions are needed to prevent the worst consequences of continued global warming? In Table 4.11 possible responses are summarized.

■ **Table 4.11**
Global warming – appropriate actions?

Effective actions that may combat threats from global warming
conserve fossil fuel stocks, using them only sparingly, and only when there are no apparent alternatives (such as oils from biofuel sources)
develop nuclear power sources to supply electricity for industrial, commercial and domestic needs
develop so-called renewable sources of power, exploiting environmental energy sources, such as wave energy and wind power
develop biofuel sources of energy that exploit organic waste matter (which will naturally decay anyway) and biofuel crops that are renewable photosynthetic sources of energy
reduce use of fuels for heating of homes (where necessary) to minimum levels by economical designs of (well-insulated) housing and reduce use of fuels in more efficient transport systems
prevent the destruction of forests in general and of rainforest all around the tropical regions of the Earth in particular, since these are a major CO_2-sumps.

To be effective, any such actions taken in response to environmental challenges need to:

■ be agreed internationally as acceptable to all nations, and to be acted on by each and every one, simultaneously

■ recognize that existing developed countries have previously experienced their Industrial Revolutions (that largely initiated these processes of environmental damage). Surely these countries must fully share the benefits with less developed countries that would be required to otherwise forgo some of the benefits of development. Figure 4.21 summarizes the issue.

In less-developed countries
natural resources such as rainforest may be the only product people can market to earn foreign currency to purchase goods and services, often at developed-world prices – and there is huge demand for their resources from developed and developing countries.

In developed countries
the destruction of natural resources like rainforest elsewhere in the world is seen as a threat to stable environmental conditions essential for an established, comfortable way of life. This way of life is often based on earlier exploitation of similar resources.

■ **Figure 4.21** The environmental conundrum for today's world!

16 Explain fully the ways in which the destruction of rainforests affects the concentration of carbon dioxide in the atmosphere.

Is effective international agreement possible?

An international agreement to limit release of greenhouse gases by all industrial countries and emerging industrial countries was first agreed at the **Earth Summit** in Rio de Janeiro in 1992. An initiative, known as **Agenda 21**, was launched. Subsequently, at the **Kyoto Conference** in Japan a first attempt was made to meet the pledges made at Rio. Carbon dioxide emission targets for the industrial nations were set for the period 2009–2012. Now, discussions continue annually.

However, 'polluting' nations (and there are many) have been allowed to offset emissions with devices such as carbon sinks – mechanisms by which atmospheric carbon dioxide is removed from the air either permanently or on a long-term basis. For example, carbon dioxide may be absorbed by additional forest trees, becoming the carbon of wood that is not harvested and burnt. However, many of the arrangements are so complex as to be difficult to enforce. Real progress is extremely slow.

Individual responses – 'Think globally, act locally'?

A point for everyone to consider is what practical actions they, as individuals, can take to reduce their 'carbon footprint'. Like all resolutions, ideas may be easier to espouse than to carry through. Perhaps this is an issue to work at with a group of peers, within the context of your own environment, spheres of influence, and evaluations of the global crisis.

Nature of Science

Assessing claims

■ Global warming – contradictory schools of thought

Long-term climate predictions depend on the accurate estimation of the impact of many uncertain factors. Owing to the complexity of this situation, it is wise to regard all predictions, whether pessimistic or optimistic, with a measure of scepticism. The distinctive schools of thought that have emerged include:

1 profoundly committed and concerned environmentalists whose views of the future are most accessible in the writings of:

   James Lovelock (2006) *The Revenge of Gaia.* Allan Lane/Penguin Books

   Al Gore (2005) *An Inconvenient Truth.* Bloomsbury Paperback

2 students of the long-term geological records of past climates and Earth conditions, who focus on the profound changes that have occurred and reoccurred in Earth's history, including through the bulk of time when humans were *not* present

3 atmospheric chemists who, for example, draw attention to the contribution of water vapour as the most influential greenhouse gas (accounting for about 95% of the Earth's greenhouse effect). Ignorance of this, they say, contributes to overestimations of human impact. Some have argued that human greenhouse gases contribute only about 0.3% of the greenhouse effect.

Evaluating the claims of those who believe human actions are *not* the cause of environmental change is difficult at this time – but they should not be ignored. There are aspects of this issue that can be summarized in the adage, 'What you see depends on where you stand', but perhaps this is always the case?

TOK Link

The Precautionary Principle

This is summed up by the expression, 'Better safe than sorry'. In effect, when an activity raises threats of harm, measures should be taken, even if a cause-and-effect relationship has not been established scientifically. In the context of 'global warming' this implies:

■ recognition of the most likely consequences that may result from failure to slow down the rising level of atmospheric carbon dioxide

■ a response – the actions needed to achieve this (see Table 4.11).

■ *Examination questions – a selection*

Questions 1–2 are taken from IB Diploma biology papers.

Q1 The diagram shows a pyramid of energy for a wetland environment. What units would be appropriate for the values shown?

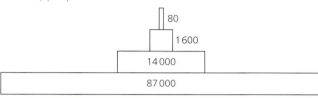

	80
	1600
14000	
87000	

A $kg\,yr^{-1}$
B $kJ\,m^{-2}\,yr^{-1}$
C $J\,m^{-2}$
D mg dry mass m^{-3}

Standard Level Paper 1, Time Zone 0, Nov 12, Q19

Q2 The energy passing from the detritivores to the predatory invertebrates in this food web is $14000\,kJ\,m^{-2}\,year^{-1}$.

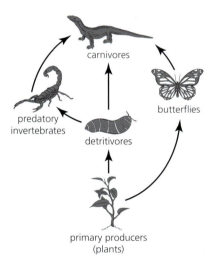

carnivores

predatory invertebrates

butterflies

detritivores

primary producers (plants)

Approximately how much energy (in $kJ\,m^{-2}\,year^{-1}$) passes from the predatory invertebrates to the carnivores?

A 140
B 1400
C 14000
D 140000

Standard Level Paper 1, Time Zone 2, May 12, Q18

Questions 3–10 cover other syllabus issues in this chapter.

Q3 When parts of a natural environment are maintained under controlled circumstances in a small area, we say that it is:
A a population
B a mesocosm
C an abiotic factor
D a small lake

Q4 What is the difference between a food chain and a food web?
A A food chain shows some relationship between animals in a community
B A food chain shows the amount of energy transfer between consumers
C A food web shows the feeding relationships between animals present in the same community
D A food chain takes place in a community while a food web takes place in a population.

Q5 To which trophic level do the butterflies from the diagram in question 2 belong to?
A producers
B primary consumers
C secondary consumers
D tertiary consumers

Q6 a Carbon is present in all organic compounds and in carbon dioxide gas in the atmosphere. Suggest where else inorganic carbon is commonly found in the biosphere, and in what forms it occurs. (4)
b Describe by what processes the carbon of organic compounds will be converted to carbon of carbon dioxide. (2)
c State where conversion of carbon dioxide to organic compounds occurs and what source of energy is involved. (2)
d Explain the fact that the average concentration of atmospheric carbon dioxide is almost stable on a daily basis. (2)
e Identify the part played by the oceans, and organisms in them, in the carbon cycle. (3)

Q7 **a** Identify within the carbon cycle, all the activities performed by living organisms that reduce the greenhouse effect. (2)

   b In an ecosystem, minerals such as nitrogen are recycled whereas energy is not. Explain the reason for this situation. (4)

Q8 Draw a labelled pyramid of energy with at least four trophic levels and label:

   i   autotrophic and heterotrophic organisms
   ii   each trophic level
   iii consumer and producer organisms
   iv   if the energy found in producers is $500\,000\,kJ\,m^{-2}\,y^{-1}$, calculate the energy transferred to the second, third and fourth trophic levels. (8)

Q9 The levels of atmospheric carbon dioxide have been increasing during the past years. Outline the reason for this increase and explain why it shows a fluctuating pattern within a year. (4)

Q10 By means of concise definition and examples, distinguish between the following terms:

   a *ecosystem* and *habitat* (4)
   b *population* and *community* (4)
   c *species* and *genus*. (4)

Evolution and biodiversity

ESSENTIAL IDEAS

- There is overwhelming evidence for the evolution of life on Earth.
- The diversity of life has evolved and continues to evolve by natural selection.
- Species are named and classified using an internationally agreed system.
- The ancestry of groups of species can be deduced by comparing their base or amino acid sequences.

5.1 Evidence for evolution – *there is overwhelming evidence for the evolution of life on Earth*

■ Evolution occurs when heritable characteristics of a species change

Today, it is generally accepted that present-day flora and fauna have arisen by change ('*descent with modification*'), probably very gradual change, from pre-existing forms of life.

By 'evolution' we mean the gradual development of life in geological time, from its earliest beginnings to the diversity of organisms we know about today, living and extinct; it is the development of new types of living organism from pre-existing types by the accumulation of genetic differences over long periods of time through the process of natural selection of chance variations. Meanwhile, the changes that an organism acquires in its lifetime are not inherited and are not transmitted to its offspring. So, evolution is the process of cumulative change in the heritable characteristics of a population. This is an organizing principle of modern biology.

Evidence for evolution comes from many sources, including from the study of fossils, from artificial selection in the production of domesticated breeds, from studies of the comparative anatomy of groups of related organisms, and from the geographical distribution of species.

■ Evidence from fossils

We learn something about the history of life from the evidence of the fossils. Fossilization is an extremely rare, chance event. Predators, scavengers and bacterial action normally break down dead plant and animal structures before they can be fossilized. Of the relatively few fossils formed, most remain buried or, if they do become exposed, are overlooked or accidentally destroyed.

Nevertheless, numerous fossils have been found – and more continue to be discovered all the time. The various types of fossil and the steps in fossil formation by petrification are illustrated in Figure 5.1. Where it is the case that the fossil, or the rock that surrounds it, can be accurately dated (using radiometric dating techniques), we have good evidence of the history of life. Radiometric dating measures the amounts of naturally occurring radioactive substances such ^{14}carbon (in relation to the amount of ^{12}carbon), or ^{40}potassium:^{40}argon.

Fossils – extinct 'life' and intermediate forms

The fossils in a rock layer (stratum) that has been accurately dated give us clues to the community of organisms living at a particular time in the past, although necessarily an incomplete picture. The fossil record may also suggest the sequence in which groups of species evolved, and the timing of the appearance of the major phyla.

■ **Figure 5.1**
Fossilization

Fossil forms

petrification – organic matter of the dead organism is replaced by mineral ions

mould – the organic matter decays, but the space left becomes a mould, filled by mineral matter

trace – an impression of a form, such as a leaf or a footprint, made in layers that then harden

preservation – of the intact whole organism; for example, in amber (resin exuded from a conifer, which then solidified) or in anaerobic, acidic peat

Steps of fossil formation by petrification

1 dead remains of organisms may fall into a lake or sea and become buried in silt or sand, in anaerobic, low-temperature conditions

2 hard parts of skeleton or lignified plant tissues may persist and become impregnated by silica or carbonate ions, hardening them

3 remains hardened in this way become compressed in layers of sedimentary rock

4 after millions of years, upthrust may bring rocks to the surface and erosion of these rocks commences

5 land movements may expose some fossils and a few are discovered by chance but, of the relatively few organisms fossilised, very few will ever be found by humans

Sedimentary rock layers (with a fault line) and the remains of extinct fossil species

1 Many fossils are preserved in sedimentary rocks. **Explain** why this is so, with reference to how sedimentary rocks form.

Intermediate forms of animals and plants may be the links between related groups of species. So, on occasions, the fossil record ought to provide evidence of the evolutionary origins of new forms of life. In fact, on this it is frustratingly incomplete – very few 'missing links' have been found. However, a possible example of an intermediate form, discovered in the fossil record, is *Archaeopteryx*, a once-living intermediate between reptiles and birds (Figure 5.2).

The fossil *Archaeopteryx* had bird-like features (e.g. feathers) together with reptile-like features (e.g. teeth and a long tail). The structure and size of the skeleton suggest that its musculature was not strong enough to allow flight.

The fossil *Archaeopteryx*, found in 1861 – a landmark triumph for the ideas of Charles Darwin

Archaeopteryx – a reconstruction
This bird-like reptile may have climbed trees using claws on its wings, and then glided from tree to tree.

■ **Figure 5.2** Archaeopteryx – a missing link

■ Evidence from selective breeding

Selective breeding (or artificial selection) is caused by humans. It is usually a deliberate and planned activity. It involves identifying the largest, the best or the most useful of the progeny, and using them as the next generation of parents. Continuous removal of progeny showing less desired features, generation by generation, leads to deliberate genetic change. Indeed, the genetic constitution of the population may change rapidly.

Charles Darwin started breeding pigeons as a result of his interest in variation in organisms (Figure 5.3). In the *Origin of Species* he noted there were more than a dozen varieties of pigeon which, had they been presented as wild birds to an expert, would have been recognized as separate species. All these pigeons were descendants of the rock dove, a common wild bird.

■ **Figure 5.3** Charles
Darwin's observation
of pigeon breeding

From his breeding of pigeons, Darwin noted that there were more than a dozen varieties that, had they been presented to an onithologist as wild birrds, would have been classsified as separate species.

jacobin

pouter

rock dove (*Columbia livia*) from which all varieties of pigeon have been derived

fantail

Darwin argued that, if so much change can be induced in so few generations, species must be able to evolve into other species by the gradual accumulation of minute alterations, as environmental conditions change – selecting some progeny and not others.

By artificial selection, the plants and animals used by humans (such as in agriculture, transport, companionship and leisure) have been bred from wild organisms. The origins of artificial selection go back to the earliest developments of agriculture, although at this stage successful practice, no doubt, evolved by accident (Figure 5.4).

■ **Figure 5.4** From
wild to domesticated
species, and the
origins of selective
breeding skills

**Wild sheep or mouflon
(*Ovis musimon*) occur today
on Sardinia and Corsica**

What domestication of wild animals involved.

- The identification of a population of a 'species' as a useful source of hides, meat, etc. and learning how to tell these animals apart from related species. Herd animals (e.g. sheep and cattle) are naturally sociable, and lend themselves to this.
- The selective killing (culling) of the most suitable members of this herd in order to meet immediate needs for food and materials for living.
- Encouraging breeding among the docile, well-endowed members of the herd, and providing protection against predators.
- Selecting from the progeny individuals with the most useful features, and making them the future breeding stock.
- Maintenance of the breeding stock during unfavourable seasons
- Ultimately, the establishment of a domesticated herd, dependent on the herdspeople rather than living wild, leading to the possibility of trading individuals of a breeding stock with neighbouring herdspeople's stocks.

2 Charles Darwin argued that the great wealth of varieties we have produced in domestication supports the concept of evolution. **Outline** how this is so.

3 Dogs of the breeds known as Alsatians, Pekinese and Dachshunds are different in appearance yet are all classified as members of the same species. **Explain** how this is justified.

**Soay sheep of the outer Hebrides
suggest to us what the earliest
domesticated sheep looked like**

**Modern selective breeding has produced shorter
animals with a woolly fleece in place of coarse
hair and with muscle of higher fat content.
Many breeds have lost their horns**

Nature of Science	Looking for patterns, trends and discrepancies

■ Evidence from comparative anatomy

The body structures of some organisms appear fundamentally similar. For example, the limbs of vertebrates seem to conform to a common plan – called the pentadactyl limb (meaning 'five fingered'). So, we describe these limbs as **homologous structures**, for they occupy similar positions in an organism, have an underlying basic structure in common, but may have evolved different functions (Figure 5.5). The fact that limbs of vertebrates conform but show modification suggests these organisms share a common ancestry. From this common origin, the vertebrates have diverged over a long period of time. This process is called **adaptive radiation**.

Another example of adaptive radiation is illustrated by the beaks of the **finches of the Galápagos Islands** (Figure 5.6). These birds mostly failed to catch Charles Darwin's attention on his visits to the Islands – the detailed study of these birds was undertaken later by the ornithologist David Lack.

The variation in finch beak morphology is a genetically controlled characteristic. It reflects differences in feeding habits. The evidence the finches provided for Darwin's theory of evolution by natural selection so impressed Lack that he coined the name 'Darwin's finches'. It has stayed with them, misleading though it is.

We can contrast homologous structures with other structures of organisms that have similar functions but fundamentally different origins. Their resemblances are superficial; these are described as **analogous structures**. The wings of an insect and a bat are analogous, for example.

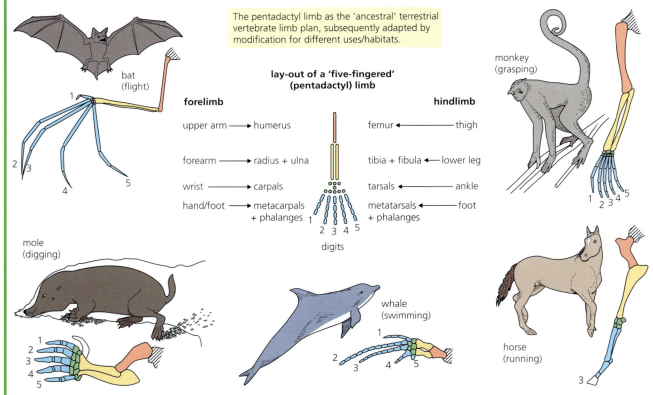

The pentadactyl limb as the 'ancestral' terrestrial vertebrate limb plan, subsequently adapted by modification for different uses/habitats.

bat (flight)

monkey (grasping)

lay-out of a 'five-fingered' (pentadactyl) limb

forelimb

upper arm ⟶ humerus
forearm ⟶ radius + ulna
wrist ⟶ carpals
hand/foot ⟶ metacarpals + phalanges

hindlimb

femur ⟵ thigh
tibia + fibula ⟵ lower leg
tarsals ⟵ ankle
metatarsals + phalanges ⟵ foot

digits

mole (digging)

whale (swimming)

horse (running)

■ **Figure 5.5** Homologous structures show adaptive radiation

Figure 5.6 Adaptive radiation in Galápagos finches

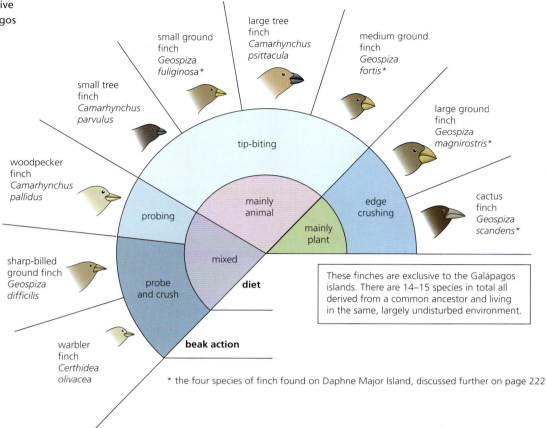

small ground finch
*Geospiza fuliginosa**

large tree finch
Camarhynchus psittacula

medium ground finch
*Geospiza fortis**

small tree finch
Camarhynchus parvulus

large ground finch
*Geospiza magnirostris**

woodpecker finch
Camarhynchus pallidus

cactus finch
*Geospiza scandens**

sharp-billed ground finch
Geospiza difficilis

warbler finch
Certhidea olivacea

tip-biting

probing

mainly animal

edge crushing

mainly plant

mixed

probe and crush

diet

beak action

These finches are exclusive to the Galápagos islands. There are 14–15 species in total all derived from a common ancestor and living in the same, largely undisturbed environment.

* the four species of finch found on Daphne Major Island, discussed further on page 222

■ Evidence from geographical distribution

Countries with similar climates and habitats might be expected to have the same flora and fauna. In fact, they are often distinctly different. The wildlife of South America and Africa are good examples. Both of these areas have a very similar range of latitudes and their habitats include tropical rainforests, savannah and mountain ranges. In addition, they share certain common fossil remains. These include a dinosaur known as *Mesosaurus*, which lived on both continents in the Jurassic period about 200 million years ago (mya).

Today the faunas (and floras) of these land masses differ profoundly. For example:

- South America now supports New World monkeys (these have tails), llamas, tapirs, pumas (a mountain lion) and jaguars.
- Africa now supports Old World monkeys, apes, African elephants, dromedaries, antelopes, giraffes and lions – but not the faunas of South America.

The explanation is that these land masses were once joined, and so shared a common flora and fauna – including the dinosaurs. However, about 150 mya they literally drifted apart (through **plate tectonics**, Figure 5.7). For the past 100 million years the organisms of South America and Africa have evolved in isolation from each other.

Species may originate in a given area and disperse out from that point, coming to occupy favourable habitats wherever they chance on them. But the huge ocean that opened up between South America and Africa formed an impossible barrier and evolution in each place took a separate path, producing different faunas and floras. This demonstrates that populations of species can gradually diverge into separate species by evolution.

■ **Figure 5.7**
Continental drift
(plate tectonics)

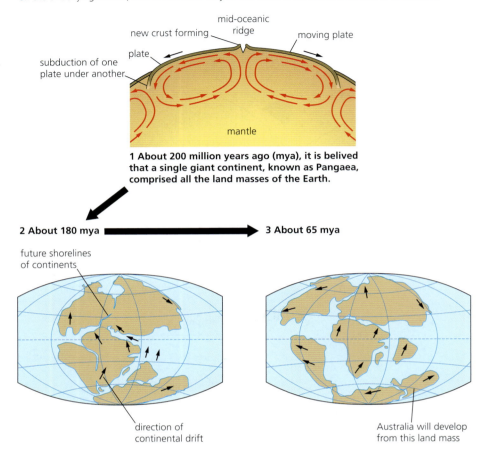

Continental drift (plate tectonics): the continents of the Earth are parts of the crust that 'float' on the underlying mantle; their movements may be the result of convection currents in the mantle.

mid-oceanic ridge

new crust forming

moving plate

plate

subduction of one plate under another

mantle

1 About 200 million years ago (mya), it is belived that a single giant continent, known as Pangaea, comprised all the land masses of the Earth.

2 About 180 mya   **3 About 65 mya**

future shorelines of continents

direction of continental drift

Australia will develop from this land mass

■ Local populations – the venues for evolution

In nature, organisms of the same species are members of a local population. Remember, a population is a group of individuals of a species, living close together, and able to interbreed. Therefore, we can look to local populations as the venues for evolution.

Look at Figure 5.8, now.

You can see that a population of garden snails might occupy a small part of a garden, say around a compost heap. A population of thrushes (snail-eating birds) might occupy some gardens and surrounding fields. So, the area occupied by a population depends on the size of the organism and on how mobile it is, for example, as well as on environmental factors (such as food supply and predation).

The boundaries of a population may be hard to define. Some populations are completely 'open', with individuals moving in from or out to nearby populations. Alternatively, some populations are more or less 'closed'; that is they are isolated communities, almost completely cut off from neighbours of the same species. The fish found in a small lake are a good example of a closed population.

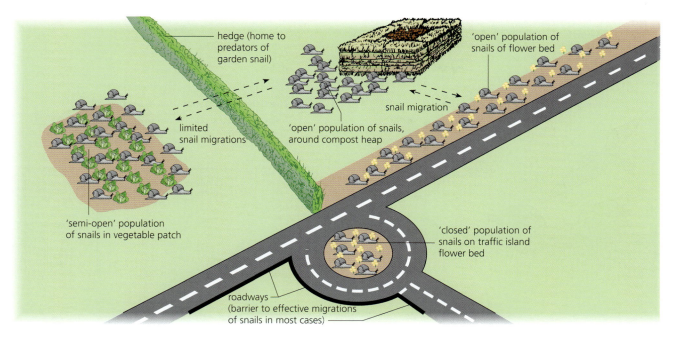

'open' population of snails of flower bed

hedge (home to predators of garden snail)

snail migration

limited snail migrations

'open' population of snails, around compost heap

'semi-open' population of snails in vegetable patch

'closed' population of snails on traffic island flower bed

roadways (barrier to effective migrations of snails in most cases)

■ **Figure 5.8** The concept of population

Divergence in local populations

Development of barriers within local populations is a possibility. Before separation, individuals share a common **gene pool** (page 430) but, after isolation, processes like mutation could trigger change in one population but not in the other.

Alternatively, a new population may form from a tiny sample that became separated from a much larger population. While the number of individuals in the new population may rapidly increase, the gene pool from which they formed might have been totally unrepresentative of the original, with many alleles lost altogether.

The outcome of these processes may be marked divergence between populations, leading to distinctly different characteristics.

The isolated islands of the Galápagos, off the coast of South America, are 500 to 600 miles from the mainland. These islands had a volcanic origin – they appeared out of the sea about 16 mya. Initially, they were uninhabited. Today, they have a rich flora and fauna which clearly relate to mainland species.

This has been brought about in a number of ways. For example, motile or mobile species are dispersed to isolated habitats when organisms are accidentally rafted from mainland territories to distant islands. The 2004 tsunami in south-east Asia demonstrated how this happens. Violent events of this type have punctuated world geological history with surprisingly frequency. The story of how the Galápagos Islands have become populated with mainland species is outlined in Figure 5.9. *Look at it now.*

The iguana lizard was one example of an animal species that came to the Galápagos from mainland South America. The iguana found no mammal competition when it arrived on the islands. Indeed, it was without any significant predators and, as a result, it became the dominant form of vertebrate life. We can assume it became extremely abundant. Today, there are *two* species of iguana present, one terrestrial and the other fully adapted to marine life. The latter is assumed to have evolved locally as a result of pressure from overcrowding and competition for food (both species are vegetarian), which drove some members of the population out of the terrestrial habitat. This is an example of a local population of a species gradually diverging into separate species – another example of evolution.

Many organisms (e.g. insects and birds) may have flown or been carried on wind currents to the Galápagos from the mainland. Mammals are most unlikely to have survived drifting there on a natural raft over this distance, but many large reptiles can survive long periods without food or water.

immigrant travel to the Galápagos

The Galápagos Islands
Today the tortoise population of each island is distinctive and identifiable.

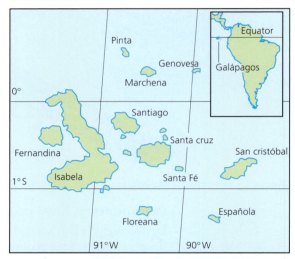

The **giant iguana lizards** on the Galápagos Islands became dominant vertebrates, and today are two distinct species, one still terrestrial, the other marine, with webbed feet and a laterally flattened tail (like the caudal fin of a fish).

Terrestrial iguana

Marine iguana

■ **Figure 5.9** The Galápagos Islands and species divergence there

■ Human activities may drive divergence – melanic insects in polluted areas

During the Industrial Revolution in Britain, in the early part of the nineteenth century, air pollution by gases (such as sulfur dioxide) and solid matter (mainly soot) was distributed over the industrial towns, cities and surrounding countryside. Here, lichens and mosses on brickwork and tree trunks were killed off and these surfaces were blackened. The numbers of dark varieties of some 80 species of moth increased in these habitats in this period. This rise in proportion of darkened forms is known as **industrial melanism**.

The dark-coloured (**melanic**) form of the peppered moth *Biston betularia* tended to increase in these industrialized areas, but their numbers were low in unpolluted countryside, where **pale, speckled forms** of moths were far more common (Figure 5.10). The melanic form was effectively camouflaged from predation by insectivorous birds in sooty areas, and became the dominant species.

Biston betularia

pale form observed
in non-polluted
habitats

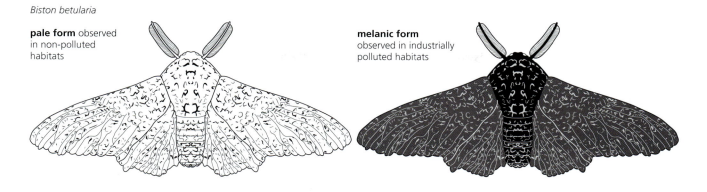

melanic form
observed in industrially
polluted habitats

experimental evidence

Key

■ melanic form

□ pale form

* evidence of
  selective
  predation

**results of frequency studies in
polluted and unpolluted habitats**

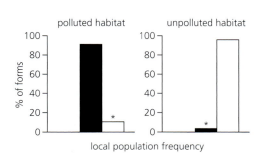

**mark–release–capture experiments using
laboratory-reared moths of both forms, in
polluted and unpolluted habitats**

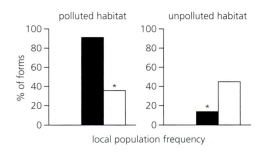

■ **Figure 5.10** The peppered moth and industrial melanism

4 **Explain** the key
difference between
natural and artificial
selection.

TOK Link

In the study of evolutionary history, do experiments have any part to play in establishing knowledge? If an
experimental approach has a limited role, is the study of evolution a 'science'?

5.2 Natural selection – *the diversity of life has evolved and continues to evolve by natural selection*

■ Evolution by natural selection – the ideas and arguments

Charles Darwin (1809–82) was a careful observer and naturalist who made many discoveries in
biology. After attempting to become a doctor (at Edinburgh University) and then a clergyman
(at Cambridge University), he became the unpaid naturalist on an Admiralty-commissioned
expedition to the Southern Hemisphere, on a ship called HMS *Beagle*. On this five-year
expedition around the world, and in his later investigations and reading, he developed the idea
of **organic evolution by natural selection**.

Darwin remained very anxious (always) about how the idea of evolution might be received,
and he made no moves to publish it until the same idea was presented to him in a letter by
another biologist and traveller, **Alfred Russel Wallace**. Only then (1859) was *On the Origin of
Species by Natural Selection* completed and published. The arguments and ideas in the *Origin of
Species* are summarized in Table 5.1.

■ **Table 5.1** Charles Darwin's ideas about the origin of species, summarized in four statements (S) and three deductions (D) from these statements

	Statements and deductions
S1	Organisms produce a far greater number of progeny than ever give rise to mature individuals.
S2	The number of individuals in species remains more or less constant.
D1	Therefore, there must be a high mortality rate.
S3	The individuals in a species are not all identical, but show variations in their characteristics.
D2	Therefore, some variants will succeed better than others in the competition for survival. So the parents for the next generation will be selected from those members of the species better adapted to the conditions of the environment.
S4	Hereditary resemblance between parents and offspring is a fact.
D3	Therefore, subsequent generations will maintain and improve in the degree of adaptation of their parents, by gradual change.

■ Neo-Darwinism

Charles Darwin (and nearly everyone else in the scientific community of his time) knew nothing about Mendel's work on genetics. Instead, biologists generally subscribed to the concept of 'blending inheritance' when mating occurred (which would *reduce* the genetic variation available for natural selection).

Neo-Darwinism is an essential restatement of the concepts of evolution by natural selection in terms of Mendelian and post-Mendelian genetics, as follows.

Organisms produce many more offspring than survive to be mature individuals

Darwin did not coin the phrase 'struggle for existence', but it does sum up the point that the over-production of offspring in the wild leads naturally to their competition for resources. Table 5.2 is a list of the normal rate of production of offspring in some common species.

■ **Table 5.2** Numbers of offspring produced

Organism	No. of eggs/seeds/young per brood or season
rabbit	8–12
great tit	10
cod	2–20 million
honey bee (queen)	120 000
poppy	6000

How many of these offspring survive to breed themselves?

In fact, in a stable population on average, a breeding pair gives rise to a single breeding pair of offspring. All their other offspring are casualties of the 'struggle'; many organisms die before they can reproduce.

So, populations do not show rapidly increasing numbers in most habitats or, at least, not for long. Population size is naturally limited by restraints that we call **environmental factors**. These include space, light and the availability of food. The never-ending competition for resources means that the majority of organisms fail to survive and reproduce. In effect, the environment can only support a certain number of organisms, and the number of individuals in a species remains more or less constant over a period of time.

Natural selection can only occur if there is variation amongst members of the same species

The individuals in a species are not all identical, but show variations in their characteristics. Today, modern genetics has shown us that there are several ways by which **genetic variations** arise.

1 Variation arises in **meiosis** in gamete formation, and in **sexual reproduction** at fertilization

We have seen that genetic variations arise via:

- **independent assortment** of paternal and maternal chromosomes in meiosis (in the process of gamete formation, page 144)
- **crossing over** of segments of individual maternal and paternal homologous chromosomes (resulting in new combinations of genes on the chromosomes of the haploid gametes produced by meiosis, pages 144 and 413)
- the **random fusion** of male and female gametes in sexual reproduction.

These processes operate each time meiosis occurs and is followed by fertilization. The results are **new combinations** of existing characteristics that may favour individuals in their lifetime – they may affect survival and opportunities to reproduce. If so, a particular individual's success in reproduction will result in certain alleles being passed on to the next generation in greater proportions than other alleles.

2   Variations arise as the product of mutation (page 132), giving entirely **new alleles**. We have seen that mutations are changes in DNA. Mutations occurring in ovaries or testes (or anthers or embryo sacs of flowering plants) are called **germ line mutations**. Since these mutations occur in the cells that give rise to gametes, they may be passed to the offspring. Meanwhile, mutations that occur in body cells of multicellular organisms (**somatic mutations**) are only passed on to the immediate descendants of those cells, and they disappear when the organism dies. So, new characteristics acquired during the lifetime of an individual are not heritable.

As a result of all these factors, the individual offspring of parents are not identical. Rather, they show variations in their characteristics.

Natural selection results in offspring with favourable characteristics

When genetic variation has arisen in organisms:

- the favourable characteristics are expressed in the phenotypes of some of the offspring
- these offspring may be better able to survive and reproduce in a particular environment; others will be less able to compete successfully to survive and reproduce.

Thus, natural selection operates to determine the survivors and the genes that are perpetuated in future progeny. In time, this selection process may lead to new varieties and new species.

The operation of natural selection is sometimes summarized in the phrase **survival of the fittest**, although these were not words that Darwin used, at least not initially.

To avoid the criticism that 'survival of the fittest' is a circular phrase (how can fitness be judged except in terms of survival?), the term 'fittest' is understood in a particular context. For example, the fittest of the wildebeest (hunted herbivores) of the African savannah may be those with the acutest senses, quickest reflexes and strongest leg muscles for efficient escape from predators. By natural selection for these characteristics, the health and survival of wildebeests is assured.

5 **Deduce** the importance of modern genetics to the theory of the origin of species by natural selection.

■ Environmental change and speciation

When the environment changes, some individuals present may be at a disadvantage. If so, natural selection is likely to operate on individuals and cause changes to gene pools. We have already noted that individuals possessing a particular allele, or combination of alleles, may be more likely to survive, breed and pass on their alleles than are other, less well-adapted individuals. This process is also referred to as **differential mortality**. Two examples follow.

Nature of Science

Use theories to explain natural phenomena

Multiple antibiotic resistance in bacteria

In this example, the majority of a population of organisms is not suited to an environment that has changed (with the addition of an antibiotic), but an unusual or abnormal form of the population is suited and, therefore, has a selective advantage.

Here we are concerned with a species of pathogenic bacteria. Patients who are infected with the bacterium are treated with an antibiotic to help them overcome the disease; antibiotics are very widely used. In a large population of that bacterium, some individuals may carry a gene for resistance to the antibiotic in question. Such genes sometimes arise by spontaneous mutation. Alternatively, the gene may have been acquired through sexual reproduction between bacteria of different populations.

The resistant bacteria in the population have no selective advantage *in the absence of the antibiotic* and must compete for resources with non-resistant bacteria. But when the antibiotic *is* present, most bacteria of the population will be killed. The resistant bacteria are very likely to survive and will be the basis of the future population. In the new population, all individuals now carry the gene for resistance to the antibiotic. The genome has changed abruptly (Figure 5.11).

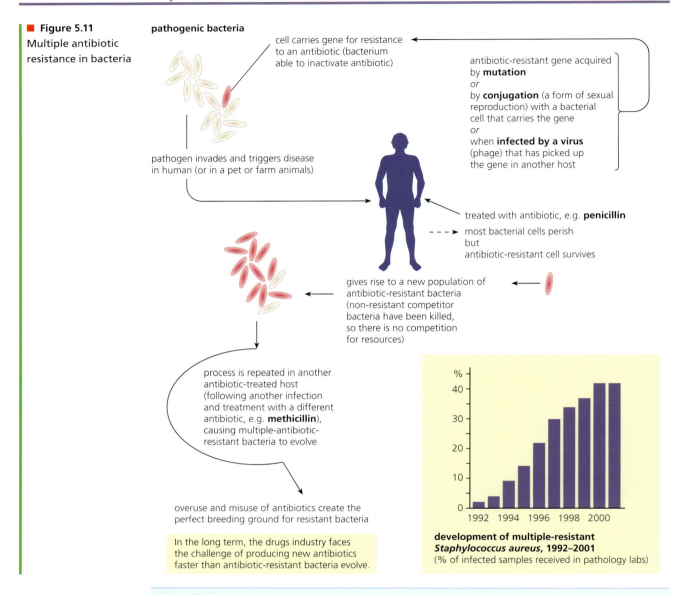

pathogenic bacteria

cell carries gene for resistance to an antibiotic (bacterium able to inactivate antibiotic)

antibiotic-resistant gene acquired by **mutation**
or
by **conjugation** (a form of sexual reproduction) with a bacterial cell that carries the gene
or
when **infected by a virus** (phage) that has picked up the gene in another host

pathogen invades and triggers disease in human (or in a pet or farm animals)

treated with antibiotic, e.g. **penicillin**

most bacterial cells perish
but
antibiotic-resistant cell survives

gives rise to a new population of antibiotic-resistant bacteria (non-resistant competitor bacteria have been killed, so there is no competition for resources)

process is repeated in another antibiotic-treated host (following another infection and treatment with a different antibiotic, e.g. **methicillin**), causing multiple-antibiotic-resistant bacteria to evolve

overuse and misuse of antibiotics create the perfect breeding ground for resistant bacteria

In the long term, the drugs industry faces the challenge of producing new antibiotics faster than antibiotic-resistant bacteria evolve.

development of multiple-resistant *Staphylococcus aureus*, 1992–2001
(% of infected samples received in pathology labs)

6 Explain:
   a why doctors ask patients to complete the full course of antibiotics, even if they start to feel better
   b why the medical profession tries to combat resistance by alternating the types of antibiotic used against an infection.

Changes in beak size of medium ground finches on Daphne Major

This example was the discovery of the evolutionary biologists Peter and Rosemary Grant. It, too, illustrates how evolution can occur surprisingly quickly, rather than *only* by slow change over long periods of time. The Grants researched the finch populations of two islands in the Galápagos, Daphne Major and Daphne Minor.

On Daphne Major, the medium ground finch (*Geospiza fortis* – Figure 5.6) tends to feed on small tender seeds, such as are available in abundance in wet years. During long dry periods (as occurred in 1977, 1980 and 1982), once the limited stocks of smaller seeds had been eaten, the surviving birds were those that could feed on larger, drier seeds that are more difficult to crack open. It was discovered in periods of drought that average beak size increased (Figure 5.12).

The explanation of this change was that when seeds were plentiful, beak size was not critical. Beaks of a range of dimensions allowed successful feeding and breeding by all pairs. At these times, a range of beak sizes was maintained in the population (but possession of a larger beak conferred no particular advantage). When only large, hard seeds were available, birds with larger, stronger beaks survived and bred, whereas those with smaller beaks mostly did not. This led to a change in the gene pool.

The Grants confirmed this hypothesis by measuring the beak size of parents and their offspring broods in this species *over several years*. They found that beak size is genetically controlled – parents with large beaks had offspring with large beaks, generation after generation. So, in drought periods differential mortality quickly changes the genetic constitution of the populations of G. *fortis*.

■ Figure 5.12 Change in mean beak size in *Geospiza fortis* populations on Daphne Major, 1975–1983, provides evidence of natural selection

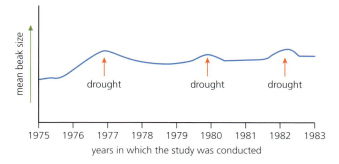

7 In Western culture, the biblical account of creation was generally accepted as authoritative until the eighteenth century. The chronology detailed in the Bible suggested that life had appeared on Earth a mere few thousand years ago. This meant the Earth was only 5000–6000 years old. James Hutton (1726–97), a doctor, farmer and experimental scientist, realized that the sedimentary rocks of many existing mountain ranges had once been the beds of lakes and seas and, before that, had been the rocks of even older mountains. He made no estimate of the age of the Earth, but he realized that, in contrast to biblical estimates, the Earth's timescale virtually had no beginning and no end.

Explain the significance for evolutionary theory of the realization by early geologists that the Earth was more than a few thousand years old.

8 **Suggest** what sort of events might, in the past, have caused violent and speedy habitat change over a substantial part of the surface of the Earth.

TOK Link

The word 'theory' comes from a Greek word meaning 'seeing'. Today, a scientific theory normally explains something, usually via some mental model. However, there is diversity of scientific theories; they are not always precise and mathematical, like those of Isaac Newton (1642–1727), which seized people's imaginations and enabled them to see a particular branch of science as extremely precise and exact.

Dip into the *Origin of Species* – say Chapter 7. Find the paragraph on the Greenland whale and the evolution of the baleen plates. How would you describe this reasoning? What sort of 'theory' is presented here?

5.3 Classification of biodiversity *– species are named and classified using an internationally agreed system*

■ The range of living organisms and their classification

There are vast numbers of different types of living organism in the world – almost unlimited diversity, in fact. We refer to this issue as '**biodiversity**', which is a contraction of the words 'biological' and 'diversity'. By biodiversity we mean '**the total number of different species living in a defined area or ecosystem**'. It is also applied to the incredible abundance of different types of species on Earth.

Up to now, about 2 million species have been described and named in total. However, until very recently, there was no international 'library of living things' where new discoveries could be checked out. Meanwhile, previously unknown species are being discovered all the time.

We might have expected all the wildlife to be known in the countries that pioneered the systematic study of plants and animals. This is not the case; previously unknown organisms are frequently found in all countries. Worldwide, the number of unknown species is estimated at between 3–5 million at the very least, and possibly as high as 100 million. So scientists are not certain just how many different types of organisms exist. Figure 5.13 is a representation of the relative numbers of known species that are estimated to exist in many of the major divisions of living things.

■ **Figure 5.13**
The relative number of animal and plant groups

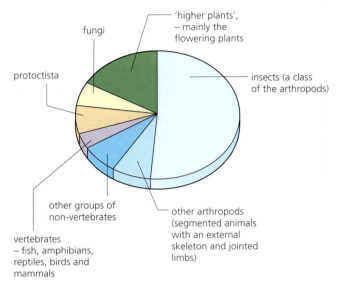

Of the 1.8 million described species, more than 50% are insects, and the higher plants, mostly flowering plants, are the next largest group. By contrast, only 4000 species of mammals are known, about 0.25% of all known species.

Taxonomy – the classification of diversity

Classification is essential to biology because there are too many different living things to sort and compare, unless they are organized into manageable categories. Biological classification schemes are the invention of biologists and are based upon the best available evidence at the time. With an effective classification system in use, it is easier to organize our ideas about organisms and make generalizations.

The science of classification is called '**taxonomy**'. The word comes from 'taxa' (sing = taxon), which is the general name for groups or categories within a classification system. The scheme of classification has to be flexible, allowing newly discovered living organisms to be added where they fit best. It should also include fossils, since we believe living and extinct species are related.

The process of classification involves:

■ giving every organism an agreed name
■ imposing a scheme upon the diversity of living things.

Nature of Science | **Cooperation and collaboration between groups of scientists**

■ The binomial system of naming

Many organisms have local names, but these often differ around the world. They do not allow observers to be confident that they are all talking about the same organism. For example, the name 'magpie' represents entirely different birds commonly seen in Europe, in Asia and in Sri Lanka (Figure 5.14). Instead, scientists use an international approach called the binomial system (meaning 'a two-part name'). It has been by international cooperation and collaboration that the principles to be followed in the classification of living organisms have been developed. Using this system, everyone everywhere knows exactly which organism is being referred to.

(a) European magpie (*Pica pica*)

(b) Asian magpie (*Platysmurus leucopterus*)

(c) Sri Lankan magpie (*Urocissa ornate*)

■ **Figure 5.14** 'Magpie' species of the world

So, each organism is given a scientific name consisting of two words in Latin. The first (a noun) designates the **genus**, the second (an adjective) describes the **species**. The generic name comes first and begins with a capital letter, followed by the specific name. Conventionally, this name is written in *italics* (or is underlined). As shown in Figure 5.15, closely related organisms have the same generic name; only their species names differ. You will see that, when organisms are referred to frequently, the full name is given initially, but thereafter the generic name is shortened to the first (capital) letter. Thus, in continuing references to humans in an article or scientific paper, *Homo sapiens* would become *H. sapiens*.

***Panthera leo* (lion)**

***Panthera tigris* (tiger)**

***Homo habilis* (handy man – extinct)**

***Homo sapiens* (modern human)**

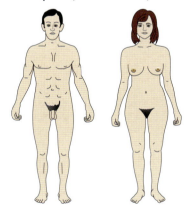

9 Scientific names of organisms are often difficult to pronounce or remember. **State** why they are used.

■ **Figure 5.15** Naming organisms by the binomial system

■ The scheme of classification

In classification, the aim is to use as many characteristics as possible when placing similar organisms together and dissimilar ones apart. Just as similar **species** are grouped together into the same **genus** (plural = genera), so too, similar genera are grouped together into **families**. This approach is extended from families to **orders**, then **classes**, **phyla** and **kingdoms**. This is the hierarchical scheme of classification, with each successive group containing more and more different kinds of organism. In a natural classification, the genus and accompanying higher taxa consist of all the species that have evolved from one common ancestral species.

The taxa used in taxonomy are given in Figure 5.16.

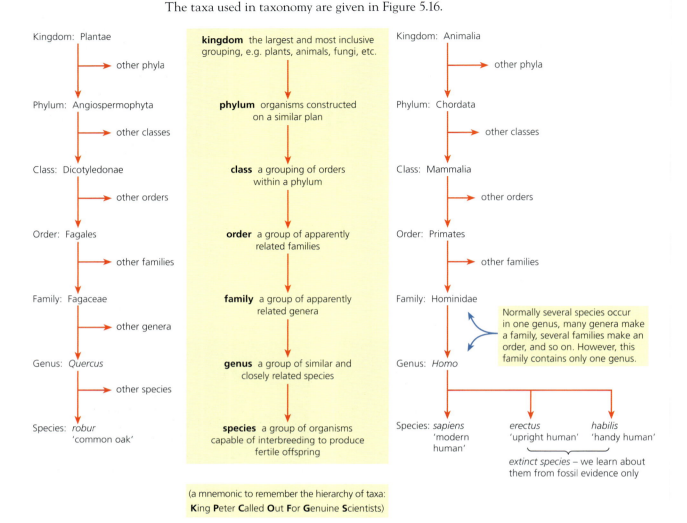

(a mnemonic to remember the hierarchy of taxa:
King **P**eter **C**alled **O**ut **F**or **G**enuine **S**cientists)

■ **Figure 5.16** The taxa used in taxonomy, applied to genera from two different kingdoms

Domains and kingdoms

At one time the living world seemed to divide naturally into two kingdoms consisting of the **plants** (with autotrophic nutrition) and the **animals** (with heterotrophic nutrition). These two kingdoms grew from the original disciplines of biology, namely **botany**, the study of plants, and **zoology**, the study of animals. Fungi and microorganisms were conveniently 'added' to botany! Initially there was only one problem; fungi possessed the typically 'animal' heterotrophic nutrition but were 'plant-like' in structure.

Then, with the use of the **electron microscope**, came the discovery of the two types of cell structure, namely **prokaryotic** and **eukaryotic** (page 29). As a result, the bacteria with their prokaryotic cells could no longer be 'plants', since plants have eukaryotic cells. The division of living things into kingdoms needed overhauling. This led to the division of living things into five kingdoms (Table 5.3). Today, taxonomists sometimes reclassify groups of species when new evidence shows that a previous taxon contains species that have evolved from different ancestral species.

■ **Table 5.3** The five kingdoms

Prokaryotae – the prokaryote kingdom, the bacteria and cyanobacteria (photosynthetic bacteria), predominately unicellular organisms.
Protoctista – the protoctistan kingdom (eukaryotes), predominately unicellular, and seen as resembling the ancestors of the fungi, plants and animals.
Fungi – the fungal kingdom (eukaryotes), predominately multicellular organisms, non-motile and with heterotrophic nutrition.
Plantae – the plant kingdom (eukaryotes), multicellular organisms, non-motile, with autotrophic nutrition.
Animalia – the animal kingdom (eukaryotes), multicellular organisms, motile, with heterotrophic nutrition.
Note that viruses are *not* classified as living organisms.

Extremophiles – and the recognition of 'domains'

Then came the discovery of the distinctive biochemistry of the bacteria found in extremely hostile environments (the **extremophiles**), such as the 'heat-loving' bacteria found in hot springs at about 70°C. This led on to a new scheme of classification. These microorganisms of extreme habitats have cells that we can identify as prokaryotic. However, the larger RNA molecules present in the ribosomes of extremophiles are different from those of previously known bacteria. Further analyses of their biochemistry in comparison with that of other groups has suggested new evolutionary relationships (Figure 5.17).

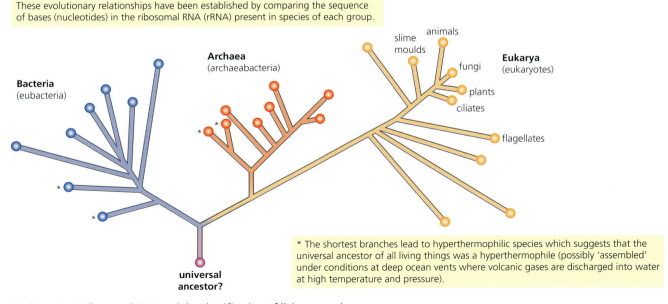

These evolutionary relationships have been established by comparing the sequence of bases (nucleotides) in the ribosomal RNA (rRNA) present in species of each group.

* The shortest branches lead to hyperthermophilic species which suggests that the universal ancestor of all living things was a hyperthermophile (possibly 'assembled' under conditions at deep ocean vents where volcanic gases are discharged into water at high temperature and pressure).

■ **Figure 5.17** Ribosomal RNA and the classification of living organisms

All organisms are now classified into three domains

As a result, we now recognize three major forms of life, called **domains**. The organisms of each domain share a distinctive, unique pattern of ribosomal RNA and there are other differences which establish their evolutionary relationships (Table 5.4). These **domains** are:

- the **Archaea** (the extremophile prokaryotes)
- the **Eubacteria** (the true bacteria)
- the **Eukaryota** (all eukaryotic cells – the protoctista, fungi, plants and animals).

Actually, the Archaea have been found in a broader range of habitats than merely extreme environments. Today, some occur in the oceans and some in fossil-fuel deposits deep underground. Some species occur at deep ocean vents, in high temperature habitats such as geysers, in salt pans and in polar environments. Some species occur only in anaerobic enclosures, such as in the guts of termites and cattle, and at the bottom of ponds among the rotting plant remains. Here, they breakdown organic matter and release methane – with important environmental consequences.

■ **Table 5.4**
Biochemical differences between the domains

Biochemical features	Domains		
	Archaea	Eubacteria	Eukaryota
DNA of chromosome(s)	circular genome	circular genome	chromosomes
bound protein (histone) present in DNA	present	absent	present
Introns in genes	typically absent	typically absent	frequent
Cell wall	present – not made of peptidoglycan	present – made of peptidoglycan	sometimes present – never made of peptidoglycan
Lipids of cell membrane bilayer:	Archaeal membranes contain lipids that differ from those of eubacteria and eukaryotes (Figure 5.18).		

■ **Figure 5.18** Lipid structure of cell membranes in the three domains

Phospholipids of archaeal membranes

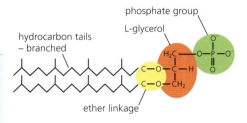

Phospholipids of eubacteria and eukaryote membranes

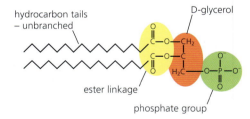

10 In Figure 5.16, one plant and one animal species are classified from kingdom to species level. **Suggest** how these flow diagrams require to be modified to show their classification from 'domain' level.

TOK Link

A major advance in the study of bacteria was the recognition, in 1977 by Carl Woese, that Archaea have a separate line of evolutionary descent to bacteria. Famous scientists, including Luria and Mayr, objected to his division of the prokaryotes. To what extent is conservatism in science desirable?

Classification of the plant kingdom

The green plants are terrestrial organisms, adapted to life on land, although some do occur in aquatic habitats. They are eukaryotic organisms, with a wall containing cellulose around each cell. In their nutrition, green plants are autotrophic organisms, manufacturing sugar by photosynthesis in their chloroplasts. The sugar is then stored or used immediately to sustain the whole of the metabolism.

A distinctive feature of green plants is their rather **complex lifecycle**: there are two stages or generations – a gametophyte generation that produces gametes and a sporophyte generation that forms spores. *The details of this feature do not concern us here, but it does account for some otherwise puzzling aspects of green plant structures and lifecycles.*

The phyla that make up the green plants

Green plants range from simple, tiny mosses to huge trees – incidentally, trees include some of the largest and oldest living things. The four main phyla of green plants are the mosses and liverworts (Phylum Bryophyta), ferns (Phylum Filicinophyta), conifers (Phylum Coniferophyta) and flowering plants (Phylum Angiospermophyta). Figures 5.19–5.22 present the **simple recognition features of the four phyla**, in turn.

■ **Figure 5.19** Phylum Bryophyta – the bryophytes

the moss *Funaria* is an early colonizer of land after a fire has killed larger plants in woods or heathland

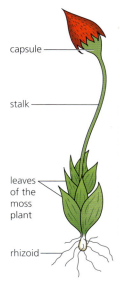

capsule

stalk

leaves of the moss plant

rhizoid

Introducing the bryophytes
- All are land plants, yet poorly adapted to terrestrial conditions – typically restricted to damp environments.
- Plant constructed of a tiny stem and leaves arranged radially. Roots absent, but stem anchored by hair-like rhizoids.
- Leaves not protected from water loss by a waxy cuticle, and stem contains no water-conducting cells or supporting 'fibres' (no vascular tissue at all).
- Spore-containing capsule grows on the main (haploid) plant on a long stalk with its 'foot' in the moss stem.
- Spore capsule may have an elaborate spore-dispersing valve mechanism.
- Alternatively, some bryophytes are flat leaf-like structures on the soil surface and are known as liverworts.

■ **Figure 5.20** Phylum Filicinophyta – the filicinophytes

the fern *Dryopteris*

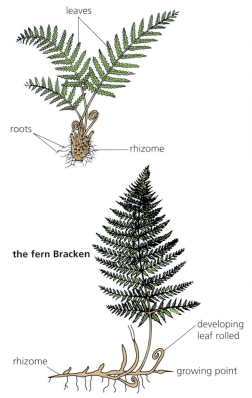

leaves

roots

rhizome

the fern Bracken

rhizome

developing leaf rolled

growing point

Introducing the filicinophytes
- Ferns are green plants with stems, leaves and roots and are well-adapted to terrestrial conditions. (Stems growing just below ground are called rhizomes.)
- Within stem, leaves and roots is vascular tissue for conducting water and nutrients around the plant.
- The leaves are elaborate structures that form tightly coiled up, and uncoil in early growth.
- Leaves are covered by a waxy cuticle that protects against water loss by evaporation.
- Spore-producing structures (sporangia) occur in clusters on the under-surface of the leaves, protected below a flap of tissue.
- Spores are released explosively, thrown some distance, and then germinate to produce a tiny, independent leaf-like plant.
- This tiny gamete-forming plant (haploid) is where the zygote is formed that then grows into a new fern plant.
- Present day ferns are relatively small survivors of an ancient group, dominant within the Carboniferous period (about 355 million years ago). Huge forests grew across a swampy land, with tree-like ferns – now present in today's fossil fuels.

■ **Figure 5.21** Phylum Coniferophyta – the coniferophytes

the conifer tree *Pinus*

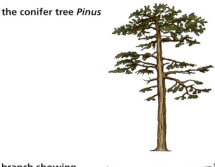

branch showing position of female cones

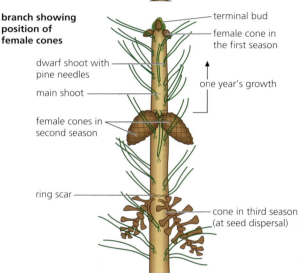

- terminal bud
- female cone in the first season
- dwarf shoot with pine needles
- main shoot
- one year's growth
- female cones in second season
- ring scar
- cone in third season (at seed dispersal)

Introducing the coniferophytes
- Cone-bearing trees, usually large with strong stem (trunk). They grow well, even on poor soils, and are the dominant plants in northern (boreal) forests, for example.
- The main trunk continues to grow straight, and side branches are typically formed in whorls, giving young trees simple cone-shaped outlines. Within stem, leaves and roots is vascular tissue for conducting water and nutrients around the plant.
- Leaves are usually waxy and needle-shaped. They are mostly evergreen, the leaves able to resist damage in low temperatures and heavy snow fall, when water supplies are locked away as ice.
- Seed-bearing plants, the seeds form in female cones. Male and female cones often occur on the same tree. Typically, the seed takes 2–3 years to form from arrival of pollen (from male cones) to release and dispersal of seeds.
- Survival on poor soils is aided by their mutually advantageous relationship with soil fungi in modified roots (mycorrhiza) in which fungal threads (hyphae) 'trade' ions from the soil for excess sugar that forms in the tree.
- Since confers mostly grow fast and straight, they form economically important crops of timbers known as 'soft woods'.

■ **Figure 5.22** Phylum Angiospermophyta – the angiospermophytes

the monocotyledon grass *Poa*

the dicotyledon *Ranunculus*

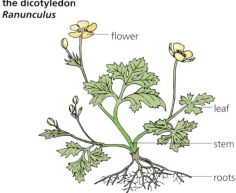

- flower
- leaf
- stem
- roots

Introducing the angiospermophytes
- The angiospermophytes are the dominant group of land plants. Many are herbaceous (non-woody) plants; others are trees (hard woods) or shrubs.
- The stem, leaves and roots contain vascular tissue (xylem and phloem) that delivers water and nutrients all around the plant.
- Leaves are elaborate structures with a waxy, waterproof covering and pores (stomata) in the surface.
- Flowers are unique to the angiospermophytes, and from them seeds are formed. Seeds are enclosed in an ovary and, after fertilization, the ovary develops into a fruit.
- With development of the flower have come complex mechanisms of pollen transfer and seed dispersal, often involving insects, birds, mammals or wind and water.
- The angiospermophytes are divided into the monocotyledons and dicotyledons.

  The monocotyledons (e.g. the grasses) mostly have parallel veins in their leaves and have a single seed-leaf in the embryo in the seed.
- The dicotyledons (the broad-leaved plants) have net veins in their leaves and two seed leaves in the embryo.

Classification of the animal kingdom

Animals are multicellular, eukaryotic organisms with heterotrophic nutrition. Typically, the body of an animal has cells that are highly specialized by their structure and physiology to perform particular functions, such as muscle cells for movement. Specialized tissues often occur together to form organs, which carry out particular functions in the body. Most animals have some form of nervous system to coordinate their body actions and responses. These features of animal structure and function are mostly explored in relation to the human mammal, in Chapter 6.

A point of contrast with plants is the simplicity of the lifecycle of animals (although some parasitic animals are an exception to this). Their lifecycle is diploid, with the adult producing haploid gametes (sperms and ova) by meiosis. After fertilization, the zygote divides to produce an embryo, which, early in development, becomes a characteristic hollow ball of cells, called a **blastula**.

An animal's lifestyle may be reflected in its body plan – many animals are in more or less constant movement, often in search of food, for example. The symmetry of the body of a motile organism is typically **bilateral**, meaning that there is only one plane that cuts the body into two equal halves. Also, with motility comes a compact body, elongated in the direction of movement, so shaped to offer least resistance to the surrounding medium – air or water. Since the anterior (front) end experiences the changing environment first, sense organs become located there. The result is the evolution of a head which is distinct from the rest of the body, a developmental process called **cephalization**.

However, the body of an animal with a non-motile or sessile lifestyle may show radial symmetry. The body is approximately cylindrically organized, with body parts arranged around a central axis. There are many planes by which the body can be cut into equal halves. No head is formed.

Some non-vertebrate phyla of the animal kingdom

The phyla of animals is introduced here, with a selection of non-vertebrate groups, namely the sponges (Phylum Porifera), jellyfishes and sea anemones (Phylum Cnidaria), flatworms (Phylum Platyhelminthes), roundworms (Phylum Nematoda), the molluscs (Phylum Mollusca), and the jointed-limbed animals (Phylum Arthropoda). Figures 5.23–5.28 present the **simple recognition features of these six phyla**, in turn.

■ **Figure 5.23**
Phylum Porifera – the sponges

the sponge *Leucosolenia*

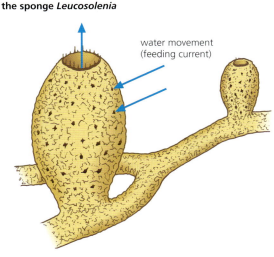

water movement (feeding current)

Leucosolenia (1–2 cm wide)

Introducing the porifera
- The porifera are the simplest multicellular animals, structurally little more than colonies of cells. They are aquatic and mostly marine animals (once thought to be plants).

- Sponges are formed into simple sac-like structures of cells in two layers, arranged around a central (gastric) cavity.

- Cells of the walls of the sponge specialize for feeding, structural support or reproduction. They are the only multicellular animals to entirely lack any nervous system.

- Feeding is on tiny suspended particles – plankton – which are drawn in through pores in the walls and are taken up by individual cells to be digested.

- Sponges may reproduce asexually by budding and also sexually – forming free-swimming larvae which is the dispersal stage for this sedentary animal.

■ **Figure 5.24**
Phylum Cnidaria –
the jellyfishes and
sea anemones

Hydra
a fresh water hydroid cnidarian

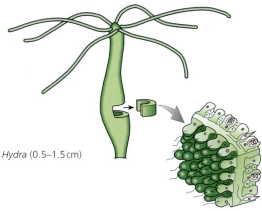

Hydra (0.5–1.5 cm)

Aurelia
a marine jellyfish
where the medusoid form
is the dominant stage

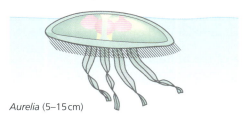

Aurelia (5–15 cm)

Introducing the cnidaria
- The cnidaria or coelenterates (meaning 'hollow gut') are aquatic animals, mostly marine, with radially symmetrical body plans.
- The body cavity is a gut with a single opening for ingestion of food and egestion of waste matter.
- The body wall is of two layers of cells – an outer **ectoderm** and an inner **endoderm** – separated by a layer of jelly, the **mesoglea**.
- The ectoderm includes **stinging cells** which may be triggered by passing prey. With these cells, prey is poisoned, paralysed and held until pushed into the gut for digestion.
- The stinging cells are found especially on the tentacles.
- Behaviour of the body is coordinated by a nerve net in the mesoglea, in contact with the bases of all wall cells.
- There are alternative body forms in the cnidaria – either a sessile *hydroid* form (illustrated by *Hydra*) or a floating *medusa* form (illustrated by the jellyfish). These body forms typically alternate in the lifecycle.
- The cnidaria also include the sea anemones and corals.

■ **Figure 5.25**
Phylum
Platyhelminthes –
the flatworms

Planaria
a free-living flatworm

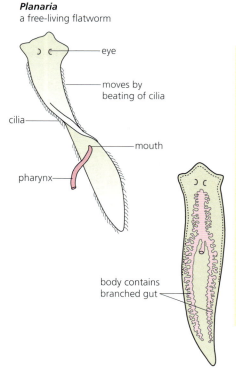

eye

moves by
beating of cilia

cilia

mouth

pharynx

body contains
branched gut

Introducing the platyhelminthes
- The platyhelminthes are flat, unsegmented animals with a body built from three cell layers (**triploblastic organisation**).
- There is no cavity in the middle layer. They have a mouth and a gut with numerous branches, but there is no anus. Feeding is by scavenging or predating on other small animals.
- There is no circulatory system in the platyhelminth body, but the generally small, thin, often flat body means that oxygen can diffuse easily to most cells.
- There are 'flame cells' present for excretion and regulation of water and ions in the body.
- Platyhelminthes often have both male and female reproductive organs in one individual (hermaphrodite organization), but the chances of self-fertilization are minimal.
- Some are free-living flatworms, but others are parasitic flukes or tapeworms. The phylum contains many important parasites.

■ **Figure 5.26**
Phylum Annelida –
the segmented
worms

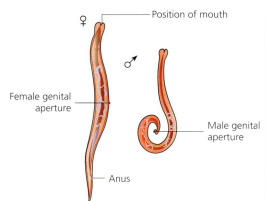

Introducing the annelids
- These are worm-like animals with a soft body (without a rigid skeleton), built of a fixed number of **similar segments** (visible externally as rings). Each segment contains the same pattern of nerves, blood vessels and excretory organs. This arrangement is known as **metameric segmentation**.

- Internally, the segments are separated by septa that divide the **body cavity** (**coelom**) into compartments, each surrounded by the muscular body wall. The coelom is fluid-filled and muscles of the body wall bring about movements by working against the pressure of the fluid in the coelom compartments. This forms a flexible **hydrostatic skeleton**.

- Gaseous exchange occurs through the whole body surface, which is kept moist. The blood contains an oxygen-transporting pigment.

- The collection of sense organs and the feeding structures at the anterior end, known as **cephalization**, modifies the segmentation pattern somewhat. A solid (non-tubular) ventral nerve cord runs the length of the body.

■ **Figure 5.27**
Phylum Mollusca –
the molluscs

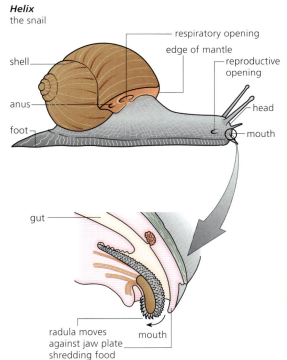

Introducing the molluscs
- The molluscs are the slugs, snails, limpets, mussels and octopuses. They are a huge and diverse group of organisms, the second largest phylum in terms of number of species. Most are aquatic and found in fresh water or marine habitats, but a few are terrestrial.

- They are animals with generally soft, flexible bodies, which show little or no evidence of segmentation. The body is divided into a head, a flattened muscular foot and a hump or visceral mass often covered by a shell, secreted by a layer of tissue called the mantle.

- The compact body shape of molluscs means that diffusion is not effective for the transport of nutrients, and molluscs have gills, or occasionally lungs, for gaseous exchange, as well as a well-developed blood circulation.

- Most molluscs have a rasping, tongue-like radula used for feeding.

■ **Figure 5.28**
Phylum Arthropoda –
the jointed-limbed
animals

the locust

adult

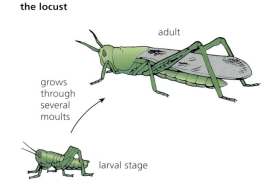

grows
through
several
moults

larval stage

the butterfly

pupa

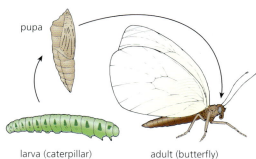

larva (caterpillar)   adult (butterfly)

Introducing the arthropods

- The arthropoda are the most numerically successful of all animals, and are divided into five distinct groups: the **crustaceans**, the **arachnids**, the **centipedes**, the **millipedes** and the **insects**.

- These animals have segmented bodies, covered by a hard external skeleton made of chitin, with jointed limbs (after which they are named). Typically, there is one pair of legs per segment, but this pattern has been lost from some arthropods.

- The exoskeleton cannot grow with the animal, so it has to be shed periodically (moulting), and replaced with a larger one, into which the enlarging animal grows.

- The blood circulation is open; the blood is in a hemocoel cavity surrounding all the organs of the body. A tubular heart pumps blood into the hemocoel.

- The functioning of the body is coordinated by a ventral nerve cord with nerves running to each segment. There is a concentration of nerves at the front of the body.

- The insects are by far the most numerous of the arthropods and have:

  – a body divided into head, thorax and abdomen
  – three pairs of legs and two pairs of wings attached to the thorax
  – a pair of compound eyes and a pair of antennae on the head
  – a head with mouthparts that are modified paired limbs
  – tracheae, a system of tubes, that pipe air to the tissues.

Introducing phylum Chordata

The chordates are a phylum that includes the vertebrates, but also some non-vertebrate groups, including the tunicates (sea-squirts). They are all triploblastic, coelomate animals that have a dorsal strengthening structure (called a **notochord**) in their bodies for at least some stage of their development. Lying above the notochord is a dorsal, tubular **nerve cord**. In the pharynx region of the body are a series of **pharyngeal slits**, opening between the pharynx (between mouth and esophagus) and the outside of the animal. There is also a **post-anal tail**. All these features are present in all chordates, but not necessarily for all stages of their life. In some vertebrates, for example, some of these features are only briefly present, being lost early in embryological development.

Sub-phylum Vertebrata

The vertebrates are those members of the phylum Chordata which have a **vertebral column** in place of the supporting rod, the notochord. The vertebrates include the fish, amphibia, reptiles, birds and mammals. The features of these classes of vertebrates are introduced in Figure 5.29. Much of animal biology in this book concerns the vertebrates, particularly the human animal.

■ The construction of dichotomous keys

The process of naming unknown organisms – for example, in ecological field work – is important but time-consuming. We often attempt this by making comparisons, using identification books that are illustrated with drawings and photographs, and that provide information on habitat and habits, to give us clues to the identity of organisms.

Alternatively, the use of **keys** may assist us in the identification of unknown organisms. The advantage of using keys is that it requires careful observation. We learn a great deal about the structural features of organisms and get some understanding of how different organisms may be related.

The steps in the construction of a dichotomous key are illustrated first (Figures 5.30–5.33). *Follow the steps, then put them into practice yourself.*

Class Osteichthyes (the bony fish):
- fish with a skeleton of bone
- mouth with a terminal position on the head
- the gills are covered by a body flap, the operculum

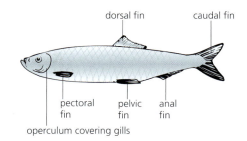

dorsal fin caudal fin

pectoral fin pelvic fin anal fin

operculum covering gills

Class Amphibia (the amphibians):
- amphibians have moist skin and use the skin for respiratory gas exchange
- largely terrestrial animals that breed in water, where fertilization is external. The larval stage (the tadpole) is aquatic
- the tadpole undergoes metamorphosis into a terrestrial adult, e.g. *Rana* (the frog)

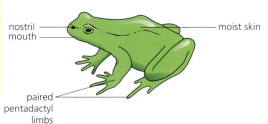

nostril
mouth
moist skin
paired pentadactyl limbs

Class Reptilia (the reptiles):
- terrestrial vertebrates, with dry, impervious skin protected by overlapping scales
- gaseous exchange occurs in lungs
- fertilization is internal, but the fertilized eggs are laid with a shell
- typically, reptiles have four pentadactyl limbs, but in the snakes the limbs are reduced or absent, e.g. *Lacerta* (lizard)

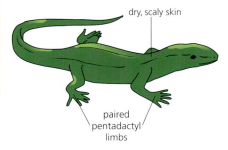

dry, scaly skin

paired pentadactyl limbs

Class Aves (the birds):
- birds have a strong, light skeleton; limb bones are hollow with internal strutting for strength
- the skin of the body is covered by scales
- the forelimbs are modified as wings; massive flight muscles are anchored to the pectoral girdle
- birds are endothermic, with a high and constant body temperature
- fertilization is internal; eggs are laid with a food store (yolk) and a hard calcareous shell, e.g. *Columba* (pigeon)

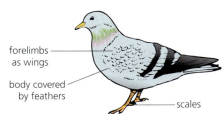

forelimbs as wings

body covered by feathers

scales

Class Mammalia (the mammals)
Subclass Eutheria (the placental mammals):
- the skin of mammals has hair
- mammals have four pentadactyl limbs
- the body cavity is divided by a muscular diaphragm between thorax and abdomen
- mammals are endothermic, with a relatively high and constant body temperature
- fertilisation is internal; the young complete their early development in the uterus; when born they are fed on milk produced in mammary glands, e.g. *Homo sapiens* (human)

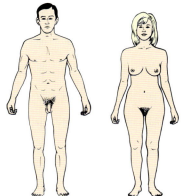

■ **Figure 5.29** Introducing the vertebrates

Steps in key construction

The steps in key construction are illustrated using eight different tree leaves, as shown in Figure 5.30. When selecting a leaf, care must be taken to ensure that it is entirely representative of the majority of the tree's leaves.

First, each leaf is carefully examined and the most significant features of structure are listed in a matrix, where their presence (or absence) is recorded against each specimen, as shown in Figure 5.31.

Then, from the matrix, a characteristic shown by half (or thereabouts) of the leaves is selected. This divides the specimens into two groups. A dichotomous flow diagram is constructed, progressively dividing the specimens into smaller groups. Each division point is labelled with the critical diagnostic feature(s), as shown in Figure 5.32.

Finally, a dichotomous key is constructed, reducing the dichotomy points in the flow diagram to alternative statements to which the answer is either 'yes' or 'no'. Each alternative leaf is given a number to which the reader must refer to carry on the identification, until all eight leaves have been identified (Figure 5.33).

■ **Figure 5.30**
Collection of leaves for the construction of a dichotomous key

A Macedonian Oak (Italy/Balkans)

B Picrasma (China)

C Horse Chestnut (Greece)

D Sweet Chestnut (N. Africa)

E Pignut Hickory (N. America)

F Osier (Central and W. Europe)

G Southern Catalpa (N. America)

H Indian Horse Chestnut (Himalaya)

■ **Figure 5.31** Matrix of characteristics shown by one or more of the sample

Feature of leaf: present (✓) or absent (−)	Tree leaves, identified by number							
	A Macedonian Oak	**B** Picrasma	**C** Horse Chestnut	**D** Sweet Chestnut	**E** Pignut Hickory	**F** Osier	**G** Southern Catalpa	**H** Indian Horse Chestnut
leaf not divided into leaflets (leaf entire)	✓	−	−	✓	−	✓	✓	−
leaf consists of leaflets	−	✓	✓	−	✓	−	−	✓
leaf blade narrow, with almost parallel sides	✓	−	−	−	−	✓	−	−
leaf blade broad	−	−	−	✓	−	−	✓	−
leaflets radiate from one point (palmate)	−	−	✓	−	−	−	−	✓
leaflets arranged in two rows along stalk	−	✓	−	−	✓	−	−	−
leaf / leaflet margin smooth	−	−	−	−	✓	✓	✓	✓
leaf / leaflet margin toothed, like a saw	✓	✓	✓	✓	−	−	−	−
leaf blade heart-shaped	−	−	−	−	−	−	✓	−
leaf blade boat-shaped (widest in middle)	−	−	−	✓	−	−	✓	−
leaflet paddle-shaped, widest near one end	−	−	✓	−	−	−	−	−
leaflets 5 in number or less	−	−	−	−	✓	−	−	−
leaflets between 6 and 10 in number	−	−	✓	−	−	−	−	✓
leaflets more than 10 in number	−	✓	−	−	−	−	−	−

■ **Figure 5.32** Dichotomous flow diagram of leaf characteristics

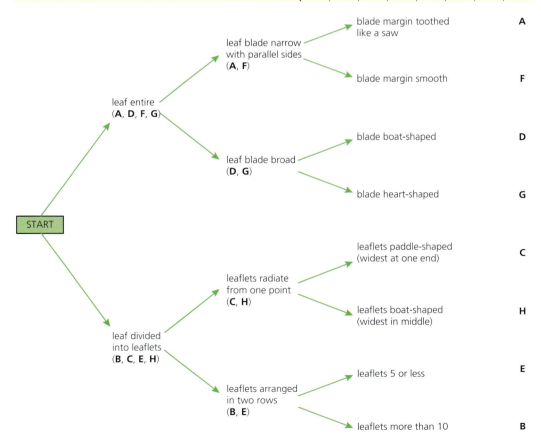

■ **Figure 5.33**
Dichotomous key to the sample of eight tree leaves

		Go to …
1	Leaves entire, not divided into leaflets	2
	Leaf blade divided into leaflets	5
2	Leaf blade narrow, with almost parallel sides	3
	Leaf blade broad rather than narrow	4
3	Blade margin toothed like a saw	**Macedonian Oak**
	Blade margin smooth	**Osier**
4	Blade boat-shaped	**Sweet Chestnut**
	Blade heart-shaped	**Southern Catalpa**
5	Leaflets radiate from one point	6
	Leaflets arranged in two rows	7
6	Leaflets paddle-shaped – widest at one end	**Horse Chestnut**
	Leaflets boat-shaped – widest in the middle	**Indian Horse Chestnut**
7	Leaflets 5 (or less) in number	**Pignut Hickory**
	Leaflets more than 10 in number	**Picrasma**

Practising key construction

Joseph Camin, a biology teacher at Kansas University, designed a family of 'animals' in order to introduce the principles of classification in a practical way. Some of his 'caminacules' are shown in Figure 5.34. Use a copy of these to construct a key, following the steps detailed on page 236. Alternatively, use a suitable selection of eight living biological specimens.

■ **Figure 5.34**
A selection of Camin's 'caminacules'

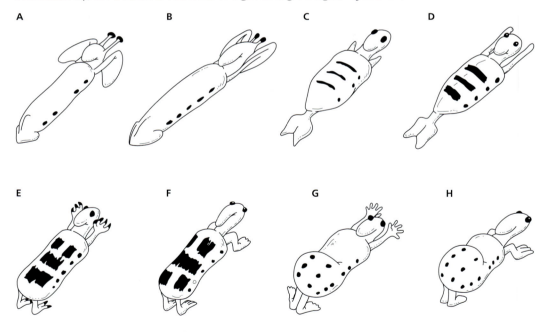

11 **Design** a method of classifying animals and plants which are commonly found in gardens or farmland that might be useful to an enthusiastic gardener or farmer.

5.4 Cladistics – *the ancestry of groups of species can be deduced by comparing their base or amino acid sequences*

■ An artificial or natural classification?

The quickest way to classify living things is on their immediate and obvious similarities and differences. For example, we might classify together animals that fly, simply because the essential organ, wings, are so easily seen. This would include almost all birds and many insects, as well as the bats and certain fossil dinosaurs. However, resemblances between the wings of the bird and the insect are superficial. Both are aerofoils (structures that generate 'lift' when moved through the air), but they are built from different tissues and have different origins in the body.

We say that the wings of birds and insects are **analogous** structures. Analogous structures resemble each other in function but differ in their fundamental structure. A classification based on analogous structures is an **artificial classification** (Table 5.5).

Alternatively, a **natural classification** is based on similarities and differences that are due to close relationships between organisms because they share common ancestors. We have already seen that the bone structure of the limbs of all vertebrates suggests they are modifications of a common plan which we call the pentadactyl limb (Figure 5.5, page 214). So, there are many comparable bones in the human arm, the leg of a horse and the limb of a mole. Structures built to a common plan, but adapted for different purposes are **homologous structures**. We see these adaptations of a common structure as evolutionary change, driven by natural selection.

■ **Table 5.5**
Analogous and homologous structures

Analogous structures	Homologous structures
resemble each other in function	are similar in fundamental structure
differ in their fundamental structure	are similar in position and development, but not necessarily in function
demonstrate only superficial resemblances	are similar because of common ancestry
for example, wings of birds and insects	for example, limbs of vertebrates

A classification based on homologous structures is believed to reflect evolutionary relationships. It is called a natural or **phylogenetic classification**. A diagram illustrating evolutionary relationships between species is called a phylogenetic tree or **cladogram**. The word 'tree' is employed because the branching pattern shows when different species 'split off' from others.

■ Clades and cladograms

Rather than any form of artificial (superficial) classification, taxonomists seek to use evolutionary relationships in the schemes of classification of living things that they propose. Phylogenetic trees or cladograms have two important features:

- **branch points** in the tree – representing the **time** at which a divide between two taxa occurred
- the **degree of divergence** between branches – representing the **differences** that have developed between the two taxa since they diverged.

So, taxonomists ponder the question of which features are the more significant in a phylogenetic taxonomy, and so should receive the greater emphasis in devising a scheme.

If the answer to this question is that the measurable similarities and differences of anatomy should be used to arrange species into dichotomously branching trees, then the product is a **dendrogram**.

In **cladistics**, classification is based upon an analysis of relatedness, and the product is a cladogram. A **clade is a group of organisms that have evolved from a common ancestor**. The differences between a dendrogram and a cladogram are illustrated by lizards, crocodiles and birds (Figure 5.35). Lizards and crocodiles resemble each other more than either resemble the birds, but crocodiles and birds share a common ancestor.

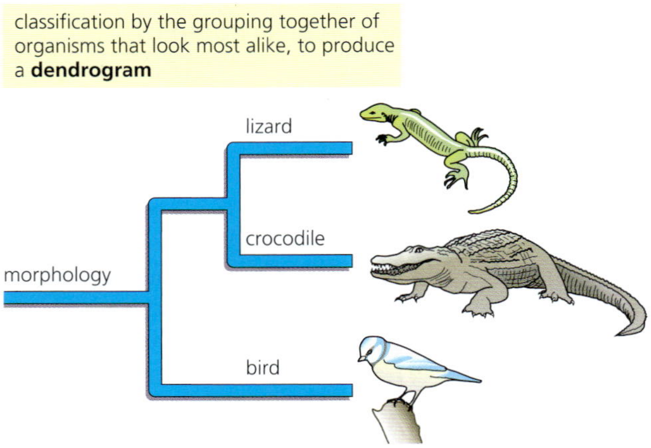

classification by the grouping together of organisms that look most alike, to produce a **dendrogram**

classification by the grouping together of organisms with a more recent common ancestor, to produce a **cladogram**

■ **Figure 5.35** Cladistics – one of two ways of classifying

■ Evidence for a natural classification

Evidence of evolutionary relationships comes from a range of studies (Table 5.6).

■ **Table 5.6** Evidence of evolutionary relatedness

Palaeontology (study of fossils)	Fossils tell us about organisms in the geological past. Sometimes, they represent intermediate forms – links between groups of organisms, such as *Archeopteryx*, a possible 'missing link' between reptiles and birds.
Comparative biochemistry	The composition of nucleic acids and cell proteins establish degrees of relatedness – closely related organisms show fewer differences in the composition of specific nucleic acids and cell proteins.
Comparative embryology	Study of the development of an organism from egg to adult may throw light on its evolution. For example, the arrangement of arteries and development of the heart in early vertebrate embryos follow similar patterns.
Comparative anatomy	Studies of comparative anatomy show that many structural features are basically similar.

All living things have DNA as their genetic material, with a genetic code that is virtually universal. The processes of 'reading' the code and protein synthesis, using RNA and ribosomes, are very similar in prokaryotes and eukaryotes. Processes such as respiration involve the same types of steps, and similar or identical intermediates and biochemical reactions, similarly catalysed. ATP is the universal energy currency. Among the autotrophic organisms the biochemistry of photosynthesis is virtually identical.

This biochemical commonality suggests a common origin for life, because the biochemical differences between the living things of today are small. Some of the earliest events in the evolution of life must have been biochemical, and the results have been inherited more or less universally.

Today, similarities and differences in the biochemistry of organisms have become extremely important in taxonomy. For example, hemoglobin, the β chain of which is built from 146 amino acid residues, shows variation in the sequence of amino acids in different species. Hemoglobin structure is determined by inherited genes, so the more closely related species are, the more likely their amino acid sequence is to match (Table 5.7). Variations are thought to have arisen by mutation of an 'ancestral' gene for hemoglobin. If so, the longer ago a species diverged from a common ancestor, the more likely it is that differences may have arisen.

■ **Table 5.7** Number of amino acid differences in β chain of hemoglobin compared to human hemoglobin

Species	Differences	Species	Differences
human	0	kangaroo	38
gorilla	1	chicken	45
gibbon	2	frog	67
rhesus monkey	8	lamprey	125
mouse	27	sea slug (mollusc)	127

Similar studies have been made of the differences in the polypeptide chains of other protein molecules, including ones common to all eukaryotes and prokaryotes. One such is the universally occurring electron transport carrier, cytochrome c (see pages 137–138).

Biochemical variation used as an evolutionary or molecular clock

Biochemical changes, like those discussed above, may occur at a constant rate and, if so, may be used as a 'molecular clock'. If the rate of change can be reliably estimated, it does record the time that has passed between the separation of evolutionary lines.

An example comes from immunological studies – a means of detecting differences in specific proteins of species, and therefore (indirectly) their relatedness.

Serum is a liquid produced from blood samples from which blood cells and fibrinogen have been removed. Protein molecules present in the serum act as antigens, if the serum is injected into an animal with an immune system that lacks these proteins (page 92).

Typically, a rabbit is used when investigating relatedness to humans. In the rabbit, the injected serum causes the production of antibodies against the human proteins. Then, serum produced from the rabbit's blood (now containing antibodies against human proteins) can be tested against serum from a range of animals. The more closely related a test animal is to humans, the greater the reaction of rabbit antibodies with human-like antigens (bringing about observable precipitation, Figure 5.36).

The precipitation produced by reaction of treated rabbit serum with human serum is taken as 100%. For each species tested, the greater the precipitation, the more recently the species shared a common ancestor with humans. This technique, called **comparative serology**, has been

■ **Figure 5.36**
Investigating evolutionary relationships by the immune reaction

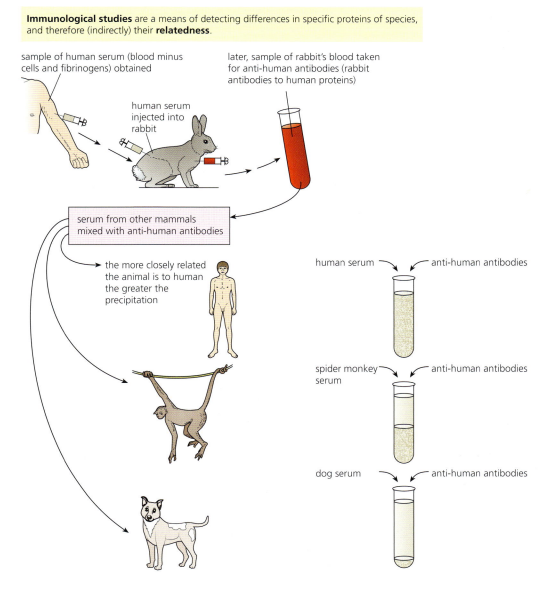

Immunological studies are a means of detecting differences in specific proteins of species, and therefore (indirectly) their **relatedness**.

sample of human serum (blood minus cells and fibrinogens) obtained

later, sample of rabbit's blood taken for anti-human antibodies (rabbit antibodies to human proteins)

human serum injected into rabbit

serum from other mammals mixed with anti-human antibodies

the more closely related the animal is to human the greater the precipitation

human serum — anti-human antibodies

spider monkey serum — anti-human antibodies

dog serum — anti-human antibodies

used by taxonomists to establish phylogenetic links in mammals and in non-vertebrates. The list of animals tested in this way is given in Table 5.8. Of course, we do not know of the common ancestor to these animals and the blood of that ancestor is not available to test.

However, if the sequence of the 584 amino acids that make up the blood protein albumin changes at a constant rate, the percentage immunological 'distance' between humans and any of these animals will be a product of the distance back to the common ancestor plus the difference 'forward' to the listed animal. Hence, the difference between a listed animal and human can be halved to gauge the difference between a modern form and the common ancestor.

The divergent evolution of the primates is known from geological (fossil) evidence. So, the forward rate of change since the lemur gives us the rate of the molecular clock – namely 35% in 60 million years, or 0.6% every million years.

This calculation has been applied to all the data (Table 5.8, column 5).

■ **Table 5.8**
Relatedness investigated via the immune reaction

1	2	3	4	5
Species	**Precipitation/%**	**Difference from human/%**	**Difference to common ancestor (half difference from human)**	**Postulated time since common ancestor/mya**
human	100	–	–	–
chimpanzee	95	5	2.5	4
gorilla	95	5	2.5	4
orang-utan	85	15	7.5	13
gibbon	82	18	9	15
baboon	73	27	13.5	23
spider monkey	60	40	20	34
lemur	35	65	32.5	55
dog	25	75	37.5	64
kangaroo	8	92	46	79

We can now construct a cladogram based on the biochemical data in Table 5.8 – Figure 5.37.

■ **Figure 5.37**
A cladogram based on immunological evidence in Table 5.8

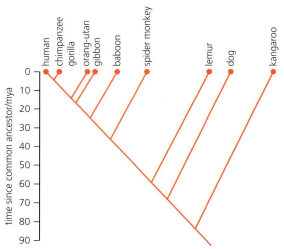

Relatedness measured from DNA samples

It is possible to measure the relatedness of different groups of organisms by the amount of difference between specific molecules, such as differences in the base sequence of genes in DNA. The genetic differences between the DNA of various organisms give us data on degrees of divergence. *Look at Figure 5.38.* Here, the degree of relatedness of the DNA of primate species suggests the number of years that have elapsed since the various primates shared a common ancestor.

■ **Figure 5.38** Genetic difference between DNA samples

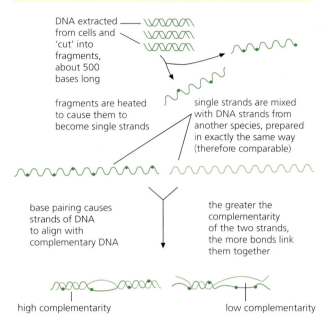

DNA hybridization is a technique that involves matching the DNA of different species, to discover how closely they are related.

DNA extracted from cells and 'cut' into fragments, about 500 bases long

fragments are heated to cause them to become single strands

single strands are mixed with DNA strands from another species, prepared in exactly the same way (therefore comparable)

base pairing causes strands of DNA to align with complementary DNA

the greater the complementarity of the two strands, the more bonds link them together

high complementarity

low complementarity

The closeness of the two DNAs is measured by finding the temperature at which they separate – the fewer bonds formed, the lower the temperature required.

The degree of relatedness of the DNA of **primate species** can be correlated with the estimated number of years since they shared a common ancestor.

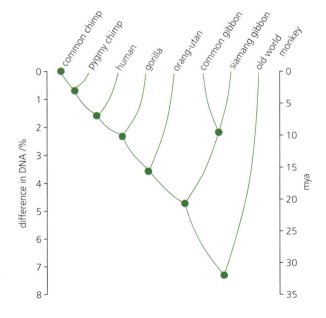

Falsification of theories

■ Reclassification of the figwort family using evidence from cladistics

The Scrophulariaceae (figwort family) – one of the groups of flowering plants that have long been recognized by plant taxonomists – was a large and varied family. Typically, genera of this family had irregular (zygomorphic) flowers, like those of the snapdragon and the foxglove, for example. The flowering plant families were designated before most biochemical studies were applied to plant taxonomy. Nevertheless, many families appear to be (largely) natural classifications, for example the rose family (Rosaceae) which includes many fruit plants and the Crucifereae which includes many economically important plants.

Today, flowering plant classification is being revisited. (Indeed, all classification is being revisited.) Most evidence for plant evolutionary relationships now comes from a comparison of DNA sequences in only 1 to 3 genes found in the chloroplasts of the plant cells. These genes are of modest length, about 1000 nucleotides long. However, they have provided more secure information on evolutionary history than the differences in anatomy and morphology on which the traditional classifications were based.

■ **Figure 5.39** Plants associated with the Scrophulariaceae, now and previously

Plants of the family Orobanchacese, previously in a large, diverse family Scrophulariaceae

Yellow-rattle (*Rhinanthus minor*)

Eye-bright (*Euphrasia micranthus*)

Plants of the modern family Scrophulariaceae

Mullein (*Verbascum lychnitis*)

Figwort (*Scrophularia nodosa*)

Plants of the family Plataginacase, previously in the family Scrophulariaceae

Antirrhinum (*Antirrhinum majus*)

Foxglove (*Digitalis purpurea*)

In the reorganization of the figwort family, three genes (totalling less than 4500 nucleotides) have been compared from many species of plants, including some formerly classified in the Scrophulariceae. The outcome has been the repositioning of several genera into other families (and the reassignment of others, previously in other families, to the newly defined Scrophulariaceae). *Examine the cladogram that sums up the evidence and identifies some major changes (Figure 5.40) and the illustrations of representative plants involved (Figure 5.39).*

A note of caution

Despite these additional, largely biochemical, sources of evidence, the evolutionary relationships of organisms are still only partly understood. Consequently, current taxonomy is only partly a phylogenetic classification.

■ **Figure 5.40**
Cladogram summarizing changes to the family Scrophulariaceae: 'bootstrap values' provide a measure of accuracy of sample estimates. On a scale of 0–100 it is a measure of support for individual branches in the cladogram

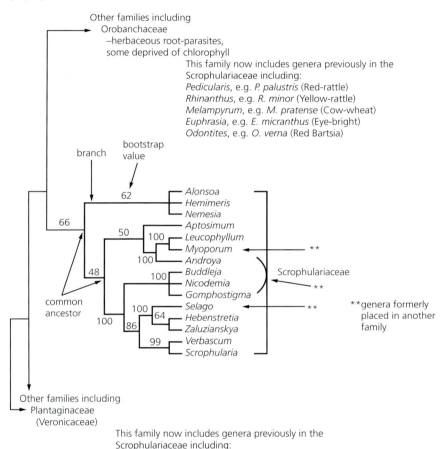

■ *Examination questions – a selection*

Questions 1–2 are taken from IB Diploma biology papers.

Q1 A biologist exploring an uninhabited island came across an unknown plant. She made the following notes:
- grows in a damp and shady corner of the island
- has large feathery leaves with spore cases (sporangia) arranged on the underside
- young leaves are tightly rolled up
- has roots.

In what phylum should she classify this plant?
- **A** Angiospermophyta
- **B** Bryophyta
- **C** Coniferophyta
- **D** Filicinophyta

Standard Level Paper 1, Time Zone 0, Nov 12, Q18

Q2 Which of the following is a characteristic of platyhelminthes?
- **A** Many pairs of legs
- **B** Flat body
- **C** Hard exoskeleton
- **D** Presence of cnidocytes

Higher Level Paper 1, Time Zone 2, May 11, Q19

Questions 3–10 cover other syllabus issues in this chapter.

Q3 Evolution by natural selection happens as result of which of the following:
- **A** the survival of the strongest organisms in a population
- **B** the variability of male gamete cells produced during meiosis
- **C** the changes in the frequency of alleles over time
- **D** when populations reach their transitional phase due to shortages in supplies

Q4 The three domain system shows the differences between Eubacteria, Archaea and Eukarya. Which of the following are characteristics used to differentiate amongst these domains?
- **A** Metabolic activities and habitats
- **B** DNA associated with proteins and presence of chlorophyll molecules
- **C** Type of locomotion and feeding activities
- **D** Size of ribosomes and materials used to make their cell wall

Q5 Identify the phylum to which the following organisms belong to. (6)

Q6 **a** Antibiotics are used to treat bacterial diseases. Outline how bacteria evolve in response to the presence of antibiotics as an example of natural selection. (4)

  b Outline how the evolution of homologous structures and selective breeding of domesticated animals can be used as evidences of evolution. (4)

Q7 Discuss how natural selection leads to evolution. (8)

Q8 The scientific name of the Arizona mud turtle is *Kinosternon arizonense* and for the yellow mud turtle it is *Kinosternon flavescens*.

  a State what is meant by the binomial system of naming living organisms. (2)

  b State if these two turtles belong to the same
    i   class
    ii genus
    iii species. (3)

  c *Sophora flavescens* is the scientific name of a different organism. State if this particular species has some evolutionary relationship with the yellow mud turtle based on their scientific names. (2)

  d Outline with a named example how scientists use the binomial system to avoid confusion. (2)

Q9 **a** Draw a graph of the growth of an animal population after introduction to a new, highly favourable environment. Annotate the stages of the curve to establish the factors influencing change in population size at each stage. (6)

  b In a natural environment the numbers of a population in the community of organisms present normally fluctuate unpredictably. Identify the natural influences that change population numbers in these circumstances. (6)

Q10 Use a table as shown to state one external feature characteristic of each group. (7)

Group	Characteristic external feature
cnidarians	
chordates	
mosses	
arthropods	
ferns	
annelids	
flowering plants	

6 Human physiology

ESSENTIAL IDEAS

- The structure of the wall of the small intestine allows it to move, digest and absorb food.
- The blood system continuously transports substances to cells and simultaneously collects waste products.
- The human body has structures and processes that resist the continuous threat of invasion by pathogens.
- The lungs are actively ventilated to ensure that gas exchange can occur passively.
- Neurons transmit the 'message', synapses modulate the message.
- Hormones are used when signals need to be widely distributed.

Physiology is about how and why the parts of the body function in the way they do. Linked with physiology is **anatomy** – the study of structure. Structure gives vital clues to function, such as when the structures of our arteries and veins are compared, for example.

Physiology is also an experimental science. For example, physiological experiments have established that cells of organisms function best in an internal environment that stays fairly constant and is maintained within quite narrow limits. We talk about a **constant internal environment** – which can be observed in mammals, such as the human, and which is our focus here.

In this chapter, we first examine how the body **obtains nutrients**, breathes and **exchanges gases** with the air, and maintains an efficient **internal transport system**.

Pathogens, many of them microorganisms, may bring disease to healthy organisms. Accordingly, cells of our blood circulation also have roles in **defence against communicable diseases**. Finally, the roles of **nerves** and **hormones**, the processes of **homeostasis**, and aspects of **sexual reproduction** are discussed.

6.1 Digestion and absorption – *the structure of the wall of the small intestine allows it to move, digest and absorb food*

An animal takes in food, which is complex organic matter, and digests it in the **alimentary canal** or gut, producing molecules that can be taken up into the body cells via the blood circulation system. This is known as **holozoic nutrition** (meaning 'feeding like an animal'). Holozoic nutrition is just one of the forms of **heterotrophic nutrition** (meaning 'feeding on complex, ready-made foods') to obtain the required nutrients. The important ecological point to remember about all heterotrophs is their dependence, directly or indirectly, on organisms that manufacture their own elaborated foods. These organisms are the photosynthetic green plants, known as autotrophs. This dependence is demonstrated in food chains.

1 State two forms of heterotrophic nutrition, other than holozoic.

■ Digestion, where and why?

The food an animal requires has to be searched out or, possibly, hunted. Some mammals eat only plant material (**herbivores**); some eat only other animals (**carnivores**); and others eat both animal and plant material (**omnivores**). Whatever is eaten, a balance of the essential nutrients is required by the body. The sum total of the nutrients eaten is the **diet**.

The mammalian gut is a long, hollow muscular tube connecting mouth to anus. Along the gut are several glands, and the whole structure is specialized for the movement and digestion of food and for the absorption of the useful products of digestion. The regions of the human gut are shown in Figure 6.1, and the steps of nutrition in Table 6.1.

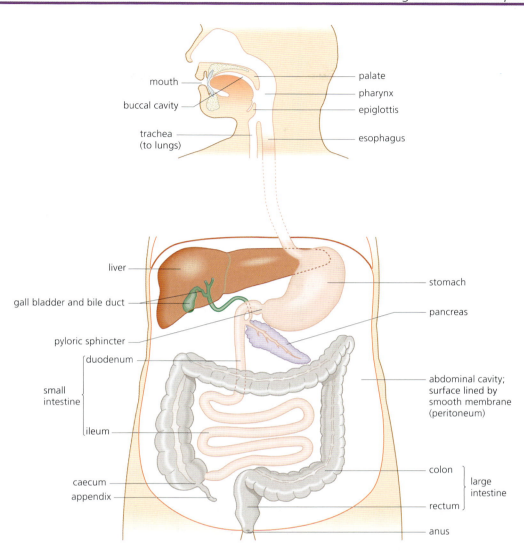

Figure 6.1
The layout of the human gut and associated organs – our digestive system

This diagram shows how the structure of the digestive system can be represented, but it is not annotated to indicate the functions of the different parts. When you are familiar with structure and function of the parts of the human gut, answer question 2.

> 2 **Construct** a table showing the regions of the human gut and its associated glands: mouth; esophagus; stomach; small intestine; pancreas; liver; gall bladder and large intestine. Complete the table by detailing the function(s) of each region. (These annotations could be added to a diagram such as Figure 6.1.)

Table 6.1 The five steps of holozoic nutrition

Step	Process
ingestion	food taken into mouth for processing in the gut
digestion	• mechanical digestion by the action of teeth and the muscular walls of the gut • chemical digestion by enzymes, mainly in the stomach and intestine • starch, glycogen, lipids and nucleic acids are digested to monomers • proteins are digested to amino acids, dipeptides (and some very short lengths of polypeptide) • cellulose remains undigested
absorption	soluble products of digestion are absorbed into the blood circulation system (into the lymphatic system in the case of fat droplets)
assimilation	products of digestion are absorbed from blood into body cells (such as liver and muscle cells) and used or stored
egestion	undigested food and dead cells from the lining of the gut, together with bacteria from the gut flora, are expelled from the body as faeces

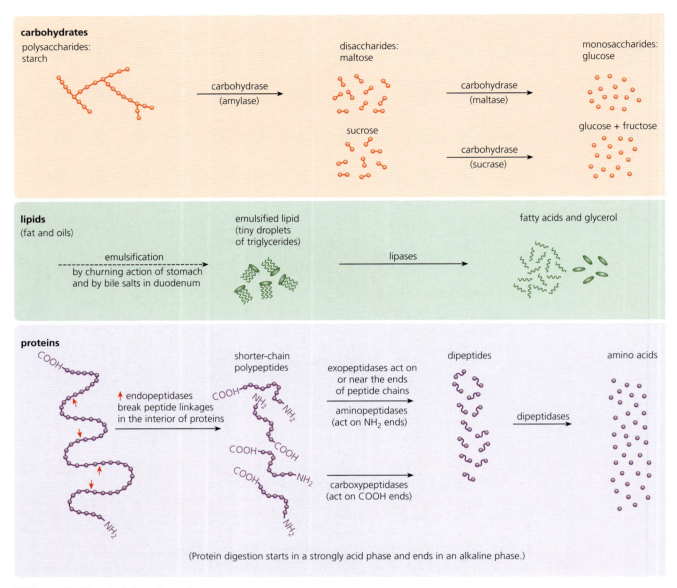

■ **Figure 6.2** Chemical digestion – the steps

The bulk of the food which is taken into the gut consists of insoluble molecules that are too large to cross the gut wall and enter the bloodstream. Our diet largely consists of carbohydrates, lipids and proteins. These must be hydrolysed to monosaccharides, fatty acids and glycerol, and free amino acids (or very small polypeptides) before they can be absorbed and, later, built up into the carbohydrates, lipids and proteins required by our bodies. The chemical structure of carbohydrates, lipids and proteins is discussed in Chapter 2. The chemical changes that occur to food during digestion are summarized in Figure 6.2. Food cannot be said to have truly entered the body until it has been digested and the products have been absorbed across the gut wall.

The first stage in the breakdown of large, insoluble food molecules is **mechanical digestion**. This occurs by the action of the jaws and teeth in the mouth, and, later, through the churning action of the muscular walls as the food is moved along the gut, particularly in the stomach. Throughout the gut, waves of contraction and relaxation of the circular and longitudinal muscles of the wall propel food along. This process is known as **peristalsis**. (Gut muscles are described as involuntary muscles because they are not under conscious control.)

The **chemical digestion** that follows is brought about by enzymes. Digestive enzymes are protein catalysts that are produced in specialized cells in glands associated with the gut.

Several different enzymes are secreted onto the food as it passes along the gut, and peristalsis mixes together enzymes and food. Most macromolecules are digested into monomers by enzyme action in the small intestine. Enzymes work efficiently at the relatively low temperature at which the body is maintained. Digestion is completed by the enzymes that were secreted onto the food, working together with those held in the plasma membranes of cells of the gut lining. Examples of digestive enzymes are listed in Table 6.2.

> **3** **a** **Explain** why digestion is an essential stage in nutrition.
> **b** **State** two essential foods that do not require digestion.
> **c** **Identify** a polysaccharide that the human gut cannot digest.

■ **Table 6.2**
Examples of digestive enzymes

Enzyme	Source	Substrate	Product	Optimum pH
amylase	salivary glands	starch	maltose	6.5–7.5
pepsin	gastric glands	protein	polypeptides	2.0
lipase	pancreas	triglyceride	fatty acids and glycerol	7.0

Note: Similar enzymes, obtained from other sources (including microorganisms), have applications in the industrial production of sugar from starch and in the brewing of beer (page 102).

■ Digestion in the stomach

The stomach, a J-shaped muscular bag, is located high in the abdominal cavity, below the diaphragm and liver. The stomach wall contains gastric glands. These secrete **hydrochloric acid** – sufficiently acidic to create an environment of pH 1.5–2.0. This is the optimum pH for protein digestion by **pepsin**, a protease enzyme in the gastric juice (Table 6.2). Pepsin is secreted in an inactive state and is activated by the hydrochloric acid. This acid also kills many of the bacteria present in the incoming food. Mucus is also secreted by the stomach lining and bathes the wall. This prevents **autolysis** (self-digestion) by the action of hydrochloric acid and pepsin. The food is now mixed with gastric juice and churned by muscle action, becoming a semi-liquid called **chyme**. This churning action is an important part of the mechanical digestion process. A typical meal may spend up to four hours in the stomach.

> **4** **Explain** what type of chemical reaction is catalysed by the enzymes involved in digestion, such as those that convert proteins to amino acids and starch to sugars.

■ The roles of the small intestine

The structure of the wall of the small intestine enables the movement, digestion and absorption of food. It is in the small intestine that completion of digestion of carbohydrates, lipids and proteins occurs, and the useful products of digestion are absorbed. In humans, the small intestine is about 5 m long. Within the wall are layers of **circular** and **longitudinal muscle**. Alternating contraction and relaxation of these muscle fibres mixes the food with enzymes and moves it along the gut. Throughout its length, the innermost layer of the wall is formed into vast numbers of finger-like or leaf-like projections called **villi**.

Digestion in the small intestine

Food enters the first part of the small intestine (known as the **duodenum**) a little at a time. Here the chyme meets bile from the bile duct and the pancreatic juice from the **pancreas**. Bile is strongly alkaline and neutralizes the acidity of the chyme. It also lowers the surface tension of large fat globules, causing them to break into tiny droplets, a process called **emulsification**. This speeds digestion by the enzyme lipase, later on. Bile itself contains no enzymes.

The pancreas secretes enzymes into the lumen of the small intestine

Pancreatic juice contains several enzymes, including an **amylase** that catalyses the hydrolysis of starch to maltose, a **lipase** that catalyses the hydrolysis of fats to fatty acids and glycerol, and an **endopeptidase**. An endopeptidase (endo means '*within*') catalyses the breakdown of proteins at numerous points *within* the long protein chain. The products are shorter-length peptides and, eventually, free amino acids (Table 6.3). Also present in the pancreatic juice are nucleotides that hydrolyse nucleic acids to nucleotides.

All these enzymes act as the chyme, bile and pancreatic juice are mixed together by a churning action (a form of peristalsis) called **segmentation** (Figure 6.3).

the duodenum

mixing the chyme

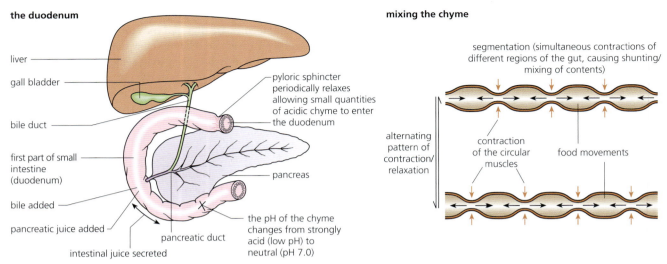

liver

gall bladder

bile duct

first part of small intestine (duodenum)

bile added

pancreatic juice added

intestinal juice secreted

pyloric sphincter periodically relaxes allowing small quantities of acidic chyme to enter the duodenum

pancreas

pancreatic duct

the pH of the chyme changes from strongly acid (low pH) to neutral (pH 7.0)

segmentation (simultaneous contractions of different regions of the gut, causing shunting/mixing of contents)

alternating pattern of contraction/relaxation

contraction of the circular muscles

food movements

■ **Figure 6.3** Chyme meets bile and pancreatic juice

■ **Table 6.3**
Hydrolytic enzymes operating in the small intestine

Enzyme	Substrate	Products
amylases	starch	maltose
lipases	lipids	fatty acids and glycerol
proteases, e.g. polypeptidese	proteins + polypeptides	shorter peptides + amino acids

Absorption in the small intestine

The products of digestion, mostly monosaccharide sugars, amino acids, fatty acids and glycerol, together with various vitamins and mineral ions, are absorbed as they make contact with the epithelial cells of the villi of the small intestine (Table 6.4). The process is efficient because the small intestine has a huge surface area, due to the vast number of **villi** (Figure 6.4).

The methods of membrane transport required to absorb different nutrients are **diffusion**, **facilitated diffusion** and **active transport** (Table 6.5). Epithelial cells transfer energy in the active transport process by which most of the products of digestion are taken into the cells. Transport involves protein pump molecules in the plasma membrane, activated by reaction with ATP.

■ **Figure 6.4**
The small intestine – the absorption surface

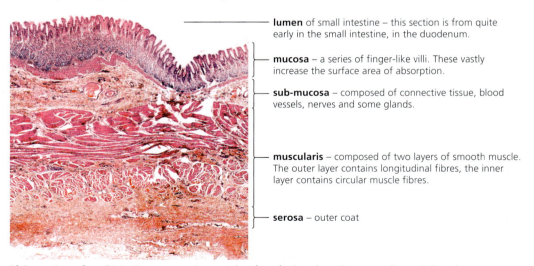

lumen of small intestine – this section is from quite early in the small intestine, in the duodenum.

mucosa – a series of finger-like villi. These vastly increase the surface area of absorption.

sub-mucosa – composed of connective tissue, blood vessels, nerves and some glands.

muscularis – composed of two layers of smooth muscle. The outer layer contains longitudinal fibres, the inner layer contains circular muscle fibres.

serosa – outer coat

(If the sections of small intestine you examine are taken from further along the mammalian gut, the sub-mucosa and muscularis layers are likely to be much thinner.)

Examine a prepared slide of a transverse section of the small intestine, first using low power and then high power, to identify the tissue layers present. Similarly, examine the photomicrograph in Figure 6.4 to identify the tissue layers labelled.

■ **Table 6.4**
Functions of the absorption surface of the small intestine

Feature	Description/function
villi	provide a huge surface area for absorption
epithelium cells	single layer of small cells, packed with mitochondria – the source of ATP (metabolic energy) for active uptake across the plasma membrane
pump proteins in the plasma membrane of epithelial cells	actively transport nutrients across the plasma membrane into the villi
network of capillaries	large surface area for uptake of amino acids, monosaccharides, and fatty acids and glycerol into blood circulation
lacteal	branch of the lymphatic system into which triglycerides (combined with protein) pass for transport to body cells
mucus from goblet cells	lubricates movement of digested food among the villi and protects plasma membranes

■ **Table 6.5**
The mechanisms of absorption in the villi

Products of digestion	Mechanism of transport
monosaccharide sugars	• Glucose enters the epithelial cells by sodium-glucose co-transporter protein in the cell membrane of the microvilli. (Sodium ions are pumped back out from the epithelial cells by sodium-potassium pumps in the same membrane). • Glucose channels then allow glucose to pass by facilitated diffusion from the epithelial cells into the blood capillary in the villus.
amino acids and small peptides	• pumped against concentration gradients across the epithelial cells of the villus into the capillaries
fatty acids and monoglycerides	• diffuse into epithelial cells where fatty acids and monoglycerides reform into triglycerides • then, as tiny droplets they are transported out of the epithelial cells by exocytosis (page 46), into the lacteals

Case study: Digestion of starch and the fate of the products of digestion

The chemical structure of starch (a mixture of amylase and amylopectin) is illustrated in Figure 2.19 (page 80). *Remind yourself of the structures of these polymers, now.*

Digestion of starch is catalysed by enzymes in the mouth, by enzymes from the pancreas that are secreted into the lumen of the small intestine, and finally by enzymes in the plasma membranes of the epithelia cells of the villi of the wall of the small intestine (Table 6.6).

■ **Table 6.6**
Steps in the digestion of starch

Location	Enzyme(s)	Chemical change
mouth	amylase from salivary glands	begins to convert starch to maltose (a disaccharide), by hydrolysis of 1,4 bonds in amylase and amylopectin
lumen of small intestine	amylase from pancreas	continues conversion of starch to maltose by hydrolysis of 1,4 bonds in amylase and amylopectin to form maltose (larger fragments containing 1,6 bonds of amylopectin (which amylase cannot hydrolyse) remain (called dextrins)
plasma membranes of villi	two enzymes + maltase	hydrolyse dextrin fragments to maltose hydrolyses maltose to glucose

5 **Draw** a large-scale diagram of an epithelial cell of a villus of the small intestine (as shown in part of Figure 6.4). **Annotate** your diagram to show the mechanisms of absorption of glucose, amino acids and small peptides, and fatty acids and glycerol, into the body.

Glucose enters the epithelial cells by co-transport with sodium ions, and then travels on into the tissue fluid beyond by facilitated diffusion. From there, it diffuses across the walls of the capillary network into the blood circulation. The hepatic portal vein (Figure 6.9) carries the products of digestion to the liver, where glucose may enter the liver cells and be converted to glycogen (page 80). Glucose remaining in the blood may be converted to glycogen in muscle cells around the body. Brain cells depend on a continuous supply of glucose from the blood circulation – they cannot store glycogen.

Assimilation follows absorption

The fate of absorbed nutrients is called **assimilation**. In the first stage of assimilation, absorbed nutrients are transported from the intestine. In the villi, **sugars** are passed into the capillary network and, from here, they are transported to the liver. The liver maintains a constant level of blood sugar (page 301). The **amino acids** are also passed into the capillary network and transported to the liver. Here, they contribute to the pool or reserves of amino acids from which new proteins are made in cells and tissues all over the body.

Finally, **lipids** are absorbed as fatty acids and glycerol and are largely absorbed into the lacteal vessels. From there, they are carried by the lymphatic system to the blood circulation outside the heart.

6 Distinguish between absorption and assimilation.

■ Role of the large intestine

The large intestine wall has no villi, but the surface area for absorption is increased by many folds of the inner lining. At this point in the gut, most of the useful products of digestion have been absorbed. What remains is the undigested matter (such as plant fibre), with mucus, dead intestinal cells, bacteria, some mineral ions and water. Water is an important component of our diet, and many litres of water are also secreted into the chyme in the form of digestive juices.

In the colon, water and mineral salts (such as Na^+ and Cl^- ions) are absorbed. What remains of the meal is now the faeces. Bacteria compose about 50% of faeces. Bile pigments (excretory products formed from the routine breakdown of red blood cells), which were added in the duodenum, uniformly colour the faeces.

The rectum is a short muscular tube which terminates at the anus. Discharge of faeces from the body at the anus is controlled by sphincter muscles.

Nature of Science

Using models as representation of the real world

■ Use of dialysis tubing to model absorption of digested foods in the intestine

Suppose you are provided with $10\,cm^3$ of a 1% starch solution, $10\,cm^3$ of a 1% amylase solution, distilled water, beakers, test tubes, a $10\,cm^3$ graduated pipette and dialysis tubing. Dialysis tube is an artificial selectively permeable membrane made from cellulose or cellophane.

You are required to use the dialysis tubing to model absorption of digested food in the intestine. You could tackle this task collaboratively – by group discussion with peers and tutor or teacher.

Design an experiment to:

- demonstrate whether a carbohydrate food item, such as starch, is able to pass across a selective permeable membrane, and
- to compare the outcome with the behaviour of the products of starch digestion under similar conditions.

Evaluate how appropriate this investigation is as a model of the membranes in the villi that are traversed in the absorption processes.

6.2 The blood system – *the blood system continuously transports substances to cells and simultaneously collects waste products*

Living cells require a supply of water and nutrients, such as glucose and amino acids, and they need oxygen. The waste products of cellular metabolism have to be removed. In single-celled organisms and very small organisms, internal distances are small, so movements of nutrients can occur efficiently by **diffusion**, although some substances require to be transported across membranes by **active transport**. In larger organisms, these mechanisms alone are insufficient – an internal transport system is required to service the needs of the cells.

Internal transport systems at work are examples of **mass flow**. The more active an organism is, the more nutrients are likely to be required by cells, and so the greater is the need for an efficient system for internal transport. Larger animals have a **blood circulatory system** that links the parts of the body and makes resources available where they are required.

■ Transport in mammals

Mammals have a **closed circulation** in which blood is pumped by a powerful, muscular heart and circulated in a continuous system of tubes – the **arteries**, **veins** and **capillaries** – under pressure. The heart has four chambers, and is divided into right and left sides. Blood flows from the right side of the heart to the lungs, then back to the left side of the heart. From here, it is pumped around the rest of the body and back to the right side of the heart. As the blood passes twice through the heart in every single circulation of the body, the system is called a **double circulation**.

The circulatory system of mammals is shown in Figure 6.5. It becomes clear that the major advantages of the mammalian circulation are:

■ simultaneous high-pressure delivery of oxygenated blood to all regions of the body

■ oxygenated blood reaches the respiring tissues, undiluted by deoxygenated blood.

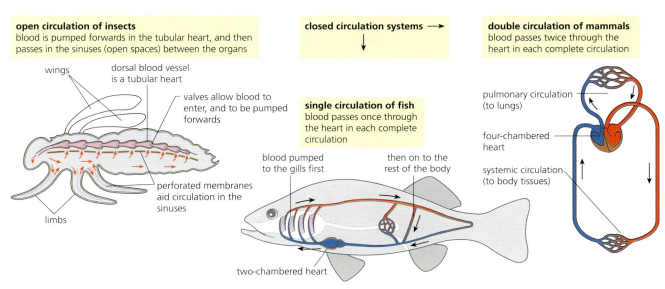

open circulation of insects
blood is pumped forwards in the tubular heart, and then passes in the sinuses (open spaces) between the organs

closed circulation systems →

double circulation of mammals
blood passes twice through the heart in each complete circulation

wings
dorsal blood vessel is a tubular heart
valves allow blood to enter, and to be pumped forwards

single circulation of fish
blood passes once through the heart in each complete circulation

perforated membranes aid circulation in the sinuses

limbs

blood pumped to the gills first

then on to the rest of the body

two-chambered heart

pulmonary circulation (to lungs)

four-chambered heart

systemic circulation (to body tissues)

■ **Figure 6.5** Open and closed circulations

> **7** In an open circulation there is 'little control over circulation'. **Suggest** what this means.

The transport medium – the blood

Blood is the medium by which the products of digestion are transported about the body (Section 6.1, page 248). It has key roles in the body's defence against disease (Section 6.3, page 270) and in transport of respiratory gases (Section 6.4, page 280). Before the circulation of blood is examined here, note (but do not memorize) the composition of the blood.

Blood is an unusual tissue in that it is made up of a liquid medium called **plasma** in which cells are suspended – the red blood cells or **erythrocytes**, white cells or **leucocytes**, and the **platelets**. The plasma, a straw-coloured fluid, supports and transports the blood cells around. It also brings about the essential exchanges of substances between cells, tissues and organ systems of the body. The chemical composition of the plasma is shown in Figure 6.6. The role of erythrocytes is the transport of respiratory gases, and the leucocytes are adapted to combat infection (Table 6.7).

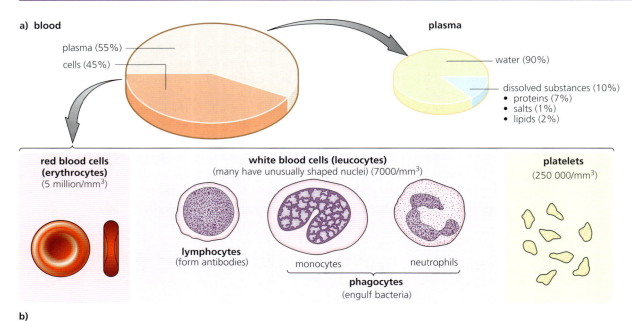

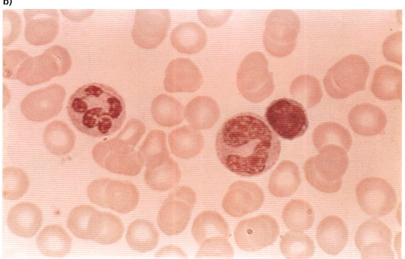

■ **Figure 6.6** The composition of the blood

■ **Table 6.7**
The components of the blood and their roles

Component	Role
plasma	transport of: • **nutrients** from gut or liver to all the cells • excretory products such as **urea** from the liver to the kidneys • **hormones** from the endocrine glands to all tissues and organs • **dissolved proteins** which have roles that include regulating the osmotic concentration (water potential) of the blood • dissolved proteins that are **antibodies** • **heat**, which is distributed to all tissues
red blood cells (erythrocytes)	transport of: • **oxygen** from lungs to respiring cell • **carbon dioxide** from respiring cells to lungs (also carried in plasma)
white cells (leucocytes)	**lymphocytes** – these have major roles in the immune system, including forming antibodies (discussed on pages 276–277) **phagocytes** – these **ingest bacteria** or cell fragments
platelets	have a role in the **blood clotting mechanism**

8 **Outline** where in the body phagocytic leucocytes function.

The plumbing of the circulation system – structure of arteries, veins and capillary walls

There are three types of vessel in the circulation system:

- **arteries**, which carry blood away from the heart
- **veins**, which carry blood back to the heart
- **capillaries**, which are fine networks of tiny tubes linking arteries and veins.

Both arteries and veins have strong, elastic walls, but the walls of the arteries are very much thicker and stronger than those of the veins. The strength of the walls comes from the collagen fibres present and the elasticity is due to the elastic and involuntary (smooth) muscle fibres. The walls of the capillaries, on the other hand, consist of endothelium only (endothelium is the innermost lining layer of arteries and veins). Capillaries branch profusely and bring the blood circulation close to cells – no cell is far from a capillary.

Blood leaving the heart is under high pressure and travels in waves or **pulses**, following each heartbeat. By the time the blood has reached the capillaries, it is under very much lower pressure, without a pulse. This difference in blood pressure accounts for the differences in the walls of arteries and veins (Table 6.8). Figure 6.7 shows an artery, vein and capillary vessel in section and details the wall structure of these three vessels. Because of the low pressure in veins there is a possibility of backflow here. Veins have **valves** at intervals that prevent this (Figure 6.8). In sectioned material (as here), veins are more likely to appear squashed, whereas arteries are circular in section. Examine this illustration carefully (in conjunction with Table 6.8) so that you can identify these vessels in cross section by their wall structure.

■ **Figure 6.7**
The walls of arteries, veins and capillaries

TS artery and vein, LP (×20)

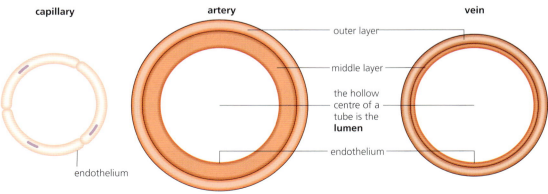

the walls of veins are thin so the blood is visible in them, under the living skin

In sectioned material (as here), veins are more likely to appear squashed, whereas arteries are circular in section.

this artery is sectioned at the point of branching

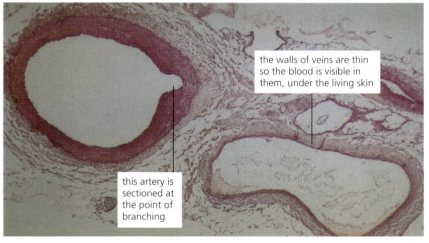

capillary

artery

vein

outer layer

middle layer

the hollow centre of a tube is the **lumen**

endothelium

endothelium

■ **Table 6.8**
Differences between arteries, veins and capillaries

	Artery	Capillary	Vein
outer layer (tunica externa) of elastic fibres and collagen	present (thick layer)	absent	present (thin layer)
middle layer (tunica media) of elastic fibres, collagen and involuntary (smooth) muscle fibres	present (thick layer)	absent	present (thin layer)
endothelium (inner lining) of pavement epithelium – single layer of cells fitting together like jigsaw pieces, with smooth inner surface that minimizes friction	present	present	present
valves	absent	absent	present

Note: Variations in thickness of muscle fibre layers and elastic fibre layers vary throughout the entire circulation system to maintain appropriate blood pressures throughout. This is detailed in Table 6.9 (page 264).

■ **Figure 6.8**
The valves in veins

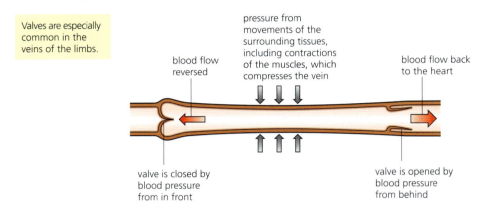

Valves are especially common in the veins of the limbs.

pressure from movements of the surrounding tissues, including contractions of the muscles, which compresses the vein

blood flow reversed

blood flow back to the heart

valve is closed by blood pressure from in front

valve is opened by blood pressure from behind

The arrangement of arteries and veins

Now, we need to look again at the mammal's double circulation. The role of the right side of the heart is to pump deoxygenated blood to the lungs. The arteries, veins and capillaries serving the lungs are known as the **pulmonary circulation**.

The left side of the heart pumps oxygenated blood to the rest of the body. The arteries, veins and capillaries serving the body are known as the **systemic circulation**.

In the systemic circulation, organs are supplied with blood by many arteries. These all branch from the main **aorta**. Within each organ, the artery divides into numerous arterioles (smaller arteries) and the smallest arterioles supply the capillary networks. Capillaries drain into venules (smaller veins) and venules join to form veins. The veins join the **vena cava** carrying blood back to the heart. The branching sequence in the circulation is, therefore:

$$\text{aorta} \rightarrow \text{artery} \rightarrow \text{arteriole} \rightarrow \text{capillary} \rightarrow \text{venule} \rightarrow \text{vein} \rightarrow \text{vena cava}$$

You can see that the arteries and veins are often named after the organs they serve (Figure 6.9). For example, the blood supply to the liver is via the hepatic artery. Notice, too, the liver also receives blood directly from the small intestine, via a vein called the **hepatic portal vein**. This brings many of the products of digestion, after they have been absorbed into the capillaries of the villi.

9 **List** the differences between the pulmonary circulation and the systemic circulation of blood.

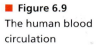

Figure 6.9
The human blood circulation

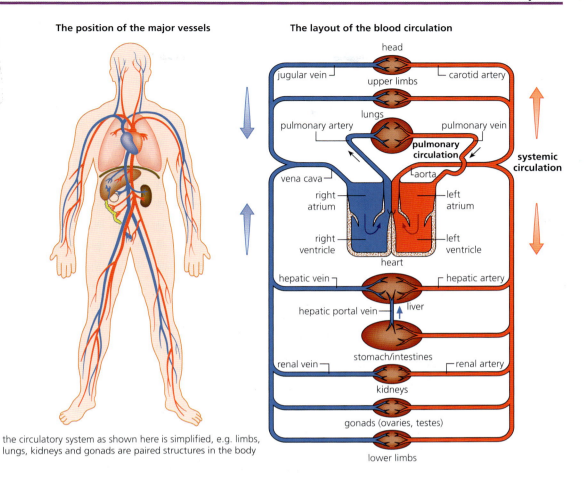

the circulatory system as shown here is simplified, e.g. limbs, lungs, kidneys and gonads are paired structures in the body

The heart as a pump

The human heart is the size of a clenched fist, and lies in the thorax between the lungs and beneath the breast bone (sternum). The heart is a hollow organ with a muscular wall and is contained in a tightly fitting membrane, the pericardium – a strong, non-elastic sac that anchors the heart within the thorax. Most importantly, the wall of the heart is well supplied with oxygenated blood from the aorta, via coronary arteries.

All muscle tissue consists of special cells called fibres that are able to shorten by a half to a third of their length. **Cardiac muscle** itself consists of cylindrical branching columns of fibres, uniquely forming a three-dimensional network (Figure 6.10). This allows contraction in three dimensions. Each fibre has a single nucleus and it appears striped or striated under the microscope. Fibres are surrounded by a special plasma membrane, called the **sarcolemma**, and all are very well supplied by mitochondria and capillaries.

Heart muscle fibres contract rhythmically from formation until they die. Muscles typically contract when stimulated to do so by an external nerve supply. However, in the heart, the impulse to contract is generated within the heart muscle itself – it is said to have a **myogenic origin**. We will return to this issue, shortly. The connection between individual cardiac muscle fibres is by special junctions called **intercalated discs**. The role of these discs is to transmit the impulse, so as to cause all cells to contract simultaneously.

TOK Link

Our current understanding is that emotions are the product of activity in the brain, rather than in the heart. Is knowledge based on science more valid than knowledge based on intuition?

■ **Figure 6.10**
Cardiac muscle

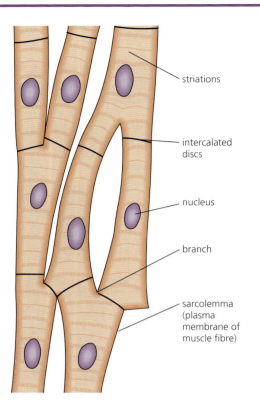

striations

intercalated discs

nucleus

branch

sarcolemma (plasma membrane of muscle fibre)

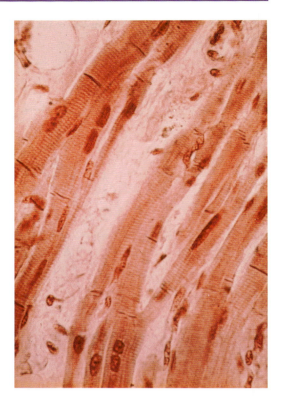

The chambers and valves of the heart

The cavity of the heart is divided into four **chambers**, with those on the right side of the heart completely separate from those on the left (Figure 6.11). The two upper chambers are thin-walled **atria** (singular, **atrium**). These receive blood into the heart. The two lower chambers are thick-walled **ventricles**, with the muscular wall of the left ventricle much thicker than that of the right ventricle. However, the volumes of the right and left sides (the quantities of blood they contain) are identical. The ventricles pump blood out of the heart.

You should be able to recognize the individual chambers of the heart in diagrams, for example. Carrying out a dissection of a fresh heart (such as a sheep's heart) is an effective aid to understanding structure and function.

The **valves** of the heart prevent backflow of the blood, so maintaining the direction of flow through the heart. You can see these valves in action, in the diagrams in Figure 6.12. Notice that the **atrioventricular valves** are large valves, in a position to prevent backflow of blood from ventricles to atria. The edges of these valves are supported by tendons, anchored to the muscle walls of the ventricles below. These tendons prevent the valves from folding back due to the (huge) pressure that develops here with each heartbeat. Actually, the atrioventricular valves are individually named: on the right side is the **tricuspid valve**; on the left is the **bicuspid** or mitral valve.

A different type of valve separates the ventricles from pulmonary artery (right side) and aorta (left side). These are pocket-like structures called **semilunar valves**, rather similar to the valves seen in veins. These cut out backflow from the aorta and pulmonary artery into the ventricles as the ventricles relax between heartbeats.

In Figure 6.11, we can see the **coronary arteries**. These arteries, and the capillaries they serve, deliver to the muscle fibres of the heart the oxygen and nutrients essential for the pumping action, and they remove the waste products. We shall return to this issue shortly and to the serious consequences for the whole body when they get blocked.

■ **Figure 6.11**
The structure of
the heart

heart viewed from the front of the body with pericardium removed

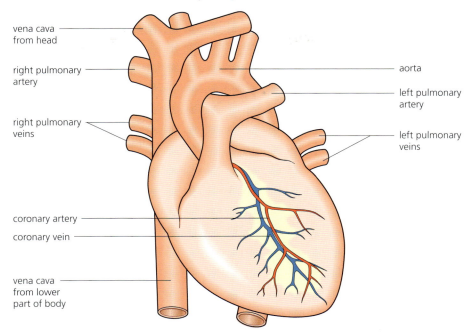

vena cava
from head

right pulmonary
artery

right pulmonary
veins

coronary artery

coronary vein

vena cava
from lower
part of body

aorta

left pulmonary
artery

left pulmonary
veins

heart in LS

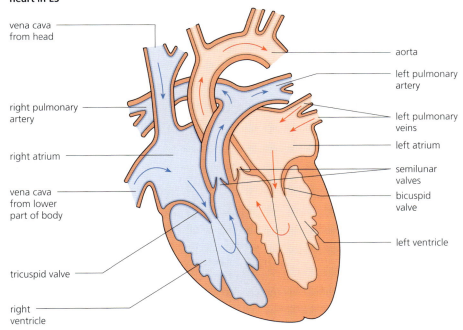

vena cava
from head

right pulmonary
artery

right atrium

vena cava
from lower
part of body

tricuspid valve

right
ventricle

aorta

left pulmonary
artery

left pulmonary
veins

left atrium

semilunar
valves

bicuspid
valve

left ventricle

Heart valves allow blood flow in one direction only. They are opened and closed by differences in blood pressure.
• They are opened when the pressure on the input side (caused by muscle contraction) exceeds that on the output side.
• They are closed when the pressure on the output side exceeds that on the input side. Typically caused by the relaxation of muscle on the input side.

Atrial systole
• Atrial walls contract – the pressure is raised in the atria
• Vena cavae and pulmonary veins are closed
• Atrioventricular valves pushed open – blood flows into the ventricles
• Ventricular walls relaxed – the pressure is low in the ventricles
• High pressure in the aorta and pulmonary arteries (due to the elastic and muscle fibres in their walls) – the semilunar valves are shut

Ventricular systole
• Ventricular walls contract strongly – the pressure is raised in the ventricles
• Atrioventricular valves are closed and semilunar valves are opened – blood is pushed out into the aorta and the arteries are stretched
• Atrial walls relax – the pressure falls in the atria
• The vena cavae and pulmonary veins are open – blood flows into the atria

Diastole – atrial and ventricular
• The muscles of the atrial and ventricular walls are relaxed – the pressure in both atria and ventricles is low
• The semilunar valves shut – there is a passive flow of blood from the veins into the atria and ventricles

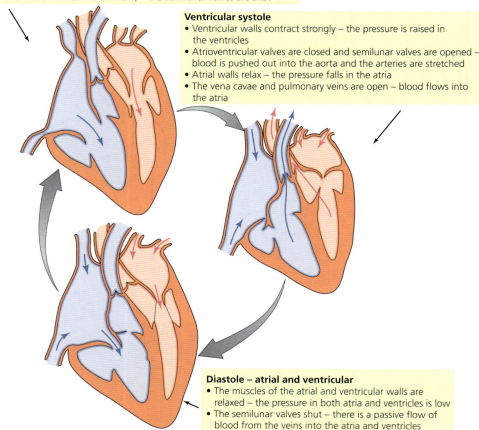

■ **Figure 6.12** The role of the heart valves

10 The edges of the atrioventricular valves have non-elastic strands attached, which are anchored to the ventricle walls (Figure 6.11). **Explain** exactly what the strands do.

The cardiac cycle

The cardiac cycle is the sequence of events of a heartbeat, by which blood is pumped all over the body. The heart beats at a rate of about 75 times per minute, so each cardiac cycle is about 0.8 s long. This period of 'heartbeat' is divided into two stages which are called **systole** and **diastole**. In the systole stage, heart muscle contracts and during the diastole stage, heart muscle relaxes. When the muscular walls of the chambers of the heart contract, the volume of the chambers is decreased. This increases the pressure on the blood contained there, forcing the blood to a region where pressure is lower. Valves prevent blood from flowing backwards to a region of low pressure, so blood always flows in one direction through the heart.

Look at the steps in Figure 6.12 and 6.13. You will see that Figure 6.13 illustrates the cycle on the left side of the heart only, but both sides function together, in exactly the same way, as Figure 6.12 makes clear.

We can start with contraction of the atrium (**atrial systole**, about 0.1 s). As the walls of the atrium contract, blood pushes past the atrioventricular valve, into the ventricles where the contents are under low pressure. At this time, any backflow of blood from the aorta, back into the ventricle chamber is prevented by the semilunar valves. Notice that backflow from the atria into the vena cavae and the pulmonary veins is prevented because contraction of the atrial walls seals off these veins. Veins also contain semilunar valves which prevent backflow here, too.

The atrium now relaxes (**atrial diastole**, about 0.7 s).

Next, the ventricle contracts (**ventricular systole**, about 0.5 s). The high pressure this generates slams shut the atrioventricular valve and opens the semilunar valves, forcing blood into the aorta. A 'pulse', detectable in arteries all over the body, is generated (see below).

This is followed by relaxation of the ventricles (**ventricular diastole**).

Each contraction of cardiac muscle is followed by relaxation and elastic recoil. The changing pressure of blood in the atria, ventricles, pulmonary artery and aorta (shown in the graph in Figure 6.13) automatically opens and closes the valves.

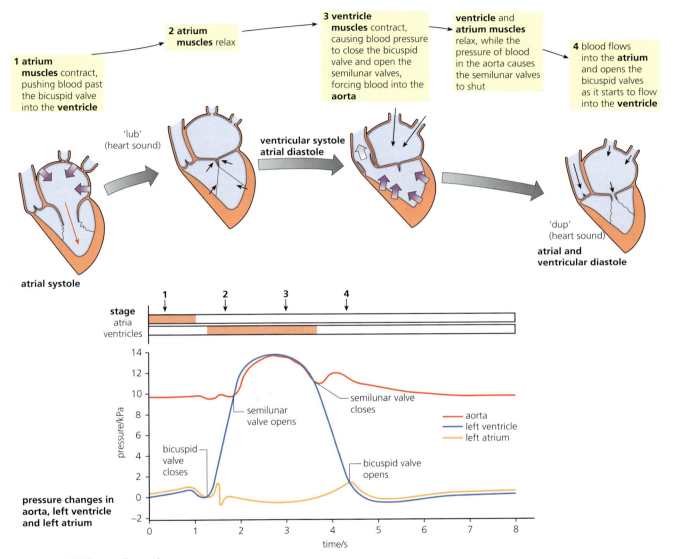

■ **Figure 6.13** The cardiac cycle

11 Examine the data on pressure change during the cardiac cycle in Figure 6.13. **Deduce** why:
 a pressure in the aorta is always significantly higher than that in the atria
 b pressure falls *most* abruptly in the atrium once ventricular systole is underway
 c the semilunar valve in the aorta does not open immediately that ventricular systole commences
 d when ventricular diastole commences, there is a significant delay before the bicuspid valve opens, despite rising pressure in the atrium
 e it is significant that about 50% of the cardiac cycle is given over to diastole.

Heart rate and the pulse

The contraction of the ventricle walls forces a surge of blood into the aorta and into the pulmonary arteries under great pressure. This volume of blood is called the **stroke volume**. Each surge stretches the elastic fibres in the artery walls. This is known as the **pulse**.

The artery walls are distended as the surge passes, followed by an elastic recoil. In this way the muscles and elastic fibres of the walls of arteries assist in maintaining blood pressure and the onward flow of the blood (Table 6.9).

■ **Table 6.9**
Blood vessels – changing structure in relation to function

Component/role	Structure in relation to function
Arteries – have the thickest and strongest walls; the tunica media is the thickest part.	
Aorta • receives blood pumped from the heart in a 'pulse' • pressure about 120 mm Hg* **Main arteries** • distribute blood under high pressure to regions of the body • arteries become wider, so lowering the pressure • supply the **arteries** that distribute blood to the main organs	• Walls stretch to accommodate the huge surge of blood when ventricles contract. The elastic and collagen fibres of the tunica externa prevent rupture as blood surges from the heart. • The high proportion of elastic fibres are first stretched and then recoil, keeping the blood flowing and propelling it forwards after each pulse passes. • With increasing distance from the heart, the tunica media progressively contains more smooth muscle fibres and fewer elastic fibres. By varying constriction and dilation of arteries, blood flow is maintained. Muscle fibres stretch and recoil, tending to even out the pressure, but a 'pulse' can still be detected.
Arterioles • deliver blood to the tissues	• High proportion of smooth muscle fibres, so able to regulate blood flow from arteries into capillaries.
Capillaries • serve the tissues and cells of the body • blood under lower pressure (about 35 mm Hg)	• Narrow tubes, about the diameter of a single red blood cell (about 7 µm), reduce the flow rate to increase exchange between blood and tissue. • Thin walls consist of a single layer of endothelial cells. • Walls have gaps between cells sufficient to allow some components of the blood to escape and contribute to tissue fluid (a liquid surrounding tissues that is the same concentration as body cells – see page 41).
Veins – have thin walls; tunica externa is the thickest layer.	
Venules • collect blood from the tissues • formed by a union of several capillaries • pressure about 15 mmHg	• Walls consist of endothelium and a very thin tunica media of a few scattered smooth muscle fibres.
Veins • receive blood from tissues under low pressure (about 5 mm Hg) • veins become wider, so lowering pressure and increasing flow rate	• tunica externa present, of elastic and collagen fibres • here tunica media contains a few elastic fibres and muscle fibres • presence of valves preventing backflow of blood (visible in LS of veins, Figure 6.8)

*Units of 'pressure'
The **pascal** (**Pa**) and its multiple the **kilopascal** (**kPa**) are generally used by scientists to measure pressure. However, in medicine the older unit of 'millimetres of mercury' (**mm Hg**) is still used (1 mm Hg = 0.13 kPa).

However, as blood is transported around the body, the pressure of blood does progressively fall. Pulsation of the blood flow has entirely disappeared by the time it reaches the capillaries. This is due to the extensive nature of the capillary networks and to the resistance the blood experiences as it flows through them.

Measuring heart rate

We can measure heart rate in the carotid artery in the neck or at the wrist, where an artery passes over a bone. Each contraction of the ventricles generates a pulse, so when we measure our pulse rate we are measuring our heart rate.

Incidentally, the amount of blood flowing from the heart is known as the **cardiac output**. At rest, our cardiac output is typically about 5 dm^3 of blood per minute.

cardiac output = stroke volume × pulse rate

12 Examine the graph showing the changing blood velocity, blood pressure and cross-sectional area of the vessels as blood circulates in the body. Then, using the information in Figure 6.14 (and in Table 6.9 where appropriate), answer the questions below.

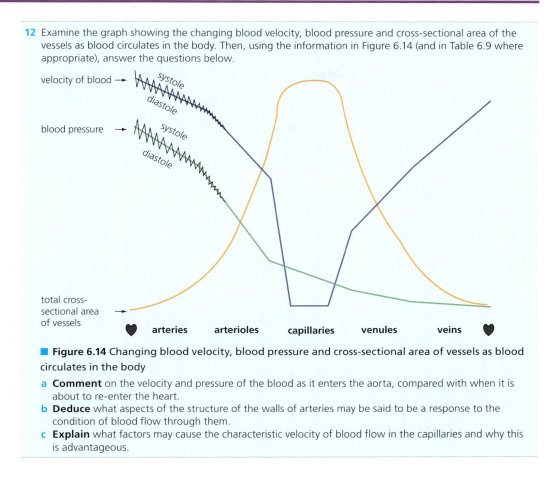

velocity of blood →
systole
diastole

blood pressure →
systole
diastole

total cross-sectional area of vessels →

arteries   arterioles   capillaries   venules   veins

■ **Figure 6.14** Changing blood velocity, blood pressure and cross-sectional area of vessels as blood circulates in the body

a **Comment** on the velocity and pressure of the blood as it enters the aorta, compared with when it is about to re-enter the heart.
b **Deduce** what aspects of the structure of the walls of arteries may be said to be a response to the condition of blood flow through them.
c **Explain** what factors may cause the characteristic velocity of blood flow in the capillaries and why this is advantageous.

Origin and control of the heartbeat

The heart beats rhythmically throughout life, without rest, apart from the momentary relaxation that occurs between each beat. Even more remarkably, heartbeats occur naturally, without the cardiac muscle needing to be stimulated by an external nerve. Since the origin of each beat is within the heart itself, we say that heartbeat is '**myogenic**' in origin. However, the alternating contractions of cardiac muscle of the atria and ventricles are controlled and coordinated precisely. Only in this way, can the heart act as an efficient pump. The positions of the structures within the heart that bring this about are shown in Figure 6.15.

The steps to control of the cardiac cycle are as follows:

■ The heartbeat originates in a tiny part of the muscle of the wall of the right atrium, called the **SAN** (sinoatrial node) or **pacemaker**.

■ From here, a wave of excitation (electrical impulses) spreads out across both atria.

■ In response, the muscle of both atrial walls contracts simultaneously (**atrial systole**).

■ This stimulus does *not* spread to the ventricles immediately, due to the presence of a narrow band of non-conducting fibres at the base of the atria. These block the excitation wave, preventing its conduction across to the ventricles. Instead, the stimulus is picked up by the **AVN** (atrioventricular node), situated at the base of the right atrium.

■ After a delay of 0.1–0.2 s, the excitation is passed from the AVN to the base of both ventricles by tiny bundles of conducting fibres, known as the **Purkyne tissue (Purkinje fibres)**. These are collectively called the **bundles of His**.

■ When stimulated by the bundles of His, the ventricle muscles start to contract from the *base* of the heart upwards (ventricular systole).

■ The delay that occurs before the AVN acts as a 'relay station' for the impulse permits the emptying of the atria into the ventricles to go to completion, and prevents the atria and ventricles from contracting simultaneously.

■ After every contraction, cardiac muscle has a period of insensitivity to stimulation, the **refractory period** (a period of enforced non-contraction – **diastole**). In this phase, the heart begins, passively, to refill with blood. This period is a relatively long one in heart muscle, and enables the heart to beat throughout life.

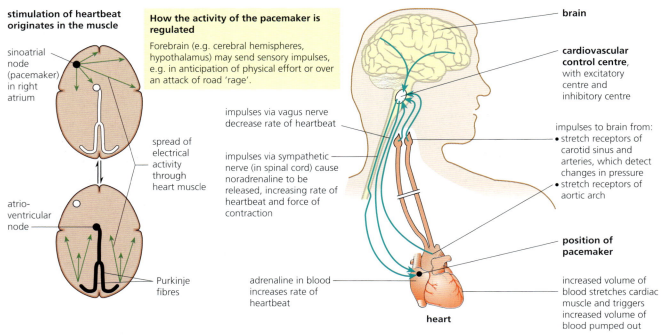

■ **Figure 6.15** Control of heart rate

The heart's own rhythm, set by the SAN, is about 50–60 beats per minute. Conditions in the body can and do override this basic rate to increase heart performance. The action of the pacemaker is modified according to the needs of the body. For example, pacemaker activity may be increased during physical activity. This occurs because increased muscle contraction causes an increased volume of blood to pass back to the heart. The response may be more powerful contractions without an increase in heart rate. Alternatively, the rate of 75 beats per minute of the heart 'at rest' may be increased to up to 200 beats a minute in very strenuous exercise.

How is the pacemaker regulated? Look at Figure 6.15 again, now.

Nervous control of the heart is by **reflex action**. The heart receives impulses from the **cardiovascular centre** in the medulla of the hind brain, via two nerves:

■ a **sympathetic nerve**, part of the sympathetic system, which speeds up the heart

■ a branch of the **vagus nerve**, part of the parasympathetic nervous system, which slows down the heart.

Since the sympathetic nerve and the vagus nerve have opposite effects, we say they are **antagonistic**.

Nerves supplying the cardiovascular centre bring impulses from stretch receptors located in the walls of the aorta, in the carotid arteries, and in the wall of the right atrium, when change in blood pressure at these positions is detected.

■ When blood pressure is high in the arteries, the rate of heartbeat is lowered by impulses from the cardiovascular centre, via the vagus nerve.

■ When blood pressure is low, the rate of heartbeat is increased.

13 Suggest likely conditions or situations in which the body is likely to secrete adrenaline.

The rate of heartbeat is also influenced by impulses from the higher centres of the brain. For example, emotion, stress and anticipation of events can all cause impulses from the sympathetic nerve to speed up heart rate. In addition, the hormone **adrenaline** (also referred to as **epinephrine**), which is secreted by the adrenal glands and carried in the blood, causes the pacemaker to increase the heart rate to prepare for vigorous physical activity.

Nature of Science

Theories are regarded as uncertain

◾ Discovery of the blood circulation

The original discovery of the circulation of mammalian blood was made in Europe by experimental scientists in the seventeenth century. At this time, much so-called medical knowledge was derived from the theories expressed in the writings of Galen, a Roman physician (AD 129–199) and on the ideas of earlier Greek writers. How William Harvey and Marcello Malpighi achieved this revolution is outlined in Figure 6.16.

◾ **Figure 6.16** William Harvey's investigation of the blood circulation

William Harvey 1578–1637
Physician – founder of modern physiology through his experimental approach.
- Harvey studied medicine at Cambridge and then at the medical school at Padua under Fabrizio (who had already discovered valves in veins but did not understand them).
- He quickly became an important physician, serving in turn, James I and later Charles I – but his main interested was research. He was an enthusiastic and skilful experimenter.
- By dissection, and by experimentation, Harvey observed and discovered:
  - the working valves in the heart and the veins, and their role in maintaining one-way flow of blood
  - in systole the heart contracts as a muscular pump
  - the right ventricle supplies the lungs
  - the left ventricle supplies the rest of the system of arteries
  - blood flow in veins was *towards* the heart.
- These discoveries and his calculation of the volume of blood in the body led to the conclusion '*therefore the blood must circulate*'.
- He concluded that the connections between arteries and veins were too small to see. Harvey had predicted the existence of capillaries before the microscope became available.
- He published 'On the Motion of the Heart and Blood' in 1628.
Working capillaries were observed after his death, by the Italian biologist Malpighi (1628–1694).

Harvey's experiment showing the veins and working valves in the arm, after ligation.

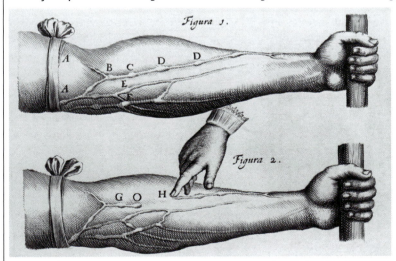

◾ The roles of the blood circulation system

The blood circulation has roles in the body's defence against diseases (pages 271 and 275) as well as being the all-important transport system of the body (Table 6.10). Nutrients from digestion, oxygen and carbon dioxide, urea, hormones and antibodies are all transported. In the tissues of the body, exchange between the blood and cells of the tissues occurs from the capillaries, the walls of which are permeable and highly 'leaky'.

■ **Table 6.10**
Transport roles of the
blood circulation

Function	Transport of
tissue respiration	• **oxygen** from lungs to all tissues • **carbon dioxide** back to the lungs
hydration	**water** to all the tissues
nutrition	**nutrients** (sugars, amino acids, lipids, vitamins) and inorganic ions to all cells
excretion	waste product, **urea**, to kidneys
development and coordination	**hormones** from endocrine glands to target organs
temperature regulation	distribution of **heat**
defence against disease	**antibodies** are circulated in the bloodstream

■ Causes and consequences of coronary heart disease

Diseases of the heart and blood vessels are primarily due to a condition called **atherosclerosis** (Figure 6.17). This is the progressive degeneration of the artery walls. The structure of a healthy artery wall is shown in Figure 6.7 (page 257).

Look up the structure of a healthy artery again, now.

■ **Figure 6.17**
Atherosclerosis,
leading to a
thrombus

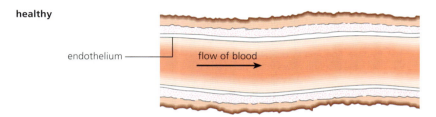

healthy

endothelium

flow of blood

diseased

blood clot = thrombus formed where atheroma has broken through the endothelium

lipid + fibre deposit = atheroma

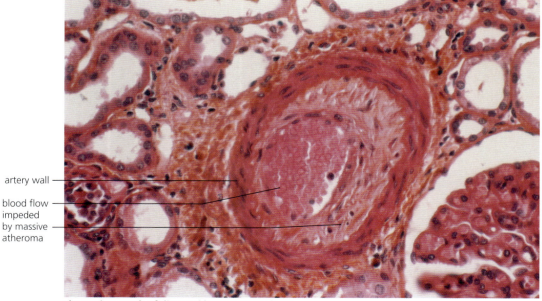

artery wall

blood flow impeded by massive atheroma

photomicrograph of diseased human artery in TS

Healthy arteries have pale, smooth linings, and the walls are elastic and flexible. The steps to atherosclerosis in arteries are:

- **Damage to the artery walls** as strands of fat (low-density lipoproteins, LDLs or 'bad cholesterol') are deposited under the endothelium. Fibrous tissue is also laid down.

- **Raised blood pressure** follows when fat deposits and fibrous tissue start to impede blood flow.

- **Lesion formation** occurs when the smooth lining of the arteries actually breaks down. These lesions are known as **atherosclerotic plaques**. Blood platelets tend to collect at the exposed, roughened surface and release factors that trigger a defensive inflammatory response that includes blood clotting. A blood clot formed within a blood vessel is known as a **thrombus**. This is until it breaks free and circulates in the bloodstream, at which point it is called an **embolus**.

The consequences of atherosclerosis

One likely outcome of atherosclerotic damage to major arteries is that an embolus may be swept into a small artery or arteriole which is narrower than the diameter of the clot, causing a blockage. Immediately, the blood supply to the tissue downstream of the block is deprived of oxygen and without oxygen tissues die. The coronary arteries are especially vulnerable, particularly those to the left ventricle. When sufficient heart muscle dies in this way, the heart may cease to be an effective pump. We say a 'heart attack' has occurred (known as a **myocardial infarction**).

Coronary arteries that have been damaged can be surgically by-passed – known as a **heart by-pass operation** (Figure 6.18 A). Typically, a blood vessel taken from the patient's leg is used. In some cases multiple by-passes are required. This operation is increasingly commonly prescribed and survival rates are high.

Coronary angioplasty or balloon angioplasty is an alternative technique for re-establishing flow in an occluded coronary artery. Here, the balloon catheter is inserted into a main artery in the patient's arm or leg, and is then guided all the way to the damaged coronary artery using X-ray observation. Here, the balloon tip is inflated to reduce the obstruction and re-establish the flow, before it is 'collapsed' and withdrawn. It may be necessary to shore up the damaged artery to maintain the subsequent flow of oxygenated blood by inserting a stainless steel spring (a stent) into the lumen of the artery (Figure 6.18 B). The risks in this procedure are outweighed by the advantages, since arteries treated to angioplasty very frequently close up again within six months without the benefit of a stent.

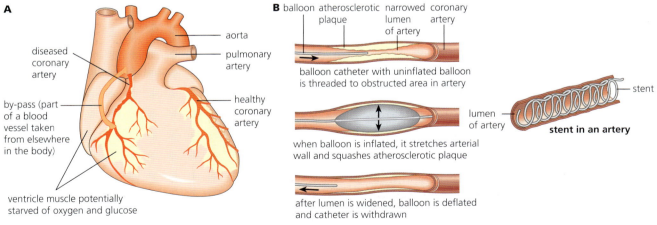

■ **Figure 6.18** Re-establishing blood flow in damaged coronary arteries

6.3 Defence against infectious disease – *the human body has structures and processes that resist the continuous threat of invasion by pathogens*

Infectious disease is caused when another organism or virus particle invades the body and lives there parasitically. The invader is known as a **pathogen** and the infected organism – a human, in this case – is the **host**. **A pathogen is an organism or virus that causes a disease.** Most, but not all, pathogens are microorganisms.

Pathogens may pass from diseased host to healthy organisms, so these diseases are known as infectious or **communicable diseases**. Disease, in general, is defined as an 'unhealthy condition of the body' and this rather broad definition includes some distinctly different forms of ill-health. For example, ill-health may be caused by unfavourable environmental conditions. Diseases of this type are non-infectious or **non-communicable** diseases and they include conditions such as cardiovascular disease, malnutrition and cancer.

WHO

Good health is more than the absence of harmful effects of a disease. This point is emphasized by the **World Health Organization (WHO)** which identifies health as '*a state of complete physical, mental and social well-being, and not merely the absence of disease or infirmity*'.

A role of the WHO is to monitor and communicate information on the spread and containment of infectious diseases, such as 'bird flu'.

Pathogens and disease

The range of **disease-causing organisms** that infect humans includes not only microorganisms such as certain **bacteria** and **fungi**, but also some **protozoa** (single-celled animals), certain non-vertebrate animals in the phyla of **flatworms** (and roundworms), and many **virus** particles,

Not all bacteria or fungi are parasitic and pathogenic – only relatively few species are, in fact. On the other hand, no virus can function in any way outside a host organism, so we can say that all viruses are parasitic. A virus particle consists of a nucleic acid core surrounded by a protein coat. It is visible only under the electron microscope (page 17). Viruses, once introduced into a host cell, may take over the machinery of protein and nucleic acid synthesis, and force their host cells to manufacture more virus components and assemble them.

14 State the differences in structure between a bacterial cell and a virus.

The body's defence against infectious disease

The skin and the mucosa (the internal linings of lungs, trachea and gut) are the primary defence against pathogens that cause infectious diseases. Not surprisingly, protective measures have evolved at these surfaces.

The **external skin** is covered by keratinized protein of the dead cells of the epidermis. This is a tough and impervious layer, and an effective barrier to most organisms unless the surface is broken, cut or deeply scratched. However, folds or creases in the skin that are permanently moist may become the home of microorganisms that degrade this barrier and cause infection, as in athlete's foot, for example (Figure 6.19).

■ **Figure 6.19**
Athlete's foot –
epidermis destruction
by a fungus

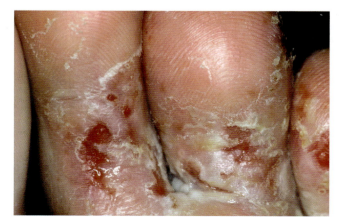

The aquatic larvae of the pathogenic flatworm *Schistosoma*, known as a blood worm, is able to burrow through the skin. This may happen when people bathe in infected water (Figure 6.20).

cycle of infection and re-infection

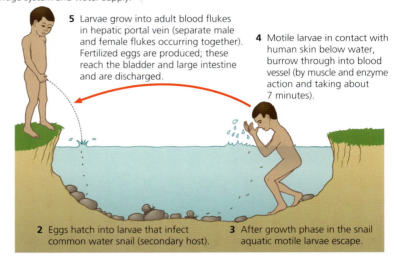

water snail
(secondary host)

half size

larva of *Schistosoma*
(at this stage in the lifecycle the parasite is equipped for entry into the human host by burrowing through the skin)

×300

glands produce an enzyme to assist in penetrating skin

tail, used to propel the larva away from the snail, breaks off here when the larva penetrates a new host

sucker used in burrowing through human skin

1 Fertilized eggs discharged by infected people (in faeces and urine) into waterways that function as sewage system and water supply.

5 Larvae grow into adult blood flukes in hepatic portal vein (separate male and female flukes occurring together). Fertilized eggs are produced; these reach the bladder and large intestine and are discharged.

4 Motile larvae in contact with human skin below water, burrow through into blood vessel (by muscle and enzyme action and taking about 7 minutes).

2 Eggs hatch into larvae that infect common water snail (secondary host).

3 After growth phase in the snail aquatic motile larvae escape.

■ **Figure 6.20** The lifecycle of *Schistosoma* and the spread of schistosomiasis

The **internal surfaces** of our breathing apparatus (the trachea, bronchi and the bronchioles) and of the gut are all lined by moist epithelial cells. These vulnerable internal barriers are protected by the secretion of copious quantities of **mucus** and by the actions of **cilia** that remove the mucus.

Cilia are organelles that project from the surface of certain cells. Cilia occur in large numbers on the lining (epithelium) of the air tubes (bronchi) serving the lungs. Here, they sweep the fluid mucus across the epithelial surface, away from the delicate air sacs of the lungs.

In the gut, digestive enzymes provide some protection, as does the strong **acid secreted in the stomach** on arrival of food (page 251).

However, all these barriers, both internal and external, may be crossed by certain pathogens – as often happens. It is fortunate there are **internal lines of defence** too. We examine these next.

15 Suggest how mucus secreted by the lungs may protect lung tissue.

1 Inflammation

Inflammation is the initial, rapid, localized response our tissues make to damage, whether due to a cut, bruising or a deep wound. The volume of blood in the damaged area is increased. The site becomes swollen, warm and painful. The increased blood flow removes toxic products that may be released by any invading microorganisms. Any local leakage of blood from the blood vessels is eventually stopped by **clotting of blood**, leading to a sealing of the wound.

Leucocytes (white cells) are present in the plasma. Many of these accumulate within and outside the enlarged capillaries. Leucocytes originate in the bone marrow and are initially distributed by the blood circulation all over the body. At the inflamed site, leucocytes become active in the resistance to infection in two ways:

■ The general phagocytic white cells engulf 'foreign' material. Ingestion of pathogens by phagocytes gives non-specific immunity to disease.

■ Other white cells produce the **antibody reaction to infection** in response to invasion by 'foreign matter'. This is the **immune response**. The production of antibodies in response to particular pathogens gives specific immunity.

You can see leucocytes in the photomicrograph of a blood smear, in Figure 6.6 on page 256.

2 Phagocytic leucocytes

Some of the leucocytes have the role of engulfing foreign material, including invading bacterial cells. Certain of these leucocytes are short-lived cells of the plasma. Others are the long-lived, rubbish-collecting cells found throughout the body tissues. Both types of cell take up material into their cytoplasm, much as the protozoan *Amoeba* is observed to feed, by a mechanism known as **phagocytosis** (Figure 6.21). Once inside the cell, the material is destroyed in a controlled way by the activity of lysosomes (page 25).

■ **Figure 6.21**
Phagocytosis of a bacterium

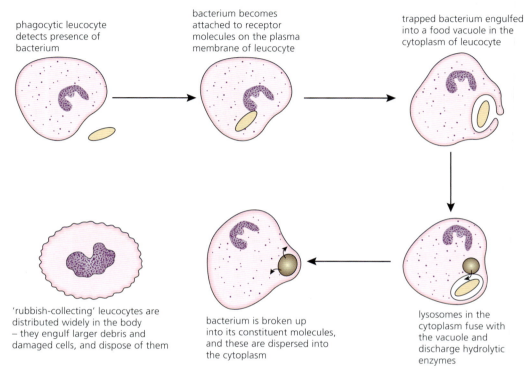

phagocytic leucocyte detects presence of bacterium

bacterium becomes attached to receptor molecules on the plasma membrane of leucocyte

trapped bacterium engulfed into a food vacuole in the cytoplasm of leucocyte

'rubbish-collecting' leucocytes are distributed widely in the body – they engulf larger debris and damaged cells, and dispose of them

bacterium is broken up into its constituent molecules, and these are dispersed into the cytoplasm

lysosomes in the cytoplasm fuse with the vacuole and discharge hydrolytic enzymes

3 Blood-clotting mechanism

When a blood vessel is ruptured, the **blood-clotting mechanism** is activated. This leads to localized clotting of blood and further blood loss is prevented. A significant fall in blood pressure is also prevented, whether at small haemorrhages, or at larger breakages or other wounds. The clot also reduces the chances of invasion by disease-causing organisms. After that, repair of the damaged tissues can get underway.

The formation of a blood clot is triggered by a 'cascade' of events at the site of a broken blood vessel (Figure 6.22). The steps are as follows.

■ Firstly, **platelets** collect at the site. These components of the blood are formed in the bone marrow, along with the red and white blood cells, and they are circulated throughout the body, suspended in the plasma. Platelets are actually cell fragments, disc-shaped and very small (only 2 µm in diameter) – too small to contain a nucleus. Each platelet consists of a sack of cytoplasm, rich in vesicles containing enzymes, and is surrounded by a plasma membrane. Platelets stick to the damaged tissues and clump together there (at this point, they change shape from sacks to flattened discs with tiny projections that interlock). This action alone seals off the smallest breaks.

■ Then, the collecting platelets release a **clotting factor** (a protein called thromboplastin – it is also released by damaged tissues at the site). This clotting factor, along with vitamin K and calcium ions (always present in the plasma), causes a soluble plasma protein called **prothrombin** to be converted to an active, proteolytic enzyme, **thrombin**.

■ The action of thrombin enzyme is to convert another soluble blood protein, **fibrinogen**, to insoluble **fibrin** fibres at the site of the cut. Within this mass of fibres, red blood cells are trapped and the blood clot has formed.

■ **Figure 6.22**
The blood-clotting
mechanism

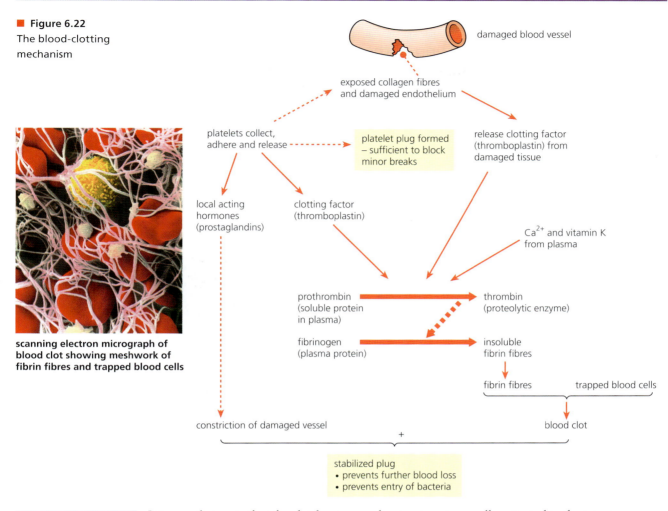

scanning electron micrograph of
blood clot showing meshwork of
fibrin fibres and trapped blood cells

16 Identify the
correct sequence
of the following
events during
blood clotting:
fibrin formation;
clotting factor
release; thrombin
formation.

It is most fortunate that the clot-forming mechanism is not normally activated in the intact circulation; clotting is only triggered by abnormal conditions, such as occur at the break. These complex steps to clotting can be seen as an essential **fail-safe mechanism**. For example, if blood clots were to form spontaneously, within the intact circulation, there would be a risk of dangerous and possibly fatal blockages in the blood supply to the capillaries in the lungs, brain or heart muscle.

Antibiotics against infection

Antibiotics are naturally occurring substances that slow down or kill microorganisms. They are obtained from fungi or bacteria and are substances which these organisms manufacture in their natural habitats. An antibiotic, when present in low concentrations, inhibits the growth of other microorganisms. Many bacterial diseases of humans and other animals can be successfully treated with them.

The discovery of penicillin – the first antibiotic

In 1929, Alexander Fleming (1881–1955), a Scottish bacteriologist, was studying *Staphylococcus*, the bacterium that causes boils and sore throats, at St. Mary's Hospital, Paddington, London. When examining some older bacteriological plates, he came across one in which a fungal colony had also become established (Figure 6.23). He noticed that the bacteria were killed in areas surrounding the mould. This he identified as *Penicillium notatum*. He cultured this mould in broth and discovered that a substance from it – he named it **penicillin** – was bactericidal. He showed that penicillin did not harm human blood cells. Fleming published these results in a scientific paper.

Nature of Science

Risks associated with scientific research

■ Early testing of the safety of penicillin as a drug

The leading figures in the development of penicillin were the Australian pathologist Harold Florey (1898–1968) and Ernest Chain (1906–79), a German biochemist who immigrated to England in 1933. This team isolated penicillin in a stable form, for therapeutic uses.

Florey and Chain used mice to test penicillin on bacterial infections in an experiment which today would not be compliant with drug-testing protocols (Figure 6.23). *Can you see why?*

Since their original discovery, over 4000 different antibiotics have been isolated, but only about 50 have proved to be safe to use as drugs. The antibiotics which are effective over a wide range of pathogenic organisms are called **broad-spectrum antibiotics**. Others are effective with just a few pathogens. Many antibiotics in use today have been synthesized.

■ Figure 6.23
Early steps in the discovery and use of penicillin as an antibiotic

17 **Suggest** possible reasons why Florey and Chain's tests on the safety of penicillin would not be compliant with current protocols on drug testing.

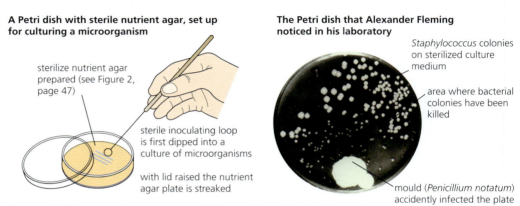

A Petri dish with sterile nutrient agar, set up for culturing a microorganism

sterilize nutrient agar prepared (see Figure 2, page 47)

sterile inoculating loop is first dipped into a culture of microorganisms

with lid raised the nutrient agar plate is streaked

The Petri dish that Alexander Fleming noticed in his laboratory

Staphylococcus colonies on sterilized culture medium

area where bacterial colonies have been killed

mould (*Penicillium notatum*) accidently infected the plate

Florey and Chain's experiment to test penicillin on a bacterial infection in mice

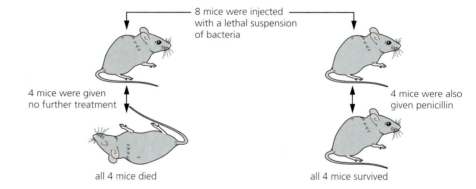

8 mice were injected with a lethal suspension of bacteria

4 mice were given no further treatment

4 mice were also given penicillin

all 4 mice died

all 4 mice survived

How antibiotics work

Most antibiotics disrupt the metabolism of prokaryotic cells – whole populations of bacteria may be quickly suppressed. Figure 1.33 (page 30) shows an actively dividing bacterial cell. It is in these division and growth phases that bacteria are vulnerable to antibiotic action (Table 6.11). At the same time, the cells of the human host organism (a eukaryote) are not affected.

■ Table 6.11
The biochemical mechanisms of antibiotic action

Mechanism targeted	Effects
Cell wall synthesis	The antibiotic interferes with the synthesis of bacterial cell walls. Once the cell wall is destroyed, the delicate plasma membrane of the bacterium is exposed to the destructive force generated by excessive uptake of water by osmosis. Several antibiotics, including penicillin, ampicillin and bacitracin, bind to and inactivate specific wall-building enzymes – the bacterium's walls fall apart. (This is the most effective mechanism.)
Protein synthesis	The antibiotic inhibits protein synthesis by binding with ribosomal RNA. The ribosomes of prokaryotes are made of particular RNA subunits. The ribosomes of eukaryotic cells are larger and are built with different types of RNA molecules. Antibiotics like streptomycin, chloramphenicol, the tetracyclines and erythromycin all bind to the prokaryotic ribosomal RNA subunits that are unique to bacteria, terminating their protein synthesis.

18 **Explain** why antibiotics are effective against bacteria but not viruses.

Viruses, on the other hand, are non-living particles and have no metabolism of their own (and so no function that can be inhibited by antibiotics). Viruses reproduce using metabolic pathways in their host cell. Antibiotics cannot be used to prevent viral diseases.

Antibiotics – 'wonder drugs' or mixed blessing?

Before antibiotics became available to treat bacterial infections, the typical UK hospital ward was filled with patients with pneumonia, typhoid fever, tuberculosis, meningitis, syphilis and rheumatic fever. These bacterial diseases claimed many lives, sometimes very quickly. Patients of all ages were affected.

Today, these infections are not the 'killers' they once were here. For example, in the 1930s about 40% of the patients with bacterial pneumonia died of the disease. Today, about 5–10% may die. Antibiotic drugs have brought about this improvement in survival rates. The viral forms of pneumonia and meningitis are not overcome by antibiotics, since antibiotics do not affect viruses.

Despite their record of successful treatment of bacterial disease, problems also arise with antibiotics over time. Sooner or later some pathogenic bacteria in a population develop genes for **resistance** to a specific antibiotic's actions (Figure 5.11, page 222). Then, when different antibiotics are used to treat infections, the pathogenic bacteria concerned slowly acquire resistances to these, too. The bacterium has then acquired multiple resistance.

For example, a strain of *Staphylococcus aureus* has acquired resistance to a range of antibiotics including **methicillin**. Now, so-called methicillin-resistant *Staphylococcus aureus* (**MRSA**) is referred to as a 'hospital superbug' because of the harm its presence has inflicted in these places. (Actually, 'superbugs' are found everywhere in the community, not just in hospitals.) MRSA presents the greatest threat to patients who have undergone surgery. With cases of MRSA, the intravenous antibiotic called **vanomycin** is prescribed, but recently there have been cases of partial resistance to this drug, too.

Similarly, a strain of the bacterium *Clostridium difficile* is now resistant to all but two antibiotics. This bacterium is a natural component of our gut 'microflora'. It is only when *C. difficile's* activities are no longer suppressed by the surrounding, hugely beneficial ('friendly') gut flora, that it may multiply to life-threatening numbers, triggering toxic damage to the colon. Suppression of beneficial gut bacteria is a typical consequence of heavy doses of broad-spectrum antibiotics, administered to overcome infections of other superbugs.

19 Antibiotics are widely used in animal farming, partly to control animal disease, especially in intensive agriculture. **Suggest** what this means, why it happens, and what possible dangers could arise.

■ **Figure 6.24**
The increasing incidence of MRSA

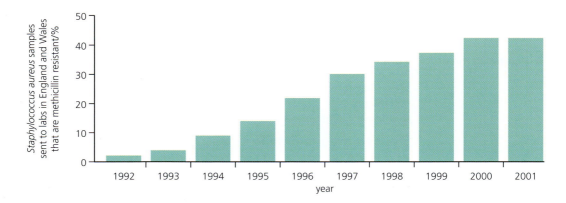

One result of bacterial resistance to antibiotics is that the pharmaceutical industry faces the challenge of producing **new antibiotics** faster than bacteria develop resistance to them. However, this is proving increasingly difficult – the number of new antibiotics being developed each year has fallen dramatically (Table 6.12).

■ **Table 6.12**
The number of new antibiotics developed annually

	1983–87	1988–92	1993–97	1998–2002
Number of new antibiotics approved (in USA)/5-year period	16	14	8	7
Source: *Clinical Infectious Diseases*, vol 38, page 179. (First seen in *New Scientist* 29/09/07, page 38.)				

20 Deduce the challenges in developing new antibiotics, today

Immunity: lymphocytes and the antigen–antibody reaction

The immune response is our main defence, once invasion of the body by harmful microorganisms or 'foreign' materials has occurred. The immune system is able to recognize 'self' – our body cells and proteins – and tell them apart from foreign or 'non-self' substances, such as those on or from an invading organism. A 'non-self' substance is called an **antigen**. It is the **lymphocytes**, particular types of white blood cells, that are able to recognize antigens and to take steps to overcome them. Lymphocytes make up 20% of the white blood cells circulating in the blood plasma.

Lymphocytes and the antigen–antibody reaction

Each type of lymphocyte in our body recognizes only one specific antigen. In the presence of that antigen (and only that antigen), the lymphocyte divides rapidly, producing many cells – known as a clone. These cloned lymphocytes then secrete an **antibody** specific to that antigen.

An **antibody** is a protein constructed in the shape of a Y. The top of each 'arm' contains an antigen-binding site (Figure 6.25). Millions of different types of antibodies may be produced by our bodies, each by a different type of lymphocyte – there are as many antibodies as there are types of foreign matter (antigens) invading the body. The amino acid sequence of the antigen-binding site differs according to the chemistry of the antibody it binds with. It is the antigen-binding site that gives each antibody its specificity.

21 Identify where antigens and antibodies may be found in the body.

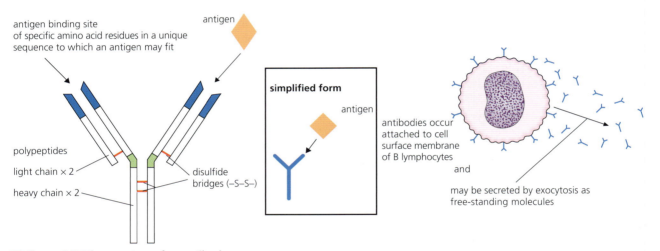

antigen binding site
of specific amino acid residues in a unique
sequence to which an antigen may fit

antigen

simplified form

antigen

antibodies occur
attached to cell
surface membrane
of B lymphocytes

polypeptides

light chain × 2

heavy chain × 2

disulfide
bridges (–S–S–)

and

may be secreted by exocytosis as
free-standing molecules

■ **Figure 6.25** The structure of an antibody

So, a huge range of **different antibody-secreting lymphocytes** exists, each type recognizing one specific antigen. The more antigens we encounter, the more antibodies we are able to form, should they be required.

Steps to the immune response

You can follow these steps in Figure 6.26.

1   When invasion of a pathogen occurs, its antigens bind to lymphocytes that recognize them.
2   Those lymphocytes then divide rapidly, producing a clone of identical **plasma cells**.
3   The plasma cells produce antibodies. These antibodies are secreted and circulate in the bloodstream.
4   When and wherever an antibody encounters the antigen (most likely on the cell membrane of the pathogen), they are destroyed (see below).
5   Sufficient antibodies to overcome the antigen 'invasion' are secreted.
6   However, this type of lymphocyte has a short lifespan. Once the harmful effects of the invading antigen are neutralized, the lymphocytes largely disappear from the blood circulation.
7   But all 'knowledge' of the antigen is *not* lost. Some lymphocytes of that type (we now call them **memory cells**) remain behind, stored in the lymph nodes.
8   With the aid of memory cells, our body can respond rapidly, if the same antigen reinvades (Figure 6.27). We say we have **immunity** to that antigen.

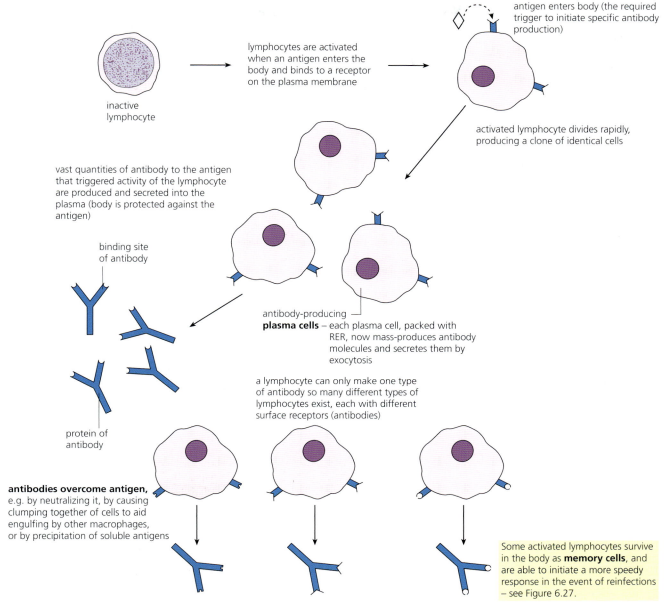

inactive
lymphocyte

lymphocytes are activated
when an antigen enters the
body and binds to a receptor
on the plasma membrane

antigen enters body (the required
trigger to initiate specific antibody
production)

activated lymphocyte divides rapidly,
producing a clone of identical cells

vast quantities of antibody to the antigen
that triggered activity of the lymphocyte
are produced and secreted into the
plasma (body is protected against the
antigen)

binding site
of antibody

antibody-producing
plasma cells – each plasma cell, packed with
RER, now mass-produces antibody
molecules and secretes them by
exocytosis

protein of
antibody

a lymphocyte can only make one type
of antibody so many different types of
lymphocytes exist, each with different
surface receptors (antibodies)

antibodies overcome antigen,
e.g. by neutralizing it, by causing
clumping together of cells to aid
engulfing by other macrophages,
or by precipitation of soluble antigens

Some activated lymphocytes survive
in the body as **memory cells**, and
are able to initiate a more speedy
response in the event of reinfections
– see Figure 6.27.

■ **Figure 6.26** Formation and function of an antibody

■ **Figure 6.27**
Profile of antibody
production on
infection and if
reinfection occurs

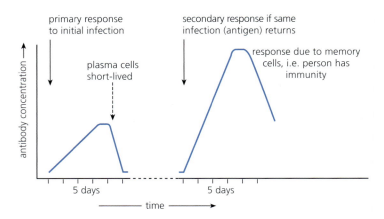

Memory cells are retained in lymph nodes. They allow a quick and specific response if the same antigen reappears.

■ **Figure 6.27**
Profile of antibody
production on
infection and if
reinfection occurs

> 22 **Deduce** which
> synthetic machinery
> in the lymphocytes
> is active in the
> production of the
> components of
> their antibodies.

Antibodies destroy antigens in different ways. Toxins may be inactivated by reaction with the antibody, and bacterial cells may be clumped together so that they 'precipitate' and can be engulfed by phagocytic cells. Antibodies also attach to foreign matter, ensuring its recognition by phagocytic cells. Antibodies also act by destroying bacterial cell walls, causing lysis of the bacterium.

■ Human immunodeficiency virus (HIV) and AIDS

Human immunodeficiency virus (**HIV**) was first identified in 1983 as the cause of a disease of the human immune system known as **acquired immune deficiency syndrome** (**AIDS**). AIDS is probably the greatest current threat to public health, because it kills people in the most economically productive stage of their lives.

HIV is a tiny virus, less than 0.1 μm in diameter (Figure 6.28). It consists of two single strands of RNA which, together with enzymes, are enclosed by a protein coat. A membrane, derived from the human host cell in which the virus was formed, encapsulates each new virus particle leaving the host cell.

■ **Figure 6.28**
The human
immunodeficiency
virus (HIV)

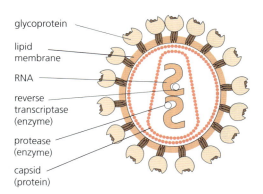

glycoprotein

lipid
membrane

RNA

reverse
transcriptase
(enzyme)

protease
(enzyme)

capsid
(protein)

election micrograph of a white cell from which HIV are budding off

HIV is a **retrovirus**. A retrovirus *reverses* the normal flow of genetic information from the DNA of genes to messenger RNA in the cytoplasm (page 111). The idea that information *always* flows in this direction in cells was called the **central dogma of cell biology** (implying it was always the case). However, in retroviruses, the information in RNA in the cytoplasm is translated *into* DNA within a host cell and then becomes attached to the DNA of a chromosome in the host's nucleus.

How a retrovirus works

The virus binds to the cell membrane of a lymphocyte, the core of the virus passes inside, and the RNA and virus enzymes are released. An enzyme from the virus, called **reverse transcriptase**, catalyses the copying of the genetic code of each of the virus's RNA strands into a DNA double helix. This DNA then enters the host nucleus and is 'spliced' into the host's chromosomal DNA. Here, it is replicated with the host's genes every time the host cell divides.

For some time, the viral genes remain latent, giving no sign of their presence in the host cells. At a later time, some event in the patient's body activates the HIV genes and the outcome is AIDS. Now, synthesis of viral messenger RNA occurs in infected lymphocytes. This then passes out into the cytoplasm and, there, codes for viral proteins (enzymes and protein coat) at the ribosomes. Viral RNA, enzymes and coat protein form into viral cores. These then move against the cell membrane and 'bud-off' new viruses, which are released, leading to death and breakdown of the lymphocyte. More lymphocytes are infected and the cycle is rapidly repeated.

■ **Figure 6.29**
Profile of an AIDS
infection

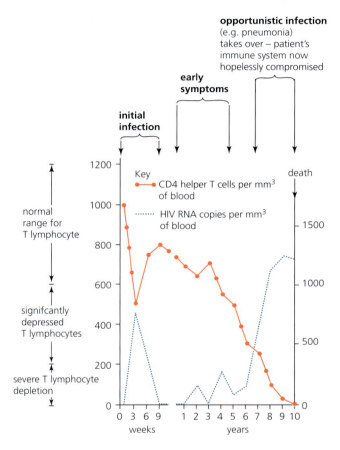

Effects of HIV on the immune system

Without treatment, this process causes the body's reserve of lymphocytes to decrease very quickly. The reduction in the number of active lymphocytes means the body loses the ability to produce antibodies. Eventually, no infection, however trivial, can be resisted; death follows.

Ideally, a vaccine against HIV would be the best solution – one designed to wipe out both infected lymphocytes and HIV particles in the patient's bloodstream. The work of several laboratories is dedicated to this solution. The problem is that in the latent state of the infection, the infected lymphocyte cells frequently change their membrane marker proteins because of the presence of the HIV genome within the cell. Effectively, HIV can hide from the body's immune response by changing its identity.

Methods of transmission

Infection with HIV is possible through contact with blood or body fluids of infected people, such as may occur during sexual intercourse, sharing of hypodermic needles by intravenous drug users and breast feeding of a newborn baby. Also, blood transfusions and organ transplants can transmit HIV, but donors are now screened for HIV infection in most countries.

HIV is *not* transferred by contact with saliva on a drinking glass, or by sharing a towel, for example. Nor does the female mosquito transmit HIV when feeding on human blood.

23 **Describe** why AIDS patients typically die from common infectious diseases that are not normally fatal.

Recent studies have shown that the prevalence of AIDS is significantly higher in sexually active uncircumcized males. Circumcized males were seven times less likely to transmit HIV to, or receive HIV from, their partner. Consequently, it is possible that HIV infection of future generations might be reduced if male circumcision were practised more widely. But health workers point out that, if this were to encourage males to participate in unsafe sex with numerous partners, it would defeat the objective.

6.4 Gas exchange – *the lungs are actively ventilated to ensure that gas exchange can occur passively*

■ All living things respire

Cellular respiration is the controlled release of energy, in the form of ATP, from organic compounds in cells. It is a continuous process in all cells.

To support aerobic cellular respiration, cells take in oxygen from their environment and give out carbon dioxide, by a process called gaseous exchange (Figure 6.30).

Gaseous exchange is the exchange of gases between an organism and its surroundings, including the uptake of oxygen and the release of carbon dioxide in animals and plants.

The exchange of gases between the individual cell and its environment takes place by **diffusion**. For example, in cells that are respiring aerobically there is a higher concentration of oxygen outside the cells than inside, and so there will be a continuous net inward diffusion of oxygen.

What speeds up diffusion?

In living things there are three factors which effectively determine the rate of diffusion in practice.

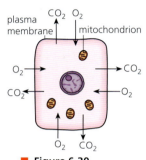

■ **Figure 6.30**
Gaseous exchange in an animal cell

- The size of the **surface area** available for gaseous exchange (the respiratory surface) – the greater this surface area, the greater the rate of diffusion. Of course, in a single cell, the respiratory surface is the whole plasma membrane.

- The **difference in concentration** – a rapidly respiring organism has a very much lower concentration of oxygen in the cells and a higher than normal concentration of carbon dioxide. The greater the gradient in concentration across the respiratory surface, the greater the rate of diffusion.

- The **length of the diffusion path** – the shorter the diffusion path, the greater the rate of diffusion, so the respiratory surface must be as thin as possible.

■ Gaseous exchange in animals

We have seen that the surface area of a single-celled organism is large in relation to the amount of cytoplasm it contains – here, the surface of the cell is sufficient for efficient gaseous exchange. On the other hand, large multicellular animals have very many of their cells too far from the body's surface to receive enough oxygen by diffusion alone.

In addition, animals often develop an external surface of tough or hardened skin that provides protection to the body, but which is not suitable for gaseous exchange. These organisms require an alternative respiratory surface.

Active organisms have an increased metabolic rate, and the demand for oxygen in their cells is higher than in sluggish and inactive organisms. So, for many reasons, large active animals, such as **mammals**, have specialized organs for gaseous exchange. In mammals, the respiratory surface consists of **lungs**. Lungs provide a large, thin surface area that is suitable for gaseous exchange. However, the lungs are in a protected position inside the thorax (chest), so air has to be brought to the respiratory surface there. The lungs must be **ventilated**.

A ventilation system is a pumping mechanism that moves air into and out of the lungs efficiently, thereby maintaining the concentration gradients of oxygen and carbon dioxide for diffusion.

24 List three characteristics of an efficient respiratory surface and explain how each influences diffusion.

In addition, in mammals the conditions for diffusion at the respiratory surface are improved by:

■ a blood circulation system, which rapidly moves oxygen to the body cells as soon as it has crossed the respiratory surface, thereby maintaining the concentration gradient in the lungs

■ a respiratory pigment, which increases the oxygen-carrying ability of the blood. This is the hemoglobin of the red blood cells, which are by far the most numerous of the cells in our blood circulation.

The working lungs of mammals

The structure of the human thorax is shown in Figure 6.31. Lungs are housed in the **thorax**, an airtight chamber formed by the **ribcage** and its muscles (**intercostal muscles**), with a domed floor, the **diaphragm**. The diaphragm is a sheet of muscle attached to the body wall at the base of the ribcage, separating thorax from abdomen. The internal surfaces of the thorax are lined by the **pleural membrane**, which secretes and maintains pleural fluid. Pleural fluid is a lubricating liquid derived from blood plasma and that protects the lungs from friction during breathing movements.

■ **Figure 6.31**
The structure of the human thorax

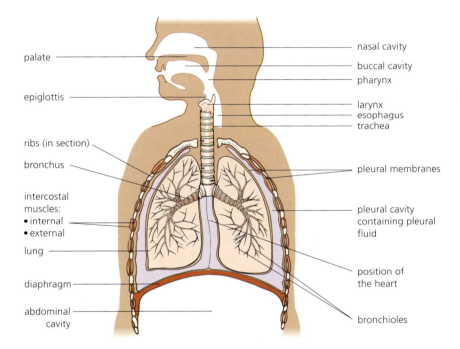

From trachea to alveoli

Lungs connect with the pharynx at the rear of the mouth by the **trachea**. Air reaches the trachea from the mouth and nostrils, passing through the larynx (voice box). Entry into the larynx is via a slit-like opening, the glottis. Above is a cartilaginous flap, the **epiglottis**. Glottis and epiglottis work to prevent the entry of food into the trachea. The trachea initially runs beside the esophagus. Incomplete rings of cartilage in the trachea wall prevent collapse under pressure from a large bolus of food passing down the esophagus.

The trachea then divides into two **bronchi**, one to each lung. Within the lungs the bronchi divide into smaller **bronchioles**. The finest bronchioles end in air sacs (**alveoli**). The walls of bronchi and larger bronchioles contain smooth muscle, and are also supported by rings or tiny plates of cartilage, preventing collapse that might be triggered by a sudden reduction in pressure that occurs with powerful inspirations of air.

Lungs are extremely efficient, but of course they cannot prevent some water loss during breathing – an issue for most terrestrial organisms.

Ventilation of the lungs

Air is drawn into the alveoli when the air pressure in the lungs is lower than atmospheric pressure, and it is forced out when pressure is higher than atmospheric pressure. Since the thorax is an airtight chamber, pressure changes in the lungs occur when the volume of the thorax changes. How the volume of the thorax is changed during breathing is illustrated in Figure 6.32 and summarized in Table 6.13.

---

25 **Compare** the roles of
   a the internal and external intercostal muscles, and
   b the diaphragm and abdominal muscles, during ventilation of the lungs (Figure 6.32 and Table 6.13).

---

inspiration:
- external intercostal muscles contract
- internal intercostal muscles relax
- diaphragm muscles contract

} ribs moved upwards and outwards, and the diaphragm down

air in

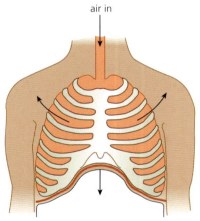

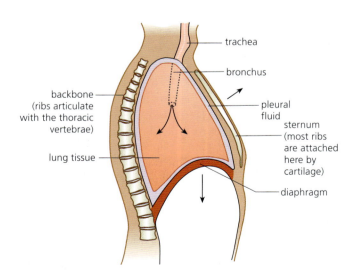

trachea

bronchus

backbone (ribs articulate with the thoracic vertebrae)

pleural fluid

sternum (most ribs are attached here by cartilage)

lung tissue

diaphragm

volume of the thorax (and therefore of the lungs) increases; pressure is reduced below atmospheric pressure and air flows in

expiration:
- external intercostal muscles relax
- internal intercostal muscles contract
- diaphragm muscles relax

} ribs moved downwards and inwards, and the diaphragm up

air out

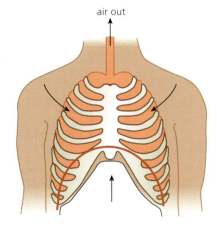

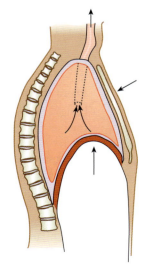

volume of the thorax (and therefore of the lungs) decreases; pressure is increased above atmospheric pressure and air flows out

■ **Figure 6.32** The ventilation mechanism of the lungs

■ **Table 6.13**
The mechanism of lung ventilation – a summary

Inspiration (inhalation)	Structure/outcome	Expiration (exhalation)
muscles contract, flattening the diaphragm and pushing down on contents of abdomen	*diaphragm	muscles relax
relax	*abdominal muscles	contract – pressure from abdominal contents pushes diaphragm into a dome shape
contract, moving ribcage up and out	*external intercostal muscles	relax
relax	*internal intercostal muscles	contract, moving ribcage down and in
increases	volume of thoracic cavity	decreases
falls below atmospheric pressure	air pressure of thorax	rises above atmospheric pressure
in	air flow	out

*Different muscles are required for inspiration and expiration because muscles only work when they contract – this is referred to as **antagonistic muscle action**.

Alveolar structure and gaseous exchange

The lung tissue consists of the alveoli, arranged in clusters, each served by a tiny bronchiole. Alveoli have **elastic connective tissue** as an integral part of their walls (Figure 6.33).

■ **Figure 6.33** Role of elastic fibres in alveoli and bronchioles

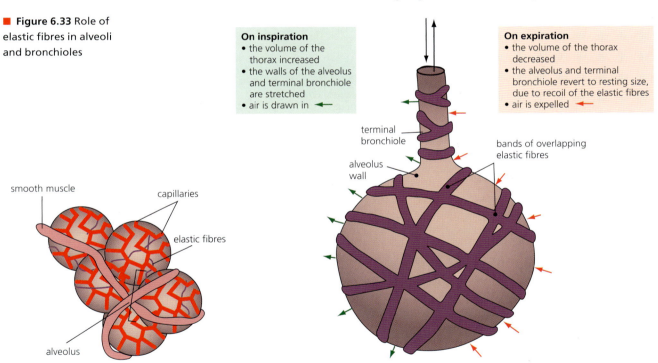

On inspiration
- the volume of the thorax increased
- the walls of the alveolus and terminal bronchiole are stretched
- air is drawn in ←

On expiration
- the volume of the thorax decreased
- the alveolus and terminal bronchiole revert to resting size, due to recoil of the elastic fibres
- air is expelled ←

terminal bronchiole

alveolus wall

bands of overlapping elastic fibres

smooth muscle

capillaries

elastic fibres

alveolus

A capillary system wraps around the clusters of alveoli (Figure 6.34). Each capillary is connected to a branch of the pulmonary artery and is drained by a branch of the pulmonary vein. The pulmonary circulation is supplied with deoxygenated blood from the right side of the heart, and returns oxygenated blood to the left side of the heart to be pumped to the rest of the body.

There are some 700 million alveoli in our lungs, providing a surface area of about $70 \, m^2$ in total. This is an area 30–40 times greater than that of the body's external skin. The wall of an alveolus is one cell thick and is formed by pavement epithelium. Lying very close is a capillary, its wall also composed of a single layer of flattened endothelium (Type I pneumocyte) cells. The combined thickness of walls separating air and blood is typically 2–4 µm thick. The capillaries are extremely narrow, just wide enough for red blood cells to squeeze through, so red blood cells are close to or in contact with the capillary walls.

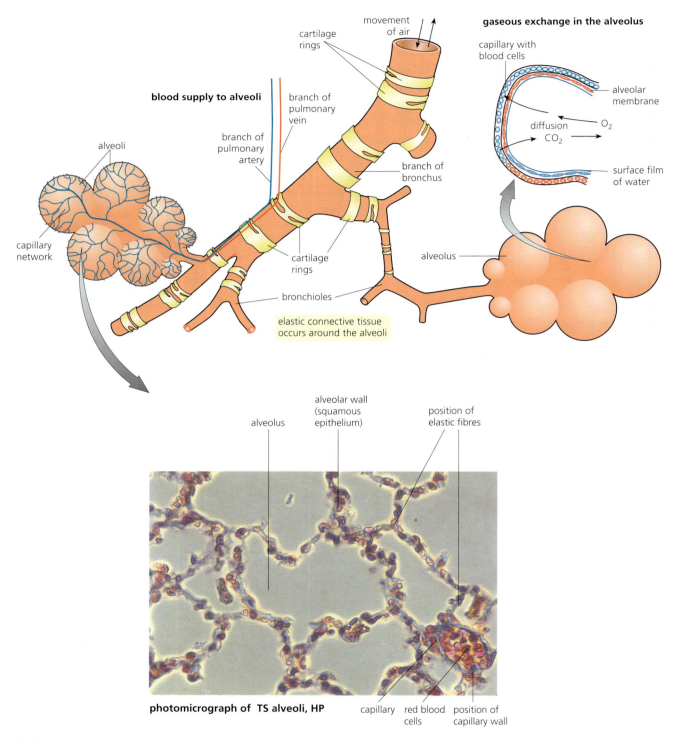

movement
of air

cartilage
rings

gaseous exchange in the alveolus

capillary with
blood cells

blood supply to alveoli

branch of
pulmonary
vein

alveolar
membrane

branch of
pulmonary
artery

diffusion

O_2

CO_2

alveoli

branch of
bronchus

capillary
network

cartilage
rings

alveolus

surface film
of water

bronchioles

elastic connective tissue
occurs around the alveoli

alveolar wall
(squamous
epithelium)

position of
elastic fibres

alveolus

photomicrograph of TS alveoli, HP

capillary

red blood
cells

position of
capillary wall

■ **Figure 6.34** Gaseous exchange in the alveoli

The extremely delicate structure of the alveoli is protected by two types of cell, present in
abundance in the surface film of moisture:

■ **macrophages** (dust cells), the main detritus-collecting cells of the body – these originate
from bone marrow stem cells and are dispersed about the body in the blood circulation.
These amoeboid cells migrate into the alveoli from the capillaries. Here, these phagocytic
white blood cells ingest any debris, fine dust particles, bacteria and fungal spores present.
They also line the surfaces of the airways leading to the alveoli;

■ **surfactant cells** (Type II pneumocytes) – these produce a detergent-like mixture of lipoproteins and phospholipid-rich secretion that lines the inner surface of the alveoli. Surface tension is a force acting to minimize the surface area of a liquid. Because of the tiny diameter of the alveoli (about 0.25 mm) they would tend to collapse under surface tension during expiration. The lung surfactant lowers surface tension, permitting the alveoli to flex easily as the pressure of the thorax falls and rises.

■ **Figure 6.35**
Keeping the alveoli flexible and clean!

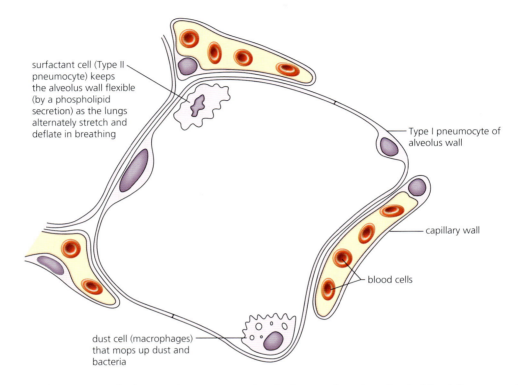

surfactant cell (Type II pneumocyte) keeps the alveolus wall flexible (by a phospholipid secretion) as the lungs alternately stretch and deflate in breathing

Type I pneumocyte of alveolus wall

capillary wall

blood cells

dust cell (macrophages) that mops up dust and bacteria

Blood arriving in the lungs is low in oxygen but high in carbon dioxide. As blood flows past the alveoli, gaseous exchange occurs by diffusion. Oxygen dissolves in the alveolar surface film of water, diffuses across into the blood plasma and into the red blood cells, where it combines with hemoglobin to form oxyhemoglobin. At the same time, carbon dioxide diffuses from the blood into the alveoli (Table 6.14).

■ **Table 6.14**
The composition of air in the lungs

Component	Inspired air/%	Alveolar air/%	Expired air/%
oxygen	20	14	16
carbon dioxide	0.04	5.5	4.0
nitrogen	79	81	79
water vapour	variable	saturated	saturated

26 Explain why, if the concentration of carbon dioxide built up in the blood of a mammal, this would be harmful.

Efficiency of lungs as organs of gaseous exchange

The lungs of mammals are just one evolutionary response to the need for efficient organs of gaseous exchange in compact multicellular animals. Alternative arrangements include the system of tubes that pipe air directly to respiring cells of insects, the internal gills of fish, and the lungs of birds in which air is drawn into air sacs that bypass the lungs. From there, it is forced out, through the lungs, which are continuously exposed to 'fresh' air as a result of this mechanism.

How effective are mammalian lungs?

Air flow in the lungs of mammals is tidal, in that air enters and leaves by the same route. Consequently, there is a residual volume of air that cannot be expelled. Incoming air mixes with and dilutes the residual air, rather than replacing it. The effect of this is that air in the alveoli contains significantly less oxygen than the atmosphere outside (Table 6.14).

Nevertheless, the lungs are efficient organs. Their success is due to numerous features of the alveoli that adapt them to gaseous exchange. These are listed in Table 6.15.

Table 6.15
Features of alveoli that adapt them to efficient gaseous exchange

Feature	Effects and consequences
surface area of alveoli	a huge surface area for gaseous exchange ($50\,m^2$ = area of doubles tennis court)
wall of alveoli (of Type I pneumocytes)	very thin, flattened (squamous) epithelium ($5\,\mu m$) means the diffusion pathway is short
capillary supply to alveoli	network of capillaries around each alveolus (supplied with deoxygenated blood from pulmonary artery and draining into pulmonary veins) maintains the concentration gradient of O_2 and CO_2
surface film of moisture	O_2 dissolves in water lining the alveoli; O_2 diffuses into the blood in solution

27 **Explain** the difference between gaseous exchange and cellular respiration.

Monitoring of ventilation in humans at rest and after exercise

One way of monitoring ventilation is by data logging with an apparatus called a **recording spirometer** (Figure 6.36). You can see this consists of a Perspex lid enclosing the spirometer chamber, hinged over a tank of water. This chamber is connected to the person taking part in the experiment via an interchangeable mouthpiece and flexible tubing. As breathing proceeds, the lid rises and falls as the chamber volume changes. With the spirometer chamber filled with air, the capacity of the lungs when breathing at different rates can be investigated. Incidentally, if the spirometer chamber is filled with oxygen and a carbon dioxide absorbing chemical, such as soda lime, is added to a compartment on the air return circuit, this apparatus can be used to measure oxygen consumption by the body, too.

Figure 6.36
Investigating breathing with a spirometer

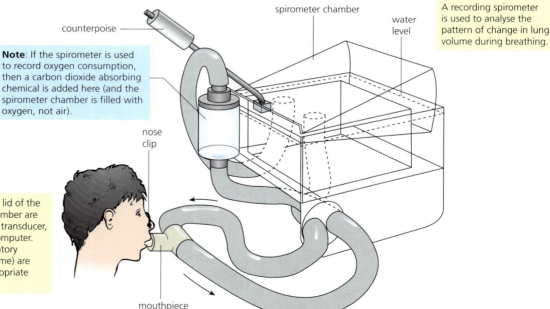

Note: If the spirometer is used to record oxygen consumption, then a carbon dioxide absorbing chemical is added here (and the spirometer chamber is filled with oxygen, not air).

A recording spirometer is used to analyse the pattern of change in lung volume during breathing.

The movements of the lid of the airtight spirometer chamber are recorded by a position transducer, connecting box and computer. The results (e.g. inspiratory capacity and tidal volume) are printed out using appropriate control software.

counterpoise

spirometer chamber

water level

nose clip

mouthpiece

The movements of the lid may be recorded by a position transducer, connecting box and computer. From 'traces' printed out from investigations of human breathing under different conditions (including at rest, and after mild and vigorous activity), the ventilation rate and tidal volume may be measured (Figure 6.37).

Examine the trace now.

■ **Figure 6.37**
A spirometer trace

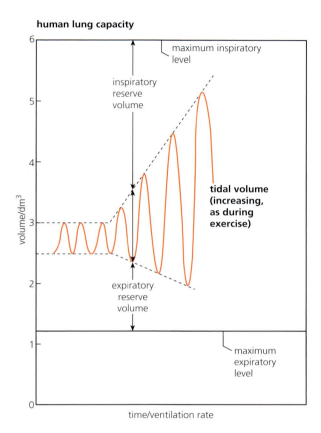

Notice that the volume of air breathed in and out during normal, relaxed, rhythmical breathing, the **tidal volume**, is typically 400–500 cm^3. However, we have the potential for an extra large intake (maximum inspiratory capacity) and an extra large expiration of air (expiratory reserve volume), when required. These, together, make up about 4.5 dm^3.

The spirometer can also be used to investigate steady breathing over a short period of time and, in these cases, the y axis of the spirometer trace is 'time'. In this way, the rate at which the lungs are ventilated can be investigated. The **ventilation rate** is the number of inhalations or exhalations per minute.

Alternative experimental approaches are possible, including use of a chest belt and pressure monitor.

■ Smoking and health

With the mass production of cigarettes and the invention of the match in the 1800s, smoking become cheap, easy and affordable. In the twentieth century, cigarette smoking in the developed world was advanced hugely by the availability of cigarettes to the troops of two World Wars. Subsequent bold and aggressive advertising campaigns, persuasive product placement in films and the generous sponsorship of sporting and cultural events by cigarette manufacturers, all encouraged greater smoking by men and persuaded women to take it up, too. Very slowly, the dangers of smoking became known, but many people doubted the evidence. By the 1950s, people started to recognize the significant dangers of cigarette smoking (and of the passive inhalation of cigarette smoke).

The composition of cigarette smoke

Cigarette smoke contains a cocktail of harmful substances – more than 4000, in fact. These include acetone, ammonia, arsenic, butane, cadmium, hydrogen cyanide, methanol, naphthalene, toluene and vinyl chloride. However, to understand the danger to health that cigarette smoke poses, we shall focus on the following components:

- **Carcinogens**, of which there are at least twenty different types, are found in the 'tar' component of smoke. Particularly harmful are certain polycyclic aromatic hydrocarbons and nitrosamines. A carcinogen is any agent that may cause cancer by damage (mutations) to the DNA molecules of chromosomes. Mutations of different types may build up in the DNA of body cells that are exposed to these substances.

- **Nicotine**, a stimulating and relaxing drug which, on entering the bloodstream, is able to cross the blood–brain barrier. In the brain, it triggers the release of dopamine, the natural neurotransmitter substance (page 296) associated with our experience of pleasure. Long-term exposure to nicotine eventually comes to have the reverse effect, actually depressing our ability to experience pleasure. So, more nicotine is needed to 'satisfy' us, and cigarettes become addictive; as addictive as heroin and cocaine, in fact. Smokers find it increasingly hard to quit the habit.

- **Carbon monoxide** is a gas that diffuses into the red blood cells and combines irreversibly with hemoglobin. In smokers, the blood is able to transport less oxygen.

Cigarette smoke reaches the smoker's lungs when it is drawn down the cigarette and inhaled, but it reaches other people, too, when it escapes from the glowing tip into the surrounding air. These latter fumes normally have a higher concentration of the toxic ingredients, and it is this mixture that others inhale. '**Passive smoking**' has, itself, been shown to be dangerous.

Lung cancer

Persistent exposure of the bronchi to cigarette smoke results in damage to the epithelium (Figure 6.38). This is progressively replaced by an abnormally thickened epithelium. With prolonged exposure to the carcinogens, permanent mutations may be triggered in the DNA of some of these cells. If this occurs in their oncogenes or tumour suppressing genes, the result is loss of control over normal cell growth.

A single mutation is unlikely to be responsible for triggering lung cancer; the danger is in the accumulation of mutations over time in a group of cells, which then divide by mitosis repeatedly, without control or regulation, forming an irregular mass of cells – a **tumour**. Tumour cells then emit signals that promote the development of new blood vessels to deliver oxygen and nutrients, all at the expense of the surrounding healthy tissues. Sometimes tumour cells break away and are carried to other parts of the body, forming a secondary tumour (a process called **metastasis**). Unchecked, cancerous cells ultimately take over the body, leading to malfunction and death.

Nature of Science

Obtain evidence for theories

■ Cigarette smoking causes lung diseases – the evidence

It was **epidemiology** (the study of the incidence and distribution of diseases, and of their control and prevention) that first identified a *likely* causal link between smoking and disease. In 1950, an American study of over 600 smokers, compared with a similar group of non-smokers, found **lung cancer** was 40 times higher among the smokers. The risk of contracting cancer increased with the number of cigarettes smoked (Table 6.16).

■ **Table 6.16** Cancer rates and numbers of cigarettes smoked/day

Number of cigarettes smoked/day	Incidence of cancer/100 000 men
0	15
10	35
15	60
20	135
30	285
40	400

Source: American Cancer Society
(http://txtwriter.com/onscience/OSpictures/smokingcancer2.2.jpg)

A survey of smoking in the UK was commenced in 1948, at which time 82% of the male population smoked, of whom 65% smoked cigarettes. This had fallen to 55% by 1970, and continued to decrease. In the same period, the numbers of females who smoked remained just above 40% until 1970, after which numbers also declined (Figure 6.39). Look carefully at the changing pattern in the incidences of lung cancer – figures are available since 1975.

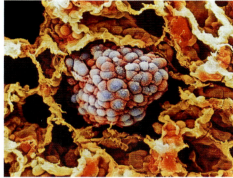

scanning electron micrograph of human lung tissue with cancer

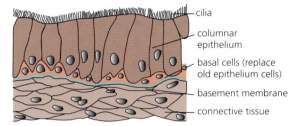

ciliated epithelium lining the bronchial tree

- cilia
- columnar epithelium
- basal cells (replace old epithelium cells)
- basement membrane
- connective tissue

in response to continuing exposure to cigarette smoke

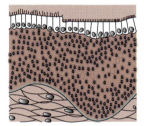   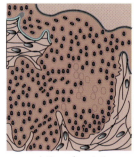

beating of cilia inhibited; excessive, abnormal proliferation of basal cells

ciliated epithelium replaced by squamous epithelium

mass of basal cells showing abnormal growth

accumulation of mutations in mass of abnormal cells; forms a tumour

■ **Figure 6.38** The development of a lung tumour

28 Comment on the incidence of lung cancer in men and women, 1975–2007 (Figure 6.39), in relation to the changing pattern of smoking since UK records began.

Meanwhile, throughout the world, there continues to be huge investment by the tobacco industry, via the growing of tobacco crops, the manufacturing and 'attractive' packaging of tobacco products, and in the placing of advertisements in countries that are open to them. While the number of people smoking cigarettes is actually falling in developed countries, smoking is on the rise in the developing world. The global health issues generated by cigarette smoking are of epidemic proportions – about one-third of the global adult male population smokes.

Among the World Health Organization areas, the East Asia and the Pacific Region has the highest smoking rate, with nearly two-thirds of men smoking. Today, apparently one in every three cigarettes is smoked in China. Globally, consumption of manufactured cigarettes continues to rise steadily. Consequently, the death toll from the use of tobacco, currently estimated to be 6 million per year worldwide, is predicted to rise steeply. This is inevitable because of the continuing popularity of cigarettes, and of the time lag between the taking up of smoking and the appearance of the symptoms of lung cancer (Figure 6.39).

■ **Figure 6.39**

Lung cancer incidence and smoking trends in the UK, 1948–2007

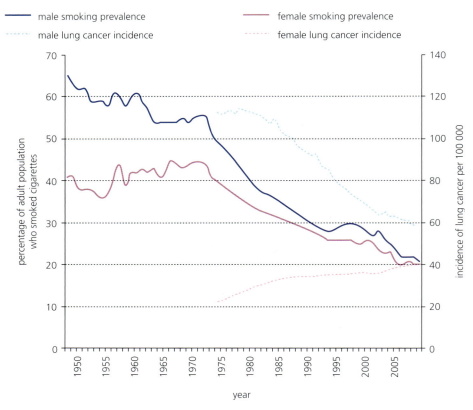

— male smoking prevalence
····· male lung cancer incidence

— female smoking prevalence
····· female lung cancer incidence

Emphysema

Emphysema is another respiratory disease that is often linked to smoking (Figure 6.40). In this disease, the walls of the alveoli lose their elasticity, resulting in the destruction of the lung tissue over time. This is because smokers' lungs typically contain large numbers of macrophages, accumulated from the blood circulation. These phagocytic cells release a high level of their natural hydrolytic enzymes and also far too little of the natural inhibitor of this enzyme. The result is the breakdown of the elastic fibres of the alveolar walls (Figure 6.33). With failing elastic fibres, the air sacs are left over-inflated when air becomes trapped in them (they fail to recoil and expire air properly). Small holes also develop in the walls of the alveoli. These begin to merge, forming huge air spaces with drastically lowered surface area for gas exchange. The patient becomes permanently breathless. The destruction of air sacs can be halted by stopping smoking, but any damage done to the lungs cannot be reversed.

■ **Figure 6.40**

Scanning electron micrograph of (A) healthy human lung tissue, and (B) lung tissue showing advanced emphysema

In (A), can you identify alveoli, the tiny capillaries and the wall of a bronchus in section?

In (B), how has the structure of the lung tissue changed?

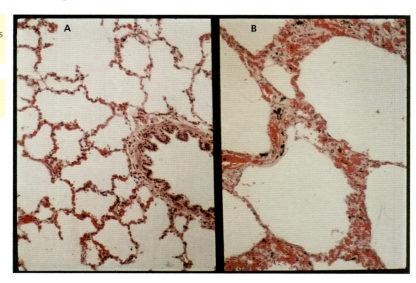

6.5 Neurons and synapses – *neurons transmit the 'message', synapses modulate the message*

Control and communication within the body involves both the nervous system and hormones from the endocrine glands. We look at the nervous system first.

■ Introducing the nervous system

The nervous system is built from nerve cells called **neurons**. A neuron has a **cell body** containing the **nucleus** and the bulk of the cytoplasm. From the cell body run fine cytoplasmic **fibres**. Most fibres are very long indeed.

Neurons are specialized for the transmission of information in the form of impulses. An **impulse** is a momentary reversal in the electrical potential difference in the membrane of a neuron. The transmission of an impulse along a fibre occurs at speeds of between 30 and 120 metres per second in mammals, so nervous coordination is extremely fast and responses are virtually immediate. Impulses travel to particular points in the body, served by the fibres. Consequently, the effects of impulses are localized rather than diffuse.

Neurons are grouped together to form the **central nervous system**, which consists of the **brain** and **spinal cord**. To and from the central nervous system run nerves of the **peripheral nervous system**. Communication between the central nervous system and all parts of the body occurs via these nerves (Figure 6.41).

■ **Figure 6.41**
The organization of the mammalian nervous system

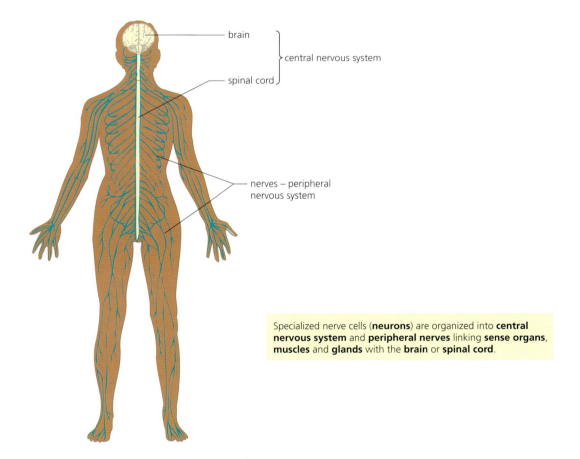

brain

central nervous system

spinal cord

nerves – peripheral nervous system

Specialized nerve cells (**neurons**) are organized into **central nervous system** and **peripheral nerves** linking **sense organs**, **muscles** and **glands** with the **brain** or **spinal cord**.

Looking at neuron structure

Three types of neuron make up the nervous system (sensory neurons, relay neurons and motor neurons). The structure of a motor neuron is shown in Figure 6.42.

The motor neurons have many fine **dendrites**, which bring impulses towards the cell body, and a single long **axon** which carries impulses away from the cell body. The function of the motor neuron is to carry impulses from the central nervous system to a muscle or gland (known as an **effector**).

■ **Figure 6.42**
Motor neuron
structure

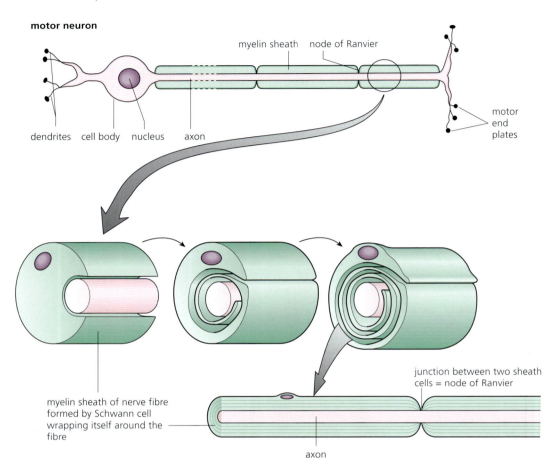

A neuron is surrounded by many supporting cells, one type of which, Schwann cells, become wrapped around the axons of motor neurons, forming a structure called a **myelin sheath**. Myelin consists largely of lipid and has high electrical resistance. Frequent junctions occur along a myelin sheath, between the individual Schwann cells. The junctions are called nodes of Ranvier.

Neurons and the transmission of an impulse

Neurons transmit information in the form of impulses. An impulse is transmitted along nerve fibres, but it is not an electrical current that flows along the 'wires' of the nerves. An impulse is a momentary reversal in electrical potential difference in the membrane – a change in the position of charged ions between the inside and outside of the membrane of the nerve fibres. This reversal flows from one end of the neuron to the other in a fraction of a second.

Between conduction of one impulse and the next, the neuron is sometimes said to be resting, but this is not the case. The 'resting' neuron membrane is actively setting up the electrical potential difference between the inside and the outside of the fibre, known as the resting potential.

The resting potential

The resting potential is the potential difference across a nerve cell membrane when it is not being stimulated. It is normally about –70 millivolts (mV).

The resting potential difference is re-established across the neuron membrane after a nerve impulse has been transmitted. We say the nerve fibre has been **repolarized**.

The resting potential is the product of two processes:

1 The active transport of potassium ions (K+) in across the membrane and sodium ions (Na^+) out across the membrane – this occurs by a K^+/Na^+ pump, using energy from ATP (Figure 1.50, page 46). The concentration of potassium and sodium ions on opposite sides of the membrane is built up, but this in itself makes no change to the potential difference across the membrane.

2 Facilitated diffusion of K^+ ions out and Na^+ ions back in – the important point here is that the membrane is far more permeable to K^+ ions flowing out than to Na^+ ions returning. This causes the tissue fluid outside the neuron to contain many more positive ions than are present in the tiny amount of cytoplasm inside. As a result, a **negative charge** is developed inside, compared to outside, and the resting neuron is said to be **polarized**. The difference in charge or potential difference (about –70 mV), is known as the **resting potential** (Figure 6.43). The next event, sooner or later, is the passage of an impulse, known as an **action potential**.

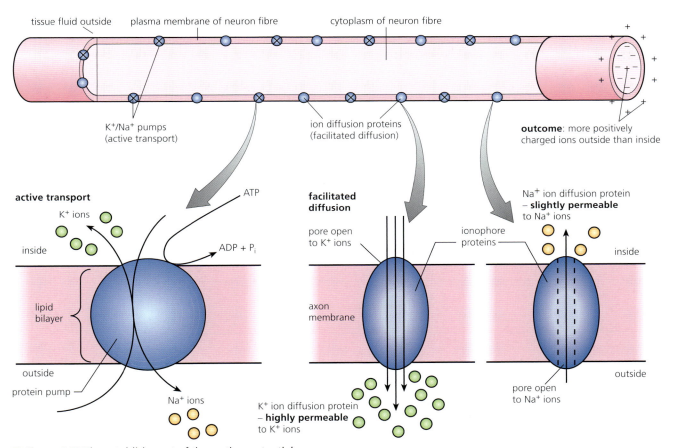

■ **Figure 6.43** The establishment of the resting potential

The action potential

The action potential is the potential difference produced across the plasma membrane of the nerve cell when stimulated, reversing the resting potential from about –70 mV to about +40 mV.

An action potential is triggered by a stimulus which is received at a receptor cell or sensitive nerve ending. The energy of the stimulus causes a temporary and local reversal of the resting potential. The result is that the membrane is briefly **depolarized**.

This change in potential across the membrane occurs because of pores in the membrane called **ion channels**. These special channels are globular proteins that span the membrane. They have a central pore with a **gate** that can open and close. One type of channel is permeable to sodium ions and another to potassium ions. During a resting potential these channels are all closed.

The transfer of energy of the stimulus first opens the gates of the **sodium channels** in the plasma membrane and sodium ions diffuse in, down their **electrochemical gradient**.

What causes an electrochemical gradient?

The electrochemical gradient of an ion is due to its electrical and chemical properties.

- **Electrical properties** are due to the charge on the ion (an ion is attracted to an opposite charge).
- **Chemical properties** are due to concentration in solution (an ion tends to move from a high to a low concentration).

With sodium channels opened, the cytoplasm of the neuron fibre (the interior) quickly becomes progressively more positive with respect to the outside. When the charge has been **reversed** from −70 mV to +40 mV (due to the electrochemical gradient), an **action potential has been created** in the neuron fibre (Figure 6.44).

The action potential then runs the length of the neuron fibre. At any one point, it exists for only two-thousandths of a second (2 milliseconds), before the resting potential starts to be re-established. So action potential transmission is exceedingly quick.

Almost immediately, an action potential has passed the sodium channels close and the **potassium channels** open. Now, potassium ions can exit the cell, again down an electrochemical gradient, into the tissue fluid outside. The interior of the neuron fibre starts to become less positive again. Then, the potassium channels also close. Finally, the resting potential is re-established by the action of the sodium–potassium pump and the process of facilitated diffusion.

29 Deduce the source of energy used to:
a establish the resting potential
b power an action potential.

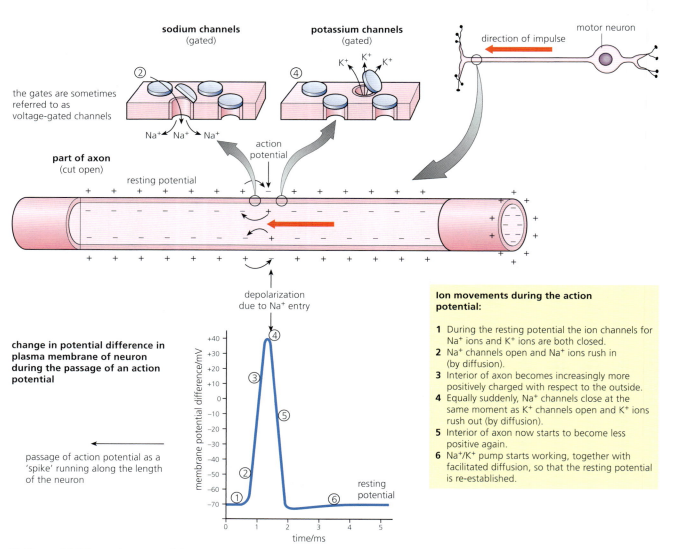

Figure 6.44 The action potential

The refractory period

For a brief period, following the passage of an action potential, the neuron fibre is no longer excitable. This is the **refractory period** and it lasts only 5–10 milliseconds in total. The neuron fibre is not excitable during the refractory period, because there is a large excess of sodium ions inside the fibre and further influx is impossible. Subsequently, as the resting potential is progressively restored, it becomes increasingly possible for an action potential to be generated again. Because of the refractory period, the maximum frequency of impulses is between 500 and 1000 per second.

The all-or-nothing principle

Obviously, stimuli are of widely different strengths: for example, contrast a light touch and the pain of a finger hit by a hammer. A stimulus must be at or above a minimum intensity, known as the **threshold of stimulation**, in order to initiate an action potential. Either the depolarization is sufficient to fully reverse the potential difference in the cytoplasm (from –70 mV to +40 mV), or it is not. If not, no action potential arises. With all sub-threshold stimuli, the influx of sodium ions is quickly reversed and the full resting potential is re-established.

However, as the **intensity of the stimulus increases**, the **frequency** at which the **action potentials pass** along the fibre **increases** (the individual action potentials are all of standard strength). For example, with a very persistent stimulus, action potentials pass along a fibre at an accelerated rate, up to the maximum possible permitted by the refractory period. This means the effector (or the brain) is able to recognize the intensity of a stimulus from the frequency of action potentials (Figure 6.45).

■ **Figure 6.45**
Weak and strong stimuli and the threshold value

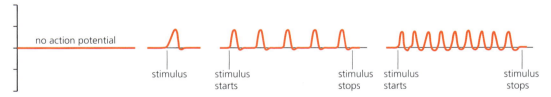

stimuli below the threshold value: not sufficient to reverse polarity of the membrane to +40 mV

brief stimulus just above threshold value: needed to cause depolarisation of the membrane of the sensory cell, and thus trigger an impulse

stronger, more persistent stimulus

much stronger stimulus: has stimulated almost the maximum frequency of impulses

no action potential

stimulus | stimulus starts | stimulus stops | stimulus starts | stimulus stops

Speed of conduction of the action potential

The presence of a myelin sheath affects the speed of transmission of the action potential. The junctions in the sheath, the nodes of Ranvier, occur at 1–2 mm intervals. Only at these nodes is the axon membrane exposed. Elsewhere along the fibre, the electrical resistance of the myelin sheath prevents depolarization of the nodes. The action potentials actually 'jump' from node to node (this is called **saltatory conduction**, meaning 'to leap', Figure 6.46). This greatly speeds up the rate of transmission.

By contrast, non-myelinated dendrons and axons are common in non-vertebrate animals. Here, step-by-step depolarization occurs as the action potential flows along the entire surface of the fibres. This is a relatively slow process, compared with saltatory conduction.

However, any non-myelinated fibre with a large diameter actually transmits an action potential much more speedily than does a narrow fibre. This is because the speed of transmission depends on resistance offered by the axoplasm within. This type of resistance is related to the diameter of the fibre; the narrower the fibre, the greater its resistance, and the lower the speed of conduction of the action potential. Some non-vertebrates, like the squid and the earthworm, have giant fibres, which allow fast transmission of action potentials (not as fast as in myelinated fibres, however). Incidentally, the original investigations of the nature of the action potential by physiologists were carried out on giant fibres.

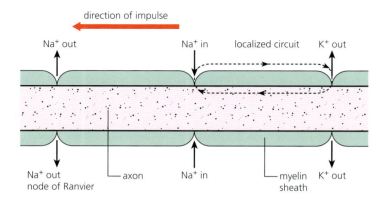

direction of impulse

Na⁺ out   Na⁺ in   localized circuit   K⁺ out

Na⁺ out   axon   Na⁺ in   myelin sheath   K⁺ out
node of Ranvier

30 **Explain** why myelinated fibres conduct faster than non-myelinated fibres of the same size.

31 Figure 6.47 shows two oscilloscope traces of specific events in an axon of a post-synaptic neuron.

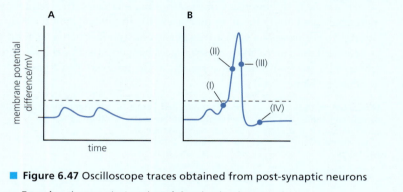

A   B

membrane potential difference/mV

(II)   (III)
(I)
(IV)

time

■ **Figure 6.47** Oscilloscope traces obtained from post-synaptic neurons

a **Examine** the trace in A and **explain** what has happened.
b In the trace in B, **outline** what specific events are occurring at the points labelled (I), (II), (III) and (IV).

■ Junctions between neurons

The synapse is the link point between neurons. A synapse consists of the swollen tip (synaptic knob) of the axon of one neuron (**pre-synaptic neuron**) and the dendrite or cell body of another neuron (**post-synaptic neuron**). At the synapse, the neurons are extremely close but they have **no direct contact**. Instead there is a tiny gap, called a **synaptic cleft**, about 20 nm wide (Figure 6.48).

The practical effect of the synaptic cleft is that an action potential can only cross it via specific chemicals, known as **transmitter substances**. Transmitter substances are all relatively small molecules that diffuse quickly. They are produced in the Golgi apparatus in the synaptic knob and are held in tiny vesicles before release.

Acetylcholine (ACh) is a commonly occurring transmitter substance (the neurons that release acetylcholine are known as cholinergic neurons). Another common transmitter substance is **noradrenalin** (from adrenergic neurons). In the brain, the commonly occurring transmitters are **glutamic acid** and **dopamine**.

Steps of synapse transmission

You may find it helpful to follow each step in Figure 6.49, the diagram of this event.

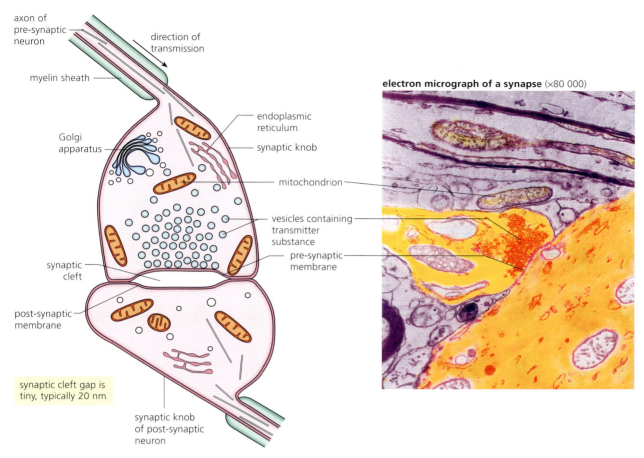

axon of
pre-synaptic
neuron

direction of
transmission

myelin sheath

Golgi
apparatus

endoplasmic
reticulum

synaptic knob

mitochondrion

vesicles containing
transmitter
substance

pre-synaptic
membrane

synaptic
cleft

post-synaptic
membrane

synaptic cleft gap is
tiny, typically 20 nm

synaptic knob
of post-synaptic
neuron

electron micrograph of a synapse (×80 000)

■ **Figure 6.48** A synapse in section

1 The arrival of an action potential at the synaptic knob opens **calcium ion channels in the pre-synaptic membrane**, and calcium ions flow in from the synaptic cleft.

2 The calcium ions cause **vesicles of transmitter substance** to fuse with the pre-synaptic membrane and release transmitter substance into the synaptic cleft.

3 The transmitter substance diffuses across the synaptic cleft and binds with a **receptor protein**.

In the **post-synaptic membrane** there are specific receptor sites for each transmitter substance. Each of these receptors also acts as a channel in the membrane which allows a specific ion (e.g. Na^+, or Cl^- or some other ion) to pass. The attachment of a transmitter molecule to its receptor instantly **opens the ion channel**.

When a molecule of ACh attaches to its receptor site, a Na^+ channel opens. As the sodium ions rush into the cytoplasm of the post-synaptic neuron, **depolarization** of the post-synaptic membrane occurs. As more and more molecules of ACh bind, it becomes increasingly likely that depolarization will reach the **threshold level**. When it does, an **action potential is generated** in the post-synaptic neuron. This process of build up to an action potential in post-synaptic membranes is called **facilitation**.

4 The transmitter substance on the receptors is immediately **inactivated** by enzyme action. For example, the enzyme cholinesterase hydrolyses ACh to choline and ethanoic acid, which are inactive as transmitters. This causes the ion channel of the receptor protein to close, and so allows the resting potential in the post-synaptic neuron to be re-established.

5 The inactivated products from the transmitter re-enter the pre-synaptic knob, are **resynthesized** into transmitter substance and packaged for reuse (Figure 6.49).

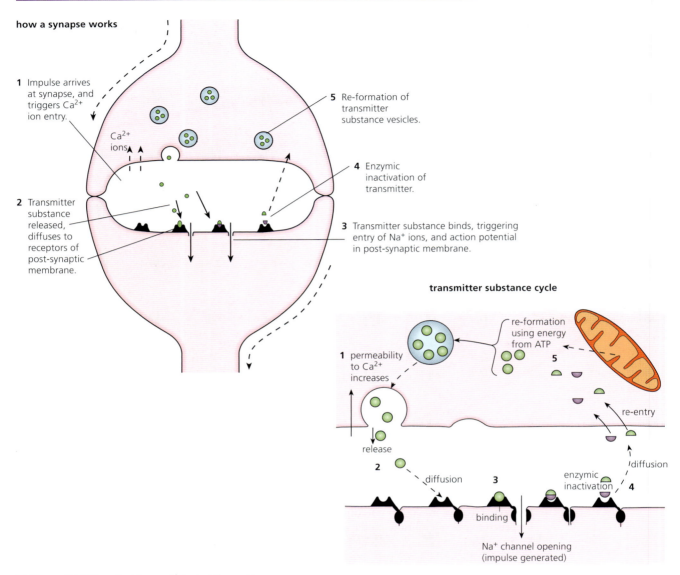

how a synapse works

1 Impulse arrives at synapse, and triggers Ca²⁺ ion entry.

Ca²⁺ ions

2 Transmitter substance released, diffuses to receptors of post-synaptic membrane.

5 Re-formation of transmitter substance vesicles.

4 Enzymic inactivation of transmitter.

3 Transmitter substance binds, triggering entry of Na⁺ ions, and action potential in post-synaptic membrane.

transmitter substance cycle

re-formation using energy from ATP

1 permeability to Ca²⁺ increases

5

re-entry

release

2

diffusion

3

binding

enzymic inactivation

diffusion

4

Na⁺ channel opening (impulse generated)

■ **Figure 6.49** Chemical transmission at the synapse

32 Identify the role of the:
  a Golgi apparatus
  b mitochondria
in the synaptic knob.

Drugs may interfere with the activity of synapses

Drugs have been discovered that interfere with the action of neurotransmitters. For example, some drugs **amplify** the processes and increase postsynaptic transmission. Nicotine and atropine have these effects.

Other drugs **inhibit** the processes of synaptic transmission, in effect decreasing synaptic transmission. Amphetamines and β-blocker drugs have these effects. The latter are used in the treatment of high blood pressure. In fact, an understanding of the workings of neurotransmitters and synapses has led to the development of numerous pharmaceuticals for the treatment of mental disorders.

Nature of Science

Cooperation and collaboration between groups of scientists

Interestingly, these and other discoveries by neurobiologists are contributing to research into memory and learning.

Another application of this technology is in the production of effective commercial pesticides. For example, chemical molecules known as neonicotinoids have been found to completely block synaptic transmission at chlorogenic synapses of insects. Since they may kill insect pests to which they are applied they are useful, but there are issues about their impact on the wider insect community.

6.6 Hormones, homeostasis and reproduction
– hormones are used when signals need to be widely distributed

■ Introducing the endocrine system

Hormones are chemical substances that are produced and secreted from the cells of the ductless or **endocrine glands**. In effect, hormones carry messages about the body – but in a totally different way from the nervous system.

Hormones are transported indiscriminately in the bloodstream, but they act only at specific sites, called **target organs**. Although present in small quantities, hormones are extremely effective messengers, helping to control and coordinate body activities. Once released, hormones may cause changes to specific metabolic reactions of their target organs. However, they circulate in the bloodstream only briefly. In the liver, hormones are broken down and the breakdown products are excreted in the kidneys. So, long-acting hormones must be secreted continuously to be effective.

An example of a hormone is insulin, released from endocrine cells in the pancreas. Insulin regulates blood glucose (page 302). The positions of endocrine glands of the body are shown in Figure 6.50.

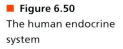
Figure 6.50
The human endocrine system

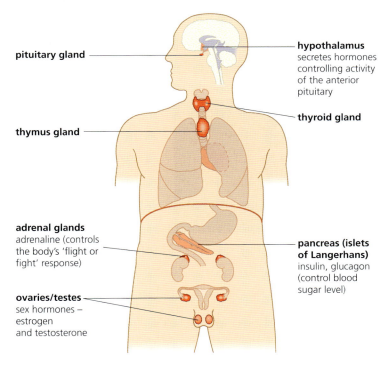

pituitary gland

hypothalamus
secretes hormones controlling activity of the anterior pituitary

thyroid gland

thymus gland

adrenal glands
adrenaline (controls the body's 'flight or fight' response)

pancreas (islets of Langerhans)
insulin, glucagon (control blood sugar level)

ovaries/testes
sex hormones – estrogen and testosterone

■ Maintaining a constant internal environment – homeostasis

Living things face changing and sometimes hostile environments; some external conditions change slowly, others dramatically. For example, temperature changes quickly on land that is exposed to direct sunlight, but the temperature of water exposed to sunlight changes very slowly (page 70).

How do organisms respond to environmental changes?

An animal that is able to maintain a constant internal environment, enabling it to continue normal activities more or less whatever the external conditions, is known as a **regulator**. For example, mammals and birds maintain a high and almost constant body temperature over a very wide range of external temperatures. Their bodies are at or about the optimum temperature for the majority of the enzymes that drive their metabolism. Their muscles contract efficiently and the nervous system coordinates responses precisely, even when external conditions are unfavourable. They are often able to avoid danger and perhaps they may also benefit from the vulnerability of prey organisms which happen to be **non-regulators**. So regulators may have greater freedom in choosing where to live. They can exploit more habitats with differing conditions than non-regulators.

Homeostasis is the name for this ability to maintain a constant internal environment. Homeostasis means 'staying the same'. The internal environment consists of the blood circulating in the body and the fluid that circulates among cells (tissue fluid that forms from blood plasma), delivering nutrients and removing waste products while bathing the cells. Mammals are excellent examples of animals that maintain remarkably constant internal conditions. They successfully regulate their blood pH, oxygen and carbon dioxide concentrations, blood glucose, body temperature and water balance at constant levels or within narrow limits (Figure 6.51).

How is homeostasis achieved?

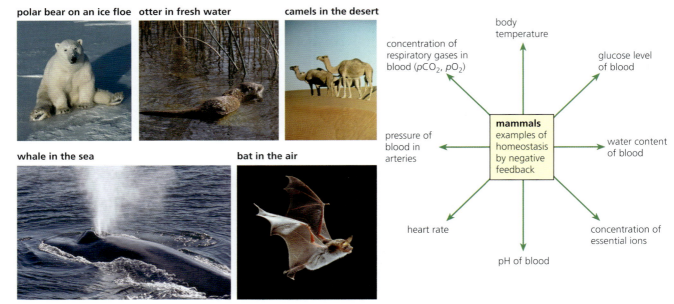

■ **Figure 6.51** Homeostasis in mammals

Negative feedback – the mechanism of homeostasis

Negative feedback is the type of control in which conditions are brought back to a set value as soon as it is detected that they have deviated from it. We see this type of mechanism at work in a system to maintain the temperature of a laboratory water bath. Analysis of this familiar example will show us the components of a negative feedback system (Figure 6.52).

In a negative feedback system, a **detector** device measures the value of the variable (the water temperature in the water bath) and transmits this information to a control unit. The **control unit** compares data from the detector with a pre-set value (the desired water temperature of the water bath). When the value from the detector is below the required value, the control unit activates an **effector** device (a water heater in the water bath) so that the temperature starts to rise. When data from the detector registers in the control box that the water has reached the set temperature, then the control box switches off the response (the water heater). How precisely the variable is maintained depends on the sensitivity of the detector, but negative feedback control typically involves some degree of 'overshoot'.

In mammals, regulation of body temperature, blood sugar level, and the amounts of water and ions in blood and tissue fluid (osmoregulation) are regulated by negative feedback. The detectors are specialized cells in either the brain or other organs, such as the pancreas. The effectors are organs such as the skin, liver and kidneys. Information passes between them via the nerves of the nervous system or via hormones (the endocrine system), or both. The outcome is an incredibly precisely regulated internal environment.

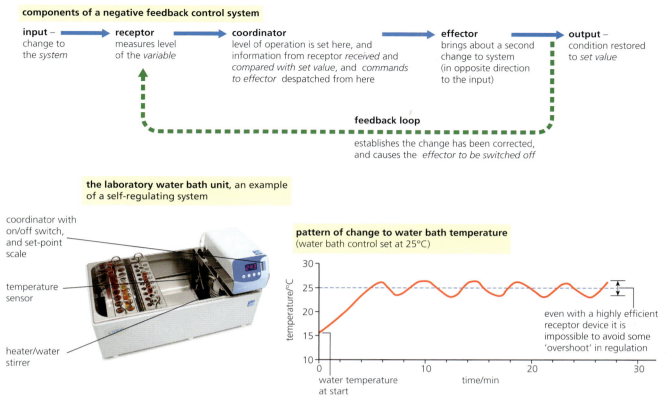

Figure 6.52 Negative feedback, the mechanism

Homeostasis in action

1 Regulation of blood glucose

Transport of glucose to all cells is a key function of the blood circulation. In humans, the **normal level of blood glucose** is about 90 mg of glucose in every 100 cm^3 of blood, but it can vary. For example, during an extended period without food, or after prolonged and heavy physical activity, blood glucose may fall to as low as 70 mg. After a meal rich in carbohydrate has been digested, blood glucose may rise to 150 mg.

The maintenance of a constant level of this monosaccharide in the blood plasma is important for two reasons.

- Respiration is a continuous process in all living cells. To maintain their metabolism, cells need a regular supply of glucose, which can be quickly absorbed across the cell membrane. Glucose is the main respiratory substrate for many tissues. Most cells (including muscle cells) hold reserves in the form of glycogen which is quickly converted to glucose during prolonged physical activity. However, glycogen reserves may be used up quickly. In the brain, glucose is the only substrate the cells can use and, here, there is no glycogen store held in reserve. If our blood glucose falls below 60 mg per 100 cm^3, we have a condition called **hypoglycemia**. If this is not quickly reversed, we may faint. If the body and brain continue to be deprived of adequate glucose levels, convulsions and coma follow.

- An abnormally high concentration of blood glucose, known as **hyperglycemia**, is also a problem. Since high concentration of any soluble metabolite lowers the water potential of the blood plasma, water is drawn from the cells and tissue fluid by osmosis, back into the blood. As the volume of blood increases, water is excreted by the kidney to maintain the correct concentration of blood. As a result, the body tends to become dehydrated and the circulatory system is deprived of fluid. Ultimately, blood pressure cannot be maintained.

For these reasons, it is critically important that the blood glucose is held within set limits.

Mechanism for regulation of blood glucose

After the digestion of carbohydrates in the gut, glucose is absorbed across the epithelial cells of the villi (Figure 6.4, page 252) into the hepatic portal vein. The blood carrying the glucose reaches the liver first. If the glucose level is too high, glucose is withdrawn from the blood and stored as glycogen. Despite this action, blood circulating in the body immediately after a meal does have a raised level of glucose. At the pancreas, the presence of an excess of blood glucose is detected in patches of cells known as the islets of Langerhans (Figure 6.53). These islets are hormone-secreting glands (endocrine glands); they have a rich capillary network, but no ducts that would carry secretions away. Instead, their hormones are transported all over the body by the blood. The islets of Langerhans contain two types of cell, α **cells** and β **cells**.

■ **Figure 6.53**
Islet of Langerhans in the pancreas

TS of pancreatic gland showing an islet of Langerhans   **drawing of part of pancreatic gland**

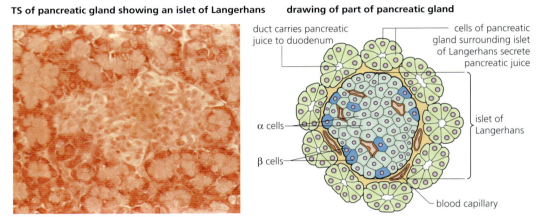

33 Predict what type of organelle you would expect to see most frequently when a liver cell is examined by electron microscopy.

A **raised blood glucose** level stimulates the β **cells**, and they secrete the hormone **insulin** into the capillary network. Insulin causes the uptake of glucose into cells all over the body, but especially by the liver and the skeletal muscle fibres (Figure 6.54). It also increases the rate at which glucose is used in respiration, in preference to alternative substrates (such as fat). Another effect of insulin is to trigger conversion of glucose to glycogen in cells (**glycogenesis**), and of glucose to fatty acids and fats, and finally the deposition of fat around the body.

As the blood glucose level reverts to normal, this is detected in the islets of Langerhans, and the β cells stop insulin secretion. Meanwhile, the hormone is excreted by the kidney tubules and the blood insulin level falls.

When the **blood glucose level falls below normal**, the α **cells** of the pancreas secrete a hormone called **glucagon**. This hormone activates the enzymes that convert glycogen and amino acids to glucose (**gluconeogenesis**). Glucagon also reduces the rate of respiration (Figures 6.55).

As the blood glucose level reverts to normal, glucagon production ceases and this hormone, in turn, is removed from the blood in the kidney tubules.

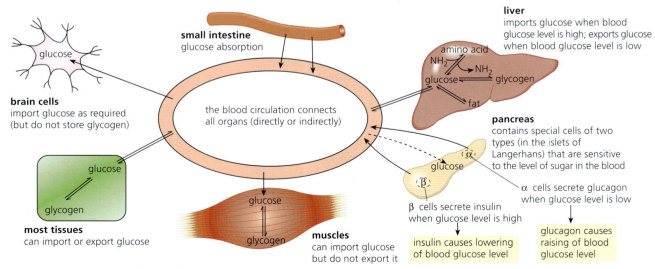

■ **Figure 6.54** The sites of blood glucose regulation

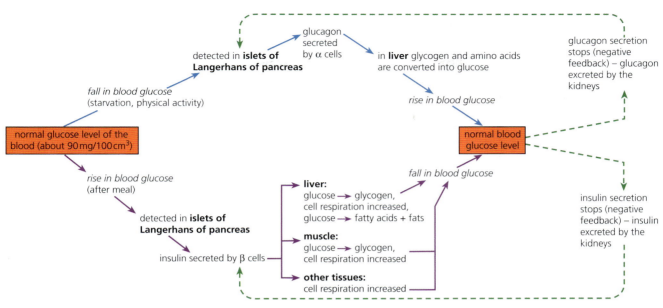

■ **Figure 6.55** Glucose regulation by negative feedback

The disease diabetes

A diabetic is a person whose body is failing to regulate blood glucose levels correctly. **Type I diabetes** is the result of a failure of insulin production by the β cells. **Type II diabetes** (diabetes mellitus) is a failure of the insulin receptor proteins on the cell membranes of target cells (Figure 6.56). As a result, blood glucose level is more erratic and, generally, permanently raised. Glucose is also regularly excreted in the urine. If this condition is not diagnosed and treated, it carries an increased risk of circulatory disorders, renal failure, blindness, strokes or heart attacks.

■ **Figure 6.56**
Diabetes, causes and treatment

**type I diabetes,
'early onset diabetes'**
affects young people, below the age of 20 years

due to the destruction of the β cells of the islets of Langerhans by the body's own immune system

symptoms:
• constant thirst
• undiminished hunger
• excessive urination

treatment:
• injection of insulin into the bloodstream daily
• regular measurement of blood glucose level

patient injecting with insulin, obtained by genetic engineering

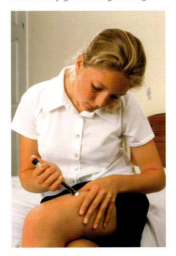

**type II diabetes,
'late onset diabetes'**
the common form (90% of all cases of diabetes are of this type)

common in people over 40 years, especially if overweight, but this form of diabetes is having an increasing effect on human societies around the world, including young people and even children in developed countries, seemingly because of poor diet

symptoms:
mild – sufferers usually have sufficient blood insulin, but insulin receptors on cells have become defective

treatment:
largely by diet alone

2 Control of body temperature

Regulation of our body temperature, known as **thermoregulation**, involves controlling both heat loss across the body surface and heat production within. Heat is lost to the environment by convection, radiation and conduction. The body also loses heat by evaporation.

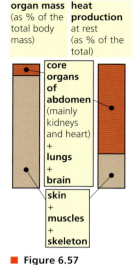

organ mass (as % of the total body mass)	heat production at rest (as % of the total)
core organs of abdomen (mainly kidneys and heart) + **lungs** + **brain**	
skin + **muscles** + **skeleton**	

■ **Figure 6.57**
Heat production in the body at rest

34 State what it means to say that most reactions in liver cells are endergonic.

Mammals maintain a high and relatively constant body temperature by:

■ generating heat within the body from the reactions of metabolism and by being able to regulate this heat production

■ subtle control of the loss of heat through the skin.

An animal with this form of thermoregulation is called an endotherm, meaning 'inside heat'. Humans hold their inner body temperature (core temperature) just below 37°C. In fact, human core temperature only varies between about 35.5 and 37.0°C within a 24-hour period, when we are in good health.

Heat production in the human body

The major sources of heat are the biochemical reactions of metabolism that generate heat as a waste product (Figure 6.57). Heat is then distributed by the blood circulation. The organs of the body vary in the amount of heat they yield. For example, the liver is extremely active metabolically, but most of its metabolic reactions require an input of energy (they are **endergonic reactions**) and little energy is lost as heat. Mostly, the liver is thermally neutral.

The bulk of our body heat (over 70%) comes from other organs, mainly from the heart and kidneys, but also from the lungs and brain (which, like a computer central processing unit, needs to be kept cool). While the body is at rest, the skeleton, muscles and skin, which make up over 90% of the body mass, produce less than 30% of the body heat. Of course, in times of intense physical activity, the skeletal muscles generate a great deal of heat as a waste product of respiration and contraction.

If the body experiences persistent cold, then heat production is increased. Under chilly conditions, heat output from the body muscles is raised by coordinated contractions of skeletal muscles, known as shivering. This raises muscles' heat production about five times above basal rate.

Thyroxin, an iodine-containing hormone produced in the thyroid gland (Figure 6.50), also plays a part in the control of temperature regulation. The presence of thyroxin in the blood circulation stimulates oxygen consumption and increases the basal metabolic rate of the body organs. Variations in secretion of thyroxin help the control of body temperature.

3 Control of appetite

Our brain houses a control centre for the appetite, located in the hypothalamus, which is a part of the floor of the forebrain (Figure 11.32, page 476). Here, our appetite is regulated. Actually, this part of the brain also regulates our thirst and body temperature, operating by means of impulses despatched to specific body organs and organ systems, via nerves and the spinal cord.

How the appetite control centre is kept informed of the 'hunger' state of the body

The appetite centre is chiefly stimulated by specific **hormones** that are secreted by tissues and organs in the body. Of the three hormones involved, one, **leptin**, plays a continuous role. In adult life, the number of fat cells does not change significantly. If we overeat they fill up with lipids; when we are short of food, reserves are used and the fat cells empty.

Now, as the fat cells fill up, they secrete more leptin. Like all hormones, leptin circulates in the blood. On reaching the appetite centre, leptin suppresses the sensation of hunger. On the other hand, when fat cells empty and shrink, they secrete less leptin, and the sensation of hunger is experienced in the brain. Clearly, leptin is associated with long-term regulation of eating.

More immediate control is exerted by hormones from parts of the gut and pancreas.

■ As the stomach empties, a hormone is secreted there that stimulates the appetite control centre and we get a hunger sensation.

■ After a meal, a hormone is released from the intestine and pancreas; on reaching the appetite centre, this suppresses the hunger sensation.

The hormone leptin and the treatment of clinical obesity

Obese people have been found to have unusually high concentrations of leptin in their bloodstream. Apparently, their bodies have become resistant to the effects of this hormone – their feeding-control mechanism having become desensitized in some way. Consequently, treatment of obesity by administration of leptin has no advantages, currently.

4 Circadian rhythms

Daily rhythms in physiological processes and behaviour are common throughout the plant and animal world. For example, animals are active for only part of the 24-hour cycle. Some function at dusk or dawn (crepuscular), some in the night (nocturnal), and many in the day (diurnal), as humans do. In fact, much of our behaviour (physical activity, sleep, body temperature, secretion of hormones, and other features) follows regular rhythms or cycles (Figure 6.58). These cycles operate over an approximately 24-hour cycle and are called **circadian rhythms** (meaning *'about a day'*).

■ **Figure 6.58**
Our physiology and behaviour show circadian rhythms

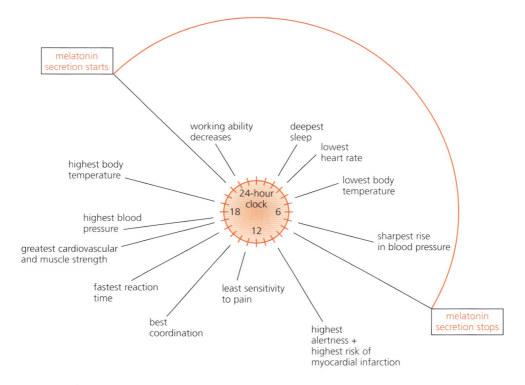

Circadian rhythms are controlled by a 'biological clock' within the brain. However, cycles are coordinated with the cycle of light and dark – with day and night.

Melatonin is a hormone that is found naturally in the body. It is produced in the brain, in the **pineal gland**. This hormone contributes to setting our biological clock. More melatonin is released in darkness and more is released in seasonal longer nights, such as in winter time. Light decreases melatonin production and this lowering of the melatonin level leads to the body's preparation for being awake. In darkness, melatonin production resumes, and sleepiness returns. The condition experienced by some people, called 'winter depression' (seasonal affective disorder, SAD) in which their mental state is affected, is associated with higher than normal melatonin levels in the body.

We see that an external stimulus, like light, has an impact on the biological clock, but the effects are not immediate. When normal patterns of light and dark are changed abruptly – for example, as a result of aircraft travel across different time zones – it may take several days before patterns of 'sleep' and 'activeness' return to normal. We say the traveller is suffering from 'jet-lag'. People who work day- and night-shift patterns that change frequently may have an even greater problem.

'Jet-lag' – does melatonin alleviate symptoms?

Melatonin tablets may prevent or reduce jet-lag symptoms, but the safe and appropriate use of this medication needs more testing.

■ Reproduction

Reproduction is the production of new individuals by an existing member or members of the same species. It is a fundamental characteristic of living things; the ability to self-replicate in this way sets the living world apart from the non-living. In reproduction, a parent generation effectively passes on a copy of itself, in the form of the genetic material, to another generation – the offspring. The genetic material of an organism consists of its chromosomes, made of nucleic acid (page 138).

Organisms reproduce either asexually or sexually and many reproduce by both these methods. However, mammals reproduce by **sexual reproduction** only.

■ **Figure 6.59**
Meiosis and the diploid lifecycle

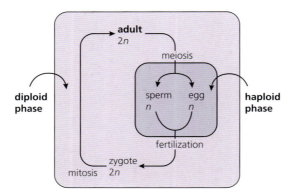

In sexual reproduction, two **gametes** (specialized sex cells) fuse to form a **zygote**, which then grows into a new individual. Fusion of gametes is called **fertilization**. In the process of gamete formation, a nuclear division by **meiosis** (page 142) halves the normal chromosome number. That is, gametes are **haploid**, and fertilization restores the **diploid** number of chromosomes (Figure 6.59). Without the reductive nuclear division in the process of sexual reproduction, the chromosome number would double in each generation. Remember, the offspring produced by sexual reproduction are unique, in complete contrast with offspring formed by asexual reproduction.

■ Sexual reproduction in mammals

In mammals, as in most of the vertebrates, the sexes are separate (unisexual individuals). In the body, the reproductive and urinary (excretory) systems are closely bound together, especially in the male, so biologists refer to these as the urinogenital system. Here, just the reproductive systems are considered (Figures 6.60 and 6.61).

Male and female gametes are produced in paired glands called **gonads**: the testes and ovaries.

The male reproductive system

- Two **testes** (singular, **testis**) are situated in the scrotal sac (**scrotum**), hanging outside the main body cavity; this allows the testes to be at the optimum temperature for sperm production, 2–3°C lower than the normal body temperature. As well as producing the male gametes, **spermatozoa** (singular, **spermatozoon**) or sperms, the testes also produce the male sex hormone, **testosterone**; the testes are, therefore, also endocrine glands (page 299).
- The **epididymis** stores the sperms and the sperm ducts carry them in a fluid, called **seminal fluid**, to the outside of the body during a process called an **ejaculation**.
- Ducted or exocrine glands secrete the nutritive seminal fluid (of alkali, proteins and fructose) in which the sperms are transported; these include the **seminal vesicles** and **prostate gland**.
- The **urethra** is a duct that carries semen during an ejaculation (and urine during urination) to the outside.
- The **penis** contains spongy erectile tissue that can fill with blood when the male is sexually stimulated. This causes the penis to enlarge, lengthen and become rigid, in a condition known as an **erection**. The erect penis penetrates the vagina in sexual intercourse.

■ **Figure 6.60**
The male urinogenital
system

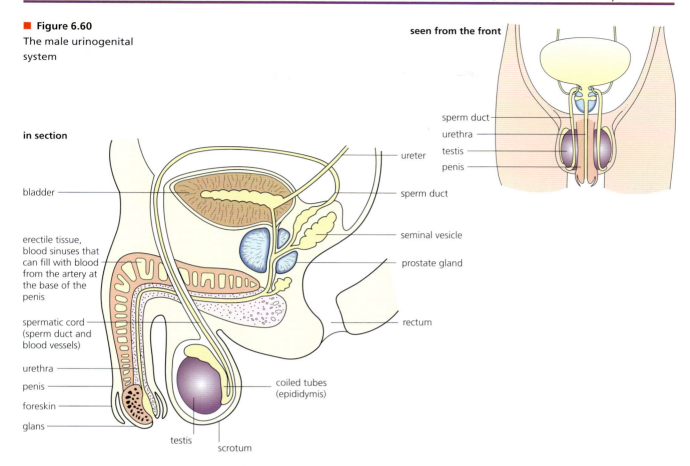

seen from the front

sperm duct
urethra
testis
penis

in section

bladder

erectile tissue,
blood sinuses that
can fill with blood
from the artery at
the base of the
penis

spermatic cord
(sperm duct and
blood vessels)

urethra

penis

foreskin

glans

testis

scrotum

ureter

sperm duct

seminal vesicle

prostate gland

rectum

coiled tubes
(epididymis)

 You should be able to annotate a diagram of the male reproductive system, noting the specific functions of testis, scrotum, epididymis, sperm duct, seminal vesicles and prostate gland, urethra and penis.

The female reproductive system

■ The **ovaries** are held near the base of the abdominal cavity. As well as producing the female gametes, ova or **egg cells**, the ovaries are also endocrine glands, secreting the female sex hormones **estrogen** (sometimes spelled 'oestrogen') and **progesterone**.

■ A pair of **oviducts** extend from the uterus and open as funnels close to the ovaries. The oviducts transport egg cells, and are the site of fertilization.

■ The **uterus**, which is about the size and shape of an inverted pear, has a thick muscular wall and an inner lining of mucous membrane that is richly supplied with arterioles. This lining, called the **endometrium**, undergoes regular change in an approximately 28-day cycle. The lining is built up each month in preparation for implantation and early nutrition of a developing embryo, should fertilization occur. If it does not occur, the endometrium disintegrates and menstruation starts.

■ The **vagina** is a muscular tube that can enlarge to allow entry of the penis and exit of a baby at birth. The vagina is connected to the uterus at the **cervix** and it opens to the exterior at the **vulva**.

■ **Figure 6.61**
The female
urinogenital system

seen from the front

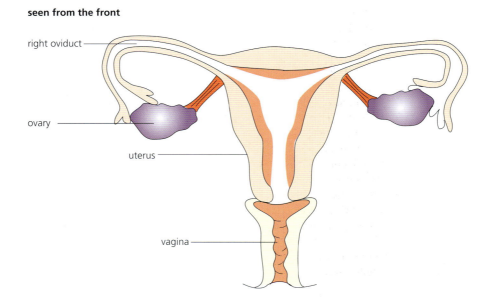

in section

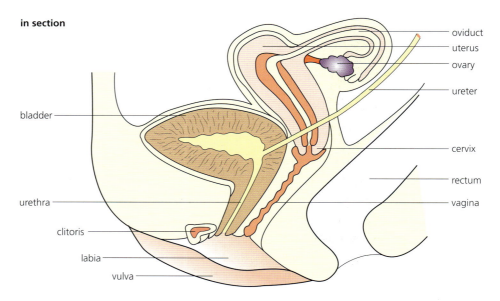

You should be able to annotate a diagram of the female reproductive system, noting the specific functions of the ovary, oviduct, uterus, cervix, vagina and vulva.

■ Hormones in reproduction

1 Hormones in embryonic development

The primary sexual characteristics of a male or a female are possession of their respective reproductive organs. We have seen that in humans, gender is primarily determined by the sex chromosomes, XX or XY (Figure 3.24, page 158). Actually, in humans the embryonic reproductive structures, the gonads and their ducts, and the structures of the external genitalia in both sexes are similar up to about the seventh week of gestation.

Then, in the male (XY), differentiation is initiated by a 'master switch' allele on the Y chromosome, named SRY (**s**ex-determining **r**egion of the **Y** chromosome). This triggers the development of primitive Sertoli cells in the embryonic gonads, followed by their secretion of **testosterone** from the eighth week. This hormone then initiates the pre-natal development of the epididymis, sperm duct, seminal vesicles, testes and prostate gland.

In the female (XX), because the SRY allele is absent (and so no testosterone is secreted), the ovary, the female ducts and genitalia develop. Subsequently, **estrogen** and **progesterone** are produced and facilitate the continuing pre-natal development of the female reproductive organs.

2 Hormones at puberty

During the first 10 years of life, the reproductive systems of girls and boys remain in a juvenile state. Then, at puberty, the secondary sexual characteristics begin to develop. This is a time of growth and development at the beginning of sexual maturation. Now, there is a significant increase in the production of the sex hormones by the gonads – testosterone in cells of the testes, and estrogen and progesterone in cells of the ovaries. These hormones are chemically very similar molecules; they are manufactured from the steroid cholesterol that is both made in the liver and absorbed as part of the diet. (Steroids are a form of lipid, page 86.)

Perhaps the most noticeable effect of the increased secretion of the sex hormones is the stimulation they cause in muscle protein formation and bone growth. Because of this effect, testosterone, estrogen and progesterone are known as **anabolic steroids** (anabolic means 'build up'). The effects of the female sex hormones are less marked in this respect than those of testosterone in the male. The onset of puberty occurs, on average, about three years earlier in girls (Figure 6.62).

■ **Figure 6.62**
The onset of the secondary sexual characteristics

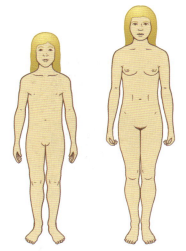

10-year-old female   13-year-old female

Secondary sexual characteristics of a human female:

- maturation of the ovaries, and enlargement of the vagina and uterus

- development of the breasts

- widening of the pelvis

- deposition of fat under the skin of the buttocks and thighs

- growth of pubic hair and hair under the armpits

- monthly ovulation and menstruation

- changes in behaviour associated with a sex drive.

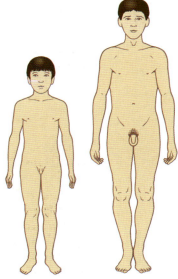

10-year-old male   16-year-old male

Secondary sexual characteristics of a human male:

- development and enlargement of the testes, scrotum, penis and glands of the reproductive tract

- increased skeletal muscle development

- enlargement of the larynx, deepening the voice

- growth of pubic hair, underarm hair and body hair

- continuous production of sperms and in the absence of sexual intercourse, occasional erections and the discharge of seminal fluid

- changes in behaviour associated with a sex drive.

The onset of puberty is triggered by a part of the brain called the hypothalamus. Here, production and secretion of a releasing hormone causes the nearby pituitary gland (the 'master' endocrine gland) to produce and release into the blood circulation two hormones, **follicle-stimulating hormone (FSH)** and **luteinizing hormone (LH)**. They are so named because their roles in sexual development were discovered in the female, although they do operate in both sexes. Their first effects are to enhance secretions of the sex hormones. Then, in the presence of FSH, LH and the respective sex hormone, there follows the development of the secondary sexual characteristics, and the preparation of the body for its role in sexual reproduction.

3 Hormones in the control of reproduction

In the female reproductive system:

■ The secretion of estrogen and progesterone is cyclical, rather than at a steady rate. Together with FSH and LH, the changing concentrations of all four hormones bring about a repeating cycle of changes that we call the menstrual cycle.

■ The menstrual cycle consists of **two cycles**, one in the **ovaries** and one in the **uterus lining**. The ovarian cycle is concerned with the monthly preparation and shedding of an egg cell from an ovary and the uterus cycle is concerned with the build-up of the lining of the uterus. 'Menstrual' means 'monthly'; the combined cycles take around 28 days (Figure 6.63).

■ **Figure 6.63**
Hormone regulation of the menstrual cycle

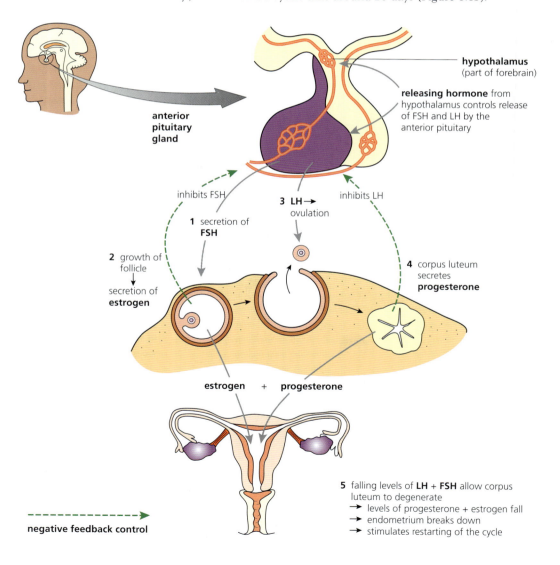

hypothalamus (part of forebrain)

releasing hormone from hypothalamus controls release of FSH and LH by the anterior pituitary

anterior pituitary gland

inhibits FSH

3 LH→ ovulation

inhibits LH

1 secretion of **FSH**

2 growth of follicle
↓
secretion of **estrogen**

4 corpus luteum secretes **progesterone**

estrogen + progesterone

5 falling levels of **LH + FSH** allow corpus luteum to degenerate
→ levels of progesterone + estrogen fall
→ endometrium breaks down
→ stimulates restarting of the cycle

- - - - - → **negative feedback control**

By convention, the start of the cycle is taken as the first day of menstruation (bleeding), which is the shedding of the endometrium lining of the uterus. The steps, also summarized in Figure 6.64, are as follows.

1 **FSH** is secreted by the pituitary gland and stimulates development of several immature egg cells (in primary follicles) in the ovary. Only one will complete development into a mature egg cell (now in the ovarian follicle).
2 The developing follicle then secretes **estrogen**. Estrogen has two targets:
   a In the uterus, it stimulates the build-up of the endometrium, the lining, for a possible implantation of an embryo should fertilization take place.
   b In the pituitary gland, estrogen inhibits the further secretion of FSH, which prevents the possibility of further follicles being stimulated to develop. This is an example of **negative feedback control** (Figure 6.64).

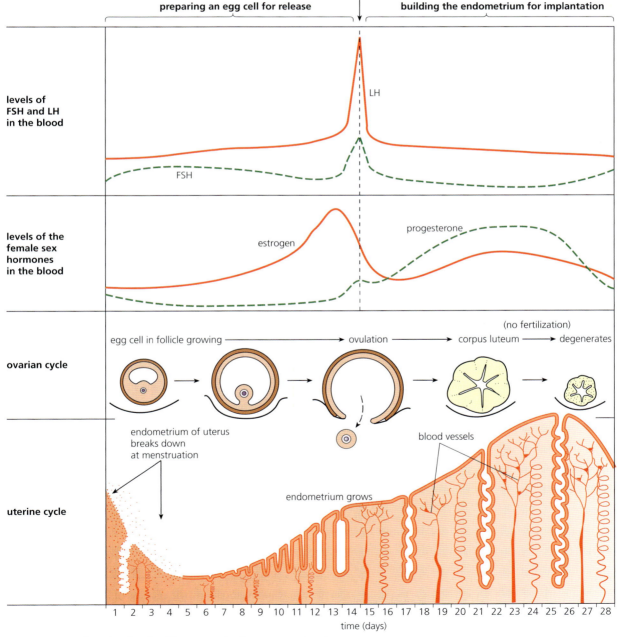

■ **Figure 6.64** Changing levels of hormones in the menstrual cycle

3   The concentration of estrogen continues to increase to a peak value just before the mid-point of the cycle. This high and rising level of estrogen suddenly stimulates the secretion of **LH** and, to a slightly lesser extent, FSH, by the pituitary gland. LH stimulates **ovulation** (the shedding of the mature egg cell from the ovarian follicle and its escape from the ovary).

   As soon as the ovarian follicle has discharged its egg, LH also stimulates the conversion of the vacant follicle into an additional temporary gland, called a **corpus luteum**.

4   The corpus luteum secretes **progesterone** and, to a lesser extent, estrogen. Progesterone has two targets:

   a   In the uterus, it continues the build-up of the endometrium, further preparing for a possible implantation of an embryo should fertilization take place.

   b   In the pituitary gland, it inhibits further secretion of LH, and also of FSH; this is a second example of **negative feedback control**.

5   The levels of FSH and LH in the bloodstream now rapidly decrease. Low levels of FSH and LH allow the corpus luteum to degenerate. As a consequence, the levels of progesterone and estrogen also fall. Soon the levels of these hormones are so low that the extra lining of the uterus is no longer maintained. The **endometrium breaks down** and is lost through the vagina in the first five days or so of the new cycle. Falling levels of progesterone again cause the secretion of FSH by the pituitary.

6   A new cycle is under way.

7   **If the egg is fertilized** (the start of a pregnancy), then the developing embryo itself immediately becomes an endocrine gland, secreting a hormone that circulates in the blood and maintains the corpus luteum as an endocrine gland for at least 16 weeks of pregnancy. When, eventually, the corpus luteum does break down, the **placenta** takes over as an endocrine gland, secreting estrogen and progesterone. These hormones continue to prevent ovulation and maintain the endometrium.

In the **male reproductive system**, the secretion of sex hormones commences in pre-natal development and is a continuous process, rather than cyclic.

   Testosterone secretion has the following roles:

■   It initiates the pre-natal development of male genitalia.

■   It triggers and regulates the development of secondary sexual characteristics.

■   It maintains the sex drive (libido) in the adult.

> **35 Identify** the critical hormone changes that respectively trigger ovulation, and cause degeneration of the corpus luteum.

4 Hormones in the treatment of infertility, and *in vitro* fertilization

In mammals, fertilization – the fusion of male and female gametes to form a **zygote** – is internal. It occurs in the upper part of the oviduct. As the zygote is transported down the oviduct, mitosis and cell division commence. By the time the embryo has reached the uterus, it is a solid ball of tiny cells. Division continues and the cells organize themselves into a fluid-filled ball, the **blastocyst**, which becomes embedded in the endometrium of the uterus, a process known as **implantation**. In humans, implantation takes from day 7 to day 14 approximately (Figure 11.48, page 491).

   Not all partners are fertile in this way. Either the male or female, or both, may be infertile, due to a number of different causes (Table 6.17).

■ Table 6.17
Causes of infertility

In males, infertility may be due to:	In females, infertility may be due to:
• failure to achieve or maintain an erect penis • structurally abnormal sperms • sperms with poor mobility • short-lived sperms • too few sperms • a blocked sperm duct, preventing semen from containing sperms	• conditions of the cervix which cause death of sperms • conditions in uterus that prevent implantation of blastocyst • eggs that fail to mature or be released • blocked or damaged oviducts, preventing egg from reaching sperms

In some cases, a couple's infertility may be overcome by the process of fertilization of eggs outside the body (*in vitro* fertilization, **IVF**, Figure 6.65). The key step in IVF is the successful removal of sufficient eggs from the ovaries. To achieve this, normal menstrual activity is temporarily suspended with hormone-based drugs.

Then, the ovaries are induced to produce a large number of eggs simultaneously, at a time controlled by doctors. In this way, the correct moment to collect the eggs can be known accurately.

■ **Figure 6.65** *In vitro* fertilization – the process

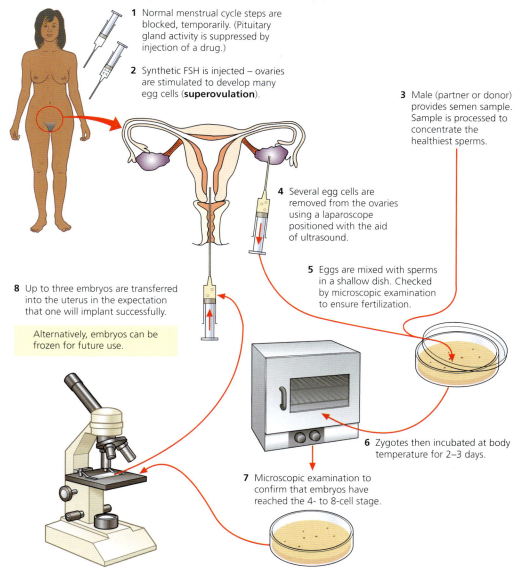

1 Normal menstrual cycle steps are blocked, temporarily. (Pituitary gland activity is suppressed by injection of a drug.)

2 Synthetic FSH is injected – ovaries are stimulated to develop many egg cells (**superovulation**).

3 Male (partner or donor) provides semen sample. Sample is processed to concentrate the healthiest sperms.

4 Several egg cells are removed from the ovaries using a laparoscope positioned with the aid of ultrasound.

5 Eggs are mixed with sperms in a shallow dish. Checked by microscopic examination to ensure fertilization.

8 Up to three embryos are transferred into the uterus in the expectation that one will implant successfully.

Alternatively, embryos can be frozen for future use.

6 Zygotes then incubated at body temperature for 2–3 days.

7 Microscopic examination to confirm that embryos have reached the 4- to 8-cell stage.

36 **Explain** what *'in vitro'* and *'in vivo'* mean.

Egg cells are then isolated from surrounding follicle cells, and mixed with sperms. If fertilization occurs, the fertilized egg cells are incubated so that embryos at the eight-cell stage may be placed in the uterus. If one (or more) imbed there, then a normal pregnancy may follow. The first 'test-tube baby' was born in 1978. Today, the procedure is regarded as a routine one.

■ William Harvey and sexual reproduction in deer

At the start of the seventeenth century, there was little understanding about how 'generation' took place or how each sex contributed. However, humans had survived and prospered since Neolithic times, sustained in large part by the products of their domestication of wild animals and the breeding of herds of sheep and cattle, for example. In all domesticated animals, breeding takes place only during estrous, and only when male and female animals were placed together. The most influential thinking on generation remained that of the Greeks – from the ideas of Aristotle. The woman (or female animal) provided the 'matter' for the baby through her menstrual blood, while the male's semen (the word means 'seed') gave the matter 'form'.

In addition to an investigation of the blood and its circulation, Harvey worked on this issue of **sexual reproduction in animals**, from 1616–1638. His results and conclusions were published in 1651. His investigations involved:

■ deer as viviparous animals and domestic fowl as oviparous animals

■ observation of the development (early embryology) of the chick within the bird's egg. He became convinced that an 'egg' was fundamental to generation. Unfortunately, the microscope was not available to use in his lifetime – Harvey was unable to see egg cells

■ dissection of the uteri of hinds (female deer) at all stages of pregnancy. He found the uterus always empty at the time of conception, so disproving Aristotle's idea that menstrual blood and semen came together there to form the fetus

■ dissection of the ovaries of hinds through the 'rutting' season, but he found no sign of an 'egg' – nor of semen, there, either. Of course, he lacked access to a microscope throughout his studies

■ establishing that the uterus was 'empty' at the time of conception, but he remained convinced that new life developed in the uterus and that an egg was involved.

Nature of Science

Developments in scientific research follow improvements in apparatus

It was later – after Harvey's death – that the Royal Society in London sent a request to van Leeuwenhoek (Figure 1.2, page 3) to examine semen (and other body fluids). Prompted in this way, he reported the presence of spermatozoa in semen in 1677. However, it was only in the nineteenth century that a union of egg and sperm was observed, and it was realized that something contained within the egg and the sperm was inherited. Only then was 'reproduction' truly understood.

■ *Examination questions – a selection*

Questions 1–4 are taken from IB Diploma biology papers.

Q1 Which enzyme is amylase?

	Source	Substrate	Product(s)
A	Pancreas	Starch	Maltose
B	Stomach	Protein	Peptides
C	Pancreas	Peptides	Amino acids
D	Small intestine	Maltose	glucose

Standard Level Paper 1, Time Zone 1, May 13, Q24

Q2 Which feature maintains a high concentration gradient of gases in the ventilation system?
A Thin-walled alveoli
B Thin-walled capillaries
C A moist lining of the alveoli
D Blood flowing in the capillaries

Higher Level Paper 1, Time Zone 0, Nov 10, Q21

Q3 Which statement describes the movements of the rib cage during inhalation of air?
A External intercostal muscles contract moving the ribs up and outwards.
B Internal intercostal muscles contract moving the ribs down and inwards.
C External intercostal muscles relax moving ribs down and inwards.
D Internal intercostal muscles relax moving ribs up and outwards.

Standard Level Paper 1, Time Zone 0, Nov 12, Q26

Q4 What causes the **rate** of heart contraction to increase or decrease?
A The heart muscle itself
B Nerve impulses from the brain
C A hormone from the thyroid gland
D The rate of return of blood to the left atrium

Standard Level Paper 1, Time Zone 2, May 13, Q24

Questions 5–11 cover other syllabus issues in this chapter.

Q5 **a** In the absorption of digested food molecules in the small intestine, explain where each of the following structures occurs and how each contributes to the absorption process:
  i   villi   (2)
  ii capillary networks   (2)
  iii microvilli   (2)
  iv protein pumps.   (2)
  b Explain the term *assimilation*, and state where it occurs in the body.   (3)

Q6 **a** Outline the way the resting potential allows a nerve cell to be ready for the transmission of a nerve impulse.   (4)
  b Describe the events that occur while a non-myelinated axon transports an impulse during an action potential.   (8)

Q7 An example of an autoimmune disease is diabetes type I. Distinguish diabetes type I from type II and explain why insulin injections are not required for patients diagnosed with diabetes type II.   (4)

Q8 State the name of the hormones that:
  i   trigger ovulation   (1)
  ii stimulate the growth of the endometrium (1)
  iii maintain the endometrium and promote its vascularisation.   (1)

Q9 The work performed by Harold Florey and Ernest Chain has a high scientific relevance in relationship with the initial use of penicillin as a possible treatment for clinic trials. Describe the tests they performed and consider why the procedures they followed were not compliant with modern investigation protocols.   (6)

Q10 a The photograph shows a heart, viewed from the ventral side.

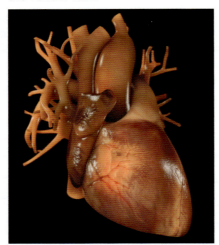

Using the following figure, label it in order to find out the following structures
- right atrium
- left ventricle
- a vein and an artery   (3)

b Explain how adrenaline and the pace maker are used to increase the heart rate.   (2)

c Distinguish between the structure of arteries and veins.   (3)

Q11 Blood has roles in transport and in the body's defences against disease. By means of a table, identify the specific roles of each component of the blood (plasma, red cells, white cells, platelets) in these functions.   (8)

7 Nucleic acids

ESSENTIAL IDEAS

- The structure of DNA is ideally suited to its function.
- Information stored as a code in DNA is copied on to mRNA.
- Information transferred from DNA to mRNA is translated into an amino acid sequence.

7.1 DNA structure and replication – *the structure of DNA is ideally suited to its function*

We have seen that **nucleic acids** are very long thread-like macromolecules of alternating sugar and phosphate molecules (forming a 'backbone') with a nitrogenous base – either cytosine (C), guanine (G), adenine (A), thymine (T) or uracil (U) – attached to each sugar molecule along the strand. Nucleic acids, as the information molecules of cells, fulfil this role throughout the living world – the genetic code is a universal one.

The structures of the two types of nucleic acid, **deoxyribonucleic acid (DNA)** and **ribonucleic acid (RNA)**, are compared in Table 2.8, page 107. In eukaryotes, while DNA occurs only in the nucleus, RNA is found in both the cytoplasm and the nucleus.

1 Distinguish between an organic base, a nucleoside, a nucleotide and a nucleic acid.

■ The packaging of DNA – nucleosomes and supercoiling

In eukaryotes, DNA occurs in the chromosomes in the nucleus, along with protein. Actually, more than 50% of a chromosome is built of **protein**. While some of the proteins of the chromosome are enzymes involved in copying and repair reactions of DNA, the bulk of chromosome protein has a **support and packaging role** for DNA.

Why is packaging necessary?

Well, take the case of human DNA. In the nucleus, the total length of the DNA of the chromosomes is over 2 m. We know this is shared out between 46 chromosomes, and that each chromosome contains one, very long DNA molecule. Chromosomes are different lengths (Figure 3.6, page 141), but we can estimate that within a typical chromosome of 5 μm length, there is a DNA molecule approximately 5 cm long. This means that about 50 000 μm of DNA is packed into 5 μm of chromosome.

This phenomenal packaging is achieved by coiling the DNA double helix and looping it around protein beads called **nucleosomes**, as illustrated in Figure 7.1.

■ **Figure 7.1**
The nucleosome and supercoiling of DNA

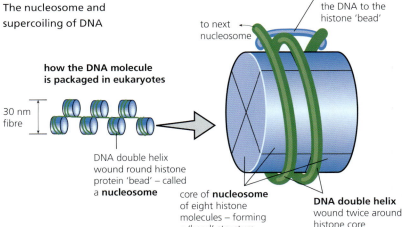

how the DNA molecule is packaged in eukaryotes

30 nm fibre

DNA double helix wound round histone protein 'bead' – called a **nucleosome**

core of **nucleosome** of eight histone molecules – forming a 'bead' structure

to next nucleosome

H1 histone binds the DNA to the histone 'bead'

DNA double helix wound twice around histone core

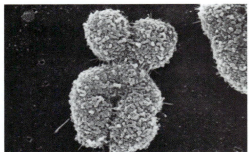

Electron micrograph of metaphase chromosome (× 40 000) – at this stage the chromosome is at maximum condensed state.

the packaging of DNA in the chromosome

2 a **Explain** a main advantage of chromosomes being 'supercoiled' in metaphase of mitosis.

b **Deduce** the significance of the positively charged histone protein and the negatively charged DNA.

The packaging protein of the nucleosome, called **histone**, is a basic (positively charged) protein containing a high concentration of amino acid residues with additional base groups ($-NH_2$), such as lysine and arginine (see also page 56). In nucleosomes, eight histone molecules combine to make a single bead. Around each bead, the DNA double helix is wrapped in a double loop.

At times of nuclear division, the whole beaded thread is, itself, coiled up, forming the chromatin fibre. The chromatin fibre is again coiled, and the coils are looped around a 'scaffold' protein fibre, made of a **non-histone protein**. This whole structure is folded (supercoiled) into the much-condensed chromosome (Figure 1.60, page 56). Clearly, the nucleosomes are the key structures in the safe storage of these phenomenal lengths of DNA that are packed in the nuclei. However, nucleosomes also allow access to selected lengths of the DNA (particular genes) during transcription – a process we will discuss shortly.

We can conclude that the much smaller genomes of prokaryotes do not require this packaging, for protein is absent from the circular chromosomes of bacteria. Here, the DNA is described as 'naked'.

■ Molecular visualization of DNA

A molecular visualization of DNA was created for the 50th anniversary of the discovery of the double helix. The dynamics and molecular shapes were based on X-ray crystallographic models and other data. You can observe an animation of the 'packaging' of the DNA molecule in nucleosomes at: www.wehi.edu.au/education/wehitv/molecular_visualizations_of_dna/

■ Discovery of the role of DNA as information molecule

Since about 50% of a chromosome consists of protein, it is no wonder that people once speculated that protein of the chromosomes might be the information substance of the cell. For example, there is more chemical 'variety' within a protein than in nucleic acid. However, this idea proved incorrect. We now know that the DNA of the chromosomes holds the information that codes for the sequence of amino acids from which the proteins of the cell cytoplasm are built.

How was this established?

The unique importance of DNA was proved by an experiment with a bacteriophage virus. A **bacteriophage** (or **phage**) is a virus that parasitizes a bacterium. A virus particle consists of a protein coat surrounding a nucleic acid core. Once a virus has gained entry to a host cell, it may take over the cell's metabolism, switching it to the production of new viruses. Eventually, the remains of the host cell breaks down (lysis) and the new virus particles escape – now able to repeat the infection in new host cells. The lifecycle of a bacteriophage, a virus with a complex 'head' and 'tail' structure, is shown in Figure 7.2.

electron micrograph of bacteriophage infecting a bacterium

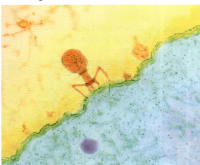

structure of the phage

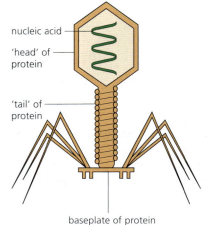

nucleic acid

'head' of protein

'tail' of protein

baseplate of protein

steps to replication of the phage

1 The phage attaches to the bacterial wall and then injects the virus DNA.

2 Virus DNA takes over the host's synthesis machinery.

3 New viruses are assembled and then escape to repeat the infection cycle.

■ **Figure 7.2** The lifecycle of a bacteriophage

Two experimental scientists, Martha Chase and Alfred Hershey, used a bacteriophage that parasitizes the bacterium *Escherichia coli* (Figure 1.32, page 29) to answer the question of whether genetic information lies in the protein (coat) or the DNA (core) (Figure 7.3).

Two batches of the bacteriophage were produced, one with radioactive phosphorus (^{32}P) built into the DNA core (so here the DNA was labelled) and one with radioactive sulfur (^{35}S) built into the protein coat (here the protein was labelled). Note that sulfur occurs in protein, but there is no sulfur in DNA. Likewise, phosphorus occurs in DNA, but there is no phosphorus in protein. So, we can be sure the radioactive labels were specific.

Two identical cultures of *E. coli* were infected, one with the ^{32}P-labelled virus and one with the ^{35}S-labelled virus. Subsequently, radioactively labelled viruses were obtained only from the bacteria infected with virus labelled with ^{32}P. In fact, the ^{35}S label did not enter the host cell at all. Chase and Hershey's experiment clearly demonstrated that it is the DNA part of the virus which enters the host cell and carries the genetic information for the production of new viruses.

■ **Figure 7.3**
The Hershey–Chase experiment

Is it the **protein coat** or the **DNA** of a bacteriophage that enters the host cell and takes over the cell's machinery, so causing new viruses to be produced?

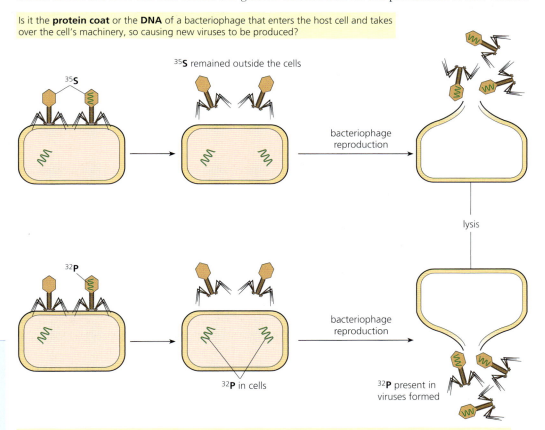

^{35}**S** remained outside the cells

bacteriophage reproduction

lysis

^{32}**P** in cells

^{32}**P** present in viruses formed

Only the DNA part of the virus got into the host cell (and radioactively labelled DNA was present in the new viruses formed). It was the virus DNA that controlled the formation of new viruses in the host, so Hershey and Chase concluded that **DNA carries the genetic message**.

3 **Deduce** what would have been the outcome of the Hershey–Chase experiment (Figure 7.3) if protein had been the carrier of genetic information.

■ Base pairing in the DNA molecule

The structure of the DNA molecule as a double helix, proposed by Watson and Crick in 1953, is illustrated in Figure 2.49 (page 107). A key feature of DNA is **base pairing**.

Discovery of the principle of base pairing by Watson and Crick was the result of their interpretation of the work of Edwin Chargaff. In 1935, Chargaff had analysed the composition of DNA from a range of organisms and found rather remarkable patterns. Apparently, the significance of these patterns was not immediately obvious to Chargaff, though.

His discoveries were:

- the numbers of purine bases (adenine and guanine) always equalled the number of pyrimidine bases (cytosine and thymine)
- the number of adenine bases equalled the number of thymine bases, and the number of guanine bases equalled the number of cytosine bases.

What does this mean?

The organic bases found in DNA are of two distinct types with contrasting shapes:

- cytosine and thymine are pyrimidines or **single-ring** bases
- adenine and guanine are purines or **double-ring** bases.

Only a **purine will fit with a pyrimidine** between the sugar–phosphate backbones, when base pairing occurs (Figure 7.4). So in DNA, adenine must pair with thymine, and cytosine must pair with guanine!

In their paper, published in *Nature* in 1953, in which the molecular structure of DNA was first announced, Crick and Watson observed that DNA was a molecule with 'novel features which are of considerable biological interest'! In a second understatement, the paper concluded, 'It has not escaped our notice that the specific pairing we have postulated immediately suggests a possible copying mechanism for the genetic material'.

We will examine the replication of DNA, shortly.

| **Nature of Science** | **Making careful observations** |

■ X-ray diffraction

Rosalind Franklin and her team at Kings College, London, found that, when X-rays are passed through crystallized DNA, they are scattered to produce a distinctive pattern (Figure 2.50, page 108). X-ray diffraction was a technique with which Franklin was already familiar, from earlier studies of the crystal structures of other molecules. It was by analysis of this X-ray diffraction pattern that the three-dimensional structure of DNA was deduced. As a result of these careful observations, Franklin was able to conclude that the cross at the centre of the X-ray pattern suggested DNA was helical in shape and also gave the pitch of the helix. Other features of the pattern indicated the dimensions of the repeating aspects of the molecule.

Watson and Crick were shown this evidence and the conclusions before Franklin had published (and without her knowledge or permission). Watson and Crick returned to their Cambridge laboratory, combined this knowledge with the fruits of their own calculations and with information from literature searches, and then built their now famous model. Subsequently, they published their paper. Rosalind Franklin was not a member of the group that was later awarded a Nobel Prize for this discovery. She had died by then, and this prize cannot be awarded posthumously.

From the model of DNA in Figure 7.4, we can also see that when A pairs with T, they are held together by two hydrogen bonds; when C pairs with G, they are held by three hydrogen bonds. Only these pairs can form hydrogen bonds. Because of base pairing and the formation of specific hydrogen bonds, the sequence of bases in one strand of the helix determines the sequence of bases in the other – a principle we know as **complementary base pairing**.

4 In base pairing, organic bases are held together (A–T, C–G) by hydrogen bonds. **State** which parts of these organic molecules form the hydrogen bonds.

■ 'Direction' in the DNA molecule

We can identify **direction** in the DNA double helix. The phosphate groups along each strand are bridges between carbon-3 of one sugar molecule and carbon-5 of the next, and one chain runs from 5' to 3' while the other runs from 3' to 5'. (Remember, the carbon atoms of organic molecules can be numbered, page 68.) That is, the two chains of DNA are **antiparallel**, as illustrated in Figure 7.4. The existence of direction in DNA strands becomes important in replication and when the genetic code is transcribed into mRNA.

■ **Figure 7.4**
Direction, base
pairing and hydrogen
bonding in the DNA
double helix

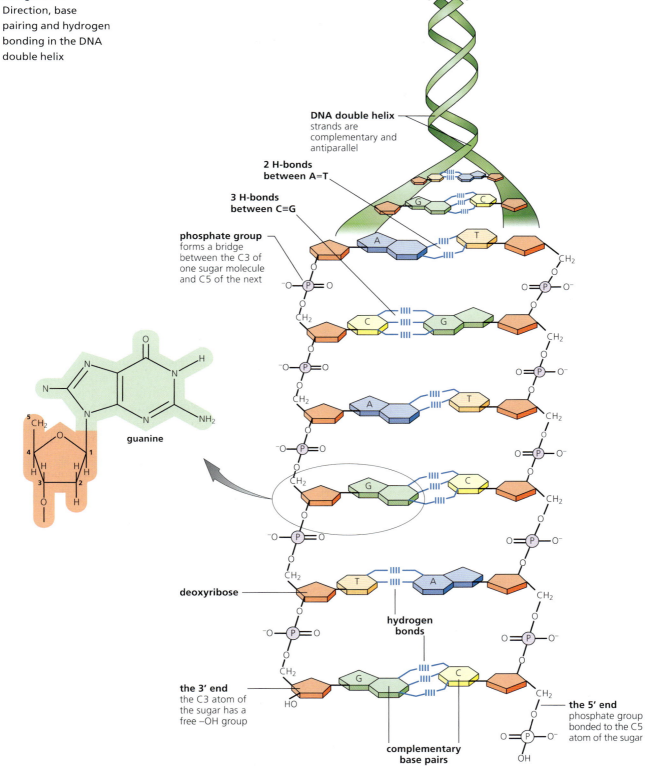

DNA double helix
strands are
complementary and
antiparallel

**2 H-bonds
between A=T**

**3 H-bonds
between C≡G**

phosphate group
forms a bridge
between the C3 of
one sugar molecule
and C5 of the next

guanine

deoxyribose

**hydrogen
bonds**

the 3' end
the C3 atom of
the sugar has a
free –OH group

**complementary
base pairs**

the 5' end
phosphate group
bonded to the C5
atom of the sugar

■ Sequencing DNA – the dideoxyribonucleotide chain termination method

The sequence of nucleotides in a gene (a DNA fragment) can now be determined by machine. This technique was devised by Fredrick Sanger – a remarkable achievement for which he was awarded a second Nobel Prize (his first was for the discovery of the structure of the protein hormone, insulin).

The method relies on introducing a nucleotide called ddNTP (dideoxyribonucleotide triphosphate), which is similar to the nucleotides used by a cell in replication but sufficiently different to stop DNA replication. So, when a ddNTP is added to a growing chain of DNA, DNA polymerase cannot add any more nucleotides; chain growth stops at that point. Since each type of ddNTP used (dd**A**TP, dd**T**TP, dd**C**TP and dd**G**TP) is tagged with a distinct fluorescent label, the identity of the nucleotide that ends each strand is automatically identified.

DNA sequencing is carried out by a machine, by a cyclic process very similar to PCR reaction (page 166). This creates many copies of the gene, but in the presence of the tagged ddNTPs and an excess of regular nucleotides. So, as the DNA polymerase continues to assemble copies, it occasionally picks up a ddNTP instead of the regular nucleotides and replication of that chain is stopped. The result is that many part copies of the gene to be sequenced are formed, all of variable length. These copies are separated in order of size by gel electrophoresis. Then a laser reads the fluorescent tag on each ddNTP to reveal the order of nucleotides in the gene.

You can see an illustration of this advanced technique at the following web address, but remember, the details do not need to be memorized: http://en.wikipedia.org/wiki/File:Sanger-sequencing.svg

■ Genes and 'nonsense' DNA sequences

The **Human Genome Project** (page 135) has established that the three million bases of our chromosomes represent far fewer genes than was originally expected. In fact, protein-coding sequences of our DNA account for only approximately 1.5%. The remainder include some DNA sequences that regulate the expression of protein-coding genes (**regulatory DNA sequences**), but that leaves 70% of our DNA with other roles or none. (At one time these regions were described as 'junk' or 'nonsense' DNA'). In fact, these extensive 'non-gene' regions of eukaryotic chromosomes also consist of:

- **introns**. These are non-coding nucleotide sequences, one or more of which interrupts the coding sequences (exons) of eukaryotic genes (page 173).

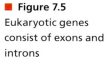

■ **Figure 7.5**
Eukaryotic genes consist of exons and introns

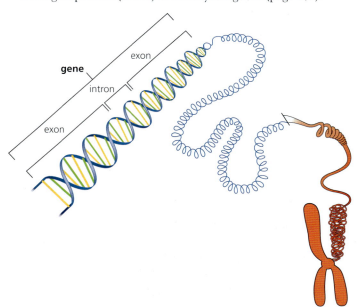

- **telomeres**. These are special nucleotide sequences, typically consisting of multiple repetitions of one short nucleotide sequence. They occur near the ends of DNA molecules and 'seal' the ends of the linear DNA. Here, they stop erosion of the genes that would occur with each repeated round of replication.

- **genes for transfer RNA (tRNA)**. These are parts of the DNA template that code for relatively short lengths of RNA that are formed in the nucleus and pass out into the cytosol. Here they transfer amino acids from the pool there, to supply a growing polypeptide in a ribosome (Figure 7.16, page 335).

- **major lengths of non-coding DNA**. Today, these are important to genetic engineers, for they are exploited in DNA profiling. They are short sequences of bases that are repeated very many times. Where they often occur together in major clusters, they are known as '**variable number tandems repeat**' regions (**VNTRs**). Their use in genetic fingerprinting (DNA profiling) is described in Chapter 3 (page 176).

TOK Link

Highly repetitive DNA sequences were once described as 'junk DNA', showing a degree of confidence that they had no role. To what extent do the labels and categories used in the pursuit of knowledge affect the knowledge that we obtain?

■ Replication – DNA copying itself

The way that DNA of the chromosome is replicated was introduced in Chapter 2 (page 109). The key features are listed here.

- Replication must be an extremely accurate process, since DNA carries the genetic message.
- Replication is quite separate from cell division – replication of DNA takes place in the **interphase** nucleus, well before the events of nuclear division.
- Strands of the DNA double helix are built up individually from free **nucleotides** (the structure of a nucleotide is shown in Figure 2.47, page 105).
- Before nucleotides can be condensed together, the DNA double helix has to **unwind**, and the hydrogen bonds holding the strands together must be broken, allowing the two strands of the helix to separate.
- The enzyme **helicase** brings about the unwinding process and holds the strands apart for replication to occur.
- **Both strands act as templates**; nucleotides with the appropriate complementary bases line up opposite the bases of the exposed strands (A with T, C with G).
- **Hydrogen bonds form** between complementary bases, holding the nucleotides in place.
- Finally, the sugar and phosphate groups of adjacent nucleotides of the new strand condense together, catalysed by the enzyme **DNA polymerase**. Replication always occurs in the $5' \rightarrow 3'$ direction.
- The two strands of each DNA molecule wind up into a double helix.
- One strand of each new double helix came from the parent chromosome and one is a newly synthesized strand, an arrangement known as **semi-conservative replication** because half the original molecule is conserved.
- DNA polymerase also has a role in '**proof-reading**' the new strands; any mistakes that start to happen (such as the wrong bases attempting to pair up) are immediately corrected, so each new DNA double helix is exactly like the original.

So, in DNA replication, both strands of the double helix serves as a template for synthesis of a new strand – that is, semi-conservatively. The evidence that DNA replication is **semi-conservative** came from the experiment conducted by Meselson and Stahl, shown in Figure 2.53, page 110). *Look at that illustration now.*

Replication forks and the system of enzymes involved in replication

Semi-conservative replication is actually initiated at many points along the DNA double helix. These points are known as **replication forks**. Here, the DNA strands separate (a 'bubble' forms), brought about by the enzyme **helicase**. (Another enzyme, **DNA gyrase**, assists in overcoming the strains that come as the double-stranded DNA is unwound.) Then, **single-strand binding proteins** attach and prevent the separated strands from repairing. The unwound sections of both strands are now ready to act as templates for the synthesis of complementary DNA strands.

Both strands are replicated simultaneously. However, since DNA polymerase (known as **DNA polymerase III**) can add nucleotides only to the free 3' end, the DNA strands can elongate only in the 5'→3' direction. Consequently, the details of the replication process differ in the two strands. Figure 7.6 illustrates the steps, and Table 7.1 lists the enzymes involved.

The leading strand

The exposed 5'→3' strand is referred to as the **leading strand**. Here, **DNA polymerase III** adds nucleotides by complementary base pairing to the free 3' end of the new strand, in the same direction as the replication fork. This process proceeds **continuously**, immediately behind the advancing helicase, as fresh template is exposed. The initial nucleotide chain formed is actually a short length of RNA called a primer. This primer is synthesized by an enzyme, **primase**. Then the new DNA starts from the 3' end of the RNA primer.

The lagging strand

In contrast to events in the leading strand, here the replication is **discontinuous**. A series of relatively short lengths of DNA (fragments), called **Okazaki fragments** are formed, each one primed separately.

So, first an RNA primer is formed by **primase**, and then **DNA polymerase III** attaches nucleotides to it, forming a fragment. Next, **DNA polymerase I** replaces the RNA nucleotides at the start of each fragment with DNA nucleotides. Finally, the enzyme **ligase** joins the Okazaki fragments together.

In this way, short lengths of DNA are synthesized and joined together.

It may help to follow these steps in Figure 7.6.

■ **Figure 7.6**
The steps of DNA replication

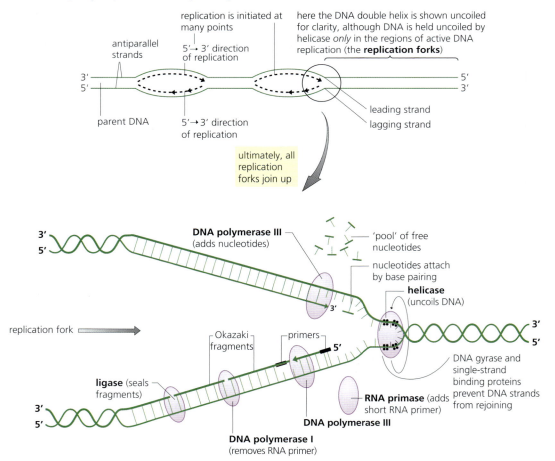

■ **Table 7.1**
The enzymes that bring about DNA replication

1 Formation of a replication fork	
helicase enzyme	separates the two strand of DNA to expose a replication fork and prevents them rejoining
DNA gyrase enzyme	
single-strand binding proteins	
2 a) DNA replication in the leading strand – a continuous process	
primase	forms a single short length of RNA primer
DNA polymerase III	forms the DNA strand, beginning at the RNA primer
2 b) DNA replication in the lagging strand – a discontinuous process	
RNA primase	forms short lengths of RNA primer at intervals along the DNA strand
DNA polymerase III	forms short DNA strands (Okazaki fragments), starting from each RNA primer
DNA polymerase I	replaces the RNA primer at the start of each Okazaki fragment with a DNA strand
ligase	joins the DNA stands together

5 Explain what is meant by antiparallel strands.

7.2 Transcription and gene expression – *information stored as a code in DNA is copied onto mRNA*

Transcription occurs in the nucleus. During transcription, a complementary copy of the information in a part of the DNA molecule (a gene) is made by the building of a molecule of **messenger RNA (mRNA)**. In the process, the DNA **triplet codes** are transcribed into **codons** in the mRNA. This process is catalysed by the enzyme **RNA polymerase** (Figure 7.7).

In transcription, only one strand of the DNA double helix serves as a **template** for synthesis of mRNA. This is called the antisense or **coding strand**. The DNA double helix first unwinds and the hydrogen bonds are broken at the site of the gene being transcribed.

Next, the enzyme RNA polymerase recognizes and binds to a **promoter region**, the 'start' signal for transcription. This is located immediately before the gene.

RNA polymerase now draws on the pool of **free nucleotides**. As with DNA replication, these nucleotides are present in the form of nucleoside triphosphates (in RNA synthesis uridine triphosphate replaces thymidine triphosphate). Transcription is a totally accurate process because of complementary base pairing.

The polymerase enzyme matches free nucleotides (A with U, C with G), working in the 5'→3' direction. Note that in RNA synthesis, it is uracil which pairs with adenine. Hydrogen bonds then form between complementary bases, holding the nucleotides in place.

Finally, each selected free nucleotide, in turn, is joined onto the growing mRNA strand by condensation reaction. It is the sugar and phosphate groups of adjacent nucleotides that are condensed together by the enzyme RNA polymerase.

The whole process continues until a base sequence known as the **transcription termination region** is reached. At this signal, both RNA polymerase and the completed new strand of mRNA are freed from the site of the gene.

Once the mRNA strand is free, it leaves the nucleus through pores in the nuclear membrane and passes to tiny structures in the cytoplasm called **ribosomes** where the information can be 'read' and is used (page 23).

The DNA double strand once more reforms into a compact helix at the site of transcription.

6 Explain when 'direction' in the DNA molecule becomes important in a replication, and b transcription.

7 Define the following terms:

transcription   translation   antisense (coding) strand   sense strand.

■ **Figure 7.7**
Gene transcription
in the eukaryote
chromosome

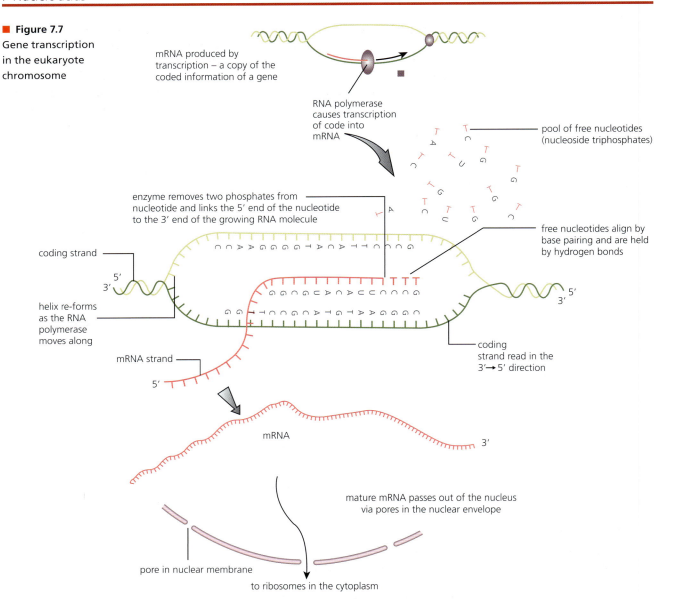

mRNA produced by
transcription – a copy of the
coded information of a gene

RNA polymerase
causes transcription
of code into
mRNA

pool of free nucleotides
(nucleoside triphosphates)

enzyme removes two phosphates from
nucleotide and links the 5′ end of the nucleotide
to the 3′ end of the growing RNA molecule

free nucleotides align by
base pairing and are held
by hydrogen bonds

coding strand

helix re-forms
as the RNA
polymerase
moves along

coding
strand read in the
3′ → 5′ direction

mRNA strand

mRNA

mature mRNA passes out of the nucleus
via pores in the nuclear envelope

pore in nuclear membrane

to ribosomes in the cytoplasm

■ Some eukaryotic genes are discontinuous

The Human Genome Project (and similar studies of the genomes of other organisms) has
established the sequence of bases in very many genes (page 135). This has led to a discovery
of some non-coding regions in the base sequences of genes. Many of the genes of eukaryotes
have non-coding DNA sequences within their boundaries. The sections of the gene that carry
meaningful information (code for amino acids) are called **exons**. The non-coding sequences that
intervene – interruptions, in effect – are called **introns**. While genes split in this way are very
common in higher plants and animals, some eukaryotic genes contain no introns at all.

When a gene consisting of exons and introns is transcribed into mRNA, the mRNA formed
contains the sequence of introns and exons, exactly as they occur in the DNA. Now, you can
see that, if this unmodified mRNA was to be 'read' and transcribed in a ribosome, it would
undoubtedly present problems in the protein-synthesis step.

In fact, an enzyme-catalysed reaction, known as **post-transcriptional modification**, removes
the introns as soon as the mRNA has been formed (Figure 7.8); the production of this enzyme

is also under the control of a gene. As a result, the short lengths of 'nonsense' transcribed into the RNA sequence of bases are removed. This is known as **RNA splicing** and the resulting shortened lengths of mRNA are described as **mature**. It this form of mRNA that passes out into the cytoplasm, to the ribosomes, where it is involved in protein synthesis (page 110).

Incidentally, the genes of prokaryotes do not have introns, and the prokaryotic cell does not have the enzyme machinery to carry out splicing, either. This means, when a genetic engineer plans to place a copy of a eukaryotic gene in the chromosome of a bacterium, that copy has to be intron-free (pages 169 and 173).

■ **Figure 7.8**
Post-transcriptional modification of mRNA

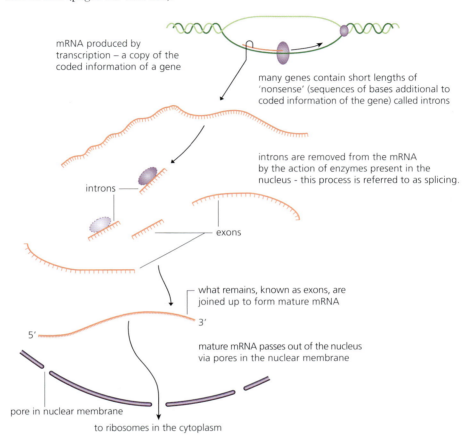

mRNA produced by transcription – a copy of the coded information of a gene

many genes contain short lengths of 'nonsense' (sequences of bases additional to coded information of the gene) called introns

introns are removed from the mRNA by the action of enzymes present in the nucleus - this process is referred to as splicing.

introns

exons

what remains, known as exons, are joined up to form mature mRNA

5'

3'

mature mRNA passes out of the nucleus via pores in the nuclear membrane

pore in nuclear membrane

to ribosomes in the cytoplasm

8 **State** the sequence of changes catalysed by RNA polymerase.

■ Regulation of gene expression

Organisms control which of their genes are **expressed** at any one moment. Remember, the cells in an organism all contain the same genome, whatever the fate of a particular cell. So, in a multicellular organism, every nucleus contains the coded information relating to the development and maintenance of *all* mature tissues and organs. Both during development and later, this genetic information is used selectively. Typically, less than 25% of the protein-coding genes in human cells are expressed at any time. The expression of genes is related to when and where the proteins they code for are needed (Table 7.2).

■ **Table 7.2**
Gene regulation

When and why genes are expressed	
expressed all the time	genes responsible for routine and continuous metabolic functions, e.g. respiration which is common to all cells and is continuous throughout life
expressed at a selected stage in cell or tissue development	e.g. as cells derived from stem cells are developing into muscle fibres or neurons
expressed only in the mature cell	e.g. genes responsible for antibody production in a mature plasma cell, after these have been cloned
expressed on receipt of an internal or external signal	e.g. when a particular hormone signal, metabolic signal or nerve impulse is received by the cell and activates a gene, such as the gene for insulin production in β cells in the islets of Langerhans

A mechanism for control of gene expression was first discovered in **prokaryotes**. However, this mechanism does not operate in eukaryotes, so the details do not concern us here.

How eukaryotic genes are regulated

1 Regulation and chromatin structure

We have seen that the DNA in the nucleus is packaged with proteins in a complex known as chromatin. Here, the much-coiled DNA double helix of each chromosome is looped around histone protein beads (**nucleosomes**, Figure 7.1). These nucleosomes are stable protein–DNA complexes, but they are not static. For example, they can inhibit or facilitate transcription. Chemical modification of histone protein plays a direct role in the regulation of gene transcription.

The ends of histone protein molecules project outwards for the nucleosome (Figure 7.9). Histone 'tails' may be chemically modified by enzymes, by the addition or removal an acetyl group ($-CO.CH_3$). It is histone acetylation, the addition of this group to a particular amino acid (lysine) at the end of the histone tails, that loosens the tight binding of the nucleosomes. Where this loosening has occurred, transcription enzymes and other proteins have access. This was previously impossible, when the nucleosomes were tightly bound together. So, histone acetylation is a first step in the initiation of transcription of a gene. However, the question remains 'What triggers acetylation in a particular part of a chromosome, in the first place?'

■ **Figure 7.9**
Histone tails, acetylation and transcription

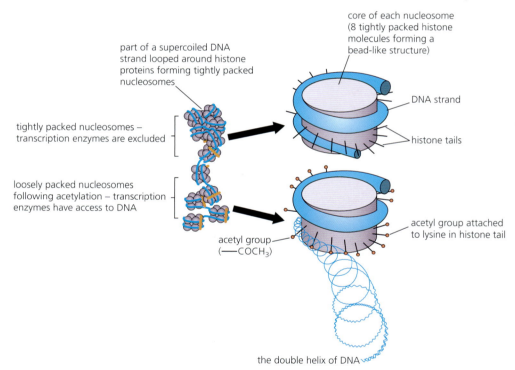

part of a supercoiled DNA strand looped around histone proteins forming tightly packed nucleosomes

core of each nucleosome (8 tightly packed histone molecules forming a bead-like structure)

tightly packed nucleosomes – transcription enzymes are excluded

DNA strand

histone tails

loosely packed nucleosomes following acetylation – transcription enzymes have access to DNA

acetyl group ($-COCH_3$)

acetyl group attached to lysine in histone tail

the double helix of DNA

2 Regulation by enhancers and transcription factors

In **eukaryotes**, before mRNA can be transcribed by the enzyme **RNA polymerase**, it first binds together with a small group of proteins called **general transcription factors** at a sequence of bases known as the **promoter**. Promoter regions occur on DNA strands just before the start of a gene's sequence of bases. (The promoter is an example of a length of non-coding DNA with a special function.) Only when this transcription complex of proteins (enzyme plus factors) has been assembled, can transcription of the template strand of the gene begin. Once transcription has been initiated, the RNA polymerase moves along the DNA, untwisting the helix as it goes and exposing the DNA nucleotides, so that RNA nucleotides can pair and the messenger RNA strand be formed and peel away.

The rate of transcription may be increased (or decreased) by the binding of **specific transition factors** on the enhancer site for the gene. The position of an **enhancer** site of a gene is shown in Figure 7.10. This is at some distance 'upstream' of the promoter and gene sequence. However, when activator proteins bind to this enhancer site, a new complex is formed and makes contact with the polymerase–transition factor complex. Then the rate of gene expression is increased.

Regulator transcription factors, activators and RNA polymerase are all proteins, and are coded for by other genes. It is clear that protein–protein interactions play a key part in the initiation of transcription.

■ **Figure 7.10**
Control sites and the initiation of transcription

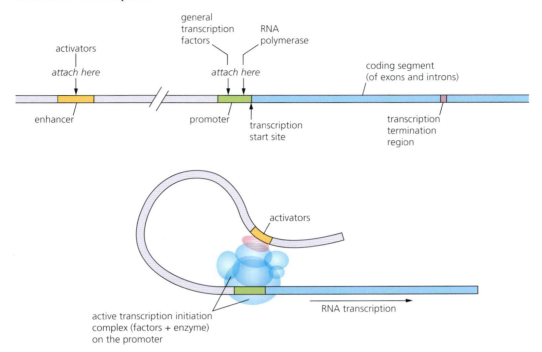

The lower part of the figure shows a portion of the upper DNA molecule with the activator (orange) region, promotor (green) region and coding (blue) region. The DNA is looped so that it comes into contact with different parts of the transcription initiation complex. The complex consists of several proteins and other factors hence the multiple overlapping spheres (blue). The activator (pink) is also part of the complex. These spheres have been drawn transparent so that contact with the DNA molecule can be shown.

3 Methylation and epigenetic inheritance – analysing patterns

Methylation is the reversible addition of a methyl group ($-CH_3$) within the chromatin, possibly to histone tails, but usually to the DNA molecule itself. Enzymes bring about this addition to the base cytosine (Figure 7.11). The addition occurs while the DNA is wrapped around histone. The effect is to change the activity of the gene – usually extensive methylation inactivates a gene. Removal of methyl groups may turn genes back on again.

Once a gene has been methylated, it may remain in this condition. The methyl groups persist *in situ* from cell division to cell division. Furthermore, external conditions adverse to the cell or organism can leave their mark on our DNA through extensive methylations, leading to switched-off genes.

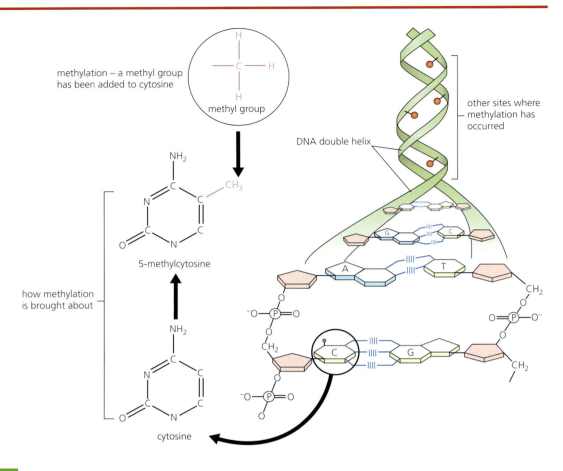

Figure 7.11
Methylation of DNA

methylation – a methyl group has been added to cytosine

methyl group

DNA double helix

other sites where methylation has occurred

NH_2

CH_3

5-methylcytosine

how methylation is brought about

NH_2

cytosine

Nature of Science	**Looking for patterns, trends and discrepancies**

A case study in epigenetic inheritance?

It now appears that, not only does methylation last a lifetime, it can be, and often is, transmitted to offspring – and sometimes to further generations. The environment of a cell and an organism may have an impact on gene expression for generations.

In 1944, communities in the western Netherlands were deprived of food supplies as a result of war-time hostilities for a period in excess of 6 months. Many people died of extreme starvation, but among the survivors were pregnant mothers who gave birth – in fact, thousands of malnourished and underweight children were born. In later years many of these survivors became parents themselves. Despite the fact these people and their offspring were continuously satisfactorily fed once hostilities were over, their own children, born many years later, were significantly underweight. It seems their parent starvation in infancy was imprinted on their children's DNA.

From the time Mendel deduced the existence of 'factors' (page 150), through to Crick and Watson's discovery of the nature of the gene in chemical terms (pages 108 and 317), genes and alleles were seen as unchangeable by external factors. The effects of environmental change were believed to be restricted to which genes (alleles) survived and contributed to the gene pool of future generations. The environment's impact was held to be on selection, not gene performance. Today, it seems bad diets (for example) can interfere with the performance of genes in succeeding generations, presumably as a result of 'markers' attached to parent DNA in earlier times.

Epigenetics is the study of heritable changes in gene activity that are *not* caused by changes in the DNA base sequences (*epi* = outside). Examples of mechanisms that produce such changes are DNA methylation and histone modifications. Notice that these modifications affect the way cells 'read' the genes, rather than alter the base sequences of DNA itself (which we would describe as a mutation).

Alternative RNA splicing, after transcription?

We have noted that change occurs to messenger RNA immediately after it has been formed by transcription, in that introns are removed. However, the remaining lengths of mRNA (exons) may be spliced together in different combinations. The consequence is that a single gene can code for more than one type of polypeptide. In fact, many genes give rise to two or more different polypeptides, depending on the order in which exons are assembled (Figure 7.12). Even more variety in gene products may result if one or more introns are treated as an exon, during RNA processing.

This process, also part of the story of how gene action is regulated, is under the control of specific genes, although these are often found on other chromosomes.

■ **Figure 7.12** Alternative RNA splicing

As a result of alternative mRNA splicing, the number of proteins produced can be greater than the number of genes present.

Alternative mRNA splicing may explain why the human genome consists of the same (low) number of genes as some small non-vertebrate animals have.

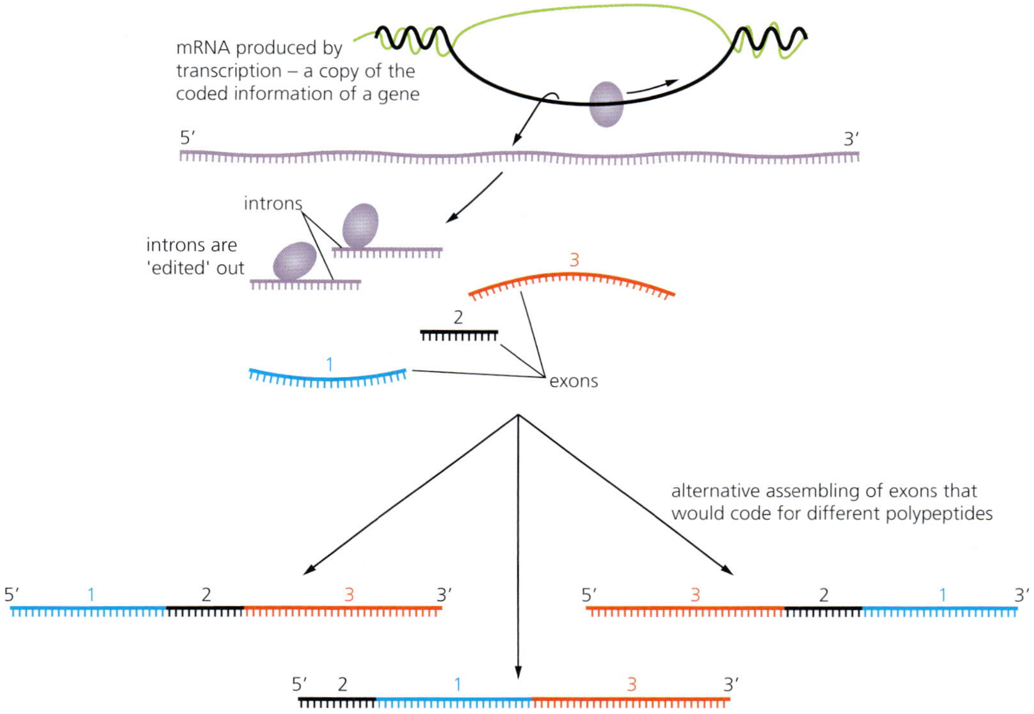

mRNA produced by transcription – a copy of the coded information of a gene

5'   3'

introns

introns are 'edited' out

3

2

1

exons

alternative assembling of exons that would code for different polypeptides

5' 1 2 3 3'   5' 3 2 1 3'

5' 2 1 3 3'

7.3 Translation – *information transferred from DNA to mRNA is translated into an amino acid sequence*

In the eukaryotic cell, the mature mRNA strand leaves the nucleus through pores in the nuclear membrane and passes to ribosomes in the cytoplasm. Here, information transferred from DNA to mRNA is translated into amino acid sequences of proteins.

■ Assembly of the components of translation

Activation of amino acids

Some 20 different amino acids are involved in protein synthesis. Amino acids are activated for protein synthesis by combination with a short length of a different sort of RNA, called transfer RNA (tRNA, Figure 7.13). The activation process involves ATP. There are 20 different tRNA molecules, one for each of the amino acids coded for in proteins.

All tRNAs molecules have a clover-leaf shape, but they differ in the sequence of bases, known as the anticodon, which is exposed on one of the 'clover leaves'. This anticodon is complementary to a codon of mRNA. The enzyme catalysing the formation of the amino acid–tRNA complex 'recognizes' only one type of amino acid and the corresponding tRNA.

■ **Figure 7.13**
Transfer RNA (tRNA) and amino acid activation

1 tRNA structure
– a 'clover-leaf' shape

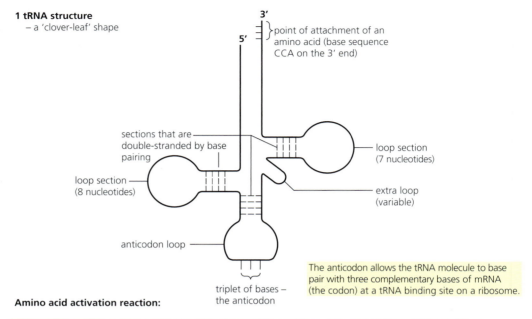

point of attachment of an amino acid (base sequence CCA on the 3′ end)

sections that are double-stranded by base pairing

loop section (7 nucleotides)

loop section (8 nucleotides)

extra loop (variable)

anticodon loop

triplet of bases – the anticodon

The anticodon allows the tRNA molecule to base pair with three complementary bases of mRNA (the codon) at a tRNA binding site on a ribosome.

Amino acid activation reaction:

Each amino acid is linked to a specific tRNA before it can be used in protein synthesis by the action of a tRNA-activating enzyme (there are 20 different tRNA-activating enzymes, one for each of the 20 amino acids).

2 role of tRNA-activating enzyme, illustrating enzyme–substrate specificity and the role of phosphorylation

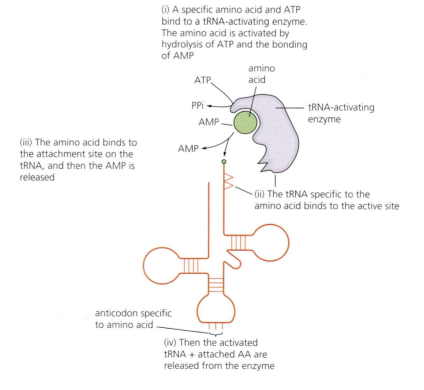

(i) A specific amino acid and ATP bind to a tRNA-activating enzyme. The amino acid is activated by hydrolysis of ATP and the bonding of AMP

ATP

amino acid

PPi

AMP

tRNA-activating enzyme

AMP

(iii) The amino acid binds to the attachment site on the tRNA, and then the AMP is released

(ii) The tRNA specific to the amino acid binds to the active site

anticodon specific to amino acid

(iv) Then the activated tRNA + attached AA are released from the enzyme

tRNA-activating enzyme, enzyme–substrate specificity and the role of ATP

Each of the amino acids is attached to the 3' terminal of its specific tRNA molecule by a tRNA-activating enzyme. Remember, this enzyme is specific to the particular amino acid, as well as to a particular tRNA molecule. This specificity is a property of the structure of the enzyme's active site. The structure of a tRNA molecule and the steps to amino acid activation are shown in Figure 7.13.

- A specific amino acid and a molecule of ATP bind to a tRNA-activating enzyme. The amino acid is activated by hydrolysis of ATP and the bonding of AMP (adenosine monophosphate). P-Pi is released.
- The tRNA specific to the amino acid binds to the active site of the enzyme.
- The amino acid binds to the attachment site on the tRNA, and then the AMP is released.
- Then the activated tRNA with attached amino acid is released from the enzyme.

Ribosomes – the site of protein synthesis

A ribosome consists of a large and a small subunit, both composed of RNA (known as rRNA) and protein (Figure 7.14). Within the ribosome are three sites where the tRNAs interact:

- **A site** – the first site. Here, a codon of the incoming mRNA binds to specific tRNA–amino acids through its anticodon (**complementary base pairing**).
- **P site** – the second site. Here, the amino acid attached to its tRNA is condensed with the growing polypeptide chain by **formation of a peptide linkage**.
- **E site** – the third site. Here, the **tRNA leaves** the ribosome, following transfer of its amino acid to the growing protein chain.

■ **Figure 7.14**
Ribosome structure and function

E site:	P site:	A site:
here the discharged tRNA exits from ribosome	holds the tRNA while peptide bonds form	this is the amino acid–tRNA binding site

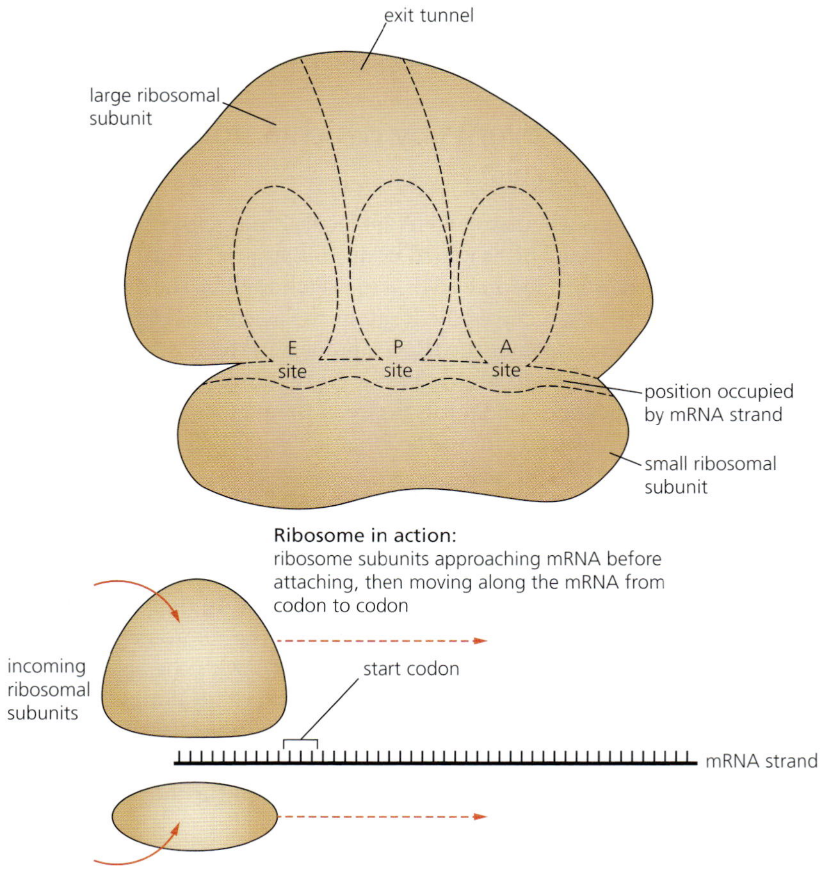

exit tunnel

large ribosomal subunit

E site   P site   A site

position occupied by mRNA strand

small ribosomal subunit

Ribosome in action:
ribosome subunits approaching mRNA before attaching, then moving along the mRNA from codon to codon

incoming ribosomal subunits

start codon

mRNA strand

The use of molecular visualization software to observe the structure of a ribosome and tRNA

Images of the three-dimensional molecular structure of biological molecules can be accessed via 'The protein data bank' (PDB): www.rcsb.org/pdh/home/home.do

Alternative sources are Proteopedia: http://proteopedia.org/wiki/index.php/Main_Page
Bioinformatics: http://bioinformatics.org/firstglance/fgij/index.htm

Figure 7.15 shows a eukaryotic ribosome and tRNA. Images may be downloaded and investigated using *Jmol*.

The ribosome model **tRNA molecule**

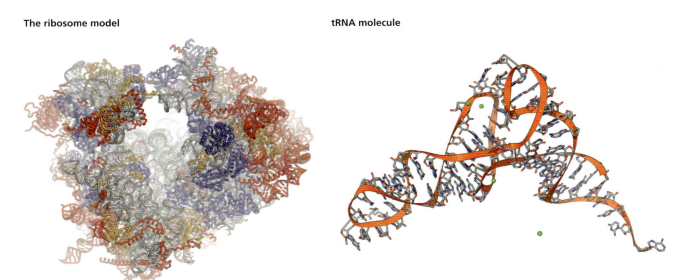

■ **Figure 7.15** Computer simulation of a eukaryotic ribosome and tRNA molecule, obtained by molecular visualization software

> **9** **Explain** in which type of RNA you would expect to find a) a codon and b) an anticodon.

■ The cycle of events by which a polypeptide is assembled

Initiation of translation

Translation begins when an mRNA molecule binds with the small ribosomal subunit at an mRNA binding site. This is joined by an initiator tRNA (to which the amino acid methionine is attached) at the start codon 'AUG'. This is followed by the attachment of a large ribosomal unit. The initiator tRNA occupies the P site in the assembled ribosome.

Now, the next codon of the tRNA, present in the A site, is available to a tRNA with the appropriate anticodon. Its arrival brings two activated amino acids (in sites P and A) into a position and a peptide bond forms between them by condensation reaction. The reaction is catalysed by enzymes present in the large subunit. A dipeptide has now been formed.

> **10** **Draw** and label the structure of a peptide linkage between two amino acids.

Elongation of the peptide

The ribosome moves three bases along the mRNA. In the process, the tRNA in the P site moves to the E site, and is released. This movement of the ribosome also brings the next codon to occupy the now vacant A site – allowing a tRNA with the appropriate anticodon to bind to that codon. This, in turn, brings a further amino acid to lie alongside and, while these amino acids are held close together, another peptide bond is formed. In this way, the ribosome progresses along the mRNA molecule in the 5'→3' direction, codon by codon. By these steps, constantly repeated, a polypeptide is formed and emerges from the large subunit (Figure 7.16).

Initiation of translation

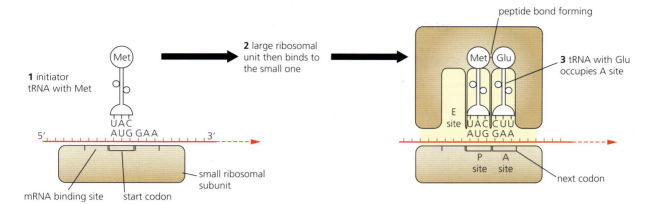

Elongation of the polypeptide

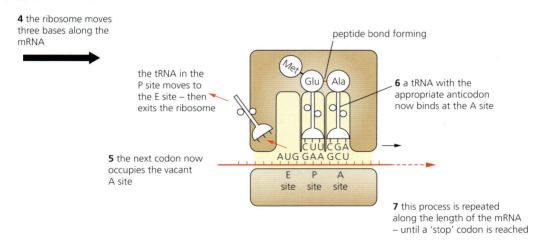

Termination of translation

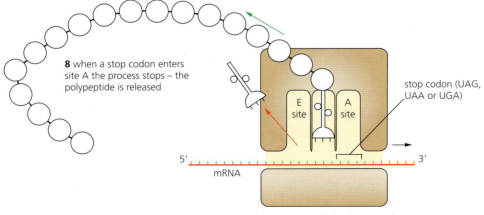

Note: the movement of the ribosome along the mRNA has been from the 5' to the 3' end

■ **Figure 7.16** Initiation, elongation and termination in translation

Termination of translation

Eventually a 'stop' codon is reached. This takes the form of one of three codons – UAA, UAG or UGA. At this point, the completed polypeptide is released from the ribosome into the cytoplasm. Disassembly of the components of the ribosome follows termination of translation.

■ Polysomes

It is common for several of the free-floating ribosomes to move along the same mRNA strand at one time. The resulting structure (mRNA, ribosomes and their growing protein chains) is called a **polysome** (Figure 7.17).

■ **Figure 7.17**
Electron micrograph of polysomes (×100 000)

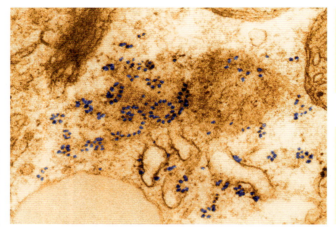

This EM of part of a leaf cell of *Vicia* (bean) shows several **polysomes**. Each polysome consists of a row of individual **ribosomes** (here stained blue) moving along a strand of spirally shaped mRNA. The outcome is that many copies of particular polypeptides are produced simultaneously.

In contrast to the way that in eukaryotes the polysomes appear linearly along the separate mRNA strands, in prokaryotes multiple ribosomes can be seen at each gene on the DNA, emerging on the mRNA thread from multiple points (genes) on the DNA. See also 'Protein synthesis in prokaryotes' on the next page.

11 Distinguish between the different forms of protein in the nucleus.

■ Post-translational modification of protein

When a protein exits from translation at a ribosome, it may take up its active three-dimensional shape and be functional immediately. For example, it may be active as an enzyme in the cytoplasm in some essential and continuous biochemical pathway. On the other hand, many proteins are produced in the form of inactive precursors, which require processing steps. These steps occur after translation and, so, are known as **post-translational modifications** (Figure 7.18). In fact, it is often important for proteins to become active only at particular sites. For example, the protein-digesting enzyme trypsin is produced in the pancreas in an inactive form (trypsinogen) which does not digest the proteins of the pancreas cells in which it is formed.

■ **Figure 7.18**
Protein synthesis and post-translational modification

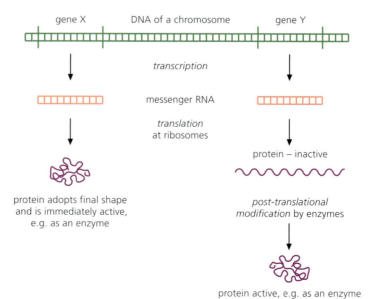

The roles of free ribosomes and those of rough endoplasmic reticulum

The free ribosomes, present in large numbers in the cytosol of the cell, synthesize the huge range of proteins that are used within the cell. However, proteins are also formed in cells for export – to be secreted outside the plasma membrane. These proteins exclusively originate in ribosomes attached to the endoplasmic reticulum and are transported in tiny vesicles budded off from the RER (see Figure 1.26, page 23). Vesicles containing polypeptides may fuse with the plasma membrane for subsequent secretion. Alternatively, they may fuse with other membranous organelles within the cytoplasm, such as the Golgi apparatus or lysosomes, where proteins are also required.

■ Protein synthesis in prokaryotes

The steps of protein synthesis in prokaryotes are very similar to those in eukaryotes, but prokaryote ribosomes are characteristically smaller than those of eukaryotes (page 28). Another difference between eukaryotes and prokaryotes is that, in the latter, there is no (nuclear) membrane between the chromosome, where mRNA is formed, and the cytoplasm. So, here, protein synthesis can begin immediately the mRNA is released. In eukaryotes the mRNA is typically modified (Figures 7.8 and 7.12) before it passes out of the nucleus via pores in the nuclear membrane and is exposed to the ribosomes.

Nature of Science **Developments in scientific research follow improvements in computing**

■ Development of the new discipline of bioinformatics

Bioinformatics is the storage, manipulation and analysis of biological information via computer science. At the centre of this development is the creation and maintenance of databases concerning nucleic acid sequences and the proteins derived from them.

Every individual has a unique **proteome** (page 92). The proteome is the entire set of proteins expressed by the genome of the individual organism. Since the genome, the whole of the genetic information of an organism, is unique to each individual, the proteome it causes to be expressed is also unique. Proteomics is the study of the structure and function of the entire set of proteins of organisms. It is a development within bioinformatics.

As an off-shoot of the Human Genome Project (page 135), the **genomes** of many prokaryotes and eukaryotes have been sequenced, as well as that of humans. The use of computers has enabled scientists to make advances such as locating genes within genomes. This huge volume of data requires organization, storage and indexing to make practical use of the subsequent analyses. These tasks involve applied mathematics, informatics, statistics and computer science.

■ Proteins

Amino acids – the building blocks

We have seen how proteins are initially built up by condensation reactions between **amino acids**, taking place within ribosomes.

The structure of amino acids was introduced in Chapter 2, page 90. The reactive or **functional groups** of the amino acid molecule are the basic amino group ($-NH_2$) and the acidic carboxyl group ($-COOH$), both attached to the same carbon atom (Figure 2.30, page 90). It is these groups that participate in the condensation reactions that form peptide bonds, linking amino acid residues in polypeptides and proteins.

The remainder of the amino acid molecule, the side chain or **–R part**, may be very variable. Something of the variety of amino acids is shown in Figure 7.19. While amino acids have the same basic structure, they are all rather different in character, because of the different R groups they carry. Categories of amino acids found in cell proteins are:

- **acidic amino acids**, having additional carboxyl groups (e.g. aspartic acid)
- **basic amino acids**, having additional amino groups (e.g. lysine)
- **amino acids with hydrophilic properties** (water soluble) have polar R groups (e.g. serine)
- **amino acids with hydrophobic properties** (insoluble) have non-polar R groups (e.g. alanine).

The combining together of different amino acids in contrasting combinations produces proteins with very different properties. The amino acid residues of proteins affect their shape, where they occur in cells and the functions they carry out.

■ **Figure 7.19**
The range of amino acids used in protein synthesis

acidic amino acids
(additional carboxyl group):
e.g. **aspartic acid**

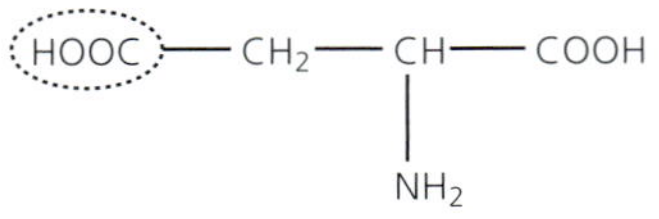

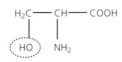

HOOC ── CH$_2$ ── CH ── COOH
                          |
                         NH$_2$

basic amino acids
(additional amino group):
e.g. **lysine**

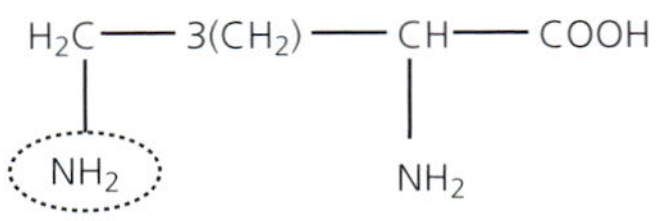

H$_2$C ── 3(CH$_2$) ── CH ── COOH
   | |
  NH$_2$ NH$_2$

acidic amino acids with hydrophilic properties
(water soluble, polar R group):
e.g. **serine**

H$_2$C ── CH ── COOH
   | |
  HO   NH$_2$

amino acids with hydrophobic properties
(insoluble, non-polar R group):
e.g. **alanine**

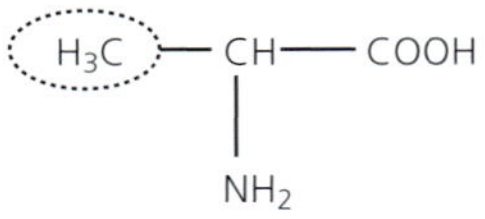

H$_3$C ── CH ── COOH
           |
          NH$_2$

The structure of proteins

There are four levels of protein structure, each of significance in biology.

- The **primary structure** of a protein is the sequence of amino acid residues linked by peptide linkages. Proteins differ in the variety, number and order of their constituent amino acids. We have seen how, in the living cell, the sequence of amino acids in the polypeptide chain is controlled by the coded instructions stored in the DNA of the chromosomes in the nucleus, mediated via mRNA. Changing just one amino acid in the sequence of a protein alters its properties, often quite drastically. This sort of mistake arises by mutation (page 132).

- The **secondary structure** of a protein develops when parts of the polypeptide chain take up a particular shape, immediately after formation at the ribosome. Parts of the chain become folded or twisted, or both, in various ways.
  The most common shapes are formed either by coiling to produce an alpha **helix** or folding into beta **sheets** (Figure 7.20). These shapes are permanent, held in place by hydrogen bonds.

- The **tertiary structure** of a protein is the precise, compact structure, unique to that protein, which arises when the molecule is further folded and held in a particular complex shape. This shape is stabilized by interactions between R groups, established between adjacent parts of the chain (Figure 7.21). The primary, secondary and tertiary structure of the protein lysozyme is shown in Figure 7.22.

- The **quaternary structure** of protein arises when two or more polypeptide chains or proteins are held together, forming a complex, biologically active molecule. An example is hemoglobin, consisting of four polypeptide chains, held around a non-protein heme group (known as a **prosthetic group**), in which an atom of iron occurs (Figure 7.23).

α **helix** (rod-like)   β **sheets**

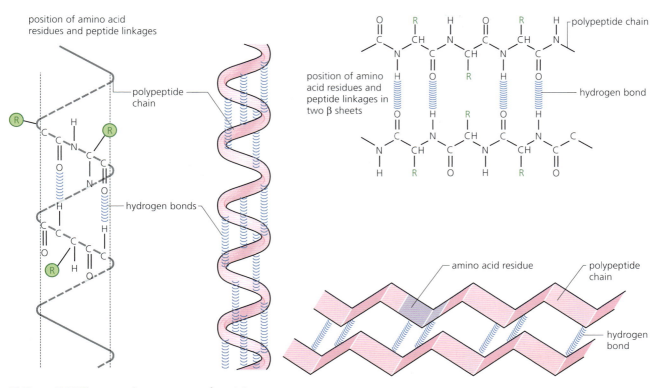

■ **Figure 7.20** The secondary structure of protein

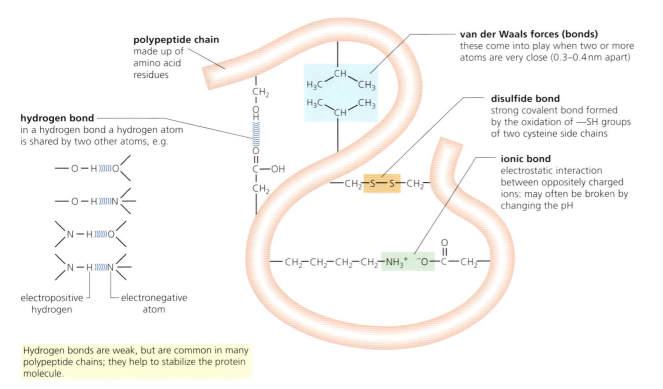

Hydrogen bonds are weak, but are common in many polypeptide chains; they help to stabilize the protein molecule.

■ **Figure 7.21** Cross-linking within a polypeptide

■ **Figure 7.22**
Lysozyme – primary,
secondary and
tertiary structure

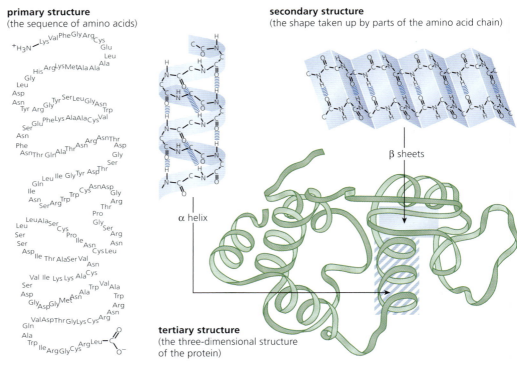

primary structure
(the sequence of amino acids)

secondary structure
(the shape taken up by parts of the amino acid chain)

β sheets

α helix

tertiary structure
(the three-dimensional structure
of the protein)

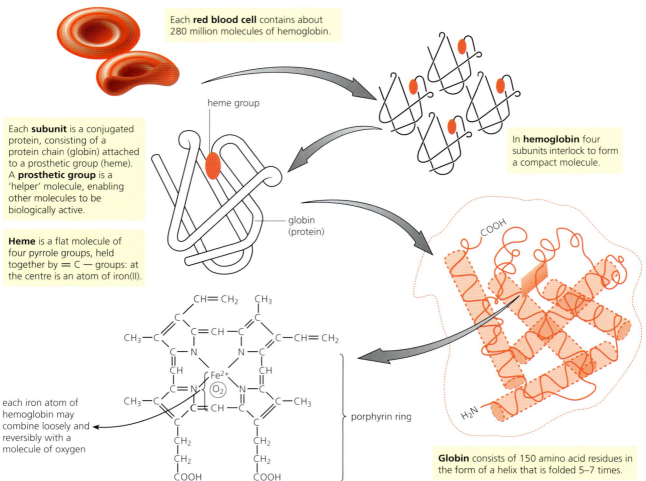

Each **red blood cell** contains about
280 million molecules of hemoglobin.

Each **subunit** is a conjugated
protein, consisting of a
protein chain (globin) attached
to a prosthetic group (heme).
A **prosthetic group** is a
'helper' molecule, enabling
other molecules to be
biologically active.

Heme is a flat molecule of
four pyrrole groups, held
together by ═ C — groups: at
the centre is an atom of iron(II).

heme group

globin
(protein)

In **hemoglobin** four
subunits interlock to form
a compact molecule.

COOH

H_2N

Globin consists of 150 amino acid residues in
the form of a helix that is folded 5–7 times.

each iron atom of
hemoglobin may
combine loosely and
reversibly with a
molecule of oxygen

porphyrin ring

■ **Figure 7.23** Hemoglobin – a quaternary protein of red blood cells

12 Describe three types of bond that contribute to a protein's secondary structure.

Polar and non-polar amino acids in proteins

Amino acids with polar R groups have hydrophilic properties. When these amino acids are built into protein in prominent positions they may influence the properties and functioning of the proteins in cells. Similarly, amino acids with non-polar R groups have hydrophobic properties.

Examples of these outcomes are illustrated in Figures 7.24 (for cell membrane proteins) and 7.25 (for an enzyme that occurs in the cytoplasm).

■ **Figure 7.24**
Polar and non-polar amino acids in membrane proteins

plasma membrane (fluid mosaic model)
(diagrammatic cross-sectional view)

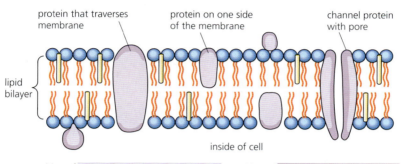

protein that traverses membrane

protein on one side of the membrane

channel protein with pore

lipid bilayer

inside of cell

position of **non-polar amino acid residues** (make bulk of protein hydrophobic – compatible with hydrocarbon tail of phospholipid molecules of bilayer)

position of **polar amino acid residues** (make surface of protein hydrophilic – this part of protein molecule protrudes or is exposed to cytosol)

■ **Figure 7.25**
Polar and non-polar amino acids in superoxide dismutase

Superoxide dismutase is an enzyme common to all cells – breaking down superoxide ions as soon as they form as by-products of the reactions of metabolism.

The molecule is approximately saucer-shaped, with active site centrally placed.

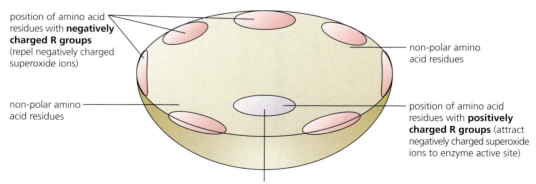

position of amino acid residues with **negatively charged R groups** (repel negatively charged superoxide ions)

non-polar amino acid residues

non-polar amino acid residues

position of amino acid residues with **positively charged R groups** (attract negatively charged superoxide ions to enzyme active site)

active site of enzyme (forms complex with superoxide ions as it disables them)

Fibrous and globular proteins – contrasting tertiary structures

Some proteins take up a tertiary structure that is a long much-coiled chain; these are called fibrous proteins. They have long, narrow shapes. Examples of fibrous proteins are **collagen**, a component of bone and tendons (Figure 7.26), and **keratin**, found in hair, horn and nails. Fibrous proteins are often insoluble.

■ **Figure 7.26**
Collagen – example of a fibrous protein

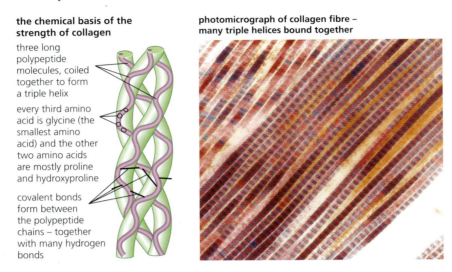

the chemical basis of the strength of collagen

three long polypeptide molecules, coiled together to form a triple helix

every third amino acid is glycine (the smallest amino acid) and the other two amino acids are mostly proline and hydroxyproline

covalent bonds form between the polypeptide chains – together with many hydrogen bonds

photomicrograph of collagen fibre – many triple helices bound together

Other proteins take up a tertiary structure that is more spherical, and are called **globular proteins** (Figure 7.27). They are mostly highly soluble in water. Enzymes, such as **lysozyme** and **catalase**, are typical globular proteins. Some of our hormones are globular proteins, too, including **insulin** (page 92).

Insulin is a hormone produced in the β cells of the islets of Langerhans in the pancreas by ribosomes of the rough endoplasmic reticulum (RER) as a polypeptide of 102 amino acid residues (preproinsulin).

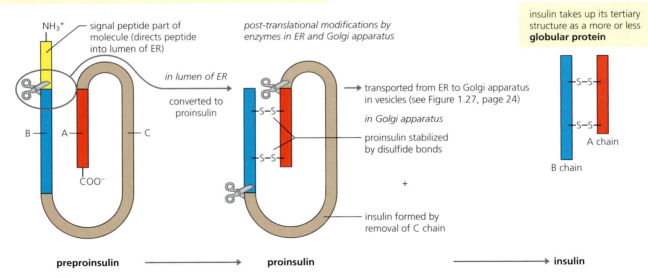

NH_3^+ — signal peptide part of molecule (directs peptide into lumen of ER)

post-translational modifications by enzymes in ER and Golgi apparatus

insulin takes up its tertiary structure as a more or less **globular protein**

in lumen of ER

converted to proinsulin

B — A — C

COO^-

transported from ER to Golgi apparatus in vesicles (see Figure 1.27, page 24)

in Golgi apparatus

proinsulin stabilized by disulfide bonds

+

insulin formed by removal of C chain

S—S

S—S

A chain

B chain

preproinsulin ⟶ proinsulin ⟶ insulin

■ **Figure 7.27** Insulin, a globular protein

13 Outline three ways in which membrane proteins are important to the functioning of a cell.

■ *Examination questions – a selection*

Questions 1–4 are taken from IB Diploma biology papers.

Q1 What is the distinction between highly repetitive DNA sequences and single-copy genes?
  A The highly repetitive sequences have greater amounts of guanine.
  B The highly repetitive sequences have greater amounts of cytosine.
  C The highly repetitive sequences are not transcribed.
  D The highly repetitive sequences are not replicated.

Higher Level Paper 1, Time Zone 2, May 09, Q26

Q2 What is a nucleosome?
  A A region in a prokaryotic cell where DNA is found
  B A DNA molecule wrapped around histone proteins
  C A ribosome of a prokaryotic cell
  D A molecule consisting of a sugar, a base and a phosphate

Higher Level Paper 1, Time Zone 2, May 12, Q25

Q3 What is the function of the tRNA activating enzyme?
  A It links tRNA to ribosomes.
  B It links tRNA to mRNA.
  C It links tRNA to a specific amino acid.
  D It links an amino acid on one tRNA to an amino acid on another tRNA.

Higher Level Paper 1, Time Zone 1, May 09, Q25

Q4 The diagram below shows the process of transcription

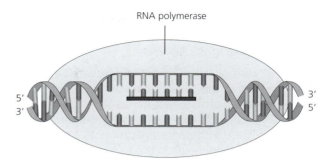

RNA polymerase

  a Label the sense and antisense strands. (1)
  b Draw an arrow on the diagram to show where the next nucleotide will be added to the growing mRNA strand. (1)

Higher Level Paper 1, Time Zone 1, May 12, Q2 ci and ii

Questions 5–10 cover other syllabus issues in this chapter.

Q5 **a** State the function of the following enzymes related with the process of duplication:
   i DNA polymerase I
   ii DNA ligase
   iii RNA primase (3)

Q6 Outline the relevance of post transcriptional processes in the removal of introns from a pre- mRNA molecule. (2)

Q7 **a** State which amino acid will be translated by the following tRNA molecules:
   i CAU
   ii UCC
   iii AGA (3)
  b Explain the relationship between transcribed genes, proteins and a functional enzyme. (4)

Q8 **a** State the **two** types of organic molecules that form the bulk of chromosomes of eukaryotes, and their approximate proportions. (3)

  b Suggest why the complex structural packaging of these molecules in chromosomes may be essential in the nucleus. (2)

Q9 **a** 'DNA replication is semi-conservative and occurs in the 5' → 3' direction.' Explain the meaning of this statement. (3)

  b By means of a fully annotated diagram, outline the steps to replication at a replication fork, making explicit the role of each of the essential enzymes. (6)

Q10 **a** Proteins are essential components of organisms. Describe the key property of the proteins specifically involved in:
  i   biological catalysis
  ii movement driven by muscle fibres
  iii structural support
  iv the movement of molecules across membranes. (4)

  b Outline the differences between the secondary and tertiary structures of proteins. (4)

  c Identify the types of bonds that maintain the tertiary structure of protein. (5)

Metabolism, cell respiration and photosynthesis

■ Metabolic reactions are regulated in response to the cell's needs.
■ Energy is converted to a usable form in cell respiration.
■ Light energy is converted into chemical energy.

8.1 Metabolism – *metabolic reactions are regulated in response to the cell's needs*

Many thousands of chemical reactions take place within cells and organisms (Figure 8.1). The molecules involved are collectively called metabolites or intermediates. Some are concerned in the formation of complex cellular components, including proteins, polysaccharides, lipids, hormones, pigments and many others. Others are molecules in the course of being broken down with the transfer of energy, including to the energy-currency molecule, ATP (page 115). Many metabolites are made in organisms but others are imported from the environment – for example, nutrients from food substances, water, and the gases carbon dioxide or oxygen.

■ **Figure 8.1**
Metabolism =
anabolism +
catabolism

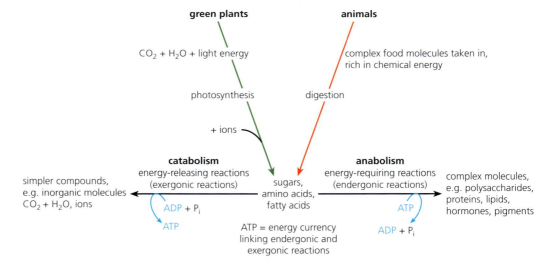

■ Metabolic pathways – chains and cycles of metabolic reactions

Metabolism consists of series of reactions in which the product of one reaction is an intermediate of the next, and so on. Many pathways consist of **straight chains** (that is, linear sequences) of reactions, whilst others are **cyclic processes** (Figure 8.2). We shall be seeing examples of both types of metabolic pathway within the reactions of respiration – a catabolic process (page 351) – and photosynthesis, an anabolic process (page 360).

In this chapter the focus is on respiration and photosynthesis – both aspects of metabolism that have traditionally been summarized in single equations.

■ Cellular respiration

$$\text{glucose} + \text{oxygen} \rightarrow \text{carbon dioxide} + \text{water} + \text{ENERGY}$$
$$C_6H_{12}O_6 + 6O_2 \rightarrow 6CO_2 + 6H_2O + \text{ENERGY}$$

■ Photosynthesis

$$\text{carbon dioxide} + \text{water} + \text{ENERGY} \rightarrow \text{glucose} + \text{oxygen}$$
$$6CO_2 + 6H_2O + \text{ENERGY} \rightarrow C_6H_{12}O_6 + 6O_2$$

1 **Define** the term
a anabolic reaction, and
b catabolic reaction, and give one example of each.

At first sight, it seems that the one process is simply the reverse of the other. In fact, these equations are merely 'balance sheets' of the inputs and outputs. Photosynthesis and respiration occur by different metabolic pathways, involving very many reactions in which each individual step is catalysed by a different enzyme. We start by looking further into the way enzymes work.

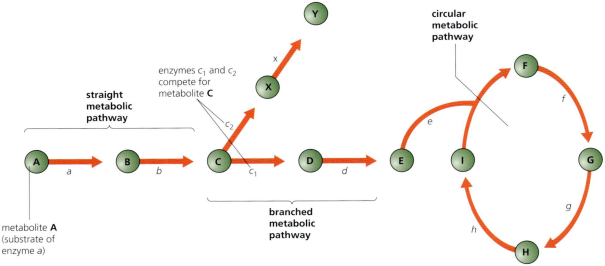

How much of metabolite **C** is converted to **X** or to **D** depends on the relative amounts of enzymes c_1 and c_2, and how readily each forms its enzyme–substrate complex.

■ **Figure 8.2** Some metabolic pathways and the roles of enzymes

■ Enzymes and activation energy

In an individual metabolic reaction, the starting substance is called the **substrate** and it is converted to the **product**. An enzyme works by binding to the substrate molecule at a specially formed pocket in the enzyme – the **active site** (page 94). So we visualize an enzyme (**E**) as a large molecule that works by reacting with another compound or compounds, the substrate (**S**). Initially, a short-lived enzyme–substrate complex (**ES**) is formed at the active site. This unstable complex exists only momentarily – it is a **transitional state**. Almost instantly, the product (**P**) is formed and the enzyme is released unchanged. The enzyme immediately takes part in another reaction. We represent this reaction as follows:

$$E + S \rightarrow [ES] \rightarrow P + E$$

Energy is released when 'substrate' becomes 'product'. However, to bring about the reaction, a small amount of energy is needed initially to break or weaken bonds in the substrate, bringing about the transitional state. This energy input is called the **activation energy** (Figure 8.3). It is a small but significant energy barrier that has to be overcome before the reaction can happen. Enzymes work by lowering the amount of energy required to activate the reacting molecules.

Another model of enzyme catalysis includes a boulder (substrate) perched on a slope, prevented from rolling down by a small hump (representing activation energy) as in Figure 8.3. The boulder can be pushed over the hump, or the hump can be dug away (= lowering the activation energy), allowing the boulder to roll down and shatter at a lower level (giving products).

■ Enzyme inhibitors

The actions of enzymes may be inhibited by other molecules, some formed in the cell and others absorbed from the external environment. These substances are known as **inhibitors**, since their effect is generally to lower the rate of reaction.

Studying the effects of inhibitors has helped our understanding of:

■ the chemistry of the active site of enzymes

■ the natural regulation of metabolism and which pathways operate

■ the ways certain commercial pesticides and drugs work, namely by inhibiting specific enzymes and preventing particular reactions.

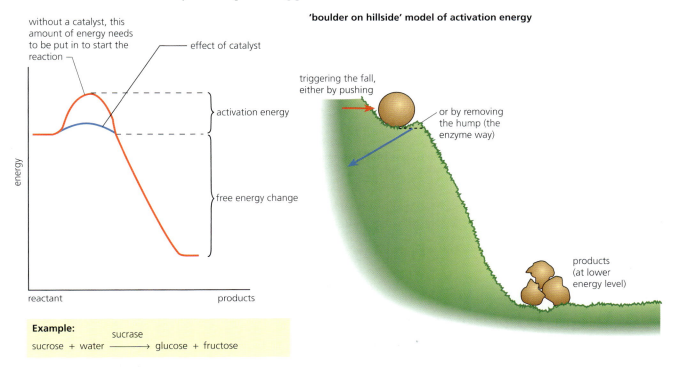

'boulder on hillside' model of activation energy

Example:

sucrose + water $\xrightarrow{\text{sucrase}}$ glucose + fructose

■ **Figure 8.3** Activation energy

For example, molecules that sufficiently resemble the substrate in shape may compete to occupy the active site. These are known as **competitive inhibitors**. The enzyme that catalyses the reaction between carbon dioxide and the acceptor molecule in photosynthesis is known as ribulose bisphosphate carboxylase (RuBisCo, page 366) and is competitively inhibited by oxygen in the chloroplasts.

Because competitive inhibitors are not acted on by the enzyme and turned into products, as normal substrate molecules are, they tend to remain attached. However, when the concentration of the substrate molecule is raised to a sufficiently high level, the inhibitor molecules are progressively displaced from the active sites.

Alternatively, an inhibitor may be unlike the substrate molecule, yet still combine with the enzyme. In these cases, the attachment occurs at some other part of the enzyme, probably quite close to the active site. Here, the inhibitor either partly blocks access to the active site by substrate molecules, or it causes the active site to change shape and so be unable to accept the substrate. These are called **non-competitive inhibitors**, since they do not compete for the active site.

Adding excess substrate does not overcome their inhibiting effects – effectively the action is non-reversible (Figure 8.4). Cyanide ions combine with cytochrome oxidase but not at the active site. Cytochrome oxidase is a respiratory enzyme present in all cells, and is a component in a sequence of enzymes and carriers that oxidize the hydrogen removed from a respiratory substrate such as glucose, forming water.

Distinguishing different types of inhibition from graphs at specified substrate concentration

Features and examples of competitive and non-competitive inhibition of enzymes are set out in Table 8.1.

■ **Table 8.1**
Competitive and non-competitive inhibition of enzymes compared

Competitive inhibitors	Non-competitive inhibitors
bind to the active site	bind to other parts of the enzyme, other than the active site
chemically resemble the substrate molecule and occupy (block) the active site	are chemically unlike the substrate molecule, but react with bulk of enzyme, either reducing accessibility of active site, or distorting the shape of the active site
at low concentration, increasing concentration of substrate eventually overcomes inhibition as substrate molecules displace inhibitor	at low concentration, increasing concentration of substrate cannot prevent binding – some inhibition remains at high substrate concentration
Examples: • O_2 competing with CO_2 for active site of RuBisCo • malonate competing with succinate for the active site of succinate dehydrogenase	Examples: • cyanide ions blocking cytochrome oxidase in terminal oxidation in cell aerobic respiration • nerve gas Sarin blocking acetyl cholinesterase in synapse transmission

Examine the graph in Figure 8.4 carefully.

You can see that, when the initial rates of reaction of an enzyme are plotted against substrate concentration, the effects of competitive and non-competitive inhibitors are clearly different.

■ **Figure 8.4**
Competitive and non-competitive inhibitors

When the initial rates of reaction of an enzyme are plotted against substrate concentration, the effects of competitive and non-competitive inhibitors are seen to be different.

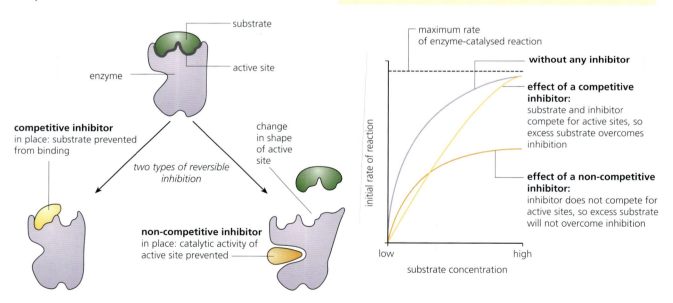

An investigation of the initial rate of reaction of the enzyme catalase in the presence and absence of heavy metal ions (Cu^{2+})

2 Explain why the shape of globular proteins that are enzymes is important in enzyme action.

The enzyme catalase catalyses the breakdown of hydrogen peroxide:

$$2H_2O_2 \xrightarrow{\text{catalase}} 2H_2O + O_2$$

Catalase occurs very widely in cells. This enzyme functions as a protective mechanism for the delicate biochemical machinery of cells. Hydrogen peroxide is a common by-product of some of the reactions of metabolism and it is a very toxic substance (being a powerful oxidizing agent, page 352). Catalase inactivates hydrogen peroxide as soon as it forms, before damage can occur.

In practical terms, the **rate of an enzyme-catalysed reaction** is taken as the amount of substrate that has disappeared from a reaction mixture, or the amount of product that has accumulated, in a period of time. For example, working with catalase, it is convenient to measure the rate at which the product (oxygen) accumulates. In this experiment, an indication of the volume of oxygen that was released at half-minute intervals was recorded at each substrate concentration.

When these 'raw' results for each concentration of hydrogen peroxide were plotted on a graph, it was found that, over a period of time, the **initial rate** of reaction was not maintained; rather, it fell off quite sharply. This is typical of enzyme actions studied outside their location in the cell. The reduction in rate can be due to a number of reasons, but most commonly it is because the concentration of the substrate in the reaction mixture has fallen. Consequently, it is the initial rate of reaction that is determined. This is the slope of the tangent to the curve in the initial stage of reaction. How this is calculated is shown in Figure 2.41 (page 100).

Look at this account of how an initial reaction rate is calculated now.

The apparatus used is illustrated in Figure 8.5 and the reaction mixtures used in this experiment are tabulated below. Note that:

1 A yeast suspension was used as the source of the enzyme. The number of bubbles released at half-minute intervals was counted and recorded, for six minutes, with each reaction mixture. (The initial rush of bubbles when the yeast was injected was discounted. The depth of the bubble nozzle was the same in each experiment.)

■ **Figure 8.5**
Apparatus for monitoring the effects of substrate concentration on the action of catalase

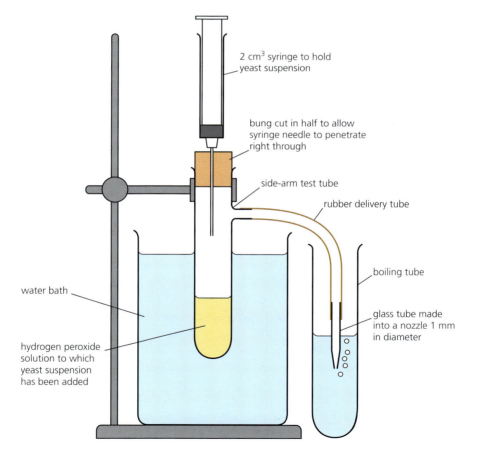

2 cm³ syringe to hold yeast suspension

bung cut in half to allow syringe needle to penetrate right through

side-arm test tube

rubber delivery tube

boiling tube

glass tube made into a nozzle 1 mm in diameter

water bath

hydrogen peroxide solution to which yeast suspension has been added

2 The concentration of a hydrogen peroxide solution is given as the volume of oxygen that can be released. For example, a 20-volume solution will, when completely decomposed, give 20 times its own volume of oxygen.

3 A duplicate investigation was carried out in the presence of a dilute solution of copper (Cu^{2+}) ions.

The reaction mixtures used and the results obtained are shown in Tables 8.2 and 8.3. Examine them and then answer question 3.

■ **Table 8.2** Effect of substrate on enzyme-catalysed reaction

Expt	1	2	3	4	5	6
distilled water (cm^3)	4.0	3.5	3.0	2.5	2.0	1.5
20-vol H_2O_2 (cm^3)	1.0	1.5	2.0	2.5	3.0	3.5
concentration of H_2O_2 (vol)	4	6	8	10	12	14
yeast suspension	1.0	1.0	1.0	1.0	1.0	1.0
initial rate of reaction (bubbles/30 s)	**0.25**	**8.50**	**12.00**	**13.50**	**15.00**	**16.00**

■ **Table 8.3** Effect of substrate on enzyme-catalysed reaction in presence of heavy metal ions

Expt	1	2	3	4	5	6
distilled water (cm^3)	3.9	3.4	2.9	2.4	1.9	1.4
0.1 M copper (Cu^{2+}) solution (cm^3)	0.1	0.1	0.1	0.1	0.1	0.1
20-vol H_2O_2 (cm^3)	1.0	1.5	2.0	2.5	3.0	3.5
concentration of H_2O_2 (vol)	4	6	8	10	12	14
yeast suspension	1.0	1.0	1.0	1.0	1.0	1.0
initial rate of reaction (bubbles/30 s)	**0.10**	**4.00**	**7.00**	**7.50**	**7.70**	**7.80**

3 a **Construct** a graph showing the initial rates of reaction of the enzyme catalase over the substrate concentration range of 4–14 vol hydrogen peroxide in the presence and absence of heavy metal ions.
  b **Explain** to what extent this data supports the hypothesis that copper ions are a non-competitive inhibitor of the enzyme that decomposes hydrogen peroxide.

■ Allosteric regulation of enzymes

Allosteric regulators are molecules that change the shape and activity of an enzyme by *reversibly* binding at a site on the enzyme, typically some distance from the active site. Binding of an allosteric **activator** temporarily stabilizes the enzyme shape as an active and effective catalyst. Binding of an allosteric **inhibitor** changes the enzyme shape into an inactive form.

We can see allosteric regulation as a form of **reversible, non-competitive inhibition or activation** of an enzyme. Subtle fluctuations in the concentrations of activators and inhibitors may fine-tune the activity of a critical pathway as constantly changing conditions or requirements of cell metabolism demands.

End-product inhibition

Individual pathways in metabolism may be switched off by the final product, acting as a reversible inhibitor of the enzyme that catalyses the first step in the pathway.

In end-product inhibition, as the product molecules accumulate, the steps in their production are switched off (Figure 8.6). But these product molecules may now become the substrates in subsequent metabolic reactions. If so, the accumulated product molecules will be removed and production of new product molecules will recommence.

regulation of a metabolic pathway by end-product inhibition

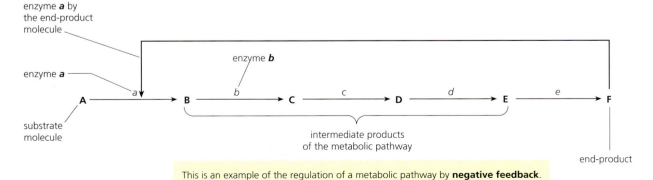

This is an example of the regulation of a metabolic pathway by **negative feedback**.

■ **Figure 8.6** End-product inhibition

■ Enzyme inhibitors as anti-malarial drugs

Malaria, a major world health problem, is caused by *Plasmodium* (a protoctistan) which is transmitted from an infected person to another by blood-sucking mosquitoes of the genus *Anopheles*. Within the human host, the parasite mainly feeds on the contents of red blood cells.

Several anti-malarial drugs currently in use are based on derivatives of quinine, such as the drug chloroquine. *How do they work?*

The parasite feeds by digesting the protein part of hemoglobin (Figure 7.23, page 340) within the red cell it has invaded. This releases amino acids for its own growth and metabolism. The residual heme is potentially toxic to *Plasmodium*, so the parasite converts heme into a harmless insoluble precipitate. Chloroquine specifically inhibits the *Plasmodium* enzyme that is involved in this conversion. Now, free heme accumulates within the cell and the parasite is killed.

A second type of drug inhibits a specific *Plasmodium* enzyme involved in DNA replication and growth.

Nature of Science

Developments in scientific research follow improvements in computing

Bioinformatics – the role of databases

Today, the *Plasmodium* parasite is increasingly developing resistance to these existing anti-malarial drugs. The search is on in laboratories worldwide for molecules that will be effective agents against the parasite, but which will be harmless to the patient. For example, anti-malarial drugs that target specific protein–protein interactions by which blood cells are attacked are sought, as are other inhibitors of enzymes unique to the parasite's metabolism. The approaches to the problem by research teams include:

1. testing of commercially available drugs that have been approved for human use for other diseases
2. chemical modification of current anti-malarials, for example by combining existing less-effective drugs to produce hybrid molecules with enhanced impacts
3. molecular modelling of target enzymes in *Plasmodium* and computer design of molecules that may specifically block their active sites
4. the application of theoretical molecular chemistry by screening of databases for new compounds with the potential for anti-malarial activity, followed by their further testing and possible drug trials.

The work continues with great urgency.

TOK Link

Many metabolic pathways have been described following a series of carefully controlled and repeated experiments. To what degree can looking at component parts give us knowledge of the whole?

8.2 Cell respiration – *energy is converted to a usable form in cell respiration*

In cellular respiration, the chemical energy of organic molecules, such as glucose, is made available for use in the living cell. Much of the energy transferred is lost in the form of heat energy, but cells are able to retain significant amounts of chemical energy in **adenosine triphosphate** (**ATP**, page 115). ATP, found in all cells, is the universal energy currency in living systems. ATP is a relatively small, soluble molecule. It is able to move, by facilitated diffusion, from the mitochondria where it is synthesized to all the very many sites where energy is required, such as in muscles for contraction movements (page 463), in membranes for active transport (page 44) and in ribosomes for protein synthesis (page 333).

How is energy transferred from respiratory substrates like glucose?

■ Cell respiration involves the oxidation and reduction of compounds

Glucose is a relatively large molecule containing six carbon atoms, all in a reduced state. During aerobic cellular respiration, glucose undergoes a series of enzyme-catalysed oxidation reactions and de-carboxylation reactions (Figure 8.7). These reactions are grouped into three major phases and a link reaction:

- **glycolysis**, in which glucose is converted to pyruvate
- a **link reaction**, in which pyruvate is converted to acetyl coenzyme A (acetyl CoA). CO_2 is given off
- the **Krebs cycle**, in which acetyl coenzyme A is converted to carbon dioxide
- the **electron-transport system**, in which hydrogen that is removed in the oxidation reactions of glycolysis and the Krebs cycle is converted to water. The bulk of the ATP is synthesized here.

■ **Figure 8.7**
The stages of aerobic cellular respiration

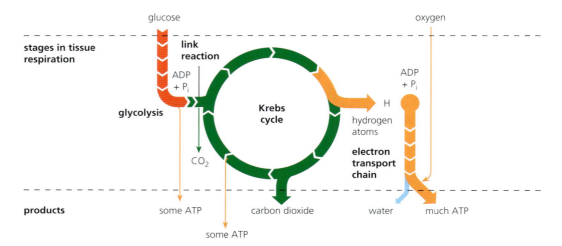

■ Respiration as a series of redox reactions.

The terms '**reduction**' and '**oxidation**' recur frequently in respiration.
What do these terms mean?

In cellular respiration, glucose is oxidized to carbon dioxide but, at the same time, oxygen is reduced to water (Figure 8.8). In fact, tissue respiration is a series of oxidation–reduction reactions, so described because when one substance in a reaction is oxidized another is automatically reduced. The short-hand name for **red**uction–**oxi**dation reactions is **redox reactions**.

In biological oxidation, oxygen atoms may be added to a compound but, alternatively, hydrogen atoms may be removed. In respiration, all the hydrogen atoms are gradually removed from glucose. They are added to hydrogen acceptors, which are themselves reduced.

Since a hydrogen atom consists of an electron and a proton, gaining hydrogen atom(s) (a case of reduction) involves gaining one or more electrons. In fact, the best definition of **oxidation is the loss of electrons**, and **reduction is the gain of electrons**. Remembering this definition has given countless people problems, so a mnemonic has been devised:

OIL RIG = **O**xidation **I**s **L**oss of electrons; **R**eduction **I**s **G**ain of electrons

Redox reactions take place in biological systems due to the presence of a compound with a strong tendency to take electrons from another compound (an **oxidizing agent**) or the presence of a compound with a strong tendency to donate electrons to another compound (a **reducing agent**).

Another feature of oxidation and reduction is **energy change**. When reduction occurs, energy is absorbed (an **endergonic reaction**, Figure 2.7, page 67). When oxidation occurs, energy is released (an **exergonic reaction**). An example of energy release in oxidation is the burning of a fuel in air. Here, energy is given out as heat. In fact, the amount of energy in a molecule depends on its degree of oxidation. An oxidized substance has less stored energy than a reduced substance. Take, for example, the fuel molecule methane (CH_4) which we know has more stored chemical energy than carbon dioxide (CO_2).

■ **Figure 8.8**
Respiration as a redox reaction

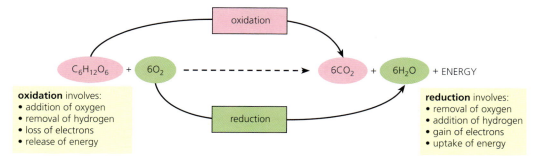

■ Cellular respiration

Glycolysis

Glycolysis is a linear series of reactions in which a 6-carbon sugar molecule is broken down to two molecules of the 3-carbon pyruvate ion (Figure 8.9). The enzymes of glycolysis are located in the cytoplasm outside organelles (known as the cytosol), rather than in the mitochondria. Glycolysis occurs by four stages:

■ **Phosphorylation** by reaction with ATP is the way glucose is first activated, forming glucose phosphate. Phosphorylation of molecules makes then less stable, meaning more reactive. Conversion to fructose phosphate follows, and a further phosphate group is then added at the expense of another molecule of ATP. So, two molecules of ATP are *consumed* per molecule of glucose respired, *at this stage of glycolysis*.

■ **Lysis** (splitting) of the fructose bisphosphate now takes place, forming two molecules of 3-carbon sugar, called **triose phosphate**.

■ **Oxidation** of the triose phosphate molecules occurs by removal of hydrogen. The enzyme for this reaction (a dehydrogenase) works with a coenzyme, **nicotinamide adenine dinucleotide** (**NAD**). NAD is a molecule that can accept hydrogen ions (H^+) and electrons (e^-). In this reaction, the NAD is reduced to NADH and H^+ (reduced NAD):

$$NAD^+ + 2H^+ + 2e^- \rightarrow NADH + H^+ \text{ (sometimes represented as } NADH_2\text{)}$$

(Reduced NAD can pass hydrogen ions and electrons on to other acceptor molecules (see below) and, when it does, it becomes oxidized back to NAD.)

■ **ATP formation** occurs twice in the reactions by which each triose phosphate molecule is converted to pyruvate. This form of ATP synthesis is described as being **at substrate level**, in order to differentiate it from the bulk of ATP synthesis that occurs later in cell respiration – during operation of the electron-transport chain (see below). As two molecules of triose phosphate are converted to pyruvate, four molecules of ATP are synthesized *at this stage of glycolysis*. So in total, there is a **net gain of two ATPs in glycolysis**.

Once pyruvate has been formed from glucose in the cytosol, the remainder of the pathway of aerobic cell respiration is located in the **mitochondria** (singular, **mitochondrion**). This is where the enzymes concerned with the **link reaction**, **Krebs cycle** and **electron transport chain** are all located.

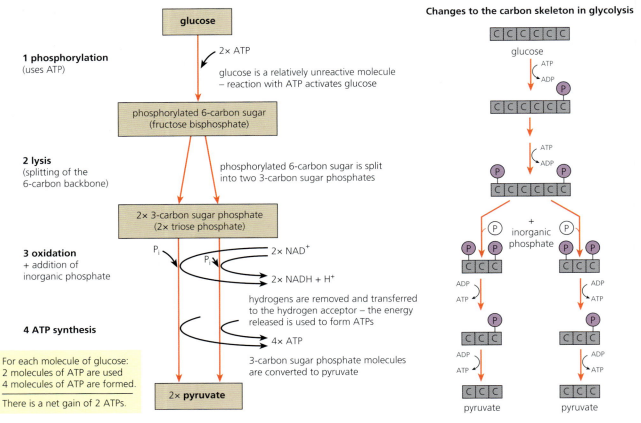

1 phosphorylation
(uses ATP)

2 lysis
(splitting of the
6-carbon backbone)

3 oxidation
+ addition of
inorganic phosphate

4 ATP synthesis

For each molecule of glucose:
2 molecules of ATP are used
4 molecules of ATP are formed.

There is a net gain of 2 ATPs.

glucose is a relatively unreactive molecule
– reaction with ATP activates glucose

phosphorylated 6-carbon sugar is split
into two 3-carbon sugar phosphates

hydrogens are removed and transferred
to the hydrogen acceptor – the energy
released is used to form ATPs

3-carbon sugar phosphate molecules
are converted to pyruvate

Changes to the carbon skeleton in glycolysis

■ **Figure 8.9** Glycolysis: a summary

---

4 **State** which of the following are produced during glycolysis:

carbon dioxide   NADH   ATP   pyruvate   lactate   glycogen   NAD⁺   glucose

---

Link reaction

Pyruvate diffuses into the matrix of the mitochondrion as it forms and is metabolized there.

First, the 3-carbon pyruvate is decarboxylated by removal of carbon dioxide and, at the same time, oxidized by removal of hydrogen. Reduced NAD is formed. The product of this **oxidative decarboxylation** reaction is an acetyl group – a 2-carbon fragment. This acetyl group is then combined with a coenzyme called coenzyme A (CoA), forming **acetyl coenzyme A (acetyl CoA)**. The production of acetyl coenzyme A from pyruvate is known as the link reaction because it connects glycolysis to reactions of the **Krebs cycle**, which now follow.

Krebs cycle

The Krebs cycle is named after Hans Krebs who discovered it, but it is also sometimes referred to as the **citric acid cycle**, after the first intermediate acid formed.

The acetyl coenzyme A enters the Krebs cycle by reacting with a **4-carbon organic acid** (oxaloacetate, OAA). The products of this reaction are a **6-carbon acid** (citrate) and coenzyme A which is released and reused in the link reaction.

The citrate is then converted back to the 4-carbon acid (an acceptor molecule, in effect) by **the reactions of the Krebs cycle**. These involve the following changes:

- two molecules of carbon dioxide are given off, in separate decarboxylation reactions
- a molecule of ATP is formed, as part one of the reactions of the cycle; as in glycolysis, this ATP synthesis is at substrate level
- three molecules of reduced NAD are formed
- one molecule of another hydrogen acceptor, the coenzyme flavin adenine dinucleotide (FAD) is reduced (NAD is the chief hydrogen-carrying coenzyme of respiration but FAD has this role in the Krebs cycle).

The above details enable you to see what types of reaction are occurring and why. Remember, the names of the intermediate compounds in glycolysis and the Krebs cycle do not need to be memorized.

■ **Figure 8.10**
Link reaction and Krebs cycle: a summary

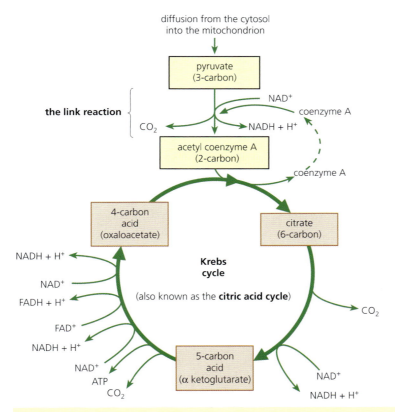

There are several other organic acid intermediates in the cycle not shown here.

Steps in aerobic respiration involve decarboxylation and oxidation. Make a copy of the pathway of the link reaction and Kreb's cycle, and highlight where these types of reaction occur. Then look again at Figure 8.9. Are both of these types of reaction observed there, too?

Because glucose is converted to two molecules of pyruvate in glycolysis, the whole Krebs cycle sequence of reactions 'turns' twice for every molecule of glucose that is metabolized by aerobic cellular respiration (Figure 8.10).

Now, we are in a position to summarize the changes to the molecule of glucose that occur in the reactions of glycolysis and the Krebs cycle. A 'budget' of the products of glycolysis and two turns of the Krebs cycle is shown in Table 8.4.

■ **Table 8.4**
Net products of aerobic respiration of glucose at the end of the Krebs cycle

Step	Product			
	CO$_2$	ATP	Reduced NAD	Reduced FAD
glycolysis	0	2	2	0
link reaction (pyruvate → acetyl CoA)	2	0	2	0
Krebs cycle	4	2	6	2
Totals:	**6 CO$_2$**	**4 ATP**	**10 reduced NAD**	**2 reduced FAD**

5 **Outline** the types of reaction catalysed by:
  a dehydrogenases
  b decarboxylases.

Fats can be respired

In addition to glucose, **fats** (lipids) are also commonly used as respiratory substrates – they are first broken down to **fatty acids** (and glycerol). Fatty acid is 'cut up' into 2-carbon fragments and fed into the Krebs cycle via **coenzyme A**. Vertebrate muscle is well adapted to the respiration of fatty acids in this way (as is our heart muscle), and they are just as likely as glucose to be the respiratory substrate.

The electron transport chain

Terminal oxidation and oxidative phosphorylation

The removal of pairs of hydrogen atoms from various intermediates of the respiratory pathway is a feature of several of the steps in glycolysis and the Krebs cycle. On most occasions, oxidized NAD is converted to reduced NAD but, once in the Krebs cycle, an alternative hydrogen-acceptor coenzyme, FAD, is reduced.

In the final stage of aerobic respiration, the hydrogen atoms (or their electrons) are transported along a **series of carriers**, from the reduced NAD (or FAD), to be combined with oxygen to form water. Oxygen is the final electron acceptor.

■ **Figure 8.11**
Terminal oxidation and the formation of ATPs

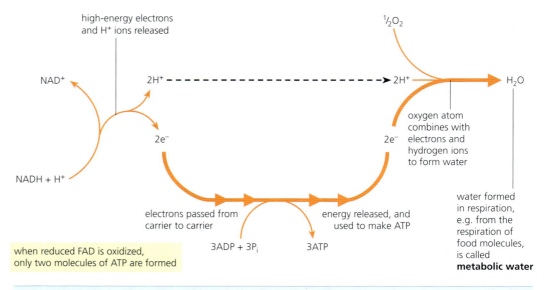

when reduced FAD is oxidized, only two molecules of ATP are formed

6 **Suggest** how the absence of oxygen in respiring tissue might 'switch off' both the Krebs cycle and terminal oxidation.

As electrons are passed between the carriers in the series, **energy is released**. Release of energy in this manner is controlled and can be used by the cell. The energy is transferred to ADP and P$_i$, to form ATP. Normally, for every molecule of reduced NAD which is oxidized (that is, for every pair of hydrogens) approximately just less than three molecules of ATP are produced (but less when FAD is oxidized). The process is summarized in Figure 8.11.

In total, the yield from aerobic respiration is 32 ATPs per molecule of glucose respired (Table 8.5).

■ **Table 8.5**
Yield from each molecule of glucose respired aerobically

	Reduced NAD (or FAD)	ATP
glycolysis	(substrate level)	(net) = 2
	2	2 × 2.5 = 5
link reaction	2	2 × 2.5 = 5
Krebs cycle	6	6 × 2.5 = 15
	2	2 × 1.5 = 3
	(substrate level)	2
Total		**32**

Paradigm shift

■ Phosphorylation by chemiosmosis

How could a mitochondrion use the energy that is made available in the flow of electrons between carrier molecules to drive the synthesis of ATP?

It was a biochemist, Peter Mitchell, who first suggested the chemiosmotic theory as the answer to this question, when he was at an independently funded research institute in Cornwall, UK. This was in 1961; at the time he was studying the metabolism of bacteria. His hypothesis was not generally accepted for many years – his ideas were, in some ways, regarded as too novel. Today, we describe the revolution that Mitchel's ideas started as a paradigm shift in the field of bioenergetics. Two decades later he was awarded a Nobel Prize for his discovery.

Chemiosmosis is a process by which the synthesis of ATP is coupled to electron transport via the movement of protons (Figure 8.12). The electron-carrier proteins are arranged in the inner mitochondrial wall in a highly ordered way. These carrier proteins oxidize the reduced coenzymes and energy from the oxidation process is used to pump hydrogen ions (protons) from the matrix of the mitochondrion into the space between inner and outer mitochondrial membranes.

Here, they accumulate – incidentally, causing the pH to drop. Because the inner membrane is largely impermeable to ions, a significant **gradient in hydrogen ion concentration** builds up across the inner membrane, generating a potential difference across the membrane. This represents a store of potential energy. Eventually, the protons do flow back into the matrix, but this occurs via the channels in **ATP synthetase** enzyme (ATPase), also found in the inner mitochondrial membrane. As the protons flow down their concentration gradient, through the enzyme, the energy is transferred as ATP synthesis occurs.

7 When ATP is synthesized in mitochondria, **explain** where the electrochemical gradient is set up, and in which direction protons move.

TOK Link

The time that elapsed from proposal of the chemiosmotic theory to its general acceptance illustrates that people may not always, or easily, accept that an earlier hypothesis must be rejected when evidence arises against it, and arises for an alternative concept. John Steinbeck (novelist and biologist) put it this way:

'There is one great difficulty with a good hypothesis. When it is completed and rounded, the corners smooth and the content cohesive, it is likely to become a thing in itself, a work of art... One hates to disturb it. Even if subsequent information should shoot a hole in it, one hates to tear it down because it once was beautiful and whole.'

John Steinbeck

Does this observation apply in this case?

■ **Figure 8.12** Mitchell's chemiosmotic theory

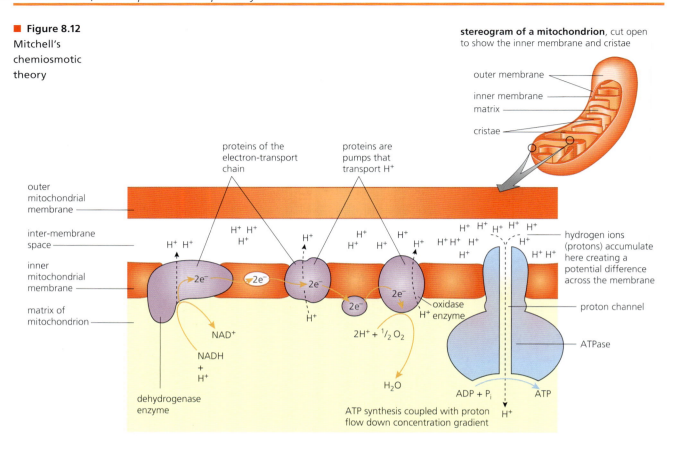

stereogram of a mitochondrion, cut open to show the inner membrane and cristae

8 Using the scale bar in Figure 8.13, **calculate** the length and width in micrometres of the mitochondrion shown in section in the electron micrograph.

■ The mitochondrion – structure in relation to function

Once pyruvate has been formed from glucose in the cytosol, the remainder of the pathway of aerobic cell respiration is located in the **mitochondria** (singular, **mitochondrion**). This is where the enzymes concerned with the link reaction, Krebs cycle and electron transport are all located.

Mitochondria are found in the cytoplasm of all eukaryotic cells, usually in very large numbers. The structure of a mitochondrion is investigated by transmission electron microscopy. The clarity of the resulting electron micrographs often makes possible a line drawing to show the mitochondrion's double membrane around the matrix, and the way the inner membrane folds to form cristae (Figure 8.13). This diagram can be annotated to show adaptation of mitochondrial structure to function (Table 8.6). Of course, the mitochondrion is a three-dimensional structure.

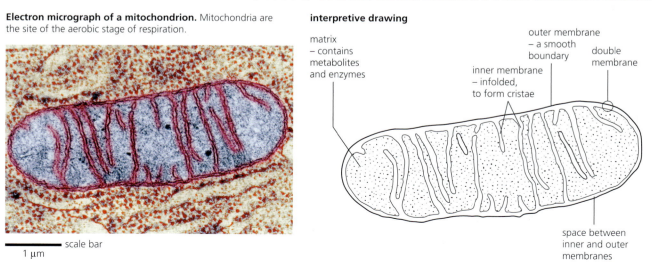

Electron micrograph of a mitochondrion. Mitochondria are the site of the aerobic stage of respiration.

interpretive drawing

■ **Figure 8.13** Electron micrograph of a mitochondrion, with an interpretive drawing

Make a line drawing of a mitochondrion, showing its internal structure. Now annotate your diagram to indicate the ways in which the structure is adapted to its function – using the information in Table 8.6.

The locations of the stages of cellular respiration are summarized in Figure 8.14. The relationship between structure and function in this organelle is summarized in Table 8.6.

■ **Table 8.6**
Mitochondrial structure in relation to function

Structure	Function/role
external double membrane	permeable to pyruvate, CO_2, O_2 and NAD/NADH + H^+
matrix	site of enzymes of link reaction and Krebs cycle
inner membrane	surface area greatly increased by folding to form cristae – since the electron-transport chain and ATP synthetase enzymes are housed here, opportunities for ATP synthesis are enhanced
	impermeable to hydrogen ions (protons) – permitting the establishment of a potential difference between the inter-membrane space and the matrix
inter-membrane space	relatively tiny space – allowing the accumulation of hydrogen ions (protons) to generate a large concentration difference with matrix, facilitating phosphorylation

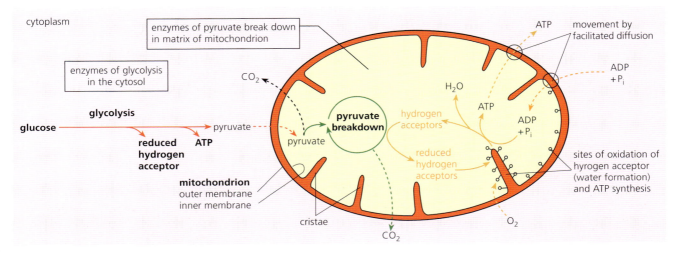

■ **Figure 8.14** The sites of respiration in the eukaryotic cell

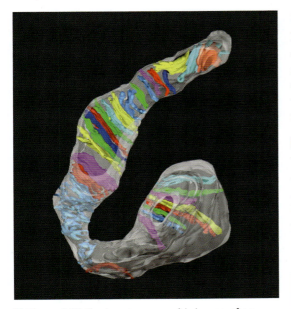

■ **Figure 8.15** Electron tomographic image of an active mitochondrion

Electron tomography

Electron tomography is used to produce three-dimensional images of the interior of active mitochondria (Figure 8.15). The technique is an extension of transmission electron microscopy. This technical development established the dynamic nature of the cristae, since they are seen to respond to changing conditions and demands of cell metabolism.

■ Respiration as a source of intermediates

We have seen that the main role of respiration is to provide a pool of ATP, and that this is used to drive the endergonic reactions of synthesis. But the compounds of the glycolysis pathway and the Krebs cycle (**respiratory intermediates**) may also serve as the **starting points of synthesis of other metabolites** needed in the cells and tissues of the organism. These include polysaccharides like starch and cellulose, glycerol and fatty acids, amino acids, and many others.

9 **Distinguish** between the following pairs:
 a substrate and intermediate
 b glycolysis and the Krebs cycle
 c oxidation and reduction.

8.3 Photosynthesis – *light energy is converted into chemical energy*

Just as the mitochondria are the site of reactions of the Krebs cycle and respiratory ATP formation, so the **chloroplasts** are the organelles where the reactions of photosynthesis occur.

Chloroplasts are found in green plants. They contain the photosynthetic pigments, along with the enzymes and electron-transport proteins, for the reduction of carbon dioxide to sugars and for ATP formation, using light energy. The detailed structure of chloroplasts is investigated with the transmission electron microscope. An electron micrograph of a thin section of a chloroplast shows the arrangement of the membranes within this large organelle (Figures 1.24, page 23, and 8.16).

There is a double membrane around the chloroplast, and the inner of these membranes folds extensively at various points to form a system of branching membranes. Here, these membranes are called **thylakoids**. Thylakoid membranes are organized into flat, compact, circular piles called **grana** (singular, **granum**), almost like stacks of coins. Between the grana are loosely arranged tubular membranes suspended in a watery **stroma**.

Chlorophyll, the photosynthetic pigment that absorbs light energy, occurs in the grana. Also suspended in the matrix are starch grains, lipid droplets and ribosomes. We shall see that the structure, composition and arrangements of membranes are central to the biochemistry of photosynthesis, just as the mitochondrial membranes are the sites of many of the reactions of aerobic cell respiration.

■ **Figure 8.16**
The ultrastructure of a chloroplast

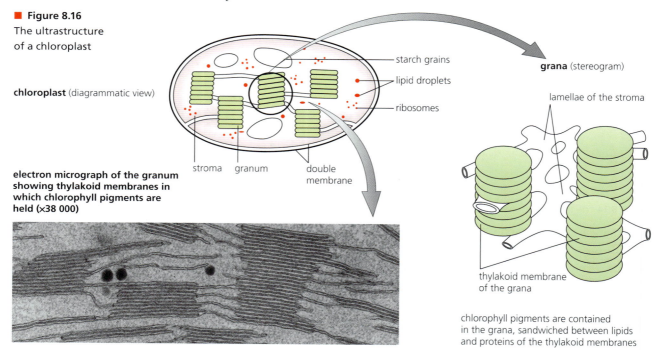

chloroplast (diagrammatic view)

starch grains

lipid droplets

ribosomes

stroma   granum   double membrane

grana (stereogram)

lamellae of the stroma

thylakoid membrane of the grana

chlorophyll pigments are contained in the grana, sandwiched between lipids and proteins of the thylakoid membranes

electron micrograph of the granum showing thylakoid membranes in which chlorophyll pigments are held (×38 000)

■ The reactions of photosynthesis

Photosynthesis is a complex set of many reactions that takes place in illuminated chloroplasts. Biochemical investigations of photosynthesis by several teams of scientists have established that the many reactions by which light energy brings about the production of sugars, using the raw materials water and carbon dioxide, fall naturally into two interconnected stages (Figure 8.17).

- **The light-dependent reactions** use light energy to split water (known as **photolysis**). Hydrogen is then removed and retained by the photosynthesis-specific hydrogen acceptor, known as NADP⁺. (NADP⁺ is very similar to the coenzyme NAD⁺ of respiration, but it carries an additional phosphate group, hence NAD**P**.) At the same time, ATP is generated from ADP and phosphate, also using energy from light. This is known as **photophosphorylation**. Oxygen is given off as a waste product of the light-dependent reactions. This stage occurs in the grana of the chloroplasts.

- **The light-independent reactions** build up sugars using carbon dioxide. Of course, the products of the light-dependent reactions (ATP and reduced hydrogen acceptor NADPH + H⁺) are used in sugar production. This stage occurs in the stroma of the chloroplast. It requires a continuous supply of the products of the light-dependent reactions, but does not directly involve light energy (hence the name). Names can be misleading, however, because this stage is an integral part of photosynthesis, and photosynthesis is a process that is powered by light energy.

We shall now look into both sets of reactions in turn to understand more about how these complex changes are brought about.

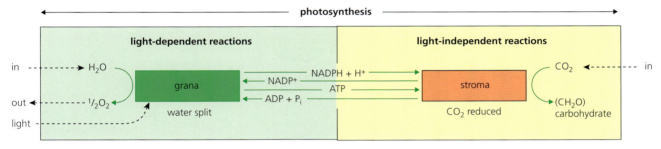

■ **Figure 8.17** The two sets of reactions of photosynthesis, inputs and outputs

■ The light-dependent reactions

In the light-dependent stage, light energy is trapped by photosynthetic pigment, chlorophyll. Chlorophyll molecules do not occur haphazardly in the grana. Rather, they are grouped together in structures called **photosystems**, held in the thylakoid membranes of the **grana** (Figure 8.18).

In each photosystem, several hundred chlorophyll molecules, plus accessory pigments (carotene and xanthophylls), are arranged. All these pigment molecules harvest light energy of slightly different wavelengths, and they funnel the energy to a single chlorophyll molecule of the photosystem, known as the **reaction centre**. The chlorophyll is then **photoactivated**.

There are **two types of photosystem** present in the thylakoid membranes of the grana, identified by the wavelength of light that the chlorophyll of the reaction centre absorbs:

- **Photosystem I** has a reaction centre that is activated by light of wavelength 700 nm. This reaction centre is referred to as P700.

- **Photosystem II** has a reaction centre that is activated by light of wavelength 680 nm. This reaction centre is referred to as P680.

Photosystems I and II have specific and differing roles, as we shall see shortly. However, they occur grouped together in the thylakoid membranes of the grana, along with proteins that function quite specifically. These consist of:

- **enzymes** catalysing
  - ☐ formation of ATP from ADP and phosphate (P_i)
  - ☐ conversion of oxidized H-carrier (NADP⁺) to reduced carrier (NADPH + H⁺)
- electron carrier molecules.

10 Construct a table that identifies the components of photosystems and the role of each one.

■ **Figure 8.18**
The structure of
photosystems

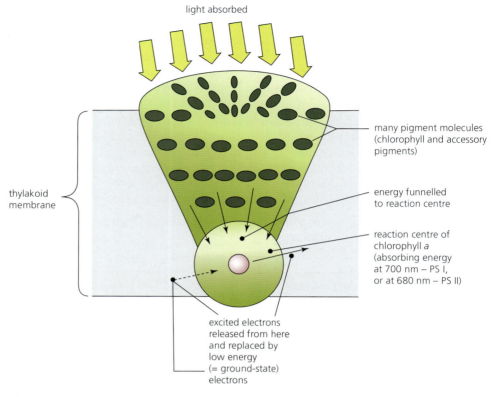

light absorbed

many pigment molecules
(chlorophyll and accessory
pigments)

thylakoid
membrane

energy funnelled
to reaction centre

reaction centre of
chlorophyll *a*
(absorbing energy
at 700 nm – PS I,
or at 680 nm – PS II)

excited electrons
released from here
and replaced by
low energy
(= ground-state)
electrons

The transfer of light energy

When light energy reaches a reaction centre, **ground-state electrons** in the key chlorophyll molecule are raised to an 'excited state' by the light energy received. As a result, **high-energy electrons** are released, and these electrons bring about the biochemical changes of the light-dependent reactions. The spaces vacated by the high-energy (excited) electrons in the reaction centres are continuously refilled by non-excited or ground-state electrons.

We will examine this sequence of reactions in the two photosystems next.

First, the excited electrons from **photosystem II** are picked up and passed along a chain of electron carriers. As these excited electrons pass, some of the energy causes the pumping of hydrogen ions (protons) from the chloroplast's matrix into the thylakoid spaces. Here, protons accumulate, causing the pH to drop. The result is a proton gradient that is created across the thylakoid membrane and which sustains the synthesis of ATP. This is another example of chemiosmosis (page 357).

As a result of these energy transfers, the excitation level of the electrons falls back to ground state and they come to fill the vacancies in the reaction centre of **photosystem I**. You can see that electrons have been transferred from photosystem II to photosystem I.

Meanwhile the 'holes' in the reaction centres of photosystem II are filled by electrons (in their ground state) from water molecules. In fact, the positively charged 'vacancies' in photosystem II are powerful enough to cause the splitting of water (photolysis). This event triggers the release of hydrogen ions and oxygen atoms, as well as ground-state electrons. The oxygen atoms combine to form molecular oxygen, the waste product of photosynthesis. The hydrogen ions are used in the reduction of $NADP^+$ (see below).

Photophosphorylation

In the grana of the chloroplasts, the synthesis of ATP is coupled to electron transport via the movement of protons by chemiosmosis, as it was in mitochondria (Table 8.7). Here, it is the hydrogen ions trapped within the thylakoid space which flow out via ATP synthase enzymes, down their electrochemical gradient. At the same time, ATP is synthesized from ADP and P_i.

We have seen that the excited electrons which provided the energy for ATP synthesis originated from water, and move on to fill the vacancies in the reaction centre of photosystem II. They are subsequently moved on to the reaction centre in photosystem I, and finally are used to reduce $NADP^+$. Because the pathway of the electrons is linear, the photophosphorylation reaction in which they are involved is described as **non-cyclic photophosphorylation**.

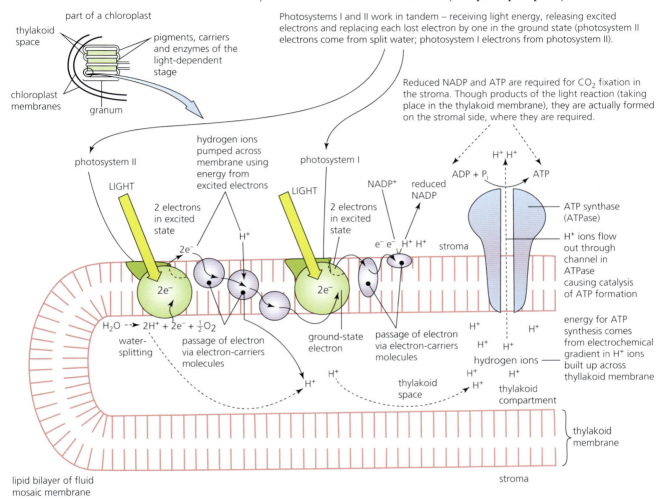

■ **Figure 8.19** The light-dependent reaction

11 In non-cyclic photophosphorylation, **deduce** the ultimate fate of electrons displaced from the reaction centre of photosystem II.

The excited electrons from photosystem I are then picked up by a different electron acceptor. Two at a time, they are passed to $NADP^+$, which, with the addition of hydrogen ions from photolysis, is reduced to form $NADPH + H^+$.

By this sequence of reactions, repeated again and again at very great speed throughout every second of daylight, the products of the light-dependent reactions ($ATP + NADPH + H^+$) are formed (Figure 8.19).

ATP and reduced NADP do not normally accumulate, however, as they are immediately used in the fixation of carbon dioxide in the surrounding stroma (light-independent reactions). Then the ADP and $NADP^+$ diffuse back into the grana for reuse in the light-dependent reactions.

12 Both reduced NADPH and ATP, products of the light stage of photosynthesis, are formed on the side of thylakoid membranes that face the stroma. **Suggest** why this fact is significant.

13 **Explain** how the gradient in protons between the thylakoid space and the stroma is generated.

■ **Table 8.7**
Chemiosmosis
in mitochondria
and chloroplasts
compared

Photosynthesis in chloroplasts		Cell respiration in mitochondria
thylakoid space in grana	**site of proton (H⁺) accumulation**	space between inner and outer membranes of mitochondria
from water molecules after photolysis has occurred	**origin of protons**	from reduced hydrogen acceptors (e.g. NADH + H⁺)
sunlight	**energy source**	glucose and respiratory intermediates
diffuses to stroma and used to sustain reduction of carbon dioxide in light-independent reactions	**fate of ATP formed**	diffuses into matrix of mitochondria and to cytosol and mainly involved in anabolic reactions of metabolism

Studying the light-dependent reactions with isolated chloroplasts

Chloroplasts can be isolated from green plant leaves and suspended in buffer solution of the same concentration as the cytosol (using an isotonic buffer). Suspended in such a buffer, it has been found that the chloroplasts are undamaged and function much as they do in the intact leaf. So, these isolated chloroplasts can be used to investigate the reactions of photosynthesis – for example, to show they evolve oxygen when illuminated (known as the **Hill reaction**). This occurs provided the natural electron-acceptor enzymes and carrier molecules are present. Note, that in the light-dependent reactions the oxygen given off in photosynthesis is derived exclusively from water (rather than from carbon dioxide).

In the research laboratory, a sensitive piece of apparatus called an **oxygen electrode** is used to detect the oxygen given off by isolated chloroplasts.

Alternatively, a **hydrogen-acceptor dye** that changes colour when it is reduced can be used. The dye known as DCPIP is an example. DCPIP does no harm when added to chloroplasts in a suitable buffer solution, but changes from blue to colourless when reduced. The splitting of water by light energy (photolysis) is the source of hydrogen that turns DCPIP colourless. The photolysis of water and the reduction of the dye are represented by the equation:

$$2DCPIP + 2H_2O \rightarrow 2DCPIPH_2 + O_2$$

In Figure 8.20 the steps for the isolation of chloroplasts and the investigation of their reducing activity are shown.

14 In the experiment shown in Figure 8.20, isolated chloroplasts are retained in isotonic buffer, standing in an ice bath. **Explain** the significance of this step.

■ The light-independent reactions

In the light-independent reactions, carbon dioxide is converted to carbohydrate. These reactions occur in the **stroma** of the chloroplasts, surrounding the grana.

How is this brought about?

The pathway by which carbon dioxide is reduced to glucose was investigated by a method we now call **feeding experiments**. In these particular experiments, radioactively labelled carbon dioxide was fed to cells. $^{14}CO_2$ is taken up by the cells in exactly the same way as non-labelled carbon dioxide and is then fixed into the same products of photosynthesis.

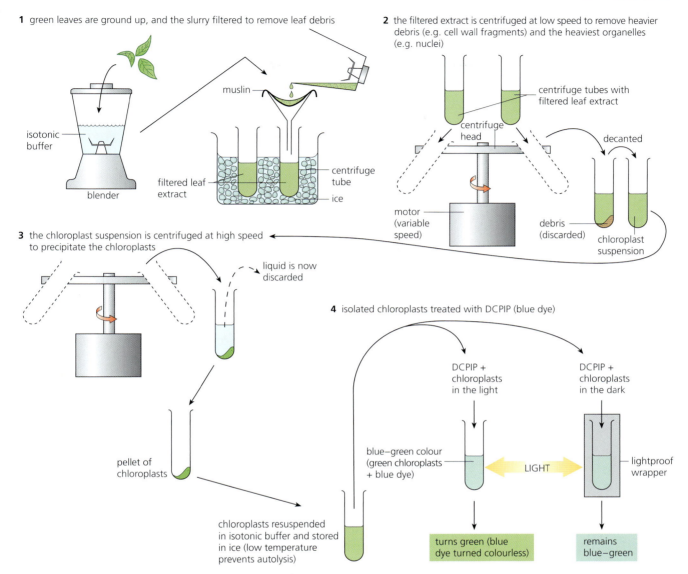

1 green leaves are ground up, and the slurry filtered to remove leaf debris

muslin

isotonic buffer

blender

filtered leaf extract

centrifuge tube

ice

2 the filtered extract is centrifuged at low speed to remove heavier debris (e.g. cell wall fragments) and the heaviest organelles (e.g. nuclei)

centrifuge tubes with filtered leaf extract

centrifuge head

decanted

motor (variable speed)

debris (discarded)

chloroplast suspension

3 the chloroplast suspension is centrifuged at high speed to precipitate the chloroplasts

liquid is now discarded

pellet of chloroplasts

chloroplasts resuspended in isotonic buffer and stored in ice (low temperature prevents autolysis)

4 isolated chloroplasts treated with DCPIP (blue dye)

DCPIP + chloroplasts in the light

DCPIP + chloroplasts in the dark

blue–green colour (green chloroplasts + blue dye)

LIGHT

lightproof wrapper

turns green (blue dye turned colourless)

remains blue–green

■ **Figure 8.20** Demonstration of the Hill reaction in isolated chloroplasts

So, a brief pulse of labelled $^{14}CO_2$ was introduced into the otherwise continuous supply of $^{12}CO_2$ to photosynthesizing cells in the light, and its progress monitored. Samples of the photosynthesizing cells, taken at frequent intervals after the $^{14}CO_2$ had been fed, contained a sequence of radioactively labelled intermediates, and (later) products, of the photosynthetic pathway. These compounds were isolated by **chromatography** from the sampled cells and identified (Figure 8.21). Chromatography was introduced in Figure 2.68, page 123.

The experimenters chose for their photosynthesizing cells a culture of *Chlorella*, a unicellular alga. They used these in place of mesophyll cells (Figure 1.13, page 13), since they have almost identical photosynthesis, but allow much easier sampling.

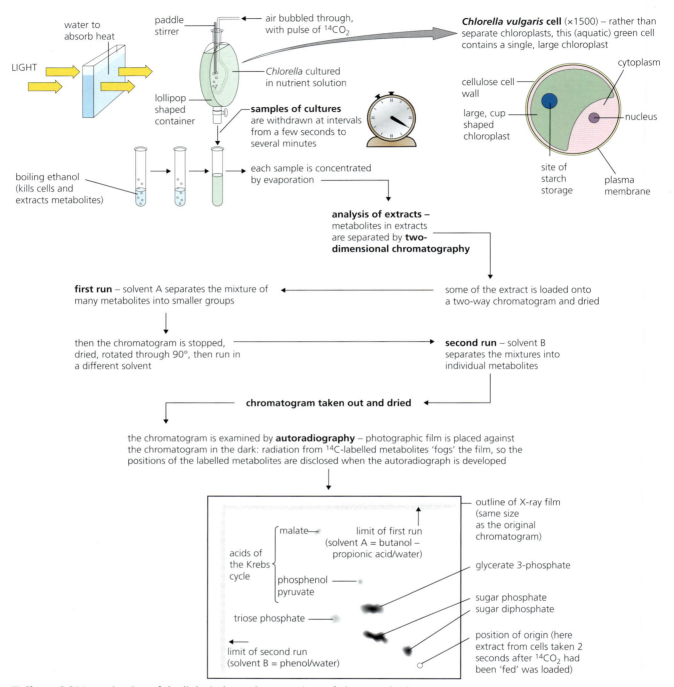

■ **Figure 8.21** Investigation of the light-independent reactions of photosynthesis

Developments in scientific research follow improvements in apparatus

■ Calvin's use of radioactive isotopes in biochemical research

The technique described above was pioneered by a team at the University of California, led by Melvin Calvin in the middle of the last century. He was awarded a Nobel Prize in 1961. The chromatography technique that the team exploited was then a relatively recent invention, and radioactive isotopes were only just becoming available for biochemical investigations. His design of 'lollipop' container is the issue in question 15.

15 Calvin used a suspension of aquatic, unicellular algal cells in his 'lollipop' container (Figure 8.21) to investigate the fixation of carbon dioxide in photosynthesis.

Suggest what advantages were obtained by this choice, compared to the use of cells of intact green leaves.

Suggest why the flask with the *Chlorella* was known as the 'lollipop' container. *Why do you think this type of container was used?*

■ **Figure 8.22**
The path of carbon in photosynthesis – Calvin cycle

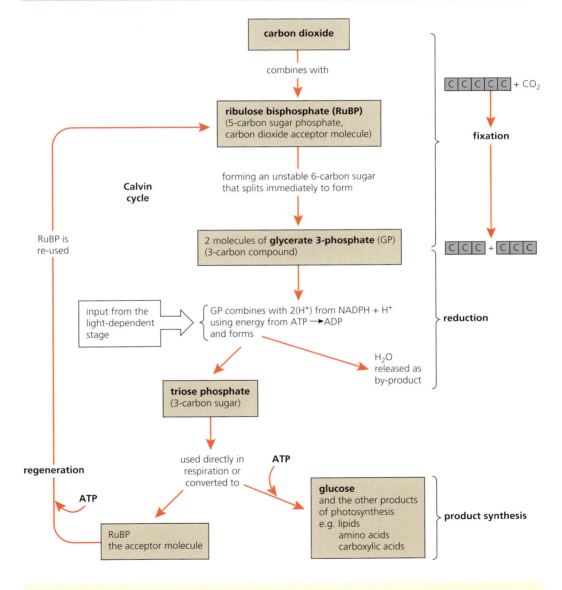

TOK Link

The lollipop experiment used to investigate the biochemical steps of the Calvin cycle shows creativity. To what extent is the creation of an elegant protocol similar to the creation of work in art?

The steps of the light-independent reactions

The experiments of Calvin's team established the details of the path of **carbon, from carbon dioxide to glucose** and other products (Figures 8.22 and 8.23). They showed that:

- The first product of the fixation of carbon dioxide is **glycerate 3-phosphate (GP)**. This is known as the **fixation step**.
- This initial product is immediately reduced to the 3-carbon sugar phosphate, **triose phosphate**, using **NADPH + H⁺** and **ATP**. This is the **reduction step**.
- Then, the triose phosphate is further metabolized to produce **carbohydrates** such as **sugars**, **sugar phosphates** and **starch**, and later **lipids**, amino acids such as **alanine**, and organic acids such as **malate**. This is the **product-synthesis step**.
- Some of the triose phosphate is metabolized to produce the molecule that first reacts with carbon dioxide (the acceptor molecule). This is the **regeneration-of-acceptor step**. The reactions of this regeneration process are today known as the **Calvin cycle**.

Even with these steps established, a problem remained for Calvin's team. *Which intermediate is the actual acceptor molecule?*

At first, a 2-carbon acceptor molecule for carbon dioxide was sought, simply because the first product of carbon dioxide fixation was known to be a 3-carbon compound. None was found.

Eventually, the acceptor molecule proved to be a 5-carbon acceptor (**ribulose bisphosphate**). When carbon dioxide has combined, the 6-carbon product immediately splits into two 3-carbon GP molecules – hence the initial confusion.

The enzyme involved is called **ribulose bisphosphate carboxylase** (commonly shortened to **RuBisCo**). RuBisCo is by far the most common protein of green plant leaves, as you would expect.

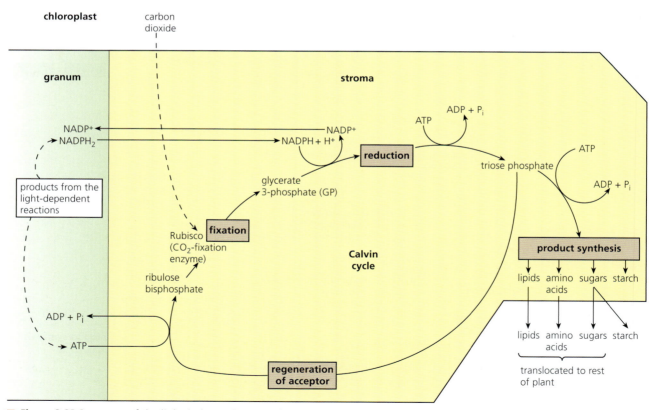

■ **Figure 8.23** Summary of the light-independent reactions

■ Photosynthesis and plant nutrition

The importance of photosynthesis to the green plant is that it provides the energy-rich sugar molecules from which the plant builds its other organic molecules. The metabolism of the green plant is sustained by the products of photosynthesis (Figure 8.24). But photosynthesis has a wider significance which was discussed on pages 127–128.

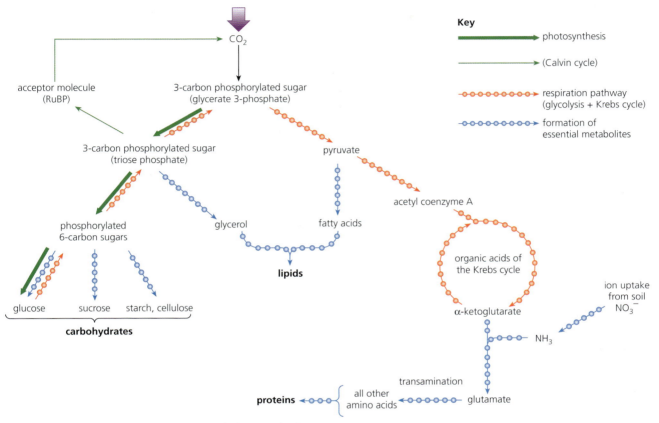

■ **Figure 8.24** The product synthesis steps of photosynthesis

Chloroplast, structure in relation to function

We have seen that the grana are the site of the light-dependent reactions and the stroma is the site of the light-independent reactions of photosynthesis. Figure 8.25 is of an electron micrograph showing the ultrastructure of a chloroplast. Table 8.8 identifies how the structure of the chloroplast facilitates function. You can make a drawing of the chloroplast ultrastructure in Figure 8.25 and annotate it, linking structures to function, guided by Table 8.8.

■ **Table 8.8**
Structure and function in chloroplasts

Structure of chloroplast	Function/role
double membrane bounding the chloroplast	contains the grana and stroma, and is permeable to CO_2, O_2, ATP, sugars and other products of photosynthesis
photosystems with chlorophyll pigments arranged on thylakoid membranes of grana	provide huge surface area for maximum light absorption
thylakoid spaces within grana	restricted regions for accumulation of protons and establishment of the gradient
fluid stroma with loosely arranged thylakoid membranes	site of all the enzymes of fixation, reduction and regeneration of acceptor steps of light-independent reactions, and many enzymes of the product synthesis steps

TEM of a thin section of a chloroplast (×15 000)

■ **Figure 8.25** Transmission electron micrograph of a chloroplast

 Make a line drawing of a chloroplast, showing its internal structure. Now annotate your diagram to indicate the ways in which the structure is adapted to its function – using the information in Table 8.8 together with that in Figure 8.16 (page 360).

16 **Distinguish** between the following:
   a light-dependent reactions and light-independent reactions
   b photolysis and photophosphorylation.

17 **Deduce** the significant difference between the starting materials and the end-products of photosynthesis.

18 **Draw** a diagram of a chloroplast to show its structure as revealed by electron microscopy. **Annotate** your diagram to indicate the adaptations of the chloroplast to its functions.

19 In cyclic photophosphorylation, **predict** which of the following occur:
   • photoactivation of photosystem I
   • reduction of $NADP^+$
   • production of ATP.

Examination questions – a selection

Questions 1–4 are taken from IB Diploma biology papers.

Q1 Which is correct for the non-competitive inhibition of enzymes?

	Inhibitor resembles substrate	Inhibitor binds to active site
A	yes	yes
B	yes	no
C	no	yes
D	no	no

Higher Level Paper1, Time Zone 1, May 10, Q27

Q2 What is an allosteric site?
  A The area on an enzyme that binds the end-product of a metabolic pathway
  B The area on a competitor molecule that inhibits an enzyme reaction
  C The site on an enzyme where the substrate binds
  D The active part of a non-competitive inhibitor of an enzyme reaction

Higher Level Paper 1, Time Zone 2, May 13, Q28

Q3 The enzyme succinic dehydrogenase catalyses the conversion of succinate to fumarate.

$$\text{succinate} \xrightarrow{\text{succinate dehydrogenase}} \text{fumarate}$$

The addition of malonate to the reaction mixture decreases the rate of the reaction. If more succinate is added, the reaction rate will increase. What is the role of malonate in this reaction?
  A End-product inhibitor
  B Non-competitive inhibitor
  C Catalyst
  D Competitive inhibitor

Higher Level Paper 1, Time Zone 1, May 13, Q27

Q4 The diagram represents components of the cristae in mitochondria. Which arrow indicates how protons (H⁺) move to generate ATP directly?

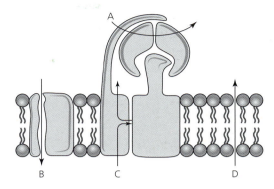

Higher Level Paper 1, Time Zone 0, Nov 12, Q29

Questions 5–10 cover other syllabus issues in this chapter.

Q5 Describe the way the rate of enzymatic reaction is affected by the interaction of the following:
  a competitive inhibitor
  b a non-competitive inhibitor
  c the presence of an alkali environment. (6)

Q6 Chemiosmosis is a process by which the synthesis of ATP is coupled to electron transport via the movement of protons. Draw and annotate a diagram to show how the structure of the membranes of a mitochondrion (given the position of electron carriers and specific enzymes within it) allows a large proton gradient to be formed, and ATP synthesis to follow. (8)

Q7 **a** Explain as concisely as you are able the distinctive nature of non-cyclic and cyclic photophosphorylation. (6)
  b Describe the specific circumstances which cause cyclic photophosphorylation to occur. (2)

Q8 **a** Explain the concept of limiting factors in photosynthesis, with reference to light intensity and the concentration of carbon dioxide. (4)

b Draw a fully annotated graph of net photosynthesis rate (*y* axis) against light intensity (*x* axis) at low and high concentrations of carbon dioxide. (4)

Q9 Distinguish between light dependent and light independent reactions and indicate the role of both water and carbon dioxide in photosynthesis. (4)

Q10 Draw and label a chloroplast indicating the place for absorption of light, where glucose is stored and where NADP and the enzyme RuBisCo are located. (6)

9 Plant biology

ESSENTIAL IDEAS

■ Structure and function are correlated in the xylem of plants.
■ Structure and function are correlated in the phloem of plants.
■ Plants adapt their growth to environmental conditions.
■ Reproduction in flowering plants is influenced by the biotic and abiotic environment.

In the long history of life, green plants (**Plantae**) evolved about 500 million years ago, from aquatic, single-celled organisms called green algae (possibly very similar to *Chlorella*, Figure 8.21, page 366). The features that make the green plants distinctly different from other organisms are:

■ a wall around each cell, the chief component of which is **cellulose**

■ possession of organelles called **chloroplasts** – the sites of photosynthesis.

The diversity of green plant life was introduced in Chapter 5, pages 228–230. Today, it is the **angiospermophytes** (flowering plants) that are the dominant plants in almost every terrestrial habitat across the world. The fossil record indicates they achieved this dominance early in their evolutionary history – about 100 million years ago.

Some flowering plants are trees and shrubs with woody stems, but many are non-woody (herbaceous) plants. Whether woody or herbaceous, the plant consists of stem, leaves and root. In this chapter it is the **internal transport** mechanisms, **growth**, and **reproduction** of plants that are discussed.

9.1 Transport in the xylem of plants – *structure and function are correlated in the xylem of plants*

■ Transpiration is the inevitable consequence of gas exchange in the leaf

Transpiration is the process by which water vapour is lost by evaporation, mainly from the leaves of the green plant. This loss of water vapour is the inevitable consequence of gas exchange in the leaf. We will see that the plant is forced to transport water from the roots to the leaves to replace losses from transpiration.

■ Structure of stem, root and leaf

■ The **stem** supports the leaves in the sunlight and transports organic materials (such as sugar and amino acids), ions and water between the roots and leaves.

■ The **root** anchors the plant and is the site of absorption of water and ions from the soil.

■ A **leaf** is an organ specialized for photosynthesis and consists of a leaf blade connected to the stem by a leaf stalk.

The internal structure of plants is seen by examination of transverse sections. A **tissue map** (sometimes called a low-power diagram) is a drawing that records the relative positions of structures within an organ or organism, seen in section; it does not show individual cells.

From the tissue map in Figure 9.1, we see that the stem surrounded or contained by a layer of cells called the **epidermis**. The stem contains **vascular** tissue (**xylem** for water transport and **phloem** for transport of organic solutes) in a discrete system of veins or **vascular bundles**. In the stem, the vascular bundles are arranged in a ring, positioned towards the outside of the stem (in a region that we call the cortex), rather like the steel girders of a ferro-concrete building.

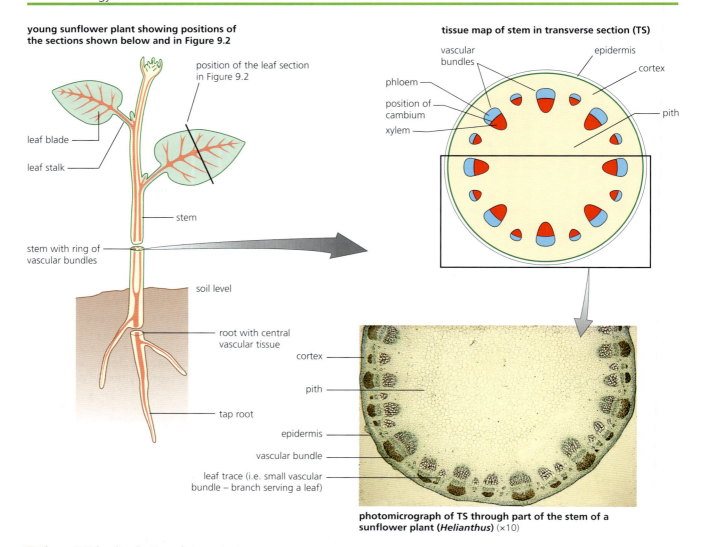

young sunflower plant showing positions of the sections shown below and in Figure 9.2

position of the leaf section in Figure 9.2

leaf blade

leaf stalk

stem

stem with ring of vascular bundles

soil level

root with central vascular tissue

tap root

tissue map of stem in transverse section (TS)

vascular bundles

epidermis

phloem

cortex

position of cambium

xylem

pith

cortex

pith

epidermis

vascular bundle

leaf trace (i.e. small vascular bundle – branch serving a leaf)

photomicrograph of TS through part of the stem of a sunflower plant (*Helianthus*) (×10)

■ **Figure 9.1** The distribution of tissues in the stem

The **leaf**, like the stem, is contained by a single layer of cells, the **epidermis**, and also contains vascular tissue in a system of vascular bundles (Figure 9.2). The vascular bundles in leaves are often referred to as **veins**. The bulk of the leaf is taken up by a tissue called **mesophyll** and the cells here are supported by veins arranged in a branching network. The roots grow through the soil, which is an entirely supporting medium. In the roots, the vascular tissue (xylem and phloem) is arranged in a central region, the stele.

■ Plants transport water from the roots to the leaves

The uptake of water occurs in the root system, but the movement of water up the stem is driven by conditions in the leaves. We need to focus on both these regions of the plant. First, we look into water absorption in the roots. The uptake of ions also occurs here.

Root system, absorption and uptake

The root system provides a huge surface area in contact with soil, because plants have a system of branching roots that continually grow at each root tip, pushing through the soil. This is important because it is the dilute solution that occurs around soil particles which the plant draws on for the large volume of **water** it requires, and also for **essential ions**.

Contact with the soil is vastly increased by the **region of root hairs** that occurs just behind the growing tip of each root (Figure 9.3). Root hairs are extensions of individual epidermal cells and are relatively short-lived. As root growth continues, fresh resources of soil solution are exploited.

■ **Figure 9.2**
The distribution
of tissues in the
leaf

tissue map of leaf in cross-section

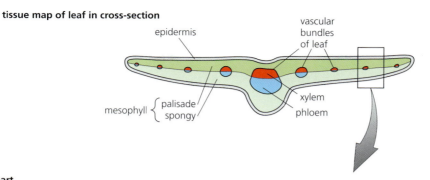

epidermis

vascular
bundles
of leaf

mesophyll { palisade
spongy

xylem

phloem

**photomicrograph of part
of a leaf in cross-section**
(×120)

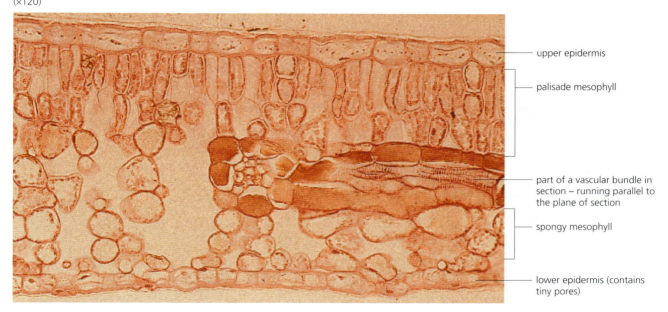

upper epidermis

palisade mesophyll

part of a vascular bundle in
section – running parallel to
the plane of section

spongy mesophyll

lower epidermis (contains
tiny pores)

So water uptake occurs from the soil solution that is in contact with the root hairs. Uptake is largely by mass flow through the interconnecting 'free spaces' in the cellulose cell walls, but there are three possible routes of water movement through plant cells and tissues (Figure 9.4).

■ **Mass flow** occurs through the interconnecting free spaces between the cellulose fibres of the plant cell walls. This free space in cellulose makes up about 50% of the wall volume. This route is a highly significant one in water movement about the plant. This pathway, entirely avoiding the living contents of cells, is called the **apoplast**. The apoplast also includes the water-filled spaces of dead cells (and the hollow xylem vessels – as we shall shortly see).

■ **Diffusion** occurs through the cytoplasm of cells and via the cytoplasmic connections between cells (called plasmodesmata). This route is called the **symplast**. As the plant cells are packed with many organelles, which offer resistance to the flow of water, this pathway is very significant.

■ **Osmosis** occurs from vacuole to vacuole of cells, driven by a gradient in osmotic pressure. Active uptake of mineral ions in the roots causes absorption of water by osmosis. This is not a significant pathway of water transport across the plant, but it is the means by which individual cells absorb water.

While all three routes are open, the bulk of water crosses from the epidermis of the root tissue to the xylem via the apoplast (Figures 9.4 and 9.8).

Ion uptake

Ion uptake by the roots from the surrounding soil solution is by active transport. In active transport, metabolic energy is used to drive the transport of molecules and ions across cell membranes. Active transport is a distinctly different process from movement by diffusion.

1 **List** the features of root hairs that facilitate absorption of water from the soil.

2 **Explain** the difference between the symplast and the apoplast.

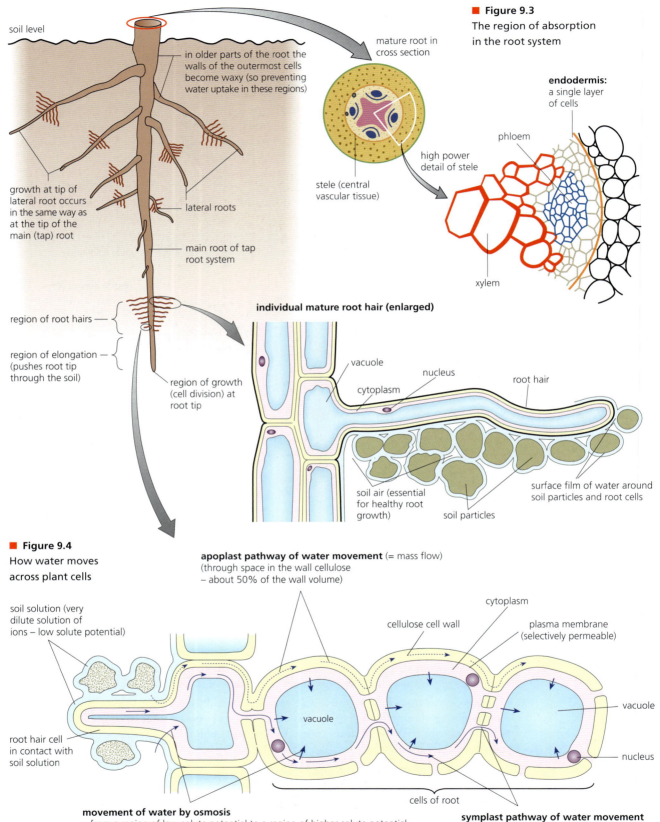

soil level

in older parts of the root the walls of the outermost cells become waxy (so preventing water uptake in these regions)

growth at tip of lateral root occurs in the same way as at the tip of the main (tap) root

lateral roots

main root of tap root system

region of root hairs

region of elongation (pushes root tip through the soil)

region of growth (cell division) at root tip

■ **Figure 9.3**
The region of absorption in the root system

mature root in cross section

endodermis:
a single layer of cells

phloem

high power detail of stele

stele (central vascular tissue)

xylem

individual mature root hair (enlarged)

vacuole

nucleus

root hair

cytoplasm

soil air (essential for healthy root growth)

soil particles

surface film of water around soil particles and root cells

■ **Figure 9.4**
How water moves across plant cells

apoplast pathway of water movement (= mass flow)
(through space in the wall cellulose – about 50% of the wall volume)

cytoplasm

cellulose cell wall

plasma membrane (selectively permeable)

soil solution (very dilute solution of ions – low solute potential)

vacuole

vacuole

nucleus

root hair cell in contact with soil solution

cells of root

movement of water by osmosis
– from a region of low solute potential to a region of higher solute potential

symplast pathway of water movement
(through living cytoplasm, by diffusion, including through the cytoplasmic connections between cells)

- **Active transport may occur against a concentration gradient** – that is, from a region of low to a region of higher concentration. The cytosol of a cell normally holds some reserves of ions that are essential to metabolism, like nitrate ions in plant cells. These useful ions do not escape; the cell membrane retains them inside the cell. Indeed, when additional ions become available to root hair cells, they are actively absorbed into the cells, too. In fact, plant cells tend to hoard valuable ions like nitrates and calcium ions, even when they are already at higher concentration inside the cytoplasm than outside the cell.
- **Active uptake is a highly selective process.** For example, in a situation where sodium nitrate (Na^+ and NO_3^- ions) is available to the root hairs, it is likely that more of the NO_3^- ions are absorbed than the Na^+, since this reflects the needs of the whole plants.
- **Active transport involves special molecules of the membrane, called pumps.** The pump molecule picks up particular ions and transports them to the other side of the membrane, where they are then released. The pump molecules are globular proteins that traverse the lipid bilayer. Movements by these pump molecules require reaction with ATP; by this reaction, metabolic energy is supplied to the process. Most membrane pumps are specific to particular molecules or ions and this is the way selective transport is brought about. If the pump molecule for a particular substance is not present, the substance will not be transported.

It is the presence of numerous specific protein pumps in the plasma membranes of all the root hair cells that makes possible the efficient way that ions are absorbed (Figure 9.5). Roots are metabolically very active and they require a supply of oxygen for aerobic cell respiration. By this process, the required supply of ATP for ion uptake is maintained.

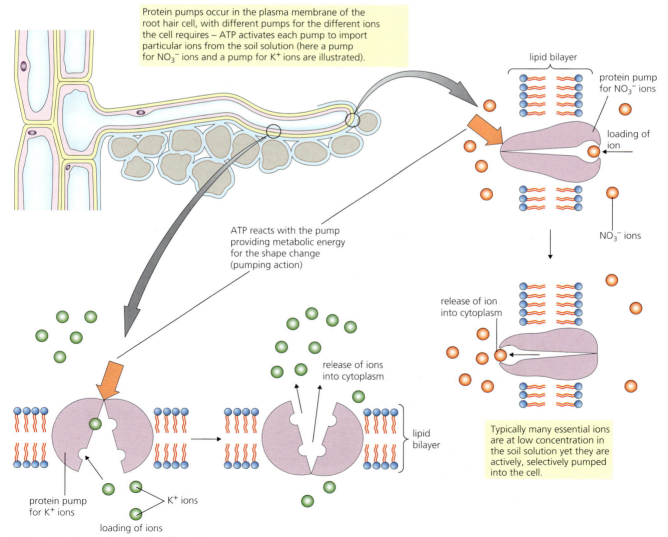

Protein pumps occur in the plasma membrane of the root hair cell, with different pumps for the different ions the cell requires – ATP activates each pump to import particular ions from the soil solution (here a pump for NO_3^- ions and a pump for K^+ ions are illustrated).

lipid bilayer

protein pump for NO_3^- ions

loading of ion

NO_3^- ions

ATP reacts with the pump providing metabolic energy for the shape change (pumping action)

release of ion into cytoplasm

release of ions into cytoplasm

Typically many essential ions are at low concentration in the soil solution yet they are actively, selectively pumped into the cell.

lipid bilayer

protein pump for K^+ ions

K^+ ions

loading of ions

■ **Figure 9.5** The active uptake of ions by protein pumps in root hair cell membranes

How ions reach the cell membranes

The soil solution contains ions at relatively low concentrations, so how do root hair cells access adequate quantities of essential ions? The mechanisms that maintain a supply of ions to the roots are as follows.

1 The **mass flow of water through the free spaces in the cellulose walls** (the apoplast pathway of water movement – Figure 9.4) continuously delivers fresh soil solution to the root hair cell plasma membranes (and to other cells of the root).

2 Active uptake of selected ions from the soil solution of the apoplast maintains the **concentration gradient**. Ions diffuse from higher concentrations outside the apoplast to the solution of lower concentration immediately adjacent to the protein pumps. Remember, active uptake of ions in the roots causes water absorption there by osmosis.

3 The **mutualistic relationship** that many plants have with species of soil-inhabiting fungi is also important. The fungal hyphae receive a supply of sugar from the plant root cells (green plants generally have an excess of sugar). In return, the fungal hyphae release to the root cells ions that have previously been taken up by the fungus as and when they became available in the soil. The spread of fungal hyphae through the surrounding soil is very extensive. The periodic death and decay of other organisms on or in the soil, releases a supply of ions – very often some distance from the plant roots, but in reach of fungal hyphae. Nevertheless, plant roots obtain many of these ions later, from the fungal hyphae, thereby establishing that the plant–fungus relationship is one of mutual benefit.

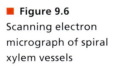

3 **Explain** the significance of root hair cells being able to take up nitrate ions from the soil solution even though their concentration in the cell is already higher than in the soil.

4 **Suggest** why plants often fail in soil which is persistently waterlogged.

Stems and water transport – transport under tension

Transport of water through the plant occurs in the **xylem** tissue. The cohesive property of water and the structure of xylem vessels allow transport under tension.

Xylem begins as elongated cells with cellulose walls and living contents, connected end to end. During development, the end walls are dissolved away so that mature xylem vessels become **long, hollow tubes**. The living contents of a developing xylem vessel are used up in the process of depositing cellulose thickening to the inside of the lateral walls of the vessel. This is hardened by the deposition of a chemical substance, **lignin**. Consequently, xylem is extremely tough tissue. Furthermore, it is strengthened internally, which means it is able to resist negative pressure (suction) without collapsing in on itself. Figure 9.6 is a scanning electron micrograph of spirally thickened xylem vessels. When you examine xylem vessels in longitudinal sections by light microscopy, you will see that some xylem vessels may have differently deposited thickening – many have rings of thickening, for example.

In a root, the xylem is centrally placed but, in the stem, xylem occurs in the ring of vascular bundles – as we saw in Figure 9.1. Nevertheless, xylem of root and stem are connected.

■ **Figure 9.6**
Scanning electron micrograph of spiral xylem vessels

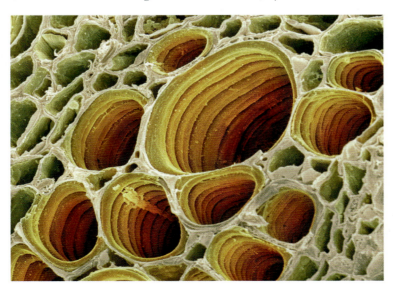

Drawing the structure of xylem vessels seen in sections of stems

The structure of xylem vessels is seen in transverse (TS) and longitudinal sections (LS) of plant stems. In prepared slides of sectioned and stained plant organs, the cell walls that have been strengthened with lignin stand out particularly clearly. You can see the xylem vessels and fibres shown *in situ* within a TS of a stem on page 374 (Figure 9.1). You can see the position of xylem in a root on page 376 (Figure 9.3). You need to be able to identify xylem and phloem in microscope images of both a stem and a root, and to be able to draw a diagram showing xylem vessels in a section of a stem.

Early growth of the plant is called primary growth (Figure 9.23, page 396). In the stem, primary xylem tissue forms on the innermost side of the developing vascular bundle. Meanwhile, primary phloem forms at the outer side of the vascular bundle. There is often a 'cap' of fibres at the outmost edge of the vascular bundle.

The very first-formed primary vessels (protoxylem) have thinner walls than the later-formed xylem. These first xylem vessels have annular (rings) or spirally-arranged thickening that permits stretching of the tissues during continuing growth in length of the stem. The late-formed metaxylem has massively thickened walls. This forms where growth in length of the stem is completed. Xylem vessels have permeable pits in their walls, too.

These differences are best seen by investigation of both transverse and longitudinal stained sections of plant stems (Figure 9.7). Examine slides carefully before you make and annotate drawings to record the structure of primary xylem vessels. The guidance on recording by drawing from microscope images (pages 7–8) should be followed.

■ **Figure 9.7**
The structure of xylem vessels seen in sections of stems

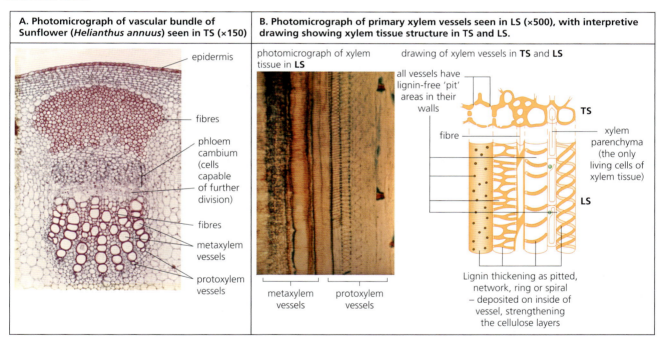

A. Photomicrograph of vascular bundle of Sunflower (*Helianthus annuus*) seen in TS (×150)	B. Photomicrograph of primary xylem vessels seen in LS (×500), with interpretive drawing showing xylem tissue structure in TS and LS.

A. labels: epidermis; fibres; phloem; cambium (cells capable of further division); fibres; metaxylem vessels; protoxylem vessels

B. photomicrograph of xylem tissue in **LS**: metaxylem vessels; protoxylem vessels

drawing of xylem vessels in **TS** and **LS**: all vessels have lignin-free 'pit' areas in their walls; fibre; **TS**; xylem parenchyma (the only living cells of xylem tissue); **LS**; Lignin thickening as pitted, network, ring or spiral – deposited on inside of vessel, strengthening the cellulose layers

Movement of water in the xylem

We have already seen that water uptake occurs from the soil solution, mainly at the root hairs (Figure 9.3). This occurs largely by mass flow through the interconnecting free spaces in the cellulose cell walls (apoplast – Figure 9.4).

In a root, the centrally placed vascular tissue is contained by the **endodermis**. The endodermis is a layer of cells that is unique to the root. At the endodermis, a waxy strip in the radial walls momentarily blocks the passage of water by the apoplast route (Figure 9.8). This waxy strip is called the **Casparian strip**. Water passes through the endodermis by osmosis. The significance of this diversion is that the cytoplasm of endodermal cells actively transports ions from the cortex to the endodermis. The result is a higher concentration of ions in the cells at the centre of the root. The resulting raised osmotic potential there causes passive water uptake

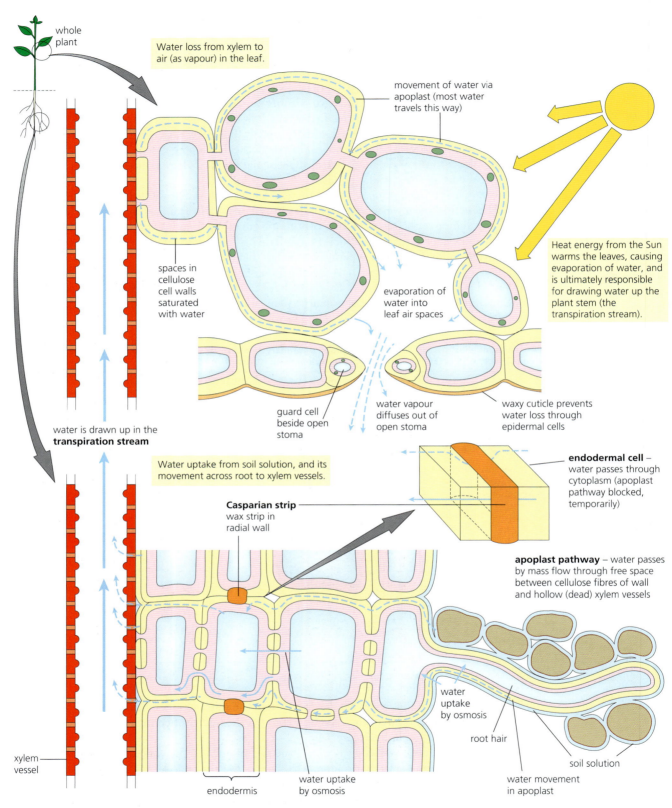

whole plant

Water loss from xylem to air (as vapour) in the leaf.

movement of water via apoplast (most water travels this way)

spaces in cellulose cell walls saturated with water

Heat energy from the Sun warms the leaves, causing evaporation of water, and is ultimately responsible for drawing water up the plant stem (the transpiration stream).

evaporation of water into leaf air spaces

water is drawn up in the **transpiration stream**

guard cell beside open stoma

water vapour diffuses out of open stoma

waxy cuticle prevents water loss through epidermal cells

Water uptake from soil solution, and its movement across root to xylem vessels.

endodermal cell – water passes through cytoplasm (apoplast pathway blocked, temporarily)

Casparian strip wax strip in radial wall

apoplast pathway – water passes by mass flow through free space between cellulose fibres of wall and hollow (dead) xylem vessels

water uptake by osmosis

root hair

xylem vessel

endodermis

water uptake by osmosis

soil solution

water movement in apoplast

■ **Figure 9.8** Water uptake and loss by a green plant – a summary

to follow, by osmosis. Once again, it is active uptake and transfer of ions within the root that causes water uptake by osmosis.

Meanwhile, in the leaves, evaporation of water occurs and water vapour diffuses out of the stomata, a process called transpiration. **Transpiration is the evaporation of water vapour through stomata of green plant leaves (and stems).**

■ The transpiration stream

The water that evaporates from the walls of the mesophyll cells of the leaf is continuously replaced. It comes, in part, from the cell cytoplasm (the symplast pathway), but mostly it comes from the water in the spaces in walls in nearby cells and then from the xylem vessels in the network of vascular bundles nearby (the apoplast pathway).

These xylem vessels are full of water. As water leaves the xylem vessels in the leaf a tension is set up on the entire water column in the xylem tissue of the plant. This tension is transmitted down the stem to the roots because of the **cohesion** of water molecules. Cohesion of water molecules is due to hydrogen bonds (page 69) – the water molecules are held together because they are polar and so are strongly attracted to each other. This same force causes the water to adhere to the sides of the xylem vessels, as well – a force called **adhesion**. Consequently, under tension, the water column does not break or pull away from the sides of the xylem vessels. The result is that water is drawn (literally 'pulled') up the stem. So, water flow in the xylem is always upwards. We call this flow of water the **transpiration stream** and the explanation of water transport up the stem is the **cohesion–tension theory**.

The tension on the water column in xylem is demonstrated experimentally when a xylem vessel is pierced by a fine needle. Immediately, a bubble of air enters the column (and will interrupt water flow). If the contents of the xylem vessel had been under pressure, a jet of water would be released from broken vessels.

Further evidence of the cohesion–tension theory comes from measurement of the diameter of a tree trunk over a 24-hour period (Figure 9.9). In a large tree, there is an easily detectable shrinkage in the diameter of the trunk during the day. Under tension, xylem vessels get narrower in diameter, although their collapse is prevented by the lignified thickening of their walls. Notice that these are on the inside of the xylem vessels (Figures 9.6 and 9.7).

The diameter of the tree trunk recovers during the night when transpiration virtually stops and water uptake makes good the earlier losses of water vapour from the aerial parts of the plant.

■ **Figure 9.9**
Diurnal changes in the circumference of a tree over a 7-day period

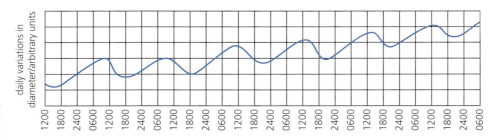

The data were obtained in early May. The tree trunk is undergoing secondary growth in girth during the experimental period. The maximum daily shrinkage amounted to nearly 5 mm.

5 **Explain** to what extent the graph in Figure 9.9 supports the cohesion–tension theory of water movement in stems.

6 **Explain** the consequence of the Casparian strip for the apoplast pathway of water movement.

Stomata and transpiration

The tiny pores of the epidermis of leaves, through which gas exchange can occur, are known as stomata (Figure 9.10). Most stomata occur in the epidermis of leaves, but some do occur in stems. In the typical broad leaf, stomata are concentrated in the lower epidermis. Each stoma consists of two elongated guard cells. These cells are attached to ordinary epidermal cells that surround them and are also securely joined together at each end. However, guard cells are detached and free to separate along the length of their abutting sides. When they separate, a pore appears between them.

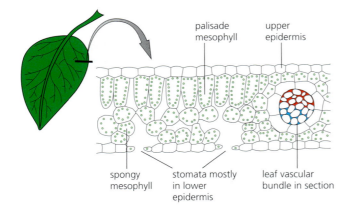

palisade mesophyll
upper epidermis
spongy mesophyll
stomata mostly in lower epidermis
leaf vascular bundle in section

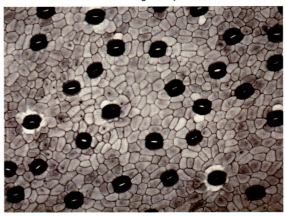

photomicrograph of lower surface of leaf – showing distribution of stomata among the epidermal cells (×100)

photomicrograph of structure of individual stoma (×500)

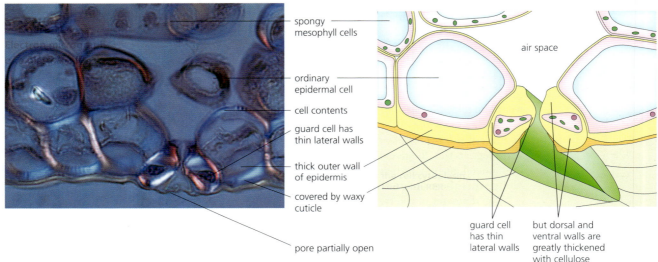

spongy mesophyll cells

ordinary epidermal cell

cell contents

guard cell has thin lateral walls

thick outer wall of epidermis

covered by waxy cuticle

pore partially open

air space

guard cell has thin lateral walls

but dorsal and ventral walls are greatly thickened with cellulose

■ **Figure 9.10** The distribution and structure of stomata

Stomata open and close due to change in turgor pressure of the guard cells. They open when water is absorbed by the guard cells from the surrounding epidermal cells. The guard cells then become fully turgid and they each push into the epidermal cell beside them (because of the way cellulose is laid down in the walls, Figure 9.10). A pore develops between the guard cells. When water is lost and the guard cells become flaccid, the pore closes again.

Stomata tend to open in daylight and be closed in the dark (but there are exceptions to this) (Figure 9.11). This diurnal pattern is overridden, however, if and when the plant becomes short of water and starts to wilt. For example, in very dry conditions when there is an inadequate water supply, stomata inevitably close relatively early in the day (turgor cannot be maintained). This curtails water vapour loss by transpiration and halts further wilting. Adequate water reserves from the soil may be taken up subsequently, thereby allowing the opening of stomata again – for example, on the following day. The effect of this mechanism is that stomata regulate transpiration by preventing excessive water loss (Figure 9.12).

7 **Deduce** why changes in the turgor of guard cells cause opening of stomata.

■ **Figure 9.11**
The site of
transpiration

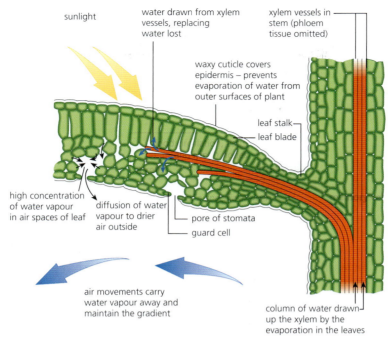

sunlight

water drawn from xylem
vessels, replacing
water lost

xylem vessels in
stem (phloem
tissue omitted)

waxy cuticle covers
epidermis – prevents
evaporation of water from
outer surfaces of plant

leaf stalk
leaf blade

high concentration
of water vapour
in air spaces of leaf

diffusion of water
vapour to drier
air outside

pore of stomata

guard cell

air movements carry
water vapour away and
maintain the gradient

column of water drawn
up the xylem by the
evaporation in the leaves

■ **Figure 9.12**
Stomatal opening
and environmental
conditions

8 Examine Figure 9.12.
Suggest why the
stomatal apertures
of the plant in very
dry conditions
differed in both
maximum size and
duration of opening
from those of the
plant with adequate
moisture.

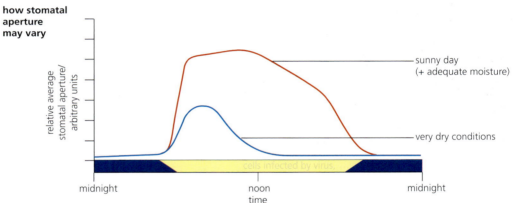

**how stomatal
aperture
may vary**

relative average
stomatal aperture/
arbitrary units

sunny day
(+ adequate moisture)

very dry conditions

midnight

noon
time

midnight

Measuring transpiration rates using a potometer

The rate of transpiration may be investigated using a potometer (Figure 9.13).

■ **Figure 9.13**
Investigating
transpiration and
the factors that
influence it

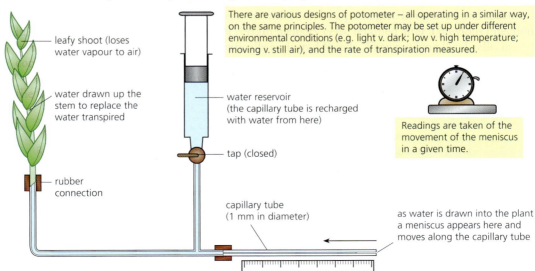

leafy shoot (loses
water vapour to air)

water drawn up the
stem to replace the
water transpired

rubber
connection

water reservoir
(the capillary tube is recharged
with water from here)

tap (closed)

capillary tube
(1 mm in diameter)

There are various designs of potometer – all operating in a similar way,
on the same principles. The potometer may be set up under different
environmental conditions (e.g. light v. dark; low v. high temperature;
moving v. still air), and the rate of transpiration measured.

Readings are taken of the
movement of the meniscus
in a given time.

as water is drawn into the plant
a meniscus appears here and
moves along the capillary tube

Transpiration is found to be dramatically affected by **environmental conditions** around the plant. *Why is this so?*

Transpiration occurs because water molecules continuously evaporate from the cellulose walls of cells in the leaf which are saturated with water. This makes the air in the air spaces between mesophyll cells more or less saturated with water vapour. If the air outside the plant is less saturated (less humid – as it very often is) *and the stomata are open*, then water vapour will diffuse out into the drier air outside.

You will need to design particular experiments to test hypotheses about the effect of temperature or humidity on transpiration rates, using a potometer. *How will changes in one of these environmental conditions affect transpiration?*

You will already appreciate these factors:

■ **Temperature** affects transpiration because it causes the evaporation of water molecules from the surfaces of the cells of the leaf. A rise in the concentration of water vapour within the air spaces increases the difference in concentration in water vapour between the leaf's interior and the air outside, and diffusion is enhanced. So, an increase in temperature of the leaf raises the transpiration rate.

■ If humid air collects around a leaf, it decreases the difference in concentration of water vapour between the interior and exterior of the leaf, so slowing diffusion of water vapour from the leaf. High **humidity** slows transpiration.

Also relevant, perhaps as factors to control in your experiments:

■ **Wind** sweeps away the water vapour molecules accumulating outside the stomata of the epidermis of the leaf surface, so enhancing the difference in concentration of water vapour between the leaf interior and the outside. Movements of air around the plant enhance transpiration.

■ **Light** affects transpiration because the stomata tend to be open in the light – essential for loss of water vapour from the leaf. Light from the Sun also contains infra-red rays which warm the leaf and raise its temperature. Light is an essential factor for transpiration.

Designing an investigation of transpiration

You have had experience of experimenting with a potometer and are, therefore, familiar with its use. The focus of this experiment is to obtain data on the effects of temperature or humidity on the rate of transpiration. In designing your approach you will need to decide and resolve:

1 the environmental factor you will investigate
2 how this factor can be varied experimentally around the leafy shoot, and how variable this condition should be
3 what other factors it is essential to control
4 how long to allow the plant material to adapt to and stabilize in the new conditions, each time a change is made
5 how to achieve and establish reproducible results.

When evaluating your approach, reflect on the fact that the potometer measures *water uptake*, and the possible significance of this fact.

Does transpiration have a role?

It is evident that transpiration is a direct result of plant structure, plant nutrition and the mechanism of gas exchange in leaves, rather than being a valuable process. In effect, the living plant is a 'wick' that steadily dries the soil around it. However, transpiration confers advantages, too:

■ Evaporation of water from the cells of the leaf in the light has a strong cooling effect.

■ The stream of water travelling up from the roots in the xylem passively carries the dissolved ions that have been actively absorbed from the soil solution in the root hairs. These are required in the leaves and growing points of the plant.

■ All the cells of a plant receive water by lateral movements of water from xylem vessels, via pits in their walls. This allows living cells to be fully hydrated. It is the turgor pressure of these cells that provides support to the whole leaf, enabling the leaf blade to receive maximum exposure to light. In fact, the entire aerial system of non-woody plants is supported by this turgor pressure.

So, transpiration does have significant roles in the life of the plant.

Using models as representation of the real world

■ Models of water transport driven by evaporation – simulations of transpiration

Models of water transport in the xylem, and demonstrations of the power and inevitability of evaporation from moist surfaces can be designed and tested using familiar laboratory equipment, after appropriate risk assessments have been carried out. Figure 9.14 illustrates two possible approaches. What does your approach establish about the mechanisms involved in transpiration?

Evaporation and the cohesive/adhesive properties of water power water movements up the fine tubes

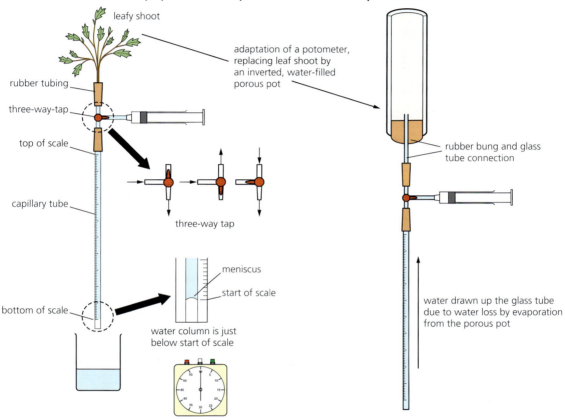

Comparison of water (mass) loss from a leaf and a comparable area of damp filter paper

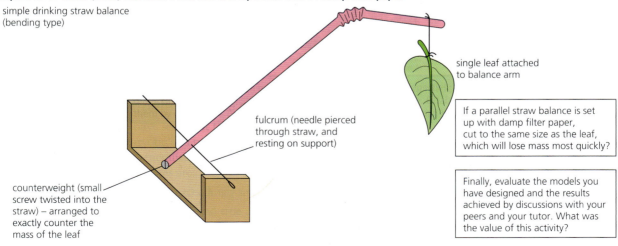

If a parallel straw balance is set up with damp filter paper, cut to the same size as the leaf, which will lose mass most quickly?

Finally, evaluate the models you have designed and the results achieved by discussions with your peers and your tutor. What was the value of this activity?

■ **Figure 9.14** Demonstrations of the power and inevitability of evaporation

■ Water scarcity and adaptations of plants

Most native plants of temperate and tropical zones, and most of our crop plants, grow best in habitats with adequate rainfall, well-drained soils and with their aerial system (stem and leaves) exposed to moderately dry air. Loss of water vapour from the leaves may be substantial in drier periods, particularly in the early part of the day, but excessive loss from the leaves is prevented by the responses of the stomata (Figure 9.12). Any deficit is normally made good by the water uptake that continues, day and night. It is the structure of these sorts of plants (known as mesophytes) that we have been considering in this chapter.

On the other hand, **xerophytes** are plants that are able to survive and grow well in habitats where water is scarce. These plants show features that directly or indirectly help to minimize water loss, due to transpiration. Their adaptations are referred to as **xeromorphic features**, and these are summarized in Table 9.1.

■ **Table 9.1**
Xeromorphic features

Structural features	Effect
exceptionally thick cuticle to leaf (and stem) epidermis	prevents water loss through the external wall of the epidermal cells
layer of hairs on the epidermis	traps moist air over the leaf and reduces diffusion
reduction in the number of stomata	reduces outlets through which moist air can diffuse
stomata in pits or grooves	moist air is trapped outside the stomata, reducing diffusion
leaf rolled or folded when short of water (cells flaccid)	reduces area from which transpiration can occur
superficial roots	exploit overnight condensation at soil surface
deep and extensive roots	exploit a deep water table in the soil

■ **Figure 9.15**
Marram grass (*Ammophilia*), a pioneer plant of sand dunes

Marram grass (*Ammophilia arenaria*) is a plant that grows in sand dunes around coasts in Western Europe. It has underground stems and roots. Above ground are cylindrical leaves, rolled up so that their lower surface (carrying the stomata) is enclosed. *Look carefully at the TS of the leaf (Figure 9.15). What other xeromorphic features does it show?*

marram grass has the ability to grow in the extremely arid environment of sand dunes, accelerating the build-up of sand

TS of leaf (×50)

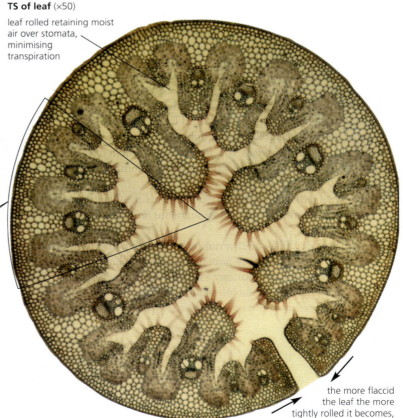

leaf rolled retaining moist air over stomata, minimising transpiration

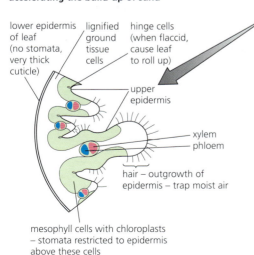

lower epidermis of leaf (no stomata, very thick cuticle)

lignified ground tissue cells

hinge cells (when flaccid, cause leaf to roll up)

upper epidermis

xylem
phloem

hair – outgrowth of epidermis – trap moist air

mesophyll cells with chloroplasts – stomata restricted to epidermis above these cells

the more flaccid the leaf the more tightly rolled it becomes, shutting off stomata from outside atmosphere

9 **Suggest** why it is that, of all the environmental factors which affect plant growth, the issue of water supply is so critical.

10 **Explain** why we can describe the external epidermis of the leaf of marram grass as botanically equivalent to the lower epidermis of a mesophyte leaf.

For the organisms living in a sand-dune habitat, water is scarce – the sand is a quick-draining soil and, here, plants experience salt spray from the nearby breaking waves in periods of high winds. Other habitats where xerophytes are common are deserts, where temperatures are high in daylight and rainfall is low, and salt marshes (Figure 9.16).

Plants found in salt marshes are known as **halophytes**. Salt marshes are periodically flooded by sea water with its high salt content (but at other times they may be exposed to fresh water, in the form of rain or river flow). So, at times, the roots may be bathed in water of higher osmotic potential than that of their cells. This generates physiological drought conditions. Halophytes respond by absorbing additional salts.

■ **Figure 9.16** Plants of deserts and salt marshes

desert plant
The saguaro cactus, Tucson, Arizona. When rain falls, this tree-like cactus soaks up the water and survives in the subsequent drought.

salt marsh plant
Salicornia europaea (glass wort) grows in intertidal salt marshes. It is rich in soda and has been used in soap and glass manufacture.

11 **Deduce** what is meant by 'physiological drought'.

9.2 Transport in the phloem of plants – *structure and function are correlated in the phloem of plants*

■ Plants transport organic compounds from sources to sinks

Translocation is the movement of manufactured food (sugars and amino acids, mainly) which occurs in the phloem tissue of the vascular bundles (Figure 9.18). Sugars are made in the leaves (in the light) by photosynthesis and transported as sucrose. So, we refer to these leaves as the **source**. The first-formed leaves, once established, transport sugars to sites of new growth (new stem, new leaves and new roots). In older plants, sucrose is increasingly transported to sites of storage, such as the cortex of roots or stems, and seeds and fruits. These sites are referred to as **sinks**.

Amino acids are mostly made in the root tips. Here, absorption of nitrates (which the plant uses in the synthesis of amino acids) occurs. So, in this case, the root tips are the source. After their manufacture, amino acids are transported to sites where protein synthesis is occurring. These are mostly in the buds, young leaves and young roots, and in developing fruits. In the case of amino acids, these sites are **sinks**.

Translocation is not restricted to organic compounds that are manufactured within the plant. Chemicals that are applied to plants, for example by spraying, and are then absorbed by the leaves, may be carried all over the organism. Consequently, pesticides of this type are called 'systemic'.

Phloem tissue structure

Phloem tissue (Figure 9.17) consists of **sieve tubes** and **companion cells**, and is served by **transfer cells** (Figure 9.20) in the leaves. Sieve tubes are narrow, elongated elements, connected end to end to form tubes. The end walls, known as **sieve plates**, are perforated by pores. The cytoplasm of a mature sieve tube has no nucleus, nor many of the other organelles of a cell. However, each sieve tube is connected to a companion cell by strands of cytoplasm, called plasmodesmata, that pass through narrow gaps (called pits) in the walls. The companion cells are believed to service and maintain the cytoplasm of the sieve tube, which has lost its nucleus.

Phloem is a living tissue and has a relatively high rate of aerobic respiration during transport. In fact, transport of manufactured food in the phloem is an active process, using energy from metabolism.

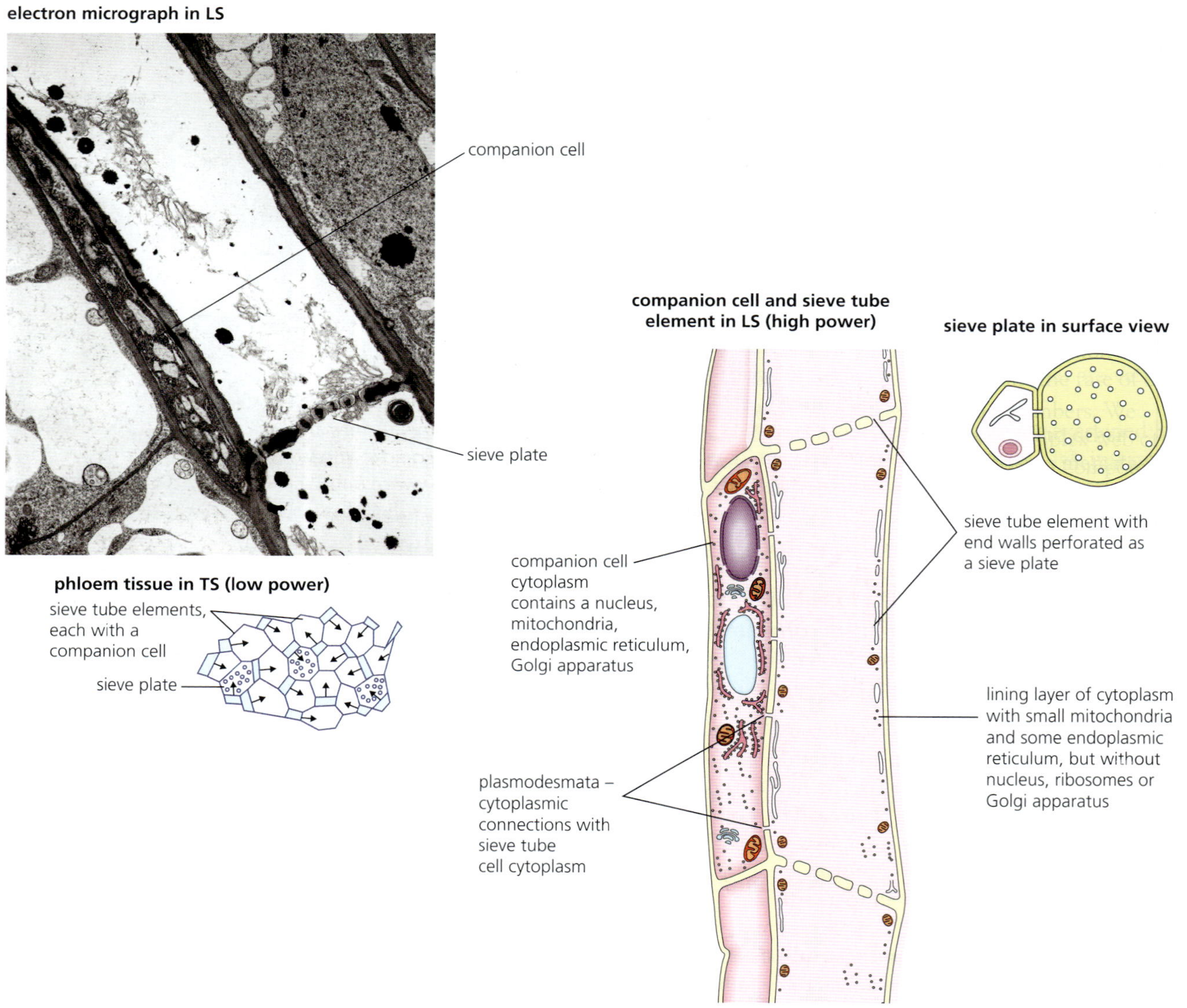

electron micrograph in LS

companion cell

sieve plate

phloem tissue in TS (low power)

sieve tube elements, each with a companion cell

sieve plate

companion cell and sieve tube element in LS (high power)

sieve plate in surface view

companion cell cytoplasm contains a nucleus, mitochondria, endoplasmic reticulum, Golgi apparatus

plasmodesmata – cytoplasmic connections with sieve tube cell cytoplasm

sieve tube element with end walls perforated as a sieve plate

lining layer of cytoplasm with small mitochondria and some endoplasmic reticulum, but without nucleus, ribosomes or Golgi apparatus

■ **Figure 9.17** The finer structure of phloem tissue

That translocation requires living cells is shown by investigation of the effect of temperature on phloem transport

Note: in neither experiment did the leaf blade wilt – xylem transport is not heat sensitive at this range of temperatures (because xylem vessels are dead, empty tubes)

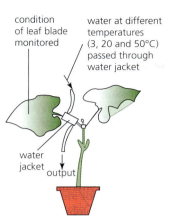

condition of leaf blade monitored

water at different temperatures (3, 20 and 50°C) passed through water jacket

water jacket output

(a) at 50°C, translocation of sugar from the leaf blade stopped – this is above the thermal death point of cytoplasm

conclusion: living cells are essential for translocation

(b) at 3°C, compared with 20°C translocation of sugar from leaf blade was reduced by almost 10% of leaf dry weight over a given time

conclusion: rate of metabolic activity of phloem cells affects rate of translocation

■ **Figure 9.18** Translocation requires living cells

 ## Identification of xylem and phloem in microscope images of stem and root

By the examination of a selection of transverse and longitudinal sections of the stems and roots of herbaceous plants, the appearance and positions of xylem vessels, phloem sieve tubes and companion cells will become clear. Ask your teacher or tutor to point out the ways that fibres differ from xylem vessels. Similarly, phloem tissue needs to be differentiated from the surrounding ground tissue (called parenchyma).

Nature of Science   **Developments in scientific research follow improvements in apparatus**

■ Investigating phloem transport

Movement of nutrients in the phloem has been investigated using radioactively labelled metabolites. This technique was pioneered between 1946 and 1953 by a team at the University of California, led by Melvin Calvin. Radioactive isotopes had become available for research in 1945, a by-product of war work on the atomic bomb. The team also exploited the relatively recent invention of paper chromatography.

[14]C-labelled sugars, manufactured in illuminated leaves 'fed' $^{14}CO_2$, can be traced during translocation. For example, the contents of individual sieve tubes can be sampled using the mouthparts of aphids as micropipettes, once these have been inserted into the plant by the insect (Figure 9.19). The techniques illustrated here can be adapted to investigate speed of phloem transport (see question 12).

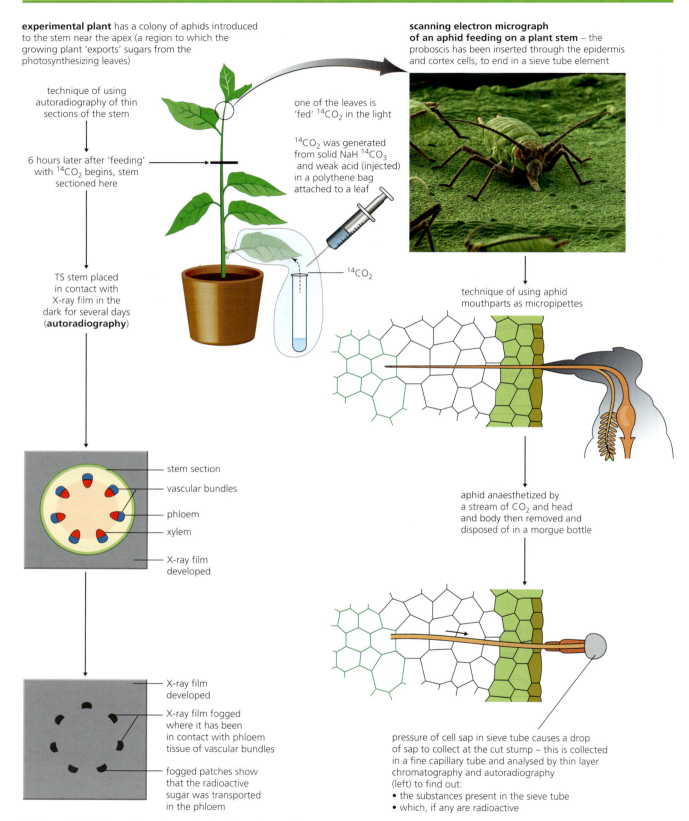

experimental plant has a colony of aphids introduced to the stem near the apex (a region to which the growing plant 'exports' sugars from the photosynthesizing leaves)

technique of using autoradiography of thin sections of the stem

6 hours later after 'feeding' with $^{14}CO_2$ begins, stem sectioned here

TS stem placed in contact with X-ray film in the dark for several days (**autoradiography**)

one of the leaves is 'fed' $^{14}CO_2$ in the light

$^{14}CO_2$ was generated from solid $NaH\,^{14}CO_3$ and weak acid (injected) in a polythene bag attached to a leaf

$^{14}CO_2$

scanning electron micrograph of an aphid feeding on a plant stem – the proboscis has been inserted through the epidermis and cortex cells, to end in a sieve tube element

technique of using aphid mouthparts as micropipettes

aphid anaesthetized by a stream of CO_2 and head and body then removed and disposed of in a morgue bottle

stem section

vascular bundles

phloem

xylem

X-ray film developed

X-ray film developed

X-ray film fogged where it has been in contact with phloem tissue of vascular bundles

fogged patches show that the radioactive sugar was transported in the phloem

pressure of cell sap in sieve tube causes a drop of sap to collect at the cut stump – this is collected in a fine capillary tube and analysed by thin layer chromatography and autoradiography (left) to find out:
• the substances present in the sieve tube
• which, if any are radioactive

■ **Figure 9.19** Using radioactive carbon to investigate phloem transport

12 This question concerns **measurement of phloem transport rates**. In a series of five investigations of the rate of movement of radioactive sucrose through phloem in the stems of willow, the mouthparts of aphids were used as micropipettes (see Figure 9.19). The time taken for a pulse of radioactive sugar to travel between sampling points A and B of known distance apart was measured and recorded.

Column	2	3	4
Experiment:	Distance between sample points A and B/mm	Time taken for sucrose to travel between A and B/ hours	Mean rate/mm hr^{-1})
1	510	2.1	
2	650	2.5	
3	480	1.6	
4	710	2.3	
5	450	1.5	
row 6	Mean distance =	Mean time taken =	Mean rate of sugar transport/mm hr^{-1} =

a Explain how radioactive sucrose may be generated in 'source' leaves close to sample point A.
b Identify two likely 'sink' sites to which phloem may transport sucrose in healthy, growing willow plants.
c Suggest why the distances between the sampling points varied in the five experiments.
d Calculate the mean distance between sampling points (column 2, row 6).
e Calculate the rate of sugar transport for each experiment (column 4).
f Calculate the mean rate of radioactive sugar transport (column 4, row 6).
g State the slowest and the fastest rates of sugar transport that were recorded (column 4).
h Suggest two possible reasons why the rate of sugar transport varied in these experiments.

■ The process of translocation

Translocation can be illustrated by examining the movement of sugar from the leaves. The story starts at the point where sugars are made and accumulate within the mesophyll in the leaf. This is the **source area**.

Sugars are loaded into the phloem sieve tubes in the leaf. This task is carried out by specialized cells called **transfer cells**, positioned between mesophyll cells (where sugar is produced) and the phloem companion cells (Figure 9.20). The transfer of sugar is driven by the combined action of **primary** and **secondary pumps**. These pumps are special proteins in the cell surface membrane. The primary pumps remove hydrogen ions (protons) *from* the cytoplasm of the companion cell into the transfer cell, so setting up a gradient in concentration of hydrogen ions. This movement requires ATP. Hydrogen ions then flow back into the companion cell, down their concentration gradient. This occurs at specific sites called secondary pumps, where their flow is linked to transport of sucrose molecules in the same direction.

electron micrograph of a leaf vein showing sieve tube elements, transfer cells, xylem vesssels and mesophyll cells (x1500)

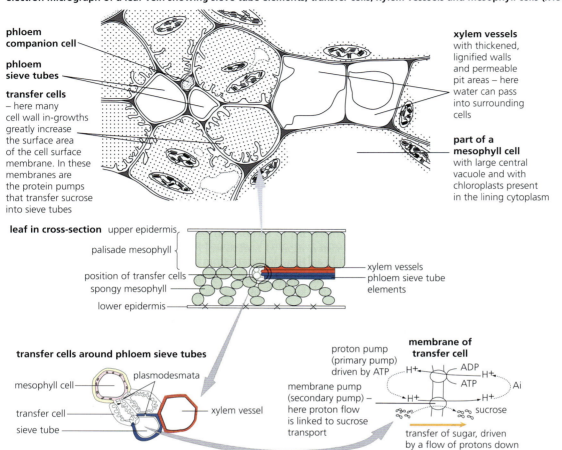

phloem companion cell

phloem sieve tubes

transfer cells
– here many cell wall in-growths greatly increase the surface area of the cell surface membrane. In these membranes are the protein pumps that transfer sucrose into sieve tubes

xylem vessels
with thickened, lignified walls and permeable pit areas – here water can pass into surrounding cells

part of a mesophyll cell
with large central vacuole and with chloroplasts present in the lining cytoplasm

leaf in cross-section
upper epidermis
palisade mesophyll
position of transfer cells
spongy mesophyll
lower epidermis

xylem vessels
phloem sieve tube elements

transfer cells around phloem sieve tubes

mesophyll cell
transfer cell
sieve tube
plasmodesmata
xylem vessel

proton pump (primary pump) driven by ATP

membrane pump (secondary pump) – here proton flow is linked to sucrose transport

membrane of transfer cell

H+
H+
ADP
ATP
Ai
H+
H+
sucrose

transfer of sugar, driven by a flow of protons down their concentration gradient

■ **Figure 9.20** Transfer cells and the loading of sieve tubes

13 Deduce what the presence of a large number of mitochondria in the companion cells implies about the role of these cells in the movement of 'sap' in the phloem.

As sucrose solution accumulates in the companion cells, it moves by diffusion into the sieve tubes, passing along the plasmodesmata (Figure 9.17). The accumulation of sugar in the phloem tissue raises the solute potential and water follows the sucrose by osmosis. This creates a high hydrostatic pressure in the sieve tubes of the source area.

Meanwhile, in living cells elsewhere in the plant – often, but not necessarily, in the roots – sucrose may be converted into insoluble starch deposits. This is a **sink area**. As sucrose flows out of the sieve tubes here, the solute potential is lowered. Water then diffuses out and the hydrostatic pressure is lowered.

These processes create the difference in hydrostatic pressures in source and sink areas that drive mass flow in the phloem.

The pressure–flow hypothesis

The principle of the pressure–flow hypothesis is that the sugar solution flows down a hydrostatic pressure gradient. There is a high hydrostatic pressure in sieve elements near mesophyll cells in the light (source area), but low hydrostatic pressure in elements near starch-storage cells of stem and root (sink area). This mass flow is illustrated in Figure 9.21, and the annotations explain the steps.

In this hypothesis, the role of the companion cells (living cells, with a full range of organelles in the cytoplasm), is to maintain conditions in the sieve tube elements favourable to mass flow of solutes. Companion cells use metabolic energy (ATP) to do this.

model demonstrating pressure flow
(A = mesophyll cell, B = starch storage cell)

In this model, the pressure flow of solution would continue until the concentration in A and B is the same.

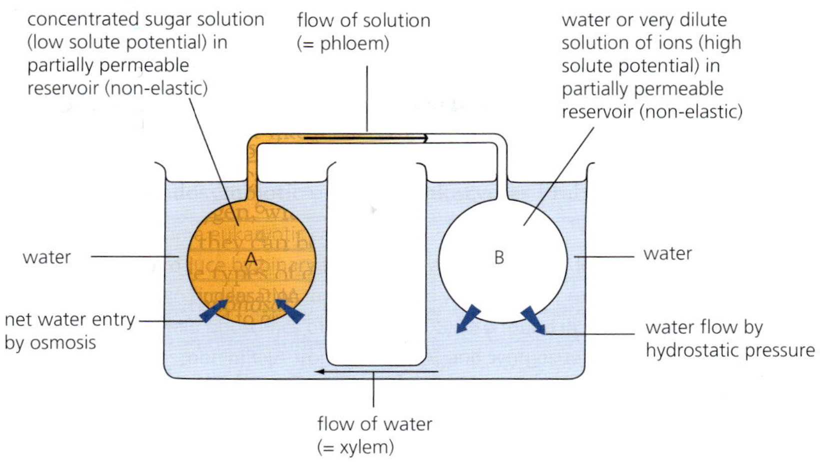

concentrated sugar solution (low solute potential) in partially permeable reservoir (non-elastic)

flow of solution (= phloem)

water or very dilute solution of ions (high solute potential) in partially permeable reservoir (non-elastic)

water

A

B

water

net water entry by osmosis

water flow by hydrostatic pressure

flow of water (= xylem)

pressure flow in the plant

sunlight

source cell, e.g. mesophyll cell of leaf where sugar is formed (= A)

chloroplast (site of sugar manufacture by photosynthesis)

water loss by evaporation

high hydrostatic pressure due to dissolved sugar

In the plant, a concentration difference between A and B is maintained by conversion of sugar to starch in cell B, while light causes production of sugar by photosynthesis in A.

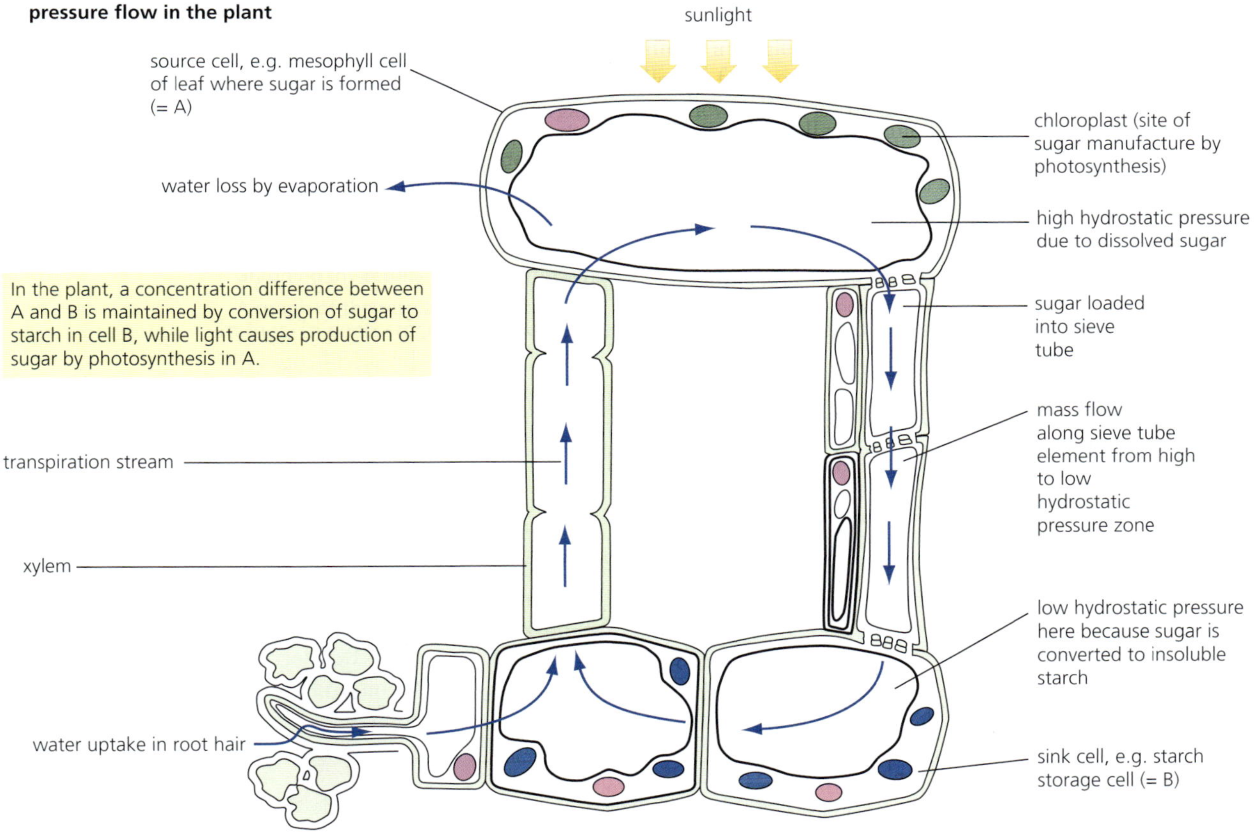

sugar loaded into sieve tube

mass flow along sieve tube element from high to low hydrostatic pressure zone

transpiration stream

xylem

low hydrostatic pressure here because sugar is converted to insoluble starch

water uptake in root hair

sink cell, e.g. starch storage cell (= B)

■ **Figure 9.21** The pressure–flow theory of phloem transport

■ **Table 9.2**
Pressure–flow
hypothesis – the
evidence

For	Against
contents of sieve tubes are under hydrostatic pressure and sugar solution exudes if phloem is cut	phloem tissue carries manufactured food to various destinations (in different sieve tubes), rather than to the greatest sink
appropriate gradients between 'source' and 'sink' tissue do exist	sieve plates are a barrier to mass flow and might be expected to have been 'lost' in the course of evolution, if mass flow is the mechanism of transport

■ 'Sinks' change in the lifecycle of the plant

The locations of the sinks change during the stages of growth of a green plant. Initially, the youngest leaves and the growing points of stems and roots are the sinks for the sugars that are exported by more matures leaves. Eventually, flower buds become the main sinks. After pollination, the developing fruits and seeds are the priority sinks. In plants that survive winter or an unfavourable season for growth, roots (and sometimes protected stems) become sinks.

14 Examine Figure 9.18 (page 389) carefully.
  a **Suggest** the sequence of events that you would anticipate in a leaf stalk as the content of the water jacket is raised to 50°C.
  b **Predict** how you would expect the phloem sap sampled from a sieve tube near leaves in the light and at the base of the same stem to differ.

15 **Explain** the processes which maintain:
  a the low solute potential of the cell of the root cortex
  b the high solute potential of the mesophyll cells of a green leaf.

16 **Describe** the differences between transpiration and translocation.

9.3 Growth in plants – *plants adapt their growth to environmental conditions*

■ Undifferentiated cells in the meristems of plants allow indeterminate growth

Initially, the plant grows from a single cell, the zygote, by repeated cell divisions, to form an embryo in the developing seed. Once a plant has grown past the early embryo stage, all later growth of the plant occurs at restricted points in the plant, called meristems. A **meristem** is a group of cells that retain the ability to divide by mitosis. These cells are small, with thin cellulose walls and dense cytoplasmic contents. Vacuoles in the cytoplasm are mostly absent, marking them apart from typical mature plant cells (with their large fluid-filled vacuoles). Meristems occur either at terminal growing points of stems and roots, or they are found laterally. In Figure 9.22, both types of meristem can be identified.

Indeterminate versus determinate growth

Growth from the apical meristem of a plant is described as **indeterminate** because it forms an unrestricted number of lateral organs, such as leaves, buds and lateral branches, more or less indefinitely. **Determinate** growth is typical of many animals, including mammals, in that growth stops when an organism or part of the organism reaches a certain size and shape.

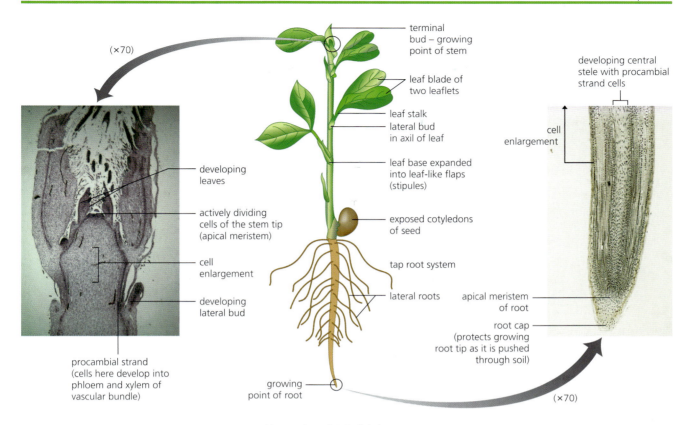

■ **Figure 9.22** Stem and root growth in the broad bean plant (*Vicia faba*)

Apical meristems occur at the tips of the stem and root and are responsible for their primary growth (Figure 9.22). Cell division and the subsequent growth of the cells produced here lead to formation of the tissues of stem (and root). First, the new cells formed by **division** rapidly increase in size. Then, this cell **enlargement** phase is followed by cell **differentiation**. The new enlarging cells become specialized. For example:

■ New cells of the **ground tissues** are formed (called collenchyma and parenchyma). These are contained within the external layer of cells known as the epidermis.

■ New cells of the **vascular tissue** form in the developing vascular bundles. These contain water-carrying cells (**xylem**) and elaborated food-carrying cells (**phloem**), both of which are assembled as extensions to the existing vascular bundles.

These are the primary tissues that make up stems (and roots), and so apical meristems are also called **primary meristems**. Between phloem and xylem of the bundles, a few meristematic cells remain after primary growth, and these form a meristematic tissue called **cambium**.

Lateral meristems (Figure 9.23) form from the cambium cells in the centre of vascular bundles, between the (outer) phloem tissue and the (inner) xylem tissue. When the lateral meristem forms and grows, it causes the secondary growth of the plant. Secondary growth involves additions of vascular tissue (secondary phloem and secondary xylem), and results in an increase in the girth of the stem. The first stage in secondary growth occurs when the cambium in the vascular bundles grows into a complete cylinder around the stem. Growth of the lateral meristem increases the circumference of the stem and also increases the strength of the stem.

Table 9.3 compares the growth due to apical and lateral meristems.

■ **Figure 9.23**
The roles of apical and lateral meristems in the growth of stems

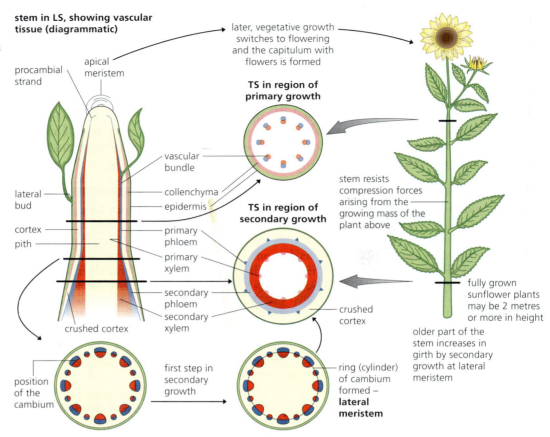

■ **Table 9.3**
Growth due to apical and lateral meristems compared

Growth due to apical meristem		Growth due to lateral meristem
occurs at tip of stems and roots	**position of meristem**	occurs laterally, between primary phloem and primary xylem
product of embryonic cells	**origin**	cambium – meristematic cells left over from primary growth
produces initial tissues of actively growing plant from the outset	**timing of activity**	functions in older stems (and roots), and in woody plants from the outset
forms epidermis, ground tissues, and primary phloem and xylem	**cell products**	forms mainly secondary phloem and xylem (and often fibres)
produces growth in length and height of plant	**outcome for stem**	produces growth in girth of stem, plus strengthening of stem

■ Plant hormones control growth in the shoot apex

Among the internal factors that play a part in control of plant growth and sensitivity are the **plant growth regulators**. These hormone-like molecules are different from animal hormones, (Table 9.4). It is accepted shorthand to refer to plant growth regulators as plant hormones, so this is what we will do.

■ **Table 9.4**
Differences between animal hormones and plant growth regulators

Plant growth regulators	Animal hormones
produced in a region of plant structure, e.g. stem or root tips, in unspecialized cells	produced in specific glands in specialized cells, e.g. islets of Langerhans in the pancreas
not necessarily transported widely, or at all, and some are active at sites of production	transported to all parts of the body by the bloodstream
not particularly specific, tend to influence different tissues and organs, sometimes in contrasting ways	effects are mostly highly specific to a particular tissues or organ, and without effects in other parts or on different processes

Plant hormones occur in low concentrations in plant tissues, which presented difficulties to earlier experimenters. Now, improvements in analytical techniques have led to the discovery of the molecules involved and of their effects on gene expression, for example. There are five major types of compound, naturally occurring in plants and classified as plant hormones. These substances tend to interact with each other in the control of growth and sensitivity, rather than working in isolation. Here, we shall illustrate the action of plant hormones largely by reference to one, known as **auxin** (Table 9.5).

■ **Table 9.5**
Introducing the plant hormone auxin

Auxin – indole acetic acid (IAA)	
discovery	initially by Charles Darwin, working with grass coleoptiles and their curvature towards a unilateral light source; later, Went devised a biological assay to find the concentrations in plant organs of 'auxin'
chief roles	cell enlargement, extension growth of stems and roots (at different concentrations), dominance of terminal buds, promotion of fruit growth and inhibition of leaf fall
synthesis	at stem and root tips, and in young leaves, from the amino acid tryptophan

Auxin is manufactured by cells undergoing repeated cell division, such as those found at the stem and root tips. Consequently, the concentration of auxin is highest there (Figure 9.24). Auxin is then transported to the region of growth behind the tip, where it causes cells to elongate.

■ **Figure 9.24**
Distribution of auxin (IAA) in a young, growing plant

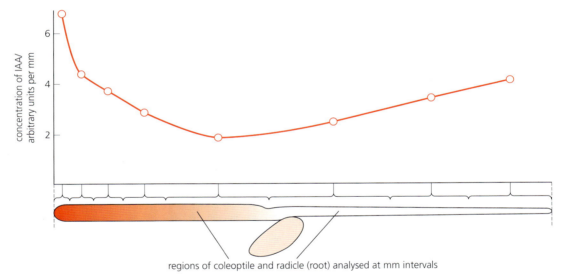

regions of coleoptile and radicle (root) analysed at mm intervals

Auxin and shoot growth

Auxin has a major role in the growth of the shoot apex, where it promotes the elongation of cells. It also inhibits growth and development of lateral buds that occur immediately below the terminal growing point. This leads to a quality known as **apical dominance**. However, a high concentration of auxin actually inhibits growth in length of the stem. Other hormones, the gibberellins, interact with auxin to enhance stem elongation. Cytokinins, from the root apex, pass back up to the stem and promote lateral bud growth – it is antagonistic to the effect of auxin in this respect. The full picture of plant hormone interactions is a complex one.

Auxin influences cell growth rates by changing the pattern of gene expression

Auxin has these effects on growth and development by direct action on the components of growing cells, including the walls, and on the gene expression mechanisms operating in the nucleus. Auxin transport across cells is polar, with its entry into the cell being passive (by diffusion) and its efflux being active (ATP-driven). This mechanism of auxin movement and the ways in which auxin may influence growth and development of the cell are outlined in

Figure 9.25. Note, particularly, the mechanisms of movement of auxin into and out of the cell; auxin efflux pumps set up concentration gradients in plant tissues.

■ **Figure 9.25**
Auxin – mechanisms
of movement and
control

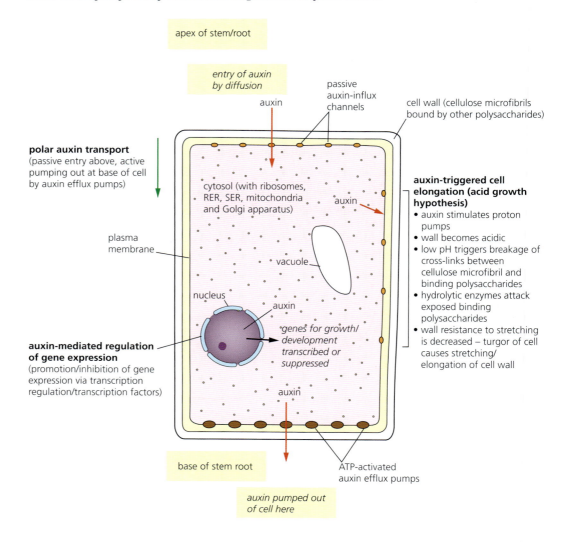

apex of stem/root

*entry of auxin
by diffusion*

auxin

passive
auxin-influx
channels

cell wall (cellulose microfibrils
bound by other polysaccharides)

polar auxin transport
(passive entry above, active
pumping out at base of cell
by auxin efflux pumps)

cytosol (with ribosomes,
RER, SER, mitochondria
and Golgi apparatus)

auxin

**auxin-triggered cell
elongation (acid growth
hypothesis)**
• auxin stimulates proton
  pumps
• wall becomes acidic
• low pH triggers breakage of
  cross-links between
  cellulose microfibril and
  binding polysaccharides
• hydrolytic enzymes attack
  exposed binding
  polysaccharides
• wall resistance to stretching
  is decreased – turgor of cell
  causes stretching/
  elongation of cell wall

plasma
membrane

vacuole

nucleus

auxin

**auxin-mediated regulation
of gene expression**
(promotion/inhibition of gene
expression via transcription
regulation/transcription factors)

*genes for growth/
development
transcribed or
suppressed*

auxin

base of stem root

ATP-activated
auxin efflux pumps

*auxin pumped out
of cell here*

Auxins and plant responses to environmental stimuli

Plant organs respond to external stimuli. A response in which the direction of the stimulus determines the direction of the response is called a tropic movement or **tropism** (Table 9.6).

■ **Table 9.6**
Tropism reviewed

Stimulus	Tropism	Example
light	phototropism	young stems are positively phototropic
gravity	geotropism	young stems are negatively geotropic; main (tap) roots are positively geotropic

Note: the auxin diffusing from a stem tip can be collected in a gelatin block and then assayed.

A. Auxin and positive phototropism in coleoptiles (stem-like caps on oat seedlings).

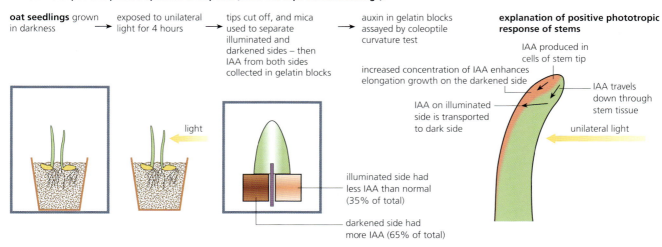

oat seedlings grown in darkness → exposed to unilateral light for 4 hours → tips cut off, and mica used to separate illuminated and darkened sides – then IAA from both sides collected in gelatin blocks → auxin in gelatin blocks assayed by coleoptile curvature test

light

illuminated side had less IAA than normal (35% of total)

darkened side had more IAA (65% of total)

explanation of positive phototropic response of stems

IAA produced in cells of stem tip

increased concentration of IAA enhances elongation growth on the darkened side

IAA on illuminated side is transported to dark side

IAA travels down through stem tissue

unilateral light

B. The role of auxin in the geotropic response.

In a horizontal seedling auxin is redistributed – a higher concentration collects on the lower surface.

The response from the plant: stem tip grows up, the root tip grows down.

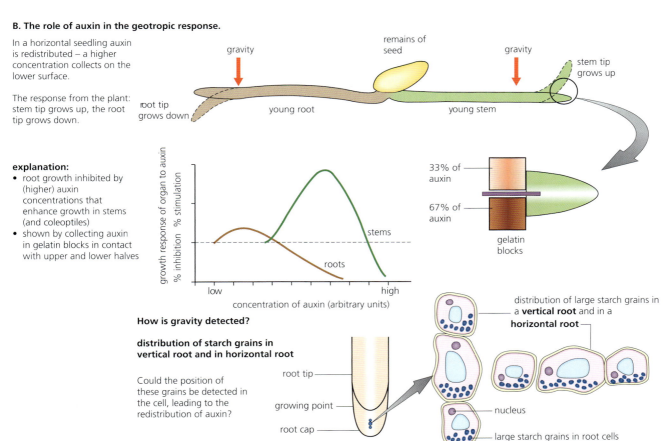

gravity

remains of seed

gravity

stem tip grows up

root tip grows down

young root

young stem

explanation:
- root growth inhibited by (higher) auxin concentrations that enhance growth in stems (and coleoptiles)
- shown by collecting auxin in gelatin blocks in contact with upper and lower halves

growth response of organ to auxin
% inhibition % stimulation

stems

roots

low   concentration of auxin (arbitrary units)   high

33% of auxin

67% of auxin

gelatin blocks

How is gravity detected?

distribution of starch grains in vertical root and in horizontal root

Could the position of these grains be detected in the cell, leading to the redistribution of auxin?

root tip

growing point

root cap

distribution of large starch grains in a **vertical root** and in a **horizontal root**

nucleus

large starch grains in root cells

■ **Figure 9.26** Investigations of the role of auxin (IAA) in tropisms

When the stem tip responds by growing towards the light, it is said to be **positively phototropic** – due to an increased concentration of auxin on the darkened side (Figure 9.26A). In a seedling that is subjected to the unilateral stimulus of gravity – that is, placed on its side (Figure 9.26B) – a higher concentration of auxin collects on the lower surface. The root tip responds by growing down (it is **positively geotropic**) but the stem tip grows up (being **negatively geotropic**).

How does auxin influence gene expression in these situations?

The current hypothesis to account for these effects is as follows:

- An environmental influence, such as unilateral light, is detected by proteins called phototropins, which respond by binding to receptors in the cell. These receptors control the transcription of specific genes. These genes may code for glycoproteins (known as PIN3 proteins) in the plasma membranes of cells that facilitate the transport of auxin.

- PIN3 proteins are involved in the lateral transport of auxin in unilaterally illuminated stems (Figure 9.26A), and lateral transport of auxin in stems and roots exposed to a unilateral gravitational stimulus (Figure 9.26B).

17 Construct a list of the various effects of light on plant growth and development.

Nature of Science   **Developments in scientific research follow improvements in analysis and deduction**

■ Developments in analytical techniques allow the detection of trace amounts of hormones

A DNA microarray consists of a collection of DNA probe sequences attached to a solid surface. The 'surface' can be a glass or silicon chip, to which the DNA is covalently bonded. One use for such microarrays is the detection and measurement of the expression of particular genes. Genes being expressed may be caused to fluoresce and, so, can be detected. In plants, the hormone auxin has been shown to influence gene expression and so regulates growth and development. Data on this has been obtained from studies on cells of a plant from the brassica family, *Arabidopsis thaliana,* when grown under the influence of unilateral environmental stimuli, such as light or gravity. A combination of several genes is typically involved.

■ Tissue culture and micropropagation

Plant tissue culture is a laboratory technique for growing new plants from blocks of undifferentiated tissue (**callus**) or from individual cells (Figure 9.27). Unlimited numbers of clones of a plant can be produced, all identical. By this technique, genetically modified cells can also be cloned and grown up into plants. So, tissue culture has increasingly important applications in agriculture, horticulture and genetic engineering. Tissue culture and micropropagation also have applications for rapid bulking up of new varieties, production of virus-free strains of existing varieties and propagation of rare species.

A practical protocol in plant tissue culture, safely and easily conducted in school laboratories, has been developed by **Science and Plants for Schools**. It can be accessed at www.saps.org.uk.

TOK Link

Plants communicate both internally and externally. *How?*

To what extent can plants be said to have 'language'. Does your answer depend on how you define 'communication'?

- new plants can be grown from mature cells, which have the necessary 'blueprint' in the DNA of their chromosomes to reproduce the complete development process by which a zygote develops into a new individual – **totipotency**
- the techniques can be adapted to allow genetic modification of the genome of the plant, leading to the production of **GM plants**

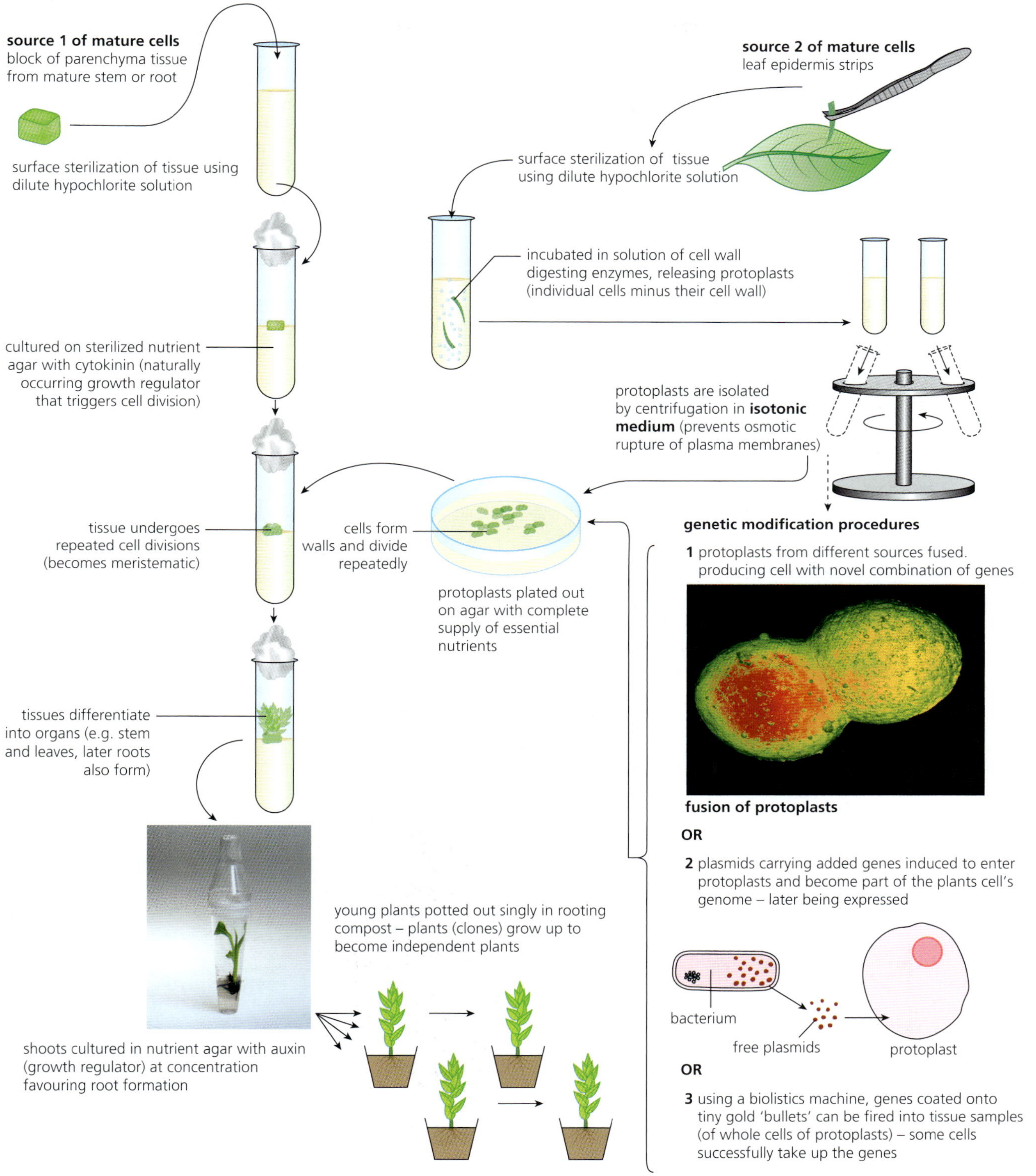

source 1 of mature cells
block of parenchyma tissue from mature stem or root

source 2 of mature cells
leaf epidermis strips

surface sterilization of tissue using dilute hypochlorite solution

surface sterilization of tissue using dilute hypochlorite solution

incubated in solution of cell wall digesting enzymes, releasing protoplasts (individual cells minus their cell wall)

cultured on sterilized nutrient agar with cytokinin (naturally occurring growth regulator that triggers cell division)

protoplasts are isolated by centrifugation in **isotonic medium** (prevents osmotic rupture of plasma membranes)

tissue undergoes repeated cell divisions (becomes meristematic)

cells form walls and divide repeatedly

protoplasts plated out on agar with complete supply of essential nutrients

genetic modification procedures

1 protoplasts from different sources fused, producing cell with novel combination of genes

fusion of protoplasts

OR

tissues differentiate into organs (e.g. stem and leaves, later roots also form)

2 plasmids carrying added genes induced to enter protoplasts and become part of the plants cell's genome – later being expressed

young plants potted out singly in rooting compost – plants (clones) grow up to become independent plants

bacterium

free plasmids

protoplast

OR

shoots cultured in nutrient agar with auxin (growth regulator) at concentration favouring root formation

3 using a biolistics machine, genes coated onto tiny gold 'bullets' can be fired into tissue samples (of whole cells of protoplasts) – some cells successfully take up the genes

■ **Figure 9.27** The techniques of tissue culture applied to flowering plants

9.4 Reproduction in plants – *reproduction in flowering plants is influenced by the biotic and abiotic environment*

Flowering plants contain their reproductive organs in the flower. Flowers are often hermaphrodite structures, carrying both male and female parts.

The parts of flowers occur in rings or whorls, attached to the swollen tip of the flower stalk, called the receptacle. The **sepals** (collectively, the calyx) enclose the flower in the bud, and are usually small, green and leaf-like. The **petals** (collectively, the corolla) are often coloured and conspicuous, and may attract insects or other small animals. The stamens are the male parts of the flower, and consist of **anthers** (housing pollen grains) and the **filament** (stalk). The carpels are the female part of the flower. There may be one or many, free-standing or fused together. Each carpel consists of an **ovary** (containing ovules), a **stigma** (a surface for receiving pollen) and a connecting style.

The buttercup flower is shown in Figure 9.28, and other flowers that are common in different parts of the world are shown in Figure 9.29.

■ **Figure 9.28**
The buttercup (*Ranunculus*) flower

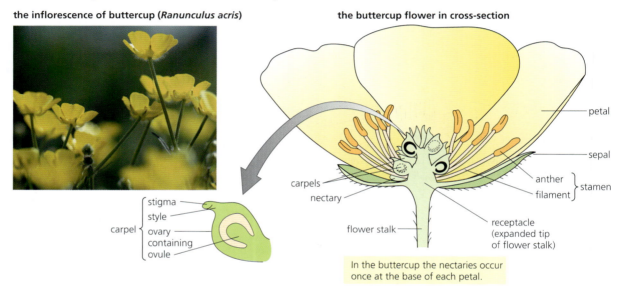

the inflorescence of buttercup (*Ranunculus acris*)

the buttercup flower in cross-section

In the buttercup the nectaries occur once at the base of each petal.

Bougainvillea rosenka Native of tropical and sub-tropical South America. The flowers are small but surrounded by brightly coloured leaves (bracts).

Hibiscus syriacus Native of warm temperate, tropical and sub-tropical regions throughout the world. The large, trumpet-shaped flowers have five petals.

■ **Figure 9.29** Other animal-pollinated flowers common in different parts of the world

■ Drawing a half-flower view of an animal pollinated flower

Look at the half-flower representations of the buttercup (Figure 9.28) and the white dead nettle (Figure 9.30). Both are insect-pollinated flowers (they have nectaries which provide a sugar solution that insects require), but the white dead nettle's structure ensures that the abdomen of a bee will be in contact with the stamens and stigma (if mature), as it pushes down to suck up the nectar. Choose an insect-pollinated flower that is available near your school or college. Study its structure and then work out its pollination mechanism. (What insects can visit and benefit?) Then create a half-flower drawing of your flower.

■ **Figure 9.30**
The white dead nettle flower

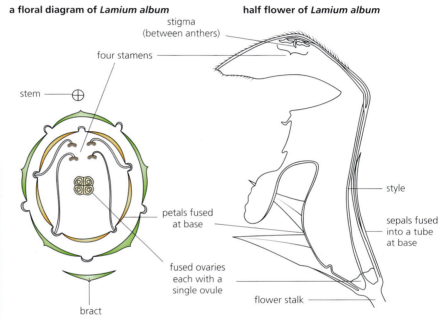

a floral diagram of *Lamium album*

half flower of *Lamium album*

stigma (between anthers)

four stamens

stem

petals fused at base

fused ovaries each with a single ovule

bract

stigma (between anthers)

style

sepals fused into a tube at base

flower stalk

■ Pollination and fertilization

Pollen grains contain the male gametes. **Pollination is the transfer of pollen from a mature anther to a receptive stigma.** The pollen may come from the anthers of the same flower or flowers of the same plant, in which case this is referred to as **self-pollination**. Alternatively, pollen may come from flowers on a different plant of the same species, which is referred to as **cross-pollination**.

Transfer of pollen is often brought about by animals (Figure 9.31). Pollinators include insects, such as butterflies or bees. In other flowers, it may be bird or bat visitors that unwittingly carry out pollination. The pollinator is typically attracted by colour or scent (or both), and is rewarded by a sugar solution, called nectar, and pollen, which usually form a key part of the diet. In return, they accidentally transfer pollen between flowers and between plants. Thus, there is a mutualistic relationship between pollinator and plant in plant sexual reproduction. Alternatively, pollen may be transferred by wind or, occasionally, by running water.

■ **Figure 9.31**
Pollinators at work

Fertilization in flowering plants can occur only after an appropriate pollen grain has landed on the stigma and germinated there. **Fertilization is the fusion of male and female gametes to form a zygote.** The pollen grain produces a pollen tube, which grows down between the cells of the style and into the ovule (Figure 9.32). Incidentally, the pollen tube delivers *two* male nuclei. One of these male nuclei then fuses with the egg nucleus in the embryo sac, forming a diploid zygote. The other fuses with another nucleus, triggering formation of the food store for the developing embryo. This 'double fertilization' is unique to flowering plants.

■ **Figure 9.32**
Fertilization in a
flowering plant

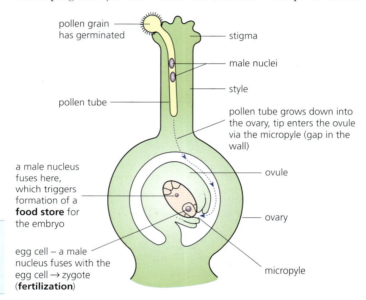

pollen grain has germinated

stigma

male nuclei

pollen tube

style

pollen tube grows down into the ovary, tip enters the ovule via the micropyle (gap in the wall)

a male nucleus fuses here, which triggers formation of a **food store** for the embryo

ovule

ovary

egg cell – a male nucleus fuses with the egg cell → zygote (**fertilization**)

micropyle

18 Explain the differences between pollination and fertilization in the flowering plant.

Nature of Science

Paradigm shift

■ The importance of pollinators – the survival of entire ecosystems

More than 85% of the world's 250 000 species of flowering plant depend on pollinators for reproduction. Without healthy populations of pollinators, plant life is threatened. Since green plants are at the heart of virtually every food chain, all life depends on plants, directly or indirectly (page 194). Terrestrial plant life, in addition to being the principal source of nutrients, determines the types of environment that are available for many other organisms. In turn, plant life is totally dependent on the activities of many other organisms. Consequently, survival of ecosystems, in general, is a product of this interdependence; this is the basis of the maintenance of biodiversity.

An example of interdependence comes from the tropical rain forests of Brazil. The Brazil nut tree (*Bertholletia excelsa*) occurs in the countries of the Amazon region, widely dispersed in the rain forests there. Key features of the Brazil nut lifecycle are as follows:

- The trees flower during the dry season, between October and December – a relatively short period in which pollination must occur.
- The large flowers can only be pollinated by orchid bees – it is the female orchid bee that is the pollinator. Only a powerful, large-bodied bee can prize open the protective 'flower hood' and access the nectar, incidentally bringing about cross-pollination.
- The chief habitat of the orchid bee is undisturbed forest. Attempts to manage and maintain colonies of orchid bees on plantations have been unsuccessful.
- The male orchid bees cannot mate successfully with the females without first visiting the flowers of small orchids that grow high on the branches of the canopy of the Brazil nut tree.
- These orchid flowers are pollinated by the male orchid bee, which visit them for exposure to a perfume present in the waxy secretion that these flowers exclusively produce.
- With this perfume, the male bees can compete for a mate, successfully breed, and so maintain orchid bee populations. Without the orchid bees and the epiphytic orchids, the Brazil nut trees would not be able to reproduce and produce seeds.
- After pollination, a further year elapses before the fruits develop. These contain 10–25 Brazil nuts (seeds) within an extremely hard shell.
- The agouti, a large rodent, is one of very few animals able to access the tree's seeds, using its tough jaws to gnaw open the hard fruit shell.
- Once the seeds have been accessed, many are hidden away in the forest floor by the agouti, to be recovered in times of food scarcity. It is the forgotten stored seeds that eventually become tree seedlings, regenerating the Brazil nut tree populations of the Amazon. So, the survival of the agouti is crucial to the survival of the Brazil nut tree, too.

In the past, conservation has focused on individual species with seriously dwindling numbers. Now it is clear that an exclusive attention to the survival of individual endangered species is mistaken – we need to protect entire ecosystems.

■ Seed formation and dispersal

The seed develops from the fertilized ovule and contains an embryo plant and a food store. After fertilization, the following occur:

- The zygote grows by repeated mitotic division to produce cells that form an **embryonic plant**, consisting of an embryo root, an embryo stem and either a single cotyledon (seed leaf) or two cotyledons. (Remember, the phylum Angiospermophyta is divided into two classes according to the number of cotyledons (seed leaves) present, page 230. The monocotyledons have a single seed leaf; the dicotyledons have two.)

- Formation of **stored food reserves** is triggered. In many seeds, the developing food store is absorbed into the cotyledons, rather than remaining as a separate store that is packed round the embryonic plant. For example, this is the case in peas and beans (Figure 9.33). Note that formation of food reserves can only occur if fertilization occurs – in the absence of fertilization, food reserves are not moved into the unfertilized ovule.

As the seed matures, the outer layers of the ovule become the protective seed coat or **testa** and the whole ovary develops into the **fruit**. Next, the water content of the seed decreases and the seed moves into a dormancy period. In a mature, fully dormant seed, water makes up only 10–15% of seed weight.

broad bean seed (*Vicia faba*)

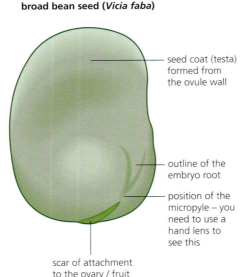

seed coat (testa) formed from the ovule wall

outline of the embryo root

position of the micropyle – you need to use a hand lens to see this

scar of attachment to the ovary / fruit

broad bean seed in section

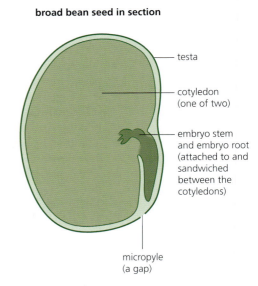

testa

cotyledon (one of two)

embryo stem and embryo root (attached to and sandwiched between the cotyledons)

micropyle (a gap)

■ **Figure 9.33**
The structure of a broad bean seed

19 **State** a fruit or vegetable which we eat that originates from:
a an ovary containing one seed
b an ovary containing many seeds
c several ovaries fused together, containing many seeds.

This drawing of a broad bean seed shows how structure can be recorded, once the seed has been examined. Now try with other seeds, such as a sunflower seed.

Seed dispersal is the carrying of the seed away from the parent plant

The **seed** is also a form in which the flowering plant may be dispersed. If offspring seeds eventually germinate some distance apart, there is more likelihood they will not be competing for the same resources of space, water and light.

Seed dispersal is the carrying of the seed away from the parent plant.

The plant structures to aid dispersal that have evolved variously exploit air currents (wind), passing animals or flowing water to transport seeds. In a few plants, seeds are flung away from the ripening fruit by an explosive mechanism. All seeds are compact, nutritious and relatively lightweight – in effect, they are food packages to a hungry animal. Many seeds taken for food are dropped and lost, or stored and forgotten. In this way, some seeds are successfully dispersed.

■ The physiology of seed germination

Many seeds do not germinate as soon as they are formed and dispersed. Such seeds are said to have a **dormant period** and germinate only when this has elapsed. Dormancy may be imposed within the seed, due to:

- **incomplete seed development** that causes the embryo to be immature, and which is overcome in time
- the **presence of a plant growth regulator** – abscisic acid, for example – that inhibits development, and which only disappears from the seed tissues with time
- an **impervious seed coat** that is eventually made permeable – for example, by abrasion with coarse soil or by the action of microorganisms
- a **requirement for pre-chilling** under moist conditions before the seed can germinate; some seeds need to be held at or below 5°C for up to 50 days (possibly the equivalent of winter in temperate climates).

Once dormancy is overcome, germination occurs if the following essential external conditions are met (Table 9.7).

■ **Table 9.7**
Conditions for germination

External	Internal
water uptake – hydration of the cytoplasm of cells of embryo	overcoming of dormancy
ambient **temperature** – within optimum range for enzyme action	production of plant growth regulator(s) by embryo cells to initiate biochemical changes of germination, leading to production of hydrolytic enzymes for mobilization of stored food
oxygen – to sustain aerobic cell respiration	

- **Water** uptake has occurred so that the seed is fully hydrated and the embryo can be physiologically active.

- **Oxygen** is present at a high enough partial pressure to sustain aerobic respiration. Growth demands a continuous supply of metabolic energy in the form of ATP that is best generated by aerobic cell respiration in all the cells.

- **A suitable temperature** exists, one that is close to the optimum temperature for the enzymes involved in the mobilization of stored food reserves, the translocation of organic solutes in the phloem, and the synthesis of intermediates for cell growth and development. For example, wheat seeds germinate in the range 1–35°C, and maize in the range 5–45°C.

The **steps of germination** are summarized in Figure 9.34. Note that a particular plant growth substance (known as gibberellic acid, **GA**) is produced by the cells of the embryo. This growth-promoting substance passes to the food stored in the cotyledons. Here, protein reserves are converted to **hydrolytic enzymes** which mobilize the stored food reserves. The main event is the production of the enzyme **amylase** which hydrolyses starch to maltose. This disaccharide is then hydrolysed to glucose. The resulting soluble sugar (and other compounds) sustain **respiration** and also provide the **building blocks for synthesis** of the intermediates essential for new cells.

■ **Figure 9.34**
Metabolic events of germination in a starchy seed

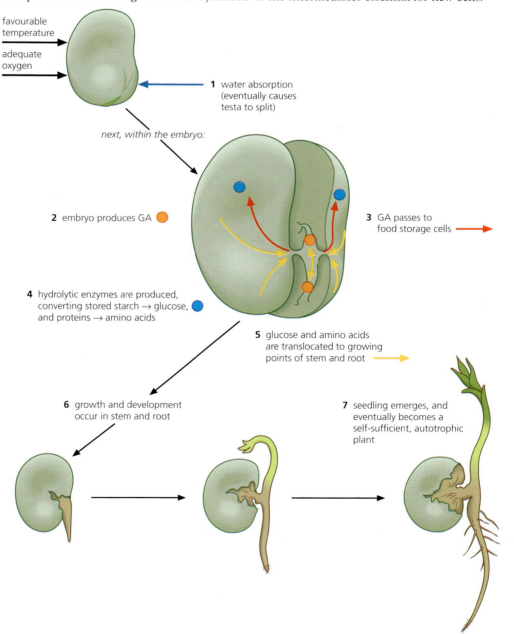

favourable temperature

adequate oxygen

1 water absorption (eventually causes testa to split)

next, within the embryo:

2 embryo produces GA

3 GA passes to food storage cells

4 hydrolytic enzymes are produced, converting stored starch → glucose, and proteins → amino acids

5 glucose and amino acids are translocated to growing points of stem and root

6 growth and development occur in stem and root

7 seedling emerges, and eventually becomes a self-sufficient, autotrophic plant

Designing an investigation of the conditions for seed germination

Samples of seeds can be set to germinate on damp filter paper in a Petri dish, but how many seeds per dish would make an appropriate sample, and how many different species might you test? Given this simple apparatus, you would be able to investigate the effect of light (presence or absence), and perhaps temperature (low and room temperatures – unless you also have access to temperature controlled cabinets). More ambitious investigations might be of the effects of intense cold treatment on newly formed seeds, or the effects of brief and prolonged pre-soaking (in effect, the degree of hydration). Whatever is investigated, there are the issues of appropriate controls, and the percentage of seeds in a sample that might be long-term dormant or non-viable. What statistical test would you apply to your results?

■ The control of flowering

You will be well aware that plants flower at different times of the year; very many species have a precise season when flowers are produced. At other times, no flowers are formed on these plants.

> *How is flowering switched on by environmental conditions?*

The answer is, in many cases, that day length provides important signals and that a plant pigment molecule is involved in the flowering process.

Plant development and phytochrome

A blue–green pigment called **phytochrome** is present in green plants in very low concentrations. The amount of phytochrome is not sufficient to mask chlorophyll, and it has been a substance that is difficult to isolate and purify from plant tissue, although this has now been done.

Phytochrome is a very large conjugated protein (protein molecule and pigment molecule combined) and it is a highly reactive molecule. It is not a plant growth substance; it is a photoreceptor pigment that is able to absorb light of particular wavelength and change its structure as a consequence. It is likely to react with different molecules around it, according to its structure.

Two forms of phytochrome

We know that phytochrome exists in two interconvertible forms.

- One form, referred to as P_R, is a blue pigment which absorbs mainly red light of wavelength 660 nm (this is what the $_R$ stands for).
- The other form is P_{FR}, a blue–green pigment which mainly absorbs far-red light of wavelength 730 nm.

When P_R is exposed to light (or red light on its own) it is converted to P_{FR}. However, in the dark (or if exposed to far-red light alone) it is converted back to P_R.

$$P_R \underset{\substack{\text{darkness} \\ \text{(slow)}}}{\overset{\substack{\text{light} \\ \text{(slow)}}}{\rightleftharpoons}} \overset{\substack{\text{(or red light)} \\ \text{(fast)}}}{\underset{\substack{\text{(or far-red light)} \\ \text{(fast)}}}{}} P_{FR}$$

Where plant growth and development are influenced by light, this is known as **photomorphogenesis**. Phytochrome is the pigment system involved in photomorphogenesis. We know this because the red–far-red absorption spectrum of phytochrome corresponds to the action spectrum of some specific effects of light on development. (See page 122 if you have forgotten the terms 'absorption spectrum' and 'action spectrum'.)

It appears that it is P_{FR} that is the active form of phytochrome, stimulating some effects in plant development and inhibiting others. In particular, P_{FR} controls of the onset of flowering.

Photoperiodism is the response of an organism to changing length of day. In fact, it is the length of the dark period in the 24-hour cycle that is important, as we shall see. The plants in which flowering is controlled by day length fall into two categories (Figure 9.35).

■ **Short-day plants** – these are plants which flower only if the period of darkness is longer than a certain critical length. If darkness is interrupted by a brief flash of red light, the plant will not flower (but this effect is reversed by a subsequent flash of far-red light).

  □ *Interpretation:* phytochrome in P_{FR} form inhibits flowering in short-day plants. The very long nights required by short-day plants allow the concentration of P_{FR} to fall to a low level, removing the inhibition. A flash of light in the darkness reverses this, but a flash of far-red light reverses the reversal and flowering still takes place.

■ **Long-day plants** – these are plants which flower only if the period of uninterrupted darkness is less than a certain critical length each day.

  □ *Interpretation:* phytochrome in P_{FR} form promotes flowering in long-day plants. The long period of daylight causes the accumulation of P_{FR}, because P_R is converted to P_{FR}.

■ **Figure 9.35**
Flowering related to day length

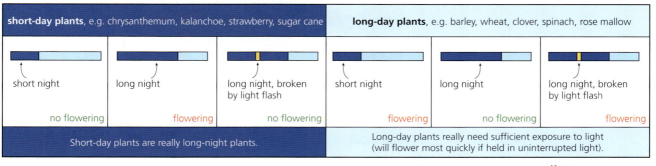

short-day plants, e.g. chrysanthemum, kalanchoe, strawberry, sugar cane			**long-day plants**, e.g. barley, wheat, clover, spinach, rose mallow		
short night	long night	long night, broken by light flash	short night	long night	long night, broken by light flash
no flowering	flowering	no flowering	flowering	no flowering	flowering
Short-day plants are really long-night plants.			Long-day plants really need sufficient exposure to light (will flower most quickly if held in uninterrupted light).		

Key 24 hours
night   day

■ Flowering involves a change in gene expression in the shoot apex

The structural switch from vegetative growth to flowering occurs in a stem apex, yet it is the leaves below that are sensitive to day length. For example, a leaf that has been exposed to the correct photoperiod, if immediately grafted onto a non-induced plant of the same type, will cause flowering there. Consequently, it was earlier assumed that a growth regulator substance is formed in leaves under the correct regime of light and dark, and is transported to the stem apex where it causes the switch in development. This growth substance was named 'florigen', but it was never detected.

Today it is known that mRNA molecules and proteins, coded for by specific genes, can also function as growth substances. It is molecules of this sort that appear to be transported about the plant via the plasmodesmata and the symplast pathway (page 375).

Currently, it is suggested that a gene ('flowering locus' – FT) is activated in leaves of photoperiodically induced plants. As a consequence, it is FT mRNA that then travels from induced leaves to stem apex. In the cells of the apex, the FT mRNA is translated into FT protein. This protein, bonded to a transcription factor, activates several flowering genes and switches off the genes for vegetative growth (Figure 9.36).

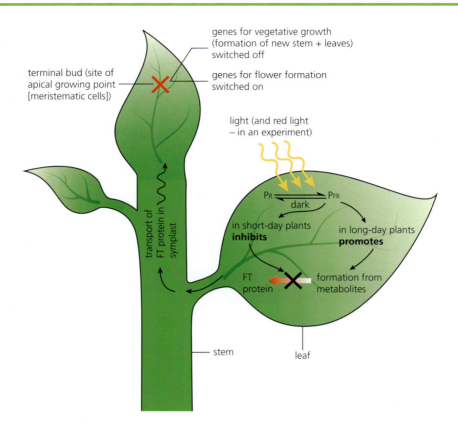

■ **Figure 9.36**
Phytochrome and the switch to flowering – a hypothesis

■ *Examination questions – a selection*

Questions 1–4 are taken from IB Diploma biology papers.

Q1 The diagram shows a section through a typical dicotyledoneous leaf.

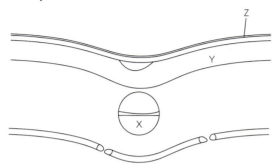

Which correctly identifies the main functions of the structures labelled X, Y and Z?

	X	Y	Z
A	Support	Gas exchange	Photosynthesis
B	Transport products of photosynthesis	Photosynthesis	Water conservation
C	Gas exchange	Water conservation	Light absorption
D	Transport water	Support	Gas exchange

Higher Level Paper 1, Time Zone 0, Nov 12, Q31

Q2 In which modified structures are sugars stored in an onion bulb?
A Stems
B Roots
C Flowers
D Leaves

Higher Level Paper 1, Time Zone 1, May 13, Q31

Q3 The diagram shows a cross-section through a leaf.

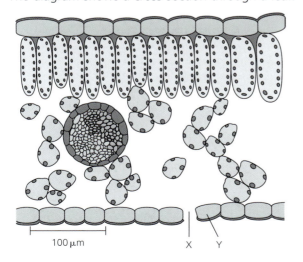

What is the relationship between structures X and Y?
A Y causes X to open allowing water to exit the leaf when water is scarce.
B Y responds to abscisic acid by closing X to prevent water loss.
C Y responds to gibberellin by opening X to allow water loss.
D Y causes X to close to increase transpiration.

Higher Level Paper 1, Time Zone 2, May 13, Q32

Q4 What is the role of P_{fr} in plants?
A To promote flowering in long-day plants
B To promote flowering in short-day plants
C To inhibit flowering in long-day plants
D To inhibit flowering in both long-day plants and short-day plants

Higher Level Paper 1, Time Zone 0, Nov 12, Q33

Questions 5–10 cover other syllabus issues in this chapter.

Q5 **a** Outline the different processes that happen during the germination of a named dicotyledoneous seed and indicate the name of the hormone that is involved in this process. (4)
b List the hormones that are involved in metabolic processes in plants with their corresponding role. (3)
c By means of a table, distinguish between the structure of monocotyledonous and dicotyledonous plants. (4)

Q6 The leaf is the site of the bulk of photosynthesis in the flowering plant. Outline, by means of concise notes, **five** structural features of leaves that specifically favour photosynthesis and the ways they may enhance this process. (5 + 5)

Q7 **a** Define *translocation* and *transpiration*. (4)
b Identify the sources of energy for movement of substances in:
 i translocation
 ii transpiration. (4)
c By means of a fully annotated drawing, describe the structure of phloem tissue. (6)
d List the structural features of phloem tissue that are not shown by xylem tissue. (3)
e Outline the essential features of the mass flow hypothesis of phloem transport. (6)

Q8 a Draw a fully labelled half-flower diagram of an insect-pollinated flower you have studied. (6)

b State what insects may visit this flower. (1)

c Identify the features of this flower that may attract insects. (3)

d Explain how pollination is brought about in this flower. (4)

Q9 a List the ways guard cells of stomata differ from the ordinary epidermal cells around them. (3)

b Explain why stomatal pores close as leaves wilt, irrespective of whether the leaf is in the light or dark. (2)

c Outline the ideal conditions for stomata to be fully open. (4)

Q10 a Describe the role of far-red absorbing phytochrome (Pfr) and how the switch to flowering is a response to the length of light and dark periods in many plants. (4)

b Distinguish between pollination, fertilization and seeds dispersal. (3)

c Describe the adaptations of plants when growing in saline soils. (4)

ESSENTIAL IDEAS

- Meiosis leads to independent assortment of chromosomes and unique composition of alleles in daughter cells.
- Genes may be linked or unlinked and are inherited accordingly.
- Gene pools change over time.

The painstaking experiments of **Gregor Mendel** (Figure 3.15, page 149) were overlooked in his lifetime. Yet, later, they provided the foundations for **modern genetics**, once the work was rediscovered. The discovery of exceptions to his 'Laws', in particular, led to major advances in genetics. Here, we learn more of Mendel's work and its rediscovery, after an examination of the behaviour of chromosomes in meiosis. Finally, the roles of genes in speciation and, hence, in evolution are discussed.

10.1 Meiosis – *meiosis leads to independent assortment of chromosomes and unique composition of alleles in daughter cells*

Meiosis is an essential event in any lifecycles that include sexual reproduction, because at fertilization the chromosome number is doubled.

The events in meiosis have already been established in Chapter 3. Meiosis is a nuclear division that is slower and more complex than mitosis, because it involves two successive divisions of the nucleus (**meiosis I** and **meiosis II**). We have seen that in meiosis I the **homologous chromosomes separate**, and that in meiosis II the **chromatids separate**.

Take a look at these points, shown in Figure 3.9 on page 143, now.

■ Chromosomes replicate in interphase *before* meiosis

The sequence of cell-cycle events of interphase (Figure 1.56, page 51) that precedes mitosis also precedes meiosis.

Remember, **chromosomes replicate to form chromatids during interphase**, well before nuclear divisions occur. Equally important is the fact that there is no interphase between meiosis I and II, so no replication of the chromosomes occurs *during* meiosis.

■ The process of meiosis

Once started, meiosis proceeds steadily as a continuous process of nuclear division. The steps of meiosis are explained in four distinct phases (**prophase**, **metaphase**, **anaphase** and **telophase**), but this is just for convenience of analysis and description – there are no breaks between the phases in nuclear division.

The behaviour of the chromosomes in the phases of meiosis is shown in Figure 10.1. For clarity, the drawings show a cell with a single pair of homologous chromosomes.

Meiosis I

Prophase I

What happens to chromosomes during prophase I is especially complex. They appear, when viewed by light microscopy, as single threads with many tiny bead-like thickenings along their length. These thickenings represent an early stage in the process of shortening and thickening by coiling that continues throughout prophase. This packaging of DNA in the chromosome is shown in Figure 1.60, page 56. Of course, each chromosome is already replicated as two chromatids, but individual chromatids are not visible as yet.

MEIOSIS I

prophase I (early)
During interphase the chromosomes replicate into chromatids held together by a centromere (the chromatids are not visible). Now the chromosomes condense (shorten and thicken) and become visible.

prophase I (mid)
Homologous chromosomes pair up (becoming **bivalents**) as they continue to shorten and thicken. Centrioles duplicate.

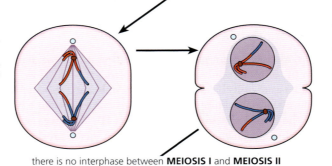

prophase I (late)
Homologous chromosomes repel each other. Chromosomes can now be seen to consist of chromatids. Sites where chromatids have broken and rejoined, causing crossing over, are visible as chiasmata.

metaphase I
Nuclear membrane breaks down. Spindle forms. Bivalents line up at the equator, attached by centromeres.

anaphase I
Homologous chromosomes separate. Whole chromosomes are pulled towards opposite poles of the spindle, centromere first (dragging along the chromatids).

telophase I
Nuclear membrane re-forms around the daughter nuclei. The chromosome number has been halved. The chromosomes start to decondense.

there is no interphase between **MEIOSIS I** and **MEIOSIS II**

MEIOSIS II

prophase II
The chromosomes condense and the centrioles duplicate.

metaphase II
The nuclear membrane breaks down and the spindle forms. The chromosomes attach by their centromere to spindle fibres at the equator of the spindle.

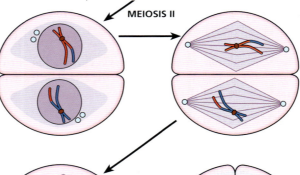

anaphase II
The chromatids separate at their centromeres and are pulled to opposite poles of the spindle.

telophase II
The chromatids (now called chromosomes) decondense. The nuclear membrane re-forms. The cells divide.

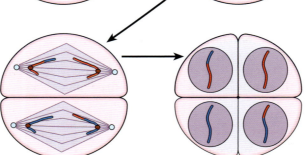

■ **Figure 10.1**
What happens in meiosis?

Formation of bivalents

As the chromosomes continue to thicken, homologous chromosomes are seen to come together in specific pairs, point by point, all along their length. The product of pairing is called a **bivalent**. Remember, in a diploid cell each chromosome has a partner that is the same length and shape, and has the same linear sequence of alleles.

The homologous chromosomes of the bivalents continue to shorten and thicken. Later in prophase, the individual chromosomes can be seen to be double-stranded, as the sister chromatids (of which each consists) become visible.

Exchange of genetic material – crossing over

Within the bivalent, during the coiling and shortening process, breakages of the chromatids occur frequently. Breaks are common in non-sister chromatids, at the same points along their lengths. Broken ends rejoin more or less immediately but, where these 'repairs' are between non-sister chromatids, swapping of pieces of the chromatids occurs, hence the term 'crossing over'. Once crossing over is complete, the non-sister chromatids continue to adhere at that point, called a **chiasma** (plural, **chiasmata**). The chiasma stabilizes the bivalent.

Exchange of alleles – the outcome of chiasmata between non-sister chromatids

Virtually every pair of homologous chromosomes forms at least one chiasma at this time, and to have two or more chiasmata in the same bivalent is very common (Figures 10.2 and 10.3).

Chiasmata increase genetic variability because the process results in the exchange of DNA between maternal and paternal chromosomes. Remember, crossing over can occur many times and between different chromatids within each bivalent. So, crossing over can produce new combinations of alleles on the chromosomes of the haploid cells that are finally formed by meiosis, followed by cytokinesis.

Then, in the later stage of prophase I, the attraction and tight pairing of the homologous chromosomes end, but the attraction between sister chromatids remains for the moment. This attraction of sister chromatids keeps the bivalents together. The chromatids are now at their shortest and thickest.

Later still, the centrioles present in animal cells (page 21) duplicate, and start to move apart as a prelude to the formation of the spindle. Plant cells are without a centriole.

Finally, the disappearance of the nucleoli and nuclear membrane marks the end of prophase I.

■ **Figure 10.2**
Formation of chiasmata

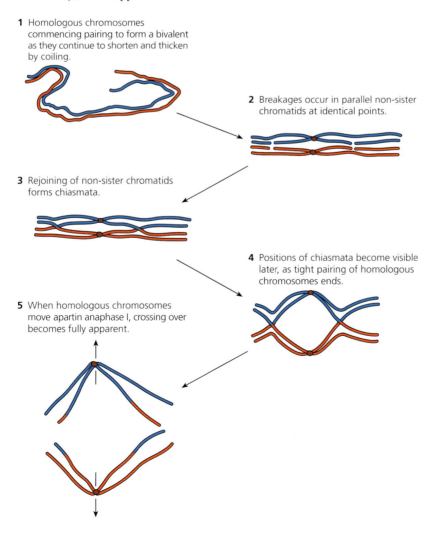

1 Homologous chromosomes commencing pairing to form a bivalent as they continue to shorten and thicken by coiling.

2 Breakages occur in parallel non-sister chromatids at identical points.

3 Rejoining of non-sister chromatids forms chiasmata.

4 Positions of chiasmata become visible later, as tight pairing of homologous chromosomes ends.

5 When homologous chromosomes move apart in anaphase I, crossing over becomes fully apparent.

Drawing diagrams to show chiasmata formed by crossing over

Diagrams of chiasmata should show sister chromatids aligned, except at the point where crossing over occurs and a chiasma forms. So chiasmata appear as X-shaped structures at one or more points between four long thin chromatids of two tightly paired homologous chromosomes. At a later stage in your sequence of drawings, the tight pairing of the homologous chromosomes ends, but the sister chromatids remain connected. Of course, the use of two colours in your diagrams is essential. Figure 10.2 shows how chiasmata formed by crossing over can be shown in a diagram, and this is further explained in Figure 10.5, page 419.

Metaphase I

Once spindle formation is complete, the bivalents become attached to individual spindle microtubules by their centromeres. The bivalents are now arranged at the equatorial plate of the spindle framework – we say that they line up at the centre of the cell. By the end of metaphase I, the members of the bivalents start to repel each other and separate.

However, at this point, they are held together by one or more chiasmata and this gives temporary but unusual shapes to the bivalents.

Anaphase I

The homologous chromosomes of each bivalent now move to opposite poles of the spindle, but with the individual chromatids remaining attached by their centromeres. The attraction of sister chromatids has lapsed and they separate slightly – both are clearly visible. However, they do not separate yet, but go to the same pole. Consequently, meiosis I has separated homologous pairs of chromosomes, but not the sister chromatids of which each is composed.

Telophase I

The arrival of homologous chromosomes at opposite poles signals the end of meiosis I. The chromosomes tend to uncoil to some extent and a nuclear membrane reforms around both nuclei. The spindle breaks down. However, these two cells do not go into interphase, but rather continue into meiosis II, which takes place at right angles to meiosis I.

Meiosis II is remarkably similar to mitosis.

Meiosis II

Prophase II

The nuclear membranes break down again, and the chromosomes shorten and rethicken by coiling. Centrioles, if present, move to opposite poles of the cell. By the end of prophase II the spindle apparatus has reformed, but is present at right angles to the original spindle.

Metaphase II

The chromosomes line up at the equator of the spindle, attached by their centromeres.

Anaphase II

The centromeres divide and the chromatids are pulled to opposite poles of the spindle, centromeres first.

Telophase II

Nuclear membranes form around the four groups of chromatids, so that four nuclei are formed. Now there are four cells, each with half the chromosome number of the original parent cell. Finally, the chromatids – now recognizable as chromosomes – uncoil and become apparently dispersed as chromatin. Nucleoli reform.

The process of meiosis is now complete, and is followed by division of the cells (cytokinesis, page 55).

■ Meiosis and genetic variation

The variation in the genetic information carried by different gametes that arises in meiosis is highly significant for the organism, as we shall see. The four haploid cells produced by meiosis differ genetically from each other because of independent assortment of chromosomes and crossing over.

Independent assortment of maternal and paternal homologous chromosomes

The way in which the bivalents line up at the equator of the spindle in meiosis I is entirely random. Which chromosome of a given pair goes to which pole is unaffected by (is independent of) the behaviour of the chromosomes in other pairs. This was introduced in Figure 3.9 (page 143), but it is represented again here in terms of the critical steps in meiosis where it occurs (Figure 10.4). These illustrations show a parent cell with only four chromosomes, for clarity. Of course, the more bivalents there are in the nucleus, the more variation is possible. In humans, there are 23 pairs of chromosomes, so the number of possible combinations of chromosomes that can be formed as a result of independent assortment is 2^{23}. This is over 8 million.

Crossing over of segments of individual maternal and paternal homologous chromosomes

Crossing over results in new combinations of genes on the chromosomes of the haploid cells produced by meiosis, as illustrated in Figure 10.5. The process generates the possibility of an almost unimaginable number of new combinations of alleles. For example, if we were to assume, for sake of discussion, that there are 30 000 individual genes on the human chromosome complement, all with at least two alternative alleles, and that crossing over was equally likely between any of these genes, there would be $2^{30\,000}$ different combinations of alleles. Of course, these assumptions could be inaccurate to varying extents, but the point that many, many allele combinations are possible is undeniable.

1 **Distinguish** the essential differences between mitosis and meiosis.

Nature of Science **Making careful observations**

■ A note on recombinants

Offspring with new combinations of characteristics, different from those of their parents, are called **recombinants**.

Recombination in genetics is the reassortment of alleles or characters into different combinations from those of the parents. We have seen that recombination occurs for genes located on separate chromosomes (unlinked genes) by chromosome assortment in meiosis (Figure 10.4), and for genes on the same chromosomes (linked genes) by crossing over during meiosis (Figure 10.5).

In Section 10.2, we look at the linkage of genes in more detail. At this stage we should just note that it was careful observation and record keeping which first disclosed anomalous data that could not be accounted for by Mendel's Law of independent assortment (page 420). It was the work of Thomas Morgan (page 423) and others that led to the notion of linked genes – a factor that accounted for many of these anomalies.

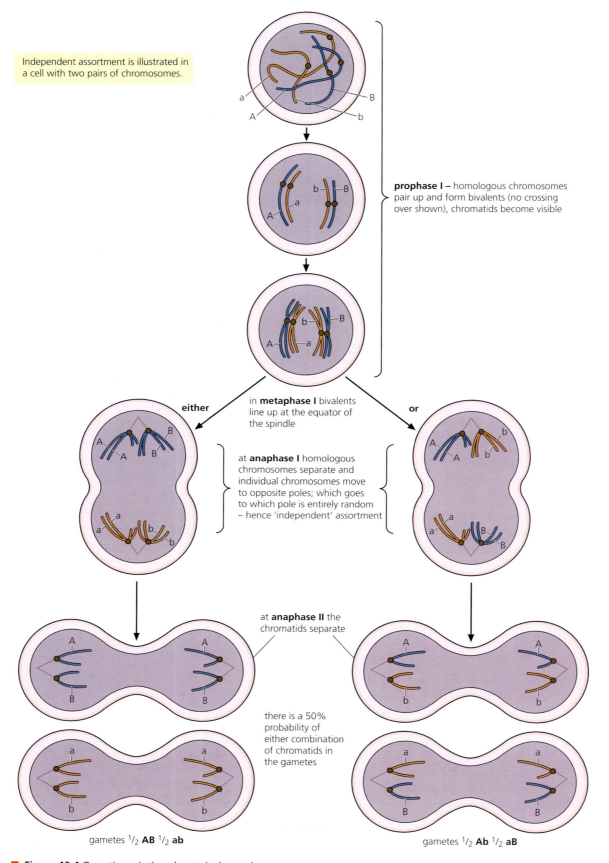

Independent assortment is illustrated in a cell with two pairs of chromosomes.

prophase I – homologous chromosomes pair up and form bivalents (no crossing over shown), chromatids become visible

in **metaphase I** bivalents line up at the equator of the spindle

either

or

at **anaphase I** homologous chromosomes separate and individual chromosomes move to opposite poles; which goes to which pole is entirely random – hence 'independent' assortment

at **anaphase II** the chromatids separate

there is a 50% probability of either combination of chromatids in the gametes

gametes ¹/₂ **AB** ¹/₂ **ab**

gametes ¹/₂ **Ab** ¹/₂ **aB**

■ **Figure 10.4** Genetic variation due to independent assortment

Figure 10.5 Genetic variation due to crossing over

The effects of chiasmata on genetic variation are illustrated in one pair of homologous chromosomes. Typically, two or three chiasmata form between the chromatids of a bivalent in prophase I.

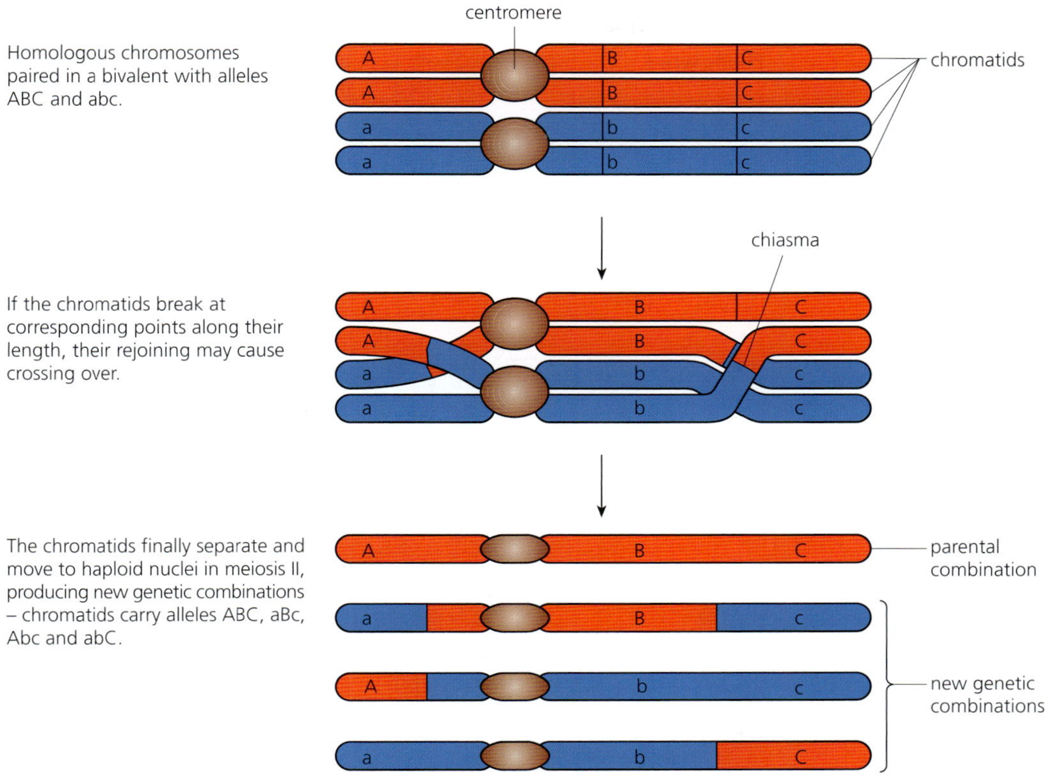
centromere
Homologous chromosomes paired in a bivalent with alleles ABC and abc.
A
A
a
a
B
B
b
b
C
C
c
c
chromatids
chiasma
If the chromatids break at corresponding points along their length, their rejoining may cause crossing over.
The chromatids finally separate and move to haploid nuclei in meiosis II, producing new genetic combinations – chromatids carry alleles ABC, aBc, Abc and abC.
parental combination
new genetic combinations

10.2 Inheritance *– genes may be linked or unlinked and are inherited accordingly*

There are many thousands of genes per cell in an organism, whereas the number of chromosomes is often less than 50 and rarely exceeds 100. Each chromosome consists of many genes – it may be thought of as a linear sequence of genes that are all linked together; gene loci are said to be **linked** if they occur on the same chromosome. Obviously, these genes tend to be inherited together.

Non-Mendelian ratios led to the discovery of linkage

This phenomenon of linkage was discovered in breeding experiments where discrepancies arose between expected results (as a consequence of Gregor Mendel's Laws) and the ratios actually obtained. We will examine the discovery and consequences of linkage, shortly.

First, we look into the inheritance of pairs of *unlinked* genes. The pioneering work of Mendel was introduced in Figure 3.15 (page 149). Mendel's experiments also included the simultaneous inheritance of two pairs of contrasting characters, using the garden pea plant. Unknown to Mendel, of course, these characters were controlled by genes on separate chromosomes – they were unlinked. Unlinked genes segregate independently as a result of meiosis. Mendel referred to this type of cross, involving two pairs of characters, as a **dihybrid cross**.

The dihybrid cross

Mendel crossed pure-breeding pea plants (P generation) from **round seeds** with **yellow cotyledons** (seed leaves) with pure-breeding plants from **wrinkled seeds** with **green cotyledons**. All the progeny (F_1 generation) were **round, yellow peas**.

When plants grown from these seeds were allowed to self-fertilize the following season, the resulting seeds (F$_2$ generation) – of which there were more than 500 to be classified and counted – were of the following four phenotypes, and they were present in the ratio shown in Table 10.1.

■ **Table 10.1**
Mendel's F$_2$
generation

Phenotypes	round seed with yellow cotyledons	round seed with green cotyledons	wrinkled seed with yellow cotyledons	wrinkled seed with green cotyledons
Ratio	9	3	3	1

■ **Figure 10.6**
Mendel's dihybrid cross

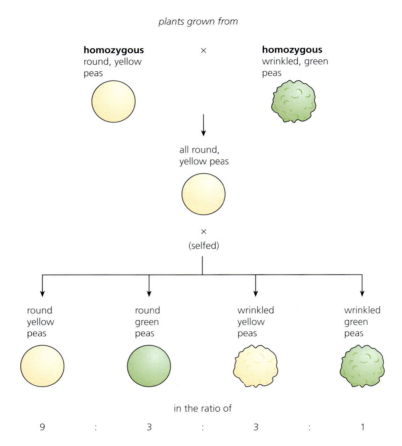

Mendel noticed that two new combinations, not represented in the parents (i.e. **recombinations**), appeared in the progeny; both round and wrinkled seeds appear with either green or yellow cotyledons. From this result, it can be seen that the two pairs of factors were inherited independently and, therefore, were on separate chromosomes. Mendel had noticed that either one of a pair of contrasting characters could be passed to the next generation. This meant that a heterozygous plant must produce four types of gametes in equal numbers (Figures 10.6 and 10.7).

Mendel did not express the outcome of the dihybrid cross as a succinct law. However, today we call Mendel's Second Law the Law of **independent assortment**. It is stated as:

Two or more pairs of alleles segregate independently of each other as a result of meiosis, provided the genes concerned are not linked by being on the same chromosome.

2 In Figure 10.6, **identify** the progeny that are
a heterozygous
b recombinants.

■ Punnett squares for dihybrid traits

Notice the use of a Punnett square diagram to predict the outcome of a breeding investigation in which independent assortment of alleles is occurring (Figure 10.7). By this device, every possible combination of maternal and paternal gametes – the product of random fertilization – is made. There are shown as many rows/columns as there are unique male and unique female gametes. Each fraction represents the probability that a particular gamete or zygote will occur.

The relationship between Mendel's Law of independent assortment and meiosis is detailed in Table 10.2.

parental (P)

phenotypes: homozygous   homozygous
              round and yellow   wrinkled and green

genotypes: **RRYY**   ×   **rryy**   ◄——— pea plants are diploid so they have two
                                                       copies of each allele

              (meiosis)   (meiosis)

gametes: (RY)   (ry)   ◄——— gametes produced by meiosis so only
                                                       have one copy of each allele

offspring (F₁)

genotypes: **RrYy**   ◄——— F₁ progeny are heterozygous for
                                                       both genes

phenotypes: heterozygous   ◄——— the alleles for 'round' and 'yellow' are
                   round and yellow   dominants; the F₁ progeny are all
                                                       round yellow peas

F₁ selfed   **RrYy**   ×   **RrYy**   ◄——— genes are on separate
                                                       chromosomes and random
              (meiosis)   (meiosis)   assortment occurs, i.e.
                                                       R may assort with **Y** or **y**
                                                       r may assort with **Y** or **y**
gametes: (RY)(Ry)(rY)(ry)   (RY)(Ry)(rY)(ry)   ◄——— so four different types of gametes
                                                       (both ♂ and ♀) are formed, in equal
                                                       proportions (**independent
                                                       assortment**)

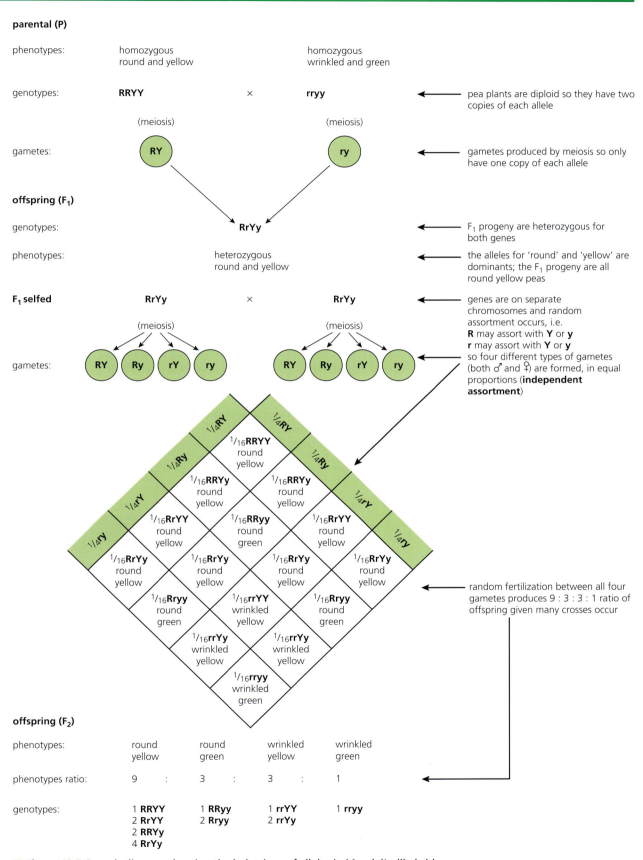

random fertilization between all four
gametes produces 9 : 3 : 3 : 1 ratio of
offspring given many crosses occur

offspring (F₂)

phenotypes: round   round   wrinkled   wrinkled
               yellow   green   yellow   green

phenotypes ratio: 9   :   3   :   3   :   1

genotypes: 1 **RRYY**   1 **RRyy**   1 **rrYY**   1 **rryy**
               2 **RrYY**   2 **Rryy**   2 **rrYy**
               2 **RRYy**
               4 **RrYy**

■ **Figure 10.7** Genetic diagram showing the behaviour of alleles in Mendel's dihybrid cross

■ **Table 10.2** How the Law of independent assortment relates to meiosis

Mendel's dihybrid cross	Feature of meiosis
Within an organism exist 'breeding factors' that control characteristics like round or wrinkled seeds, and yellow or green cotyledons. These factors remain intact from generation to generation.	Each chromosome holds a linear sequence of genes. A particular gene always occurs on the same chromosome in the same position (locus) after each nuclear division.
There are two factors for each characteristic in each cell. One factor comes from each parent. (A recessive factor is not expressed in the presence of a dominant factor.)	The chromosomes of a cell occur in pairs, called homologous pairs. One of each pair came originally from each parent.
Factors separate in reproduction; either can be passed to an offspring. Only one of the factors can be in any gamete.	At the end of meiosis, each cell (gamete) contains a single member of each of the homologous pairs of chromosomes present in the parent cell.
The factors for seed shape and seed colour segregate independently of each other as a result of meiosis.	The genes for seed shape and seed colour **are on separate chromosomes.**
The 9:3:3:1 ratio shows that all four types of gamete are equally common. The inheritance of the two characteristics is separate.	The arrangement of bivalents at the equatorial plate of the spindle is random; maternal and paternal homologous chromosomes are independently assorted. In a large number of matings, all possible combinations of chromosomes will occur in equal numbers.

3 **Deduce** the positions of the genes for round/wrinkled and for yellow/green cotyledons within the nucleus of the pea plant. **Explain** the significance of their location (which was unknown to Mendel).

Gene linkage

After the rediscovery of Mendel's work in the early 1900s, geneticists investigated other dihybrid crosses in order to confirm his results. For example, William Bateson and Reginald Punnett (who had devised the 'Punnett square') crossed pure-breeding sweet pea plants with purple flowers and long pollen grains with plants having red flowers and round pollen grains. All the F_1 plants were purple-flowered with long pollen grains. This shows that the allele for purple flower is dominant over the allele for red flower and the allele for long pollen is dominant over that for round pollen.

When the F_1 were self-crossed, however, most of the offspring resembled the parental phenotypes, but with a small number of recombinants. The actual results obtained are shown in Figure 10.8.

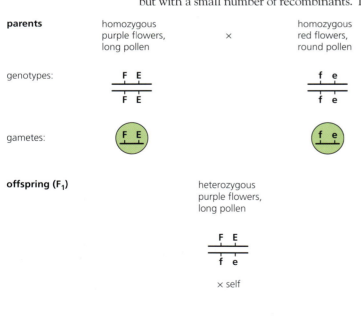

flowers of sweet pea (*Lathyrus odoratus*)

outcome	Phenotypes			
	purple flower long pollen	purple flower round pollen	red flower long pollen	red flower round pollen
expected	240	80	80	27
ratio	9	3	3	1
actual	296	19	27	85

■ **Figure 10.8** An example of linked genes in the sweet pea plant

The Mendelian ration of 9:3:3:1 had not been obtained. Since most of the F_2 offspring resembled the parental phenotypes (with a small number of recombinants) it seemed reasonable to conclude that the genes for flower colour and pollen shape were present on the same chromosome. If so, these genes were linked – they did not segregate in meiosis, but were inherited together.

Notice that in crosses involving linkage, the alleles are typically shown as vertical pairs:

$$\frac{F\ E}{f\ e}$$

rather than as FfEe, for example.

■ What caused the recombinants in the sweet pea experiment?

Well, if the genes concerned were on the same chromosome, when the F_1 plants were crossed, the appearance of the F_2 offspring depended on whether a chiasma formed between these alleles, or elsewhere along the chromosome during meiosis in gamete formation. *Look at Figure 10.9 now.* Here, the consequence of a chance chiasma – and its location – is made clear.

■ **Figure 10.9**
Chiasmata and the origin of recombinants

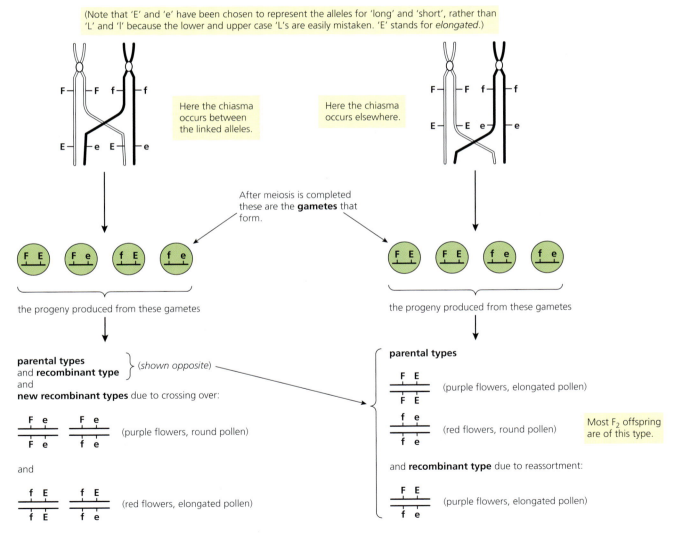

■ *Drosophila* and the work of Thomas Morgan

Drosophila melanogaster (the fruit fly) was first selected in 1908 by an American geneticist, **Thomas Morgan**, as an experimental organism for his series of investigations of Mendelian genetics, in this case in an animal. The remarkable breakthroughs in understanding that Morgan achieved resulted in the award of a Nobel Prize in 1933.

His experimental work:

■ showed that non-Mendelian ratios are commonly obtained in breeding experiments with *Drosophila* – but not always

■ established that Mendel's 'factors' are linear sequences of genes on chromosomes (this is now called the **Chromosome Theory of Inheritance**)

■ discovered sex chromosomes and sex linkage (page 158)

■ demonstrated crossing over and the exchange of alleles between chromosomes, resulting from the chiasmata that form during meiosis.

Drosophila is an organism that commonly occurs around rotting vegetable material, existing in a form called a 'wild type' (a non-mutant form) and in various naturally occurring mutant forms (Figure 10.10). This animal rapidly became a useful experimental animal in the study of genetics because:

■ *Drosophila* has only four pairs of chromosomes (Figure 10.10).

■ from mating to emergence of adult flies (generation time) takes about 10 days at 25°C

■ a single female fly produces hundreds of offspring

■ the flies are relatively easily handled, cultured on sterilized artificial medium in glass bottles (they can be temporarily anaesthetized for setting up cultures and sorting progeny).

4 **Define** what is meant by the term 'mutant'.

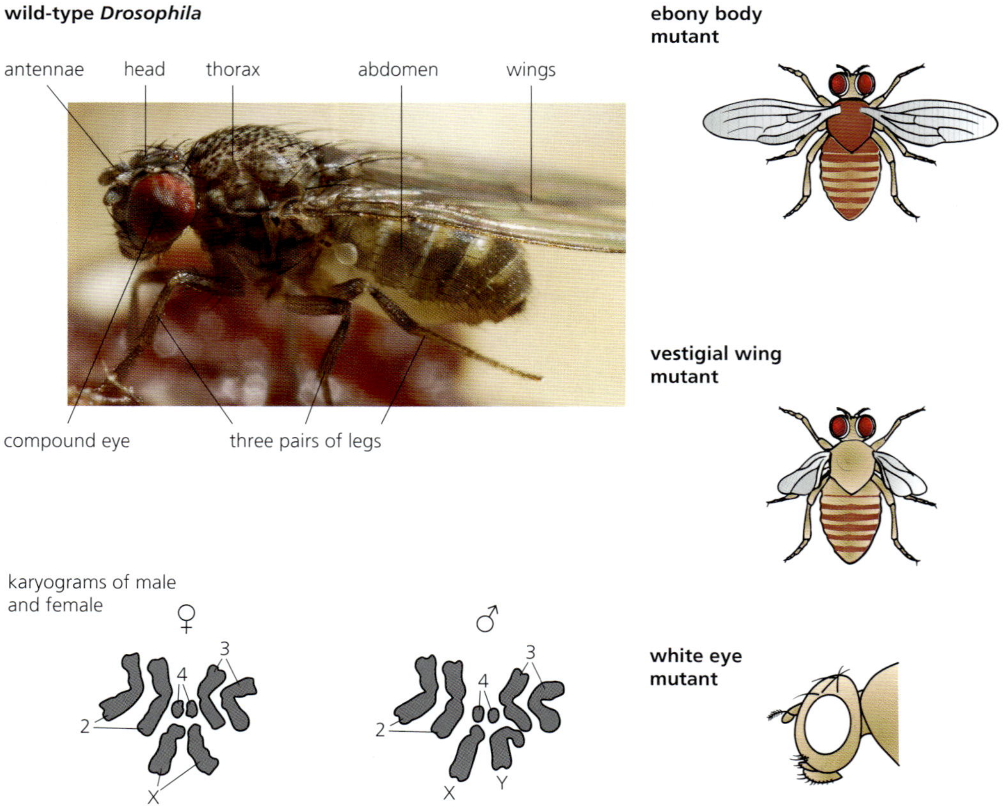

■ **Figure 10.10** Wild-type *Drosophila* and some common mutants

wild-type *Drosophila*

antennae   head   thorax   abdomen   wings

compound eye   three pairs of legs

ebony body mutant

vestigial wing mutant

white eye mutant

karyograms of male and female

Drosophila and the dihybrid cross

First, we will consider a dihybrid cross in *Drosophila*, involving unlinked genes. In this experiment we are crossing normal flies (wild type) with flies that are homozygous for vestigial wing and ebony body (Figure 10.10). These characteristics are controlled by genes that are located on separate **autosomal chromosomes** (chromosomes other than the sex chromosomes, page 158).

The prediction of genotype and phenotype ratios in a dihybrid cross involving unlinked autosomal genes

After you have examined Figure 10.11, respond to question 5. This requires you to determine the genotype and phenotype offspring of the F_2 generation raised in Figure 10.11.

■ **Figure 10.11**
A dihybrid cross in *Drosophila* – a summary

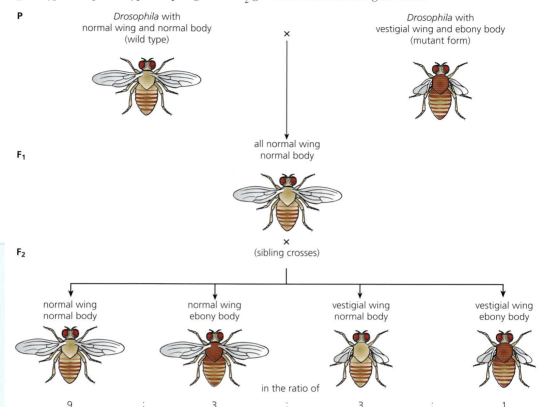

P

Drosophila with
normal wing and normal body
(wild type)

×

Drosophila with
vestigial wing and ebony body
(mutant form)

F₁

all normal wing
normal body

×
(sibling crosses)

F₂

| normal wing normal body | normal wing ebony body | vestigial wing normal body | vestigial wing ebony body |

in the ratio of

9 : 3 : 3 : 1

5 **Construct** a genetic diagram for the dihybrid cross shown in Figure 10.11, using the layout given in Figure 10.7. **Determine** the genotype and phenotype ratios of the offspring of the F_2 generation.

Probability and chance in genetic crosses – the chi-squared test

We know the expected ratio of offspring of a dihybrid cross is 9:3:3:1. Actually, the offspring produced in many dihybrid cross experiments do not *exactly* agree with the expected ratio. This is illustrated by the results of the experiment with mutant forms of *Drosophila*, shown in Table 10.3.

■ **Table 10.3** Observed and expected offspring

Offspring in F_2 generation	normal wing, grey body = 315	normal wing, ebony body = 108	vestigial wing, grey body = 101	vestigial wing, ebony body = 32	total = 556
Predicted ratio	9	3	3	1	
Expected numbers of offspring	313	104	104	35	

Clearly, these results are fairly close to (but not precisely in) the predicted ratio 9:3:3:1.

What, if anything, went wrong?

Well, we can expect precisely this ratio among the progeny *only* if three conditions are met:

■ fertilization is entirely random
■ there are equal opportunities for survival among the offspring
■ very large numbers of offspring are produced.

In the above experiment with *Drosophila*, the exact ratio may not be obtained because, for example:

■ more male flies of one type may have succeeded in fertilizing females on this occasion
■ more females of one type may have died before reaching egg laying condition
■ fewer eggs of one type may have completed their development.

Similarly, in breeding experiments with plants, such as the pea plant, exact ratios may not be obtained. This could, perhaps, be due to parasite damage to some seeds or plants, to the action of browsing predators on the anthers or ovaries in some flowers, or because some pollen types fail to be transported by pollinating insects as successfully as others. Such accidental events are quite common.

The chi-squared test on data from a dihybrid cross

Experimental geneticists often ask the question:

> **'Do the observed values differ significantly from the expected outcome?'**

For example, for the sorts of reasons discussed above.

This question is resolved by a simple statistical test, known as the **chi-squared (χ^2) test**. This is used to estimate the probability that any difference between the observed results and the expected results is due to chance. If it is not due to chance, it may be due to an entirely different explanation and the phenomenon needs further investigation.

The chi-squared (χ^2) test (also on pages 188–191)

$$\chi^2 = \Sigma \frac{(O - E)^2}{E}$$

where:
O = observed result
E = expected result
Σ = the sum of

Chi-squared applied

We can test whether the observed values obtained from the dihybrid cross between *Drosophila* of normal flies (wild type) with flies homozygous for vestigial wing and ebony body differ significantly from the expected outcome.

First we calculate χ^2 (Table 10.4).

■ **Table 10.4**
Calculating χ^2

Category	Predicted	O	E	O–E	(O – E)^2	(O – E)^2/E
normal wing, normal body	9	315	312.75	2.25	5.062	0.016
normal wing, ebony body	3	108	104.25	3.75	14.062	0.135
vestigial wing, normal body	3	101	104.25	–3.25	10.562	0.101
vestigial wing, ebony body	1	32	34.75	–2.75	7.562	0.218
Total		**556**		$\Sigma(\chi^2)$		**0.47**

In this example we have calculated χ^2 to be 0.47.

To see if this value of chi-squared represents a significant difference between observed and expected results, we now consult a table, such as that shown in Table 10.5, of the **distribution of** χ^2. We can find out the **probability (P)** of obtaining by chance alone a deviation as large as (or larger than) the one we have observed.

Note that the table takes account of the number of independent comparisons involved in our test. In our example, there were four categories and, therefore, three comparisons were made – we call this 'three **degrees of freedom (df)**'. (Another way of putting this is that for any one condition there are three alternatives.) So, we look along the row 'df = 3' to see whether 0.47 lies to the left or to the right of the 0.05 level of probability (shown in red).

■ **Table 10.5**
Table of χ^2 distribution

Degrees of freedom	Probability greater than							
	0.99	**0.95**	**0.90**	**0.50**	**0.10**	**0.05**	**0.01**	**0.001**
df = 1	0.00016	0.004	0.016	0.455	2.71	3.84	6.63	10.83
df = 2	0.0201	0.103	0.21	1.386	4.60	5.99	9.21	13.82
df = 3	0.115	0.35	0.58	1.39	6.25	7.81	11.34	16.27

Using the χ^2 distribution table, we can resolve whether the difference between the result we expected and the result we actually observed is due to chance – or whether the difference is, in fact, significant.

- If the value of χ^2 is *bigger* than the critical value highlighted in red (a probability of 0.05) then we can be at least 95% confident that the difference between the observed and expected results is *significant*.

- If the value of χ^2 is *smaller* than the critical value highlighted in red (a probability of 0.05) then we can be confident that the difference between the observed and expected results is *due to chance*.

In biological experiments we take a probability of 0.05 or larger to indicate that the difference between the observed (O) and expected (E) results is not significant. We can say it is due to chance.

In this example, the value (0.47) lies between a probability of 0.95 and 0.90. This means that a deviation of this size can be expected 90–95% of the times the experiment is carried out (due to chance). So, there is clearly *no* significant deviation between the observed (O) and the expected (E) results; the data conforms to a Mendelian ratio.

The chi-squared test is similarly applicable to the results of other test crosses.

In any chi-squared test that produces a significant deviation of observed from expected results (does *not* confirm that the results conform to the anticipated values) by giving a value for χ^2 that is *bigger* than the critical value and a probability that is *smaller* than 0.05), we must reconsider our experimental hypothesis. In this outcome, the statistical test gives no clue as to the true location or behaviour of the alleles. Further genetic investigations are required.

We will consider examples of non-Mendelian ratios, next.

---

6 In the dihybrid test cross between **homozygous dwarf** pea plants with **terminal flowers** (**ttaa**) and **heterozygous tall** pea plants with **axial flowers** (**TtAa**), the progeny were:
- tall, axial = 55 peas
- tall, terminal = 51 peas
- dwarf, axial = 49 peas
- dwarf, terminal = 53 peas.

Use the χ^2 test to **determine** whether or not the difference between these observed results and the expected results is significant

---

Looking for patterns, trends and discrepancies

■ Non-Mendelian ratios in *Drosophila*

It was Thomas Morgan who, by careful observation and record keeping, made the discovery of linkage in *Drosophila*. Like others, he was aware of Mendel's discoveries. The genetic crosses he conducted gave results at odds with the expected Mendelian ratios. The cross in Figure 10.12 is one example. Here, a fly homozygous for ebony body and curled wing was crossed with a fly heterozygous for normal body and straight wing. Note that the characteristic 'curled wing' is different from the characteristic 'vestigial wing' shown in Figure 10.10 (a cross concerned with contrasting characteristics controlled by genes on separate chromosomes).

This experiment, and many others, confirmed the existence of linked genes in *Drosophila*.

Identification of recombinants in a cross involving two linked genes

Using the annotated cross shown in Figure 10.12 as a guide, answer question 7 below.

TOK Link

The Law of independent assortment was soon found to have exceptions, such as when looking at linked genes. What is the difference between a 'law' and a 'theory' in science?

---

7 **Deduce** what recombinants may be formed in a cross involving the linked genes:
$\dfrac{tb}{tb} \times \dfrac{TB}{tb}$

■ Polygenes and continuous variation

We began the story of genetics in Chapter 3 with an investigation into the inheritance of height in the garden pea (page 149), where one gene with two alleles gave tall or dwarf plants. This clear-cut difference in an inherited characteristic is an example of discontinuous or **discrete variation** – there is no intermediate form and no overlap between the two phenotypes.

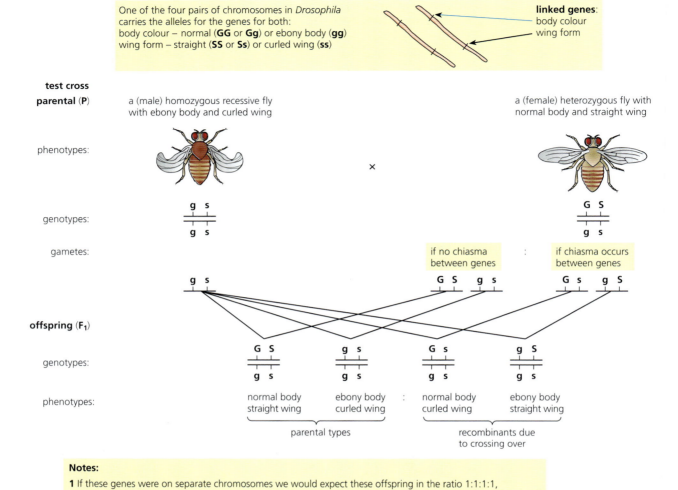

One of the four pairs of chromosomes in *Drosophila* carries the alleles for the genes for both:
body colour – normal (**GG** or **Gg**) or ebony body (**gg**)
wing form – straight (**SS** or **Ss**) or curled wing (**ss**)

linked genes:
body colour
wing form

test cross
parental (P)

a (male) homozygous recessive fly with ebony body and curled wing

a (female) heterozygous fly with normal body and straight wing

phenotypes:

×

genotypes:

$\dfrac{g \quad s}{g \quad s}$

$\dfrac{G \quad S}{g \quad s}$

gametes:

if no chiasma between genes   :   if chiasma occurs between genes

g s

G S   g s

G s   g S

offspring (F₁)

genotypes:

$\dfrac{G \quad S}{g \quad s}$ $\dfrac{g \quad s}{g \quad s}$ $\dfrac{G \quad s}{g \quad s}$ $\dfrac{g \quad S}{g \quad s}$

phenotypes:

normal body straight wing   ebony body curled wing   :   normal body curled wing   ebony body straight wing

parental types

recombinants due to crossing over

Notes:

1 If these genes were on separate chromosomes we would expect these offspring in the ratio 1:1:1:1, but these genes are linked;

2 If no crossing over occurred we would expect parental types only, in the ratio 1:1;

The outcome of this experiment was:

Offspring	Phenotypes	Genotypes	Numbers obtained
parental types	normal body straight wing	$\dfrac{G \quad S}{g \quad s}$	536
	ebony body curled wing	$\dfrac{g \quad s}{g \quad s}$	481
recombinants	normal body curled wing	$\dfrac{G \quad s}{g \quad s}$	101
	ebony body straight wing	$\dfrac{g \quad S}{g \quad s}$	152

3 The majority of offspring were parental types, so we can conclude that few chiasmata occurred *between* the gene loci of these linked genes.

■ **Figure 10.12** A non-Mendelian ratio in a *Drosophila* cross

In fact, very few characteristics of organisms are controlled by a single gene. Mostly, characteristics of organisms are controlled by a number of genes. Groups of genes which together determine a characteristic are called **polygenes**.

Polygenic inheritance is the inheritance of phenotypes that are determined by the collective effect of several genes.

The genes that make up a polygene are often (but not necessarily always) located on different chromosomes. Any one of these genes has a very small effect on the phenotype, but the combined effect of all the genes of the polygene is to produce infinite variety among the offspring. This variety we refer to as **continuous variation**.

Many features of humans are controlled by polygenes, including body weight and height. The graph of the variation in the heights of a population of 400 people, Figure 10.13, shows continuous variation in height between the shortest at 160 cm and the tallest at 186 cm, with a mean height of 173 cm.

■ **Figure 10.13**
Human height as a case of polygenic inheritance

Human height is determined genetically by interactions of the alleles of several genes, probably located at loci on different chromosomes.

Variation in the height of adult humans
The results cluster around a mean value and show a normal distribution. For the purpose of the graph, the heights are collected into arbitrary groups, each of a height range of 2 cm.

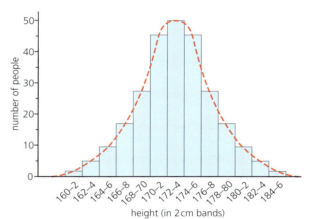

Human skin colour

The colour of human skin is due to the amount of the pigment called **melanin** produced in the skin. Melanin synthesis is genetically controlled. It seems that three, four or more separately inherited genes control melanin production. The outcome is an almost continuous distribution of skin colour from very pale (no alleles coding for melanin production) to very dark brown (all alleles for skin colour code for melanin production).

In Figure 10.14, polygenic inheritance of human skin colour involving only two independent genes is illustrated. This is because dealing with all four genes is unwieldy and the principle can be demonstrated clearly enough using just two genes.

It should be noted, too, that both human height and skin colour are characteristics that may be influenced by environmental factors.

8 **Derive** the ratio of phenotypes produced in the F_2 generation shown in Figure 10.14.

Continuous variation

There are many other examples of polygenic inheritance. In all cases, the number of genes controlling a characteristic does not have to be large before the variation in a phenotype becomes more or less continuous. The outcome is that characteristics controlled by polygenes show **continuous variation**. Nevertheless, the individual genes concerned are inherited in accordance with the principles established above. However, there are so many intermediate combinations of alleles that the discrete ratios are not observed.

Other forms of gene interaction

Another factor that may affect the appearance of the phenotype is the effect of **environmental conditions**. For example, a tall plant may appear almost dwarf if it has been consistently deprived of adequate essential mineral ions. Similarly, the physique of humans may be greatly affected by the levels of nourishment received, particularly as children. So, the phenotype of an organism is the product of both its genotype and the **influences of the environment**.

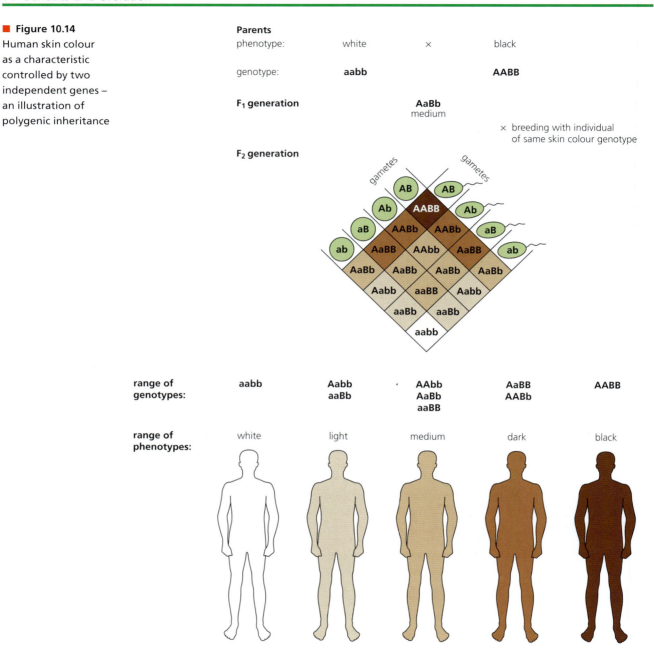

■ **Figure 10.14**
Human skin colour as a characteristic controlled by two independent genes – an illustration of polygenic inheritance

Parents

phenotype: white   ×   black

genotype: **aabb**   **AABB**

F₁ generation   AaBb
medium

× breeding with individual
   of same skin colour genotype

F₂ generation

range of genotypes:	**aabb**	**Aabb** **aaBb**	**AAbb** **AaBb** **aaBB**	**AaBB** **AABb**	**AABB**
range of phenotypes:	white	light	medium	dark	black

10.3 Gene pools and speciation – *gene pools change over time*

In nature, organisms are members of local populations. Remember, a population is a group of individuals of a species, living close enough to be able to interbreed. Species typically exist in localized populations, although the boundaries of a local population can be hard to define. (*Check this idea out by reference to Figure 5.8, page 217.*)

Individuals in local populations tend to resemble each other. They may become quite different from members of other populations. Local populations are very important because they are a potential starting point for **speciation**. Speciation is the name we give to the process by which one species evolves into another. Present-day flora and fauna have arisen by change from pre-existing forms of life. The term 'speciation' emphasizes the fact that species change. The fossil record provides evidence for the process.

■ Population genetics

Population genetics is the study of genes in populations (breeding groups). In any population, the total of all the genes located in the reproductive cells of the individuals make up a **gene pool**. A sample of the genes of the gene pool will form the genomes (gene sets of individuals) of the next generation, and so on, from generation to generation.

Now, when the breeding group is a large one and all the individuals of the population have an equal opportunity of contributing gametes, random matings will perpetuate the original proportions of alleles in the population. In *these* circumstances, the allele frequency will not change.

We can demonstrate this by setting two alleles of a gene at some arbitrary frequency in an infinitely large gene pool, and then following the frequency of possible genotypes these will produce by random matings over several generations. The example of two alleles, **A** and **a**, present in the population at frequencies of 0.8 and 0.2 is illustrated in Figure 10.15. *Look at the single mating cycle illustrated there*. You can see why the frequency of genes will not change after one cycle. In fact, this frequency will never change however many matings occur *unless* there are what geneticists call 'disturbing factors' at work.

Without change in a gene pool we can say the population is not evolving.

■ **Figure 10.15**
The fate of genes in a gene pool

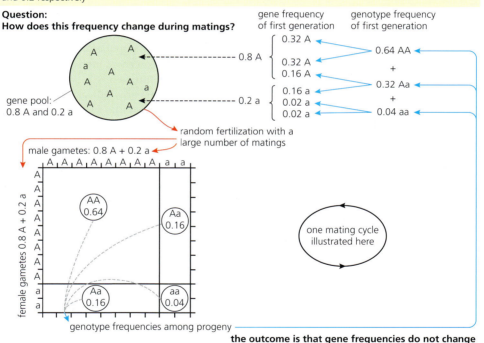

the fate of genes in a gene pool, where there are two alleles A and a, initially present in the frequencies 0.8 and 0.2 respectively

Question:
How does this frequency change during matings?

gene pool: 0.8 A and 0.2 a

random fertilization with a large number of matings

male gametes: 0.8 A + 0.2 a

gene frequency of first generation

0.8 A
0.32 A
0.32 A
0.16 A

0.2 a
0.16 a
0.02 a
0.02 a

genotype frequency of first generation

0.64 AA
+
0.32 Aa
+
0.04 aa

female gametes 0.8 A + 0.2 a

AA 0.64
Aa 0.16
Aa 0.16
aa 0.04

one mating cycle illustrated here

genotype frequencies among progeny

the outcome is that gene frequencies do not change

Changing gene pools and speciation

Studies of gene pools show that, in some populations, the composition of the gene pool does change. This is due to a range of factors that may alter the proportions of some alleles. These 'disturbing factors' are identified in Figure 10.16.

With one or more 'disturbing factors' operating, allele frequencies are likely to change from generation to generation. For example, some alleles may increase in frequency because of the advantage they give to the individuals that carry them. Because of these alleles, an organism may be more successful – producing more offspring, for example. If we can detect change in a gene pool (that is, if the proportions of particular alleles are altered) we may be seeing evolution *before a new species is observed*.

9 **Predict** the factors that may cause the composition of a gene pool to change. (Think about the changes that may go on in a population and between its members.)

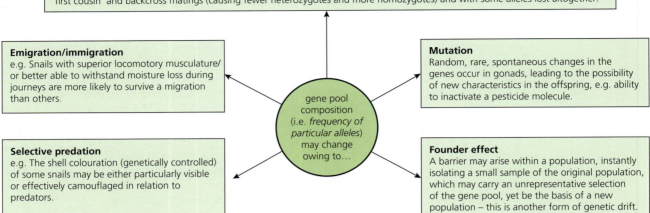

Random genetic drift
Sudden hostile physical condition, e.g. cold, flooding, drought, may sharply reduce a natural population to very few survivors.
   On the return of a favourable environment, numbers of the affected species may quickly return to normal (e.g. owing to reduced competition for food sources), but the new population has been built from a small sample of the original population, with more 'first cousin' and backcross matings (causing fewer heterozygotes and more homozygotes) and with some alleles lost altogether.

Emigration/immigration
e.g. Snails with superior locomotory musculature/ or better able to withstand moisture loss during journeys are more likely to survive a migration than others.

Mutation
Random, rare, spontaneous changes in the genes occur in gonads, leading to the possibility of new characteristics in the offspring, e.g. ability to inactivate a pesticide molecule.

Selective predation
e.g. The shell colouration (genetically controlled) of some snails may be either particularly visible or effectively camouflaged in relation to predators.

gene pool composition (i.e. *frequency of particular alleles*) may change owing to...

Founder effect
A barrier may arise within a population, instantly isolating a small sample of the original population, which may carry an unrepresentative selection of the gene pool, yet be the basis of a new population – this is another form of genetic drift.

■ **Figure 10.16** The compositions of gene pools can change

Speciation by isolation

A first step to speciation may be when a local population (particularly a small local population) becomes completely cut off in some way. Even then, many generations may elapse before the composition of the gene pool has changed sufficiently to allow us to define a different species. However, changes in local gene pools are an early indication of speciation.

Occasionally, a population is suddenly divided into two isolated populations by the appearance of a barrier (Figure 10.17). Before separation, individuals shared a common gene pool, but after isolation disturbing processes such as natural selection, mutation and random genetic drift may occur independently in both populations, causing them to diverge in their features and characteristics.

Geographic isolation between populations occurs when barriers arise and restrict the movement of individuals (and their spores and gametes in the case of plants) between the divided populations. Barriers can be natural or made by humans. An example of geographical isolation comes from the Galápagos Islands, about 500–600 miles from the South American mainland (Figure 5.9, page 218).

Reproductive isolation occurs when two potentially compatible populations are prevented from interbreeding. This may be due to geographical, seasonal or behavioural isolation.

■ Geographical isolation – organisms in different microhabitats within an area could be a good example. Think of an example in a habitat you have studied, if you can.

■ Temporal or seasonal isolation – this occurs when organisms produce gametes at different times or seasons. An example in animals is rainbow trout (spring) and brown trout (autumn).

■ Behavioural isolation – organisms acquire distinctive behaviour routines, such as in courtship or mating, not matched by other individuals of their species.

An example of behavioural isolation is seen in the Birds of Paradise (Figure 10.18). Here, bright, glittering and prominently posing males seek to secure the attention of females. Elaborate and attractive display traits have evolved because they apparently satisfy some innate preference in females. The resulting progressive elaboration of plumage and performance may be at the expense of flight ability, vulnerability to the attention of predators and, possibly, thermoregulation. Despite this, the genetic constitution of the next generation is strongly influenced by the few sexually successful males in each generation. It is the critical female selection that follows in response that leads to isolation of populations and, ultimately, reproductive isolation.

1 isolation by a new, natural physical barrier
A natural habitat became divided when a river broke its banks and took a new route

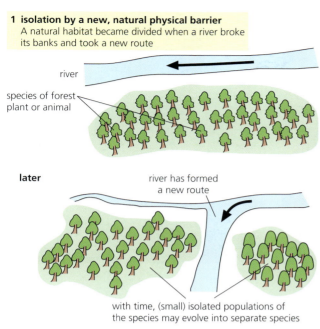

river

species of forest plant or animal

later

river has formed a new route

with time, (small) isolated populations of the species may evolve into separate species

2 isolation by a human-imposed barrier
A new road cuts through established habitats, separating local populations

■ **Figure 10.17** Geographical barriers

■ **Figure 10.18**
A male Bird of Paradise displays – a potential mate looks on!

In the examples considered up to now, speciation has been a gradual process. On the other hand, speciation can occur abruptly, too. We will examine this next.

Nature of Science   **Looking for patterns, trends and discrepancies**

■ Speciation by polyploidy

An abrupt change in the structure or number of chromosomes may lead to a new species. Such an alteration in the number of whole sets of chromosomes is known as polyploidy. Polyploids are largely restricted to plants. Some animals that reproduce asexually are polyploids, too, but the sex determination mechanism of vertebrates prevents polyploidy in these animals.

In polyploids, an additional set of chromosomes results when the spindle fails to form during meiosis, causing diploid gametes to be produced. Many crop species are polyploids. A well-known and economically very important example is the cultivated potato, *Solanum tuberosum* ($2n = 48$), a polyploid of the smaller wild variety, *Solanum brevidens* ($2n = 24$) (Figure 10.19).

wild potato (*Solanum brevidens*) (2*n* = 24)

cultivated potato (*Solanum tuberosum*) (2*n* = 48)

■ **Figure 10.19** Polyploidy in the potato plant

Speciation by polyploidy in the genus *Allium*

Allium is a genus of major economic importance. It exhibits a great diversity in morphology and in life forms (having either rhizomes or bulbs below ground). *Allium* species also show variation in basic chromosome number, ploidy level and genome size. Polyploidy can be advantageous because it increases allelic diversity and permits novel phenotypes to be generated. It also leads to hybrid vigour. An example of this is *A. porrum*, the cultivated leek. This plant is a fertile tetraploid (Figure 10.20).

■ **Figure 10.20**
Allium porrum, the cultivated leek

10 **Explain** the differences between a variety and a species. Find out about an example of each from an organism with which you are familiar.

11 **Explain** why mutation is regarded as an important force for speciation.

Speciation – a summary

Apart from the cases of instant speciation by polyploidy discussed above, species do not evolve in a simple or rapid way. The process is usually gradual, taking place over a long period of time. In fact, in many cases, speciation has occurred over several thousand years. During the process of change, there is a time when the differences between members of a species become great enough to identify separate **varieties** or **sub-species**. Eventually, these may become new species.

We can recognize that all cases of speciation require a 'isolation'. And isolation mechanisms are of two broad types:

■ Isolating mechanisms that involve spatial separation are illustrated by geographical isolation (**allopatric** ['different country'] **isolation**).

■ Isolating mechanisms that occur within the same location are illustrated by reproductive isolation (**sympatric** ['same country'] **isolation**).

Natural selection and speciation

Natural selection operates on individuals, or rather on their phenotypes. Phenotypes are the product of a particular combination of alleles, interacting with the effects of the environment of the organism. Consequently, natural selection causes changes to gene pools. For example, individuals with a particular allele or combination of alleles may be more likely to survive, breed and pass on their alleles. Individuals that are less well adapted may not survive or be reproductively successful. This process is referred to as differential mortality. Actually, whether the individual lives or dies is not important. What is relevant is that their alleles are not passed on to the next generation.

So, natural selection operates to change the composition of gene pools, but the outcomes of this vary. We can recognize different types of selection, and these are explained in Figure 10.21.

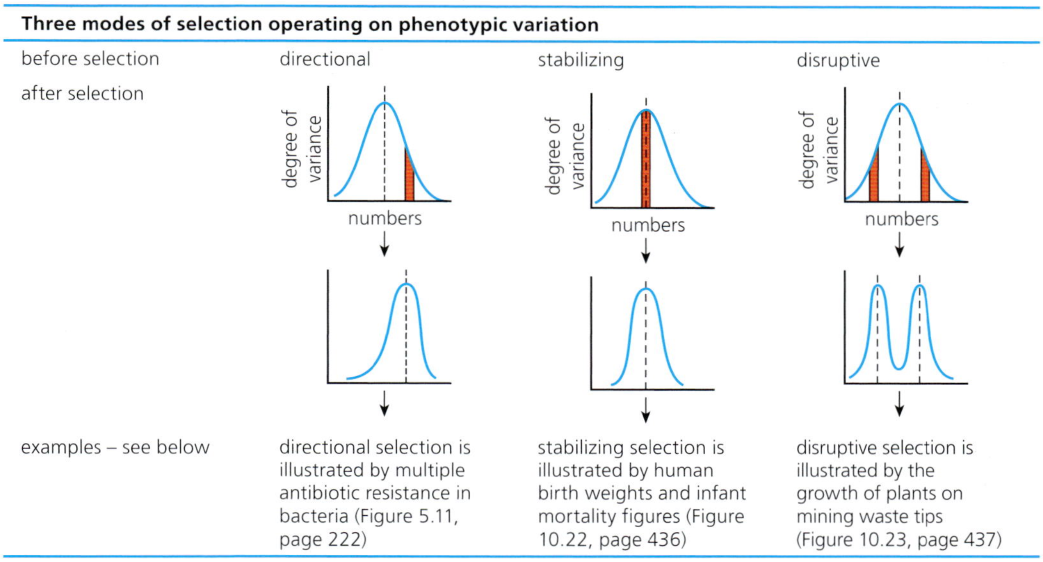

■ **Figure 10.21**
Types of selection

Three modes of selection operating on phenotypic variation

before selection / after selection — directional, stabilizing, disruptive

examples – see below

directional selection is illustrated by multiple antibiotic resistance in bacteria (Figure 5.11, page 222)

stabilizing selection is illustrated by human birth weights and infant mortality figures (Figure 10.22, page 436)

disruptive selection is illustrated by the growth of plants on mining waste tips (Figure 10.23, page 437)

Stabilizing selection occurs where environmental conditions are largely unchanging. Stabilizing selection does not lead to evolution. It is a mechanism which maintains a favourable characteristic and the alleles responsible for it, and eliminates variants and abnormalities that are useless or harmful. Probably, most populations undergo stabilizing selections. Our example (in Figure 10.22) comes from human birth records on babies born between 1935 and 1946, in London. It shows there was an optimum birth weight for babies, and those with birth weights heavier or lighter were at a selective disadvantage.

Directional selection may result from changing environmental conditions. In these situations, the majority form of an organism may become unsuited to the environment because of change. Some other alternative phenotypes may have a selective advantage.

An example of directional selection is the development of resistance to an antibiotic by bacteria (Figure 5.11, page 222). Certain bacteria cause disease, and patients with bacterial infections are frequently treated with an antibiotic to help them overcome the infection. However, in a large population of a species of bacterium, some may carry a gene for resistance to the antibiotic in question. A 'resistant' bacterium has no selective advantage in the absence of the antibiotic and must compete for resources with non-resistant bacteria. But, when the antibiotic is present, most bacteria in the population are killed. Resistant bacteria remain and create the future population, all of which now carry the gene for resistance to the antibiotic. The genome has been changed abruptly.

■ **Figure 10.22** Birth weight and infant mortality, a case of stabilizing selection

The birth weight of humans is influenced by **environmental factors** (e.g. maternal nutrition, smoking habits, etc.) and by **inheritance** (about 50%).

When more babies than average die at very low and very high birth weights, this obviously affects the gene pool because it tends to eliminate genes for low and high birth weights.

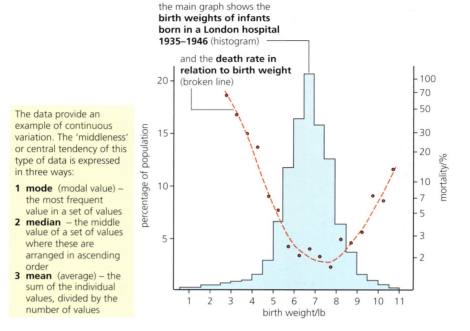

the main graph shows the **birth weights of infants born in a London hospital 1935–1946** (histogram)

and the **death rate in relation to birth weight** (broken line)

The data provide an example of continuous variation. The 'middleness' or central tendency of this type of data is expressed in three ways:

1 **mode** (modal value) – the most frequent value in a set of values
2 **median** – the middle value of a set of values where these are arranged in ascending order
3 **mean** (average) – the sum of the individual values, divided by the number of values

Disruptive selection occurs when particular environmental conditions favour the extremes of a phenotypic range over intermediate phenotypes. As a result, it is likely that the gene pool will split into two distinct gene pools; new species may be formed. This phenomenon has been illustrated by plant colonization of mine waste tips. These localized habitats often contain high concentrations of toxic metals, such as copper and lead. While several heavy metal ions are essential for normal plant growth in trace amounts, in mining spoils these levels are frequently exceeded. Seeds regularly fall on spoil heap soil, but most plants fail to establish themselves. However, spoil heaps at many locations show local populations of plants that have evolved tolerance. One example is the grass *Agrostis tenuis* (bent grass), populations of which are tolerant of otherwise toxic concentrations of copper (Figure 10.23).

Biochemical and physiological mechanisms have evolved in tolerant species, including:

- the selective ability to avoid uptake of heavy metal ions

- heavy metal ions that do enter are accumulated in insoluble compounds in cell walls by formation of stable complexes with wall polysaccharides

- transport of toxic ions into the vacuoles of cells, the membrane of which are unable to pump them out again, so avoiding interactions with cell enzymes.

The evolution of this form of tolerance has been demonstrated in several species of terrestrial plants (and also in species of seaweeds that are now tolerant of the copper-based antifouling paints frequently applied to the hulls of ships).

Grasses are wind-pollinated plants, so breeding between resistant grass plants and non-resistant grass plants goes on. When their seeds fall to the ground and attempt to germinate, disruptive selection may occur. Both the non-resistant plants germinating on contaminated soil and the resistant plants growing on uncontaminated soil may fail to survive and reproduce. The result is increasing divergence of populations, initially into two distinctly different varieties of grass plant. In time, new species may be formed.

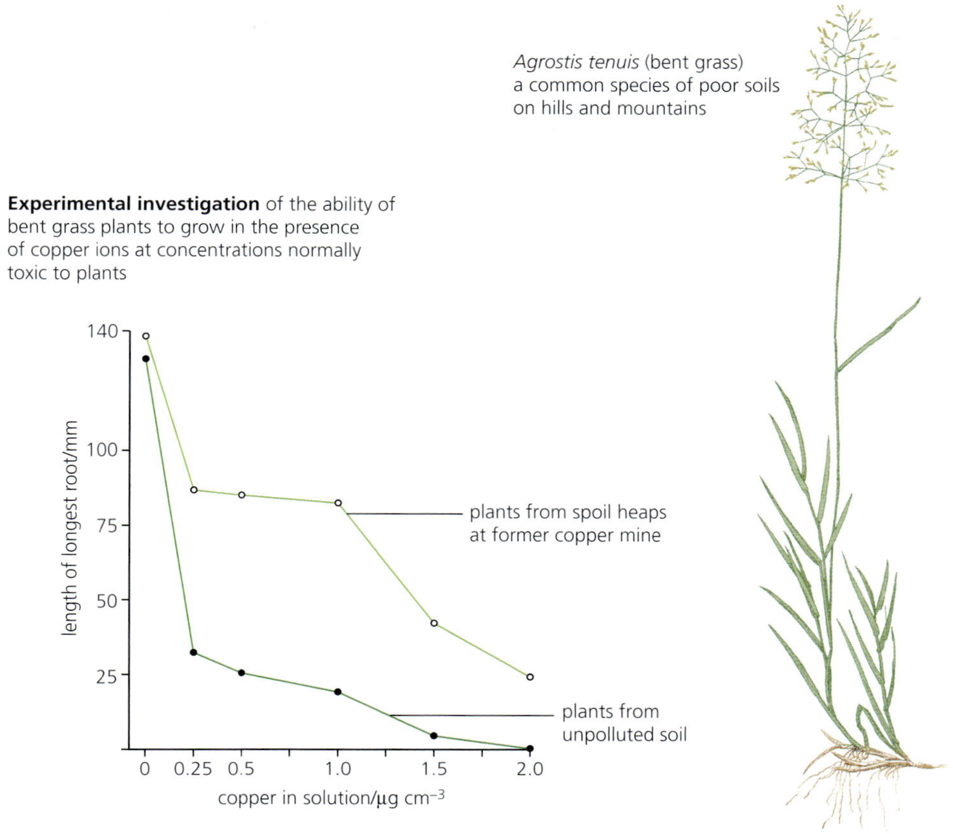

■ **Figure 10.23**
Mining waste
tip habitat, and
disruptive selection

Agrostis tenuis (bent grass)
a common species of poor soils
on hills and mountains

Experimental investigation of the ability of
bent grass plants to grow in the presence
of copper ions at concentrations normally
toxic to plants

plants from spoil heaps
at former copper mine

plants from
unpolluted soil

length of longest root/mm

copper in solution/µg cm⁻³

■ Comparisons of allele frequencies of geographically isolated populations

We have noted that in any population:

■ The total of the alleles of the genes located in the reproductive cells of the individuals makes up a **gene pool**. A sample of the alleles of the gene pool will contribute to form the genomes (gene sets of individuals) of the next generation, and so on, from generation to generation.

■ When the gene pool of a population remains more or less unchanged, that population is not evolving. If the gene pool of a population is changing (the proportions of particular alleles pair is altered), then evolution may be happening.

How can we detect change or constancy in gene pools?

The answer is by a mathematical formula called the **Hardy–Weinberg formula**. The general formula to represent the frequency of dominant and recessive alleles is:

$$p + q = 1$$

where   p = frequency of the dominant alleles
and   q = frequency of the recessive alleles.

The problem in estimating gene frequencies is that it is not possible to distinguish between homozygous dominants and heterozygotes based on their appearance or phenotype, as has been noted previously (page 152). The Hardy–Weinberg equation overcomes this problem:

$$p^2 \quad + \quad 2pq \quad + \quad q^2 \quad = \quad 1$$

| frequency of dominant homozygous individuals | frequency of heterozygous individuals | frequency of homozygous recessive individuals | TOTAL |

Using the Hardy–Weinberg formula

We can use the Hardy–Weinberg formula to find the frequency of a gene in cases of dominance in which we are unable to distinguish between the homozygous dominants and the heteroygotes on the basis of phenotype.

Example

In humans, the ability to taste the chemical phenylthiocarbamide (PTC) is conferred by the dominant allele T. Both the dominant homozygotes (**TT**) and the heterozygotes (**Tt**) are 'tasters'. The non-tasters are the homozygotes (**tt**).

In a sample of a local population in Western Europe of 200 people, 130 (65%) were tasters and 70 (35%) were non-tasters (Table 10.6).

■ **Table 10.6** Tasters and non-tasters

Phenotypes	Tasters	Non-tasters
Genotypes	TT + Tt	tt
Frequency	0.65	0.35

Applying this data to the Hardy–Weinberg formula, we know the value of q^2 to be 0.35.

Taking the square root, the value of q = 0.59.

So, the frequency of the non-tasting alleles (**t**) in this European population was 0.59.

---

12 In two small isolated populations the incidence of PCT non-tasters was as shown in the table.

Population	Sample size	Number of non-tasters
Indigenous Australians	500	245
Indigenous North Americans	500	20

Calculate the frequency of the non-tasting alleles in each of the two sample populations.

■ Pace of evolution: gradualism versus punctuated equilibria

Since geological estimates set the age of the Earth at 4500 million years, and since life originated about 3500 million years ago, the timescale over which evolution has occurred seems almost unimaginably long. Furthermore, the fossil record provides evidence of the long evolutionary history of most major groups. This definition of evolution by natural selection as an exceedingly slow process is known as '**gradualism**'. Indeed, from the theory of evolution by natural selection, we might expect species to very gradually disappear and to be replaced by new species at a similarly slow rate.

In fact, this is not necessarily the case. Some new species have appeared in the fossil record relatively quickly (in terms of geological time) and then have tended to remain *apparently* unchanged or little changed for millions of years. Sometimes, periods of stability were followed by sudden mass extinctions, all evidenced by the fossil record (Figure 10.24).

Some scientists say the fossil record is misleading because it is incomplete and does not record all the organisms that have lived. This is quite possible; we have no way of being certain the fossil record is fully representative of life in earlier times.

However, two evolutionary biologists, Niles Eldredge and Stephen Jay Gould, proposed an alternative explanation. They argued that the fossil record *for some groups* is not incomplete, but accords with their view of the origins of new species – which they called **punctuated equilibria**.

■ **Figure 10.24**
The extinction
sequence in
geological time

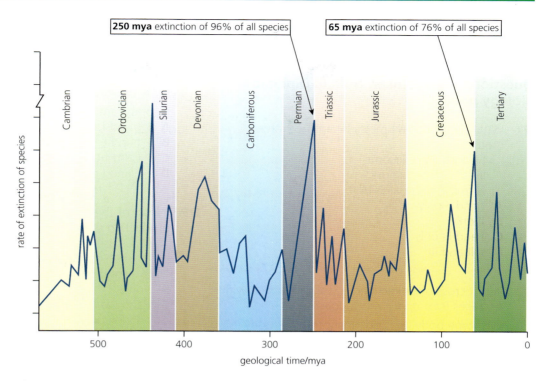

This hypothesis holds that:

- When environments become unfavourable, populations attempt to migrate to more favourable situations.

- If the switch to adverse conditions is very sudden or very violent, a mass extinction occurs. Major volcanic eruptions or meteor impacts can throw so much detritus into the atmosphere that the Earth's surface is darkened for many months, cooling the planet and killing off much plant life.

- However, populations at the fringe of a massive disturbance may be protected from the worst effects of extreme conditions, surviving into the future.

- Members of these populations may become small and isolated reproductive communities from which repopulation eventually occurs.

- This surviving group may have an unrepresentative selection of alleles from the original gene pool. If the group becomes the basis of a repopulation event, and then quickly adapts to the new conditions, abrupt genetic changes may occur. This phenomenon is known as the **'founder effect'**. In this situation, evolution can occur suddenly.

So, there are alternative proposals for the ways in which natural selection has operated in practice in the establishment of life in geological time. In fact, gradualism and punctuated equilibria may not be alternatives; both may have contributed to the pattern of life on Earth in geological time.

TOK Link

Punctuated equilibrium was long considered an alternative theory of evolution and a challenge to the long-established paradigm of Darwinian gradualism. How do paradigm shifts proceed in science and what factors are involved in their success?

■ *Examination questions – a selection*

Questions 1–5 are taken from IB Diploma biology papers.

Q1 What do **all** human males inherit from their mother?

    I an X chromosome
    II a Y chromosome
    III mitochondrial DNA

  A I and II only
  B II only
  C I and III only
  D I, II and III

High Level Paper 1, Time Zone 2, May 09, Q12

Q2 A test cross of **linked** genes was performed with fruit flies (*Drosophila melanogaster*).

Wild type body (B) is dominant to black body (b).

Normal wings (W) is dominant to vestigial wings (w).

BbWw were crossed with bbww.

The resulting offspring were
- 952 wild type body, normal wings
- 948 black body, vestigial wings
- 200 wild type body, vestigial wings
- 198 black body, normal wings

What is the most likely explanation for these results not fitting the expected ratio?

  A Crossing-over
  B Non-disjunction
  C Gene mutation
  D Random variation

Higher Level Paper 1, Time Zone 2, May 13, Q35

Q3 In fruit flies (*Drosophila melanogaster*) grey body is dominant to black body and long wings are dominant to vestigial wings. Two flies heterozygous for both genes were crossed. What proportion of the offspring would be expected to have black bodies and long wings?

  A 1/2
  B 3/16
  C 1/4
  D 1/16

Higher Level Paper 1, Time Zone 1, May 13, Q33

Q4 Flower colour and pollen grain shape are linked genes in sweet peas. Purple (F) is dominant to red (f) flower colour and long pollen grains (L) are dominant to round pollen grains (l).

If the parental genotype is $\frac{F\ L}{f\ l}$, what would be the recombinant chromosomes in the gametes?

  A $\underline{F\ L}$ and $\underline{f\ l}$
  B $\underline{F\ f}$ and $\underline{L\ l}$
  C $\underline{F\ l}$ and $\underline{f\ L}$
  D $\underline{F\ F}$ and $\underline{l\ l}$

Higher Level Paper 1, Time Zone 0, Nov 12, Q35

Q5 Which of the following processes result in the production of recombinants?

    I Crossing over between linked genes
    II Reassortment of non-linked genes
    III Mutation

  A I only
  B I and II only
  C I and III only
  D I, II and III

Higher Level Paper 1, Time Zone 1, May 09, Q35

Questions 6–10 cover other syllabus issues in this chapter.

Q6 **a** Mendel conducted many experiments with garden pea plants, but some later workers used the fruit fly *Drosophila* in experimental genetics investigations. Suggest **three** reasons why this insect was found to be useful. (3)

    b When dihybrid crosses are carried out the progeny are rarely present in the exact proportions predicted. Explain why small deviations of this sort arise in dihybrid crosses with:
      i garden pea plants (3)
      ii *Drosophila*. (3)

Q7 In *Drosophila*, mutants with scarlet eyes and vestigial wings are recessive to flies with red eyes and normal wings. (These contrasting characters are controlled by single genes on different chromosomes, i.e. they are not linked on a single chromosome.) Explain the phenotypic ratio to be expected in the F_2 generation when normal flies are crossed with scarlet eyes/vestigial wing mutants and sibling crosses of the F_1 offspring are then conducted. Show your reasoning by means of a genetic cross diagram. (8)

Q8 Distinguish between the following pairs:
  a *discontinuous variable* and *continuous variable* (4)
  b *monohybrid cross* and *dihybrid cross* (4)
  c *linkage* and *crossing over* (4)
  d *autosomes* and *sex chromosomes* (4)
  e *X chromosomes* and *Y chromosomes* (4)
  f *multiple alleles* and *polygenes*. (4)

Q9 Explain what is described as *continuous variation* by using one named example. (4)

Q10 Outline how meiosis and two other mechanisms promote genetic variability among individuals of the same species. (8)

11 Animal physiology

ESSENTIAL IDEAS

- Immunity is based on recognition of self and destruction of foreign material.
- The roles of the musculoskeletal system are movement, support and protection.
- All animals excrete nitrogenous waste products, and some animals also balance water and solute concentrations.
- Sexual reproduction involves the development and fusion of haploid gametes.

In this chapter, we look again into the body's **defence against disease**. Then, the way the skeleton, joints and muscles combine to bring about **movement** is investigated, followed by the role of the **kidneys** in excretion and **osmoregulation**. Finally, the processes of **gametogenesis**, **fertilization** and the early development of the human **embryo** are discussed.

11.1 Antibody production and vaccination –
immunity is based on recognition of self and destruction of foreign material

■ The immune response

The immune response is our main defence once invasion of the body by harmful microorganisms or 'foreign' materials has occurred. We have seen that particular leucocytes (white blood cells) called **lymphocytes** are responsible for our specific immune response. These cells make up 20% of the white blood cells circulating in the blood plasma (or in the tissue fluid – remember, white blood cells move freely through the walls of blood vessels).

Lymphocytes are able to detect as different from our own cells and proteins any 'foreign' matter that enters from outside the body (macromolecules, as well as microorganisms). Any molecule that the body recognizes as foreign or 'non-self' is known as an **antigen**.

An antigen is normally a protein (but some carbohydrates and other macromolecules can act as antigens). They are recognized by the body as foreign (non-self) and stimulate an immune response. The specificity of antigens allows for responses that are customized to specific pathogens.

How are 'self' and 'non-self' recognized?

Every organism has unique molecules on the surface of their cells. Cells are identified by these specific molecules – markers, if you like, that are lodged in the outer surface of the plasma membrane. These molecules include the highly variable **glycoproteins** on the cell surface membrane. Remember, carbohydrates occur attached to proteins here.

Remind yourself of the fluid mosaic structure of this membrane (Figure 1.34, page 31).

The glycoproteins that identify cells are known as the **major histocompatibility complex** antigens – but we can refer to them as **MHC**. In humans, the genes for MHC antigens are on chromosome 6. The MHC antigens of each individual are genetically determined – they are a feature we inherit. In inherited characteristics that are products of sexual reproduction, variations occur, so each of us has distinctive MHC antigens present on our cell surface membranes. Unless you have an identical twin, your MHC antigens are unique.

Now, lymphocytes of our immune system have **antigen receptors** that recognize our own MHC antigens, and can tell them apart from any 'foreign' antigens detected in the body. It is critically important that our own cells are not attacked by our immune system. This is the basis of the 'self' and 'non-self' recognition mechanism.

Pathogen–host specificity

An invading organism or virus particle that enters the body and causes disease is a **pathogen**. Pathogens are often specific in their choice of host. For example, humans are the only host for the pathogens that cause the diseases of syphilis, gonorrhoea, measles and poliomyelitis.

On the other hand, other pathogens are able to cross species barriers, infecting a range of hosts. For example, rabies is a viral disease that affects many carnivorous animals, including dogs, cats, foxes, skunks, jackals and wolves. Rabies is almost invariably fatal to humans, too. (**Zoonoses** are diseases of other animals that can be transmitted to humans.) Generally, infectious diseases are most often 'shared' between species that are closely related and inhabit the same geographic area, if they are shared at all.

Natural resistance to a specific disease by a species may be due to factors such as:

■ Absence of a type of cellular receptor for attachment – for example, a particular strain of *E. coli* which causes an infection in the intestines recognizes and requires a specific carbohydrate receptor on the cells of the intestine. These occur only in the gut of human infants, calves and piglets.

■ Lack of the target site for a microbial toxin – for example, diphtheria toxin, when injected into a rat, fails to kill the animal (the toxin is excreted in the urine) but it will kill a guinea pig when injected in the same way.

Lymphocytes and the antigen–antibody reaction

There are two types of lymphocyte at work in our immune system – the **B lymphocytes** (**B cells**) and **T lymphocytes** (**T cells**). Both of these cell types originate in the bone marrow, where they are formed from **stem cells** (page 15). As they mature, these cells undergo different development process in preparation for their distinctive roles (Figure 11.1).

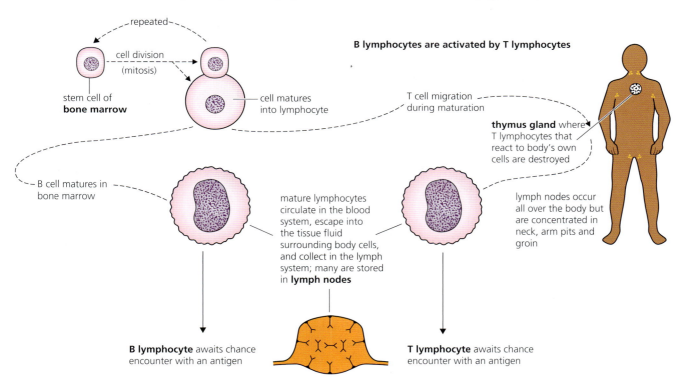

■ **Figure 11.1** T and B lymphocytes

T lymphocytes

T lymphocytes leave the bone marrow soon after they have been formed and migrate to the **thymus gland**. The thymus gland is found in the chest, just below the breast bone (sternum). It is active and enlarged during the early stages of our growth and development. While the T cells are present in the thymus gland, all of those that would react to the body's own cells are removed and destroyed. The surviving T cells are released and circulate in the blood plasma. Many are stored in lymph nodes.

The thymus gland shrinks in size after puberty is reached, its task completed.

The role of T cells is *not* to secrete antibodies. Their role is to reactivate B cells after 'activation' by contact with antigens of a particular pathogen or other foreign matter.

B lymphocytes

Meanwhile, B lymphocytes complete their maturation in the bone marrow, prior to circulating in the blood. Many of these lymphocytes are also stored in lymph nodes. The role of the majority of B cells, after activation by T cells, is to form clones of **plasma cells** that then secrete antibodies into the blood system. (In addition, **memory cells** are formed – of which, more later).

An antibody is a special protein called an **immunoglobulin** (Figure 11.2). It is made of four polypeptide chains held together by disulfide bridges (-S–S-) and forming a molecule in the shape of a **Y**. The arrangement of amino acids in the polypeptides that form the 'fork' region in this molecule is unique to that antibody. It is this region that forms the highly specific **binding site** for the antigen. Antibodies initially occur attached to the cell surface membrane of B cells, but later are mass-produced and secreted by cells derived from the B cell. This occurs after that B cell has undergone an **activation step**.

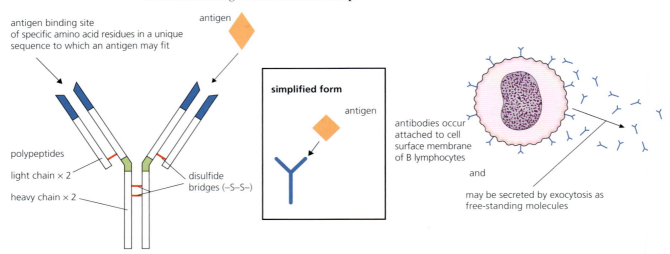

■ **Figure 11.2** The structure of an antibody

1 Explain the significance of the role of the thymus gland in destroying T cells that would otherwise react to 'self' body proteins.

In summary:

■ **B lymphocytes** form clones of plasma cells, when activated by T lymphocytes.

■ **T lymphocytes** are activated when they are in contact with an antigen from a pathogen or foreign matter. They, in turn, activate B lymphocytes.

■ The steps to an immune reaction response to an infection

When an infection occurs, the leucocyte population immediately increases and many of these cells collect at the site of the invasion. The complex response to infection has begun.

The roles of T and B cells in this response are as listed here.

You can follow these steps in Figure 11.3.

1 When a specific antigen enters the body, **B cells** with surface receptors (antibodies) that recognize the antigen **bind** to it.

2 On binding to the B cell, the antigen is taken into the cytoplasm by endocytosis. Then it is expressed and displayed on the cell surface membrane of the B cell.

3 Meanwhile phagocytic cells, the **macrophages**, engulf any antigens they encounter. (Macrophages occur in the plasma, lymph or tissue fluid.) Once antigens have been taken up, they are presented externally, attached to the MHC antigens, on the surface of the macrophages. **T cells** respond to antigens that are presented on the surface of other cells, as on the macrophages. This is called **antigen presentation** by a macrophage.

■ **Figure 11.3**

Stages in antibody production

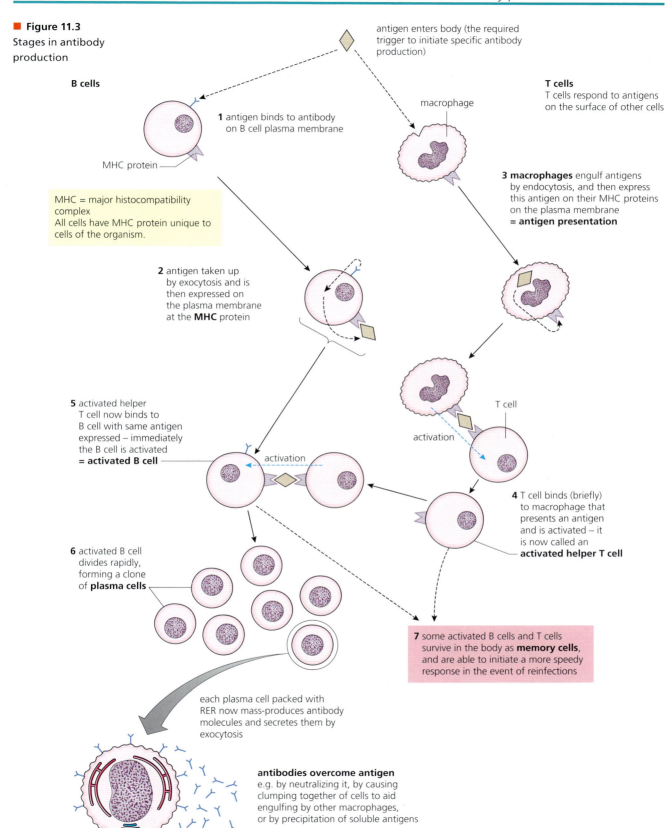

B cells

MHC protein

1 antigen binds to antibody on B cell plasma membrane

MHC = major histocompatibility complex
All cells have MHC protein unique to cells of the organism.

2 antigen taken up by exocytosis and is then expressed on the plasma membrane at the **MHC** protein

antigen enters body (the required trigger to initiate specific antibody production)

macrophage

T cells
T cells respond to antigens on the surface of other cells

3 **macrophages** engulf antigens by endocytosis, and then express this antigen on their MHC proteins on the plasma membrane
= **antigen presentation**

5 activated helper T cell now binds to B cell with same antigen expressed – immediately the B cell is activated
= **activated B cell**

activation

T cell

activation

4 T cell binds (briefly) to macrophage that presents an antigen and is activated – it is now called an **activated helper T cell**

6 activated B cell divides rapidly, forming a clone of **plasma cells**

7 some activated B cells and T cells survive in the body as **memory cells**, and are able to initiate a more speedy response in the event of reinfections

each plasma cell packed with RER now mass-produces antibody molecules and secretes them by exocytosis

antibodies overcome antigen
e.g. by neutralizing it, by causing clumping together of cells to aid engulfing by other macrophages, or by precipitation of soluble antigens

4   As T cells come into contact with these macrophages and briefly bind to them, they are immediately activated. They are now called 'armed' or **activated helper T cells**.

5   Now, activated helper T cells bind to B cells with the same antigen expressed on their cell surface membrane (*step 2 above*). As a result, the B cell is activated. It is now an 'armed' or **activated B cell**.

6   Next, activated B cells divide rapidly by mitosis, forming a clone of **plasma cells**. An electron micrograph of plasma cells shows each one is packed with endoplasmic reticulum (RER, Figure 11.4). It is in these organelles that the antibody is mass-produced and exported from the B cell by exocytosis. The generation of a large number of plasma cells that produce one specific antibody type is known as clonal selection.

The antibodies are normally produced in such numbers that the antigen is overcome. The action of antibodies is to bind to antigens, neutralizing them or making them clear targets for phagocytic cells (Table 11.1).

We have noted that the T and B cells have molecules on the outer surface of their cell surface membrane that enable them to recognize antigens, but each B and T lymphocyte has **only one type** of surface receptor. Consequently, each lymphocyte can recognize only one type of antigen.

■ **Figure 11.4**
Electron micrograph
of a plasma cell

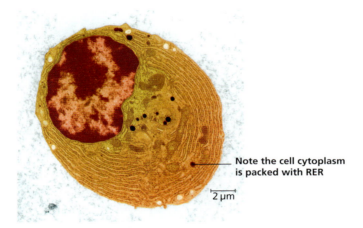

Note the cell cytoplasm
is packed with RER

2 µm

7   After antibodies have tackled the foreign matter and the disease threat is overcome, the special proteins disappear from the blood and tissue fluid. So, too, do the bulk of the specific B cells and T cells that were responsible for their formation.

It is now helpful to summarize the complex roles of B and T cells in the immune system (Figure 11.5 and Table 11.1).

■ **Table 11.1**
How antibodies aid
the destruction of
pathogens –
a summary

Agglutination	Antibodies attach to pathogens, causing them to stick together – clumped in this way they more easily ingested by phagocytic cells.
Complement activation	Complement proteins in the plasma cause cell lysis by destroying the plasma membrane of pathogen cells, after antibodies have identify them by binding.
Toxin neutralization	Antibodies bind to toxins in the plasma, preventing them from affecting susceptible cells.
Opsonization	Antibodies make pathogens instantly recognizable by binding to them, and then linking them to phagocytic cells.

■ **Figure 11.5**
The roles of B and
T cells in the immune
system – a summary

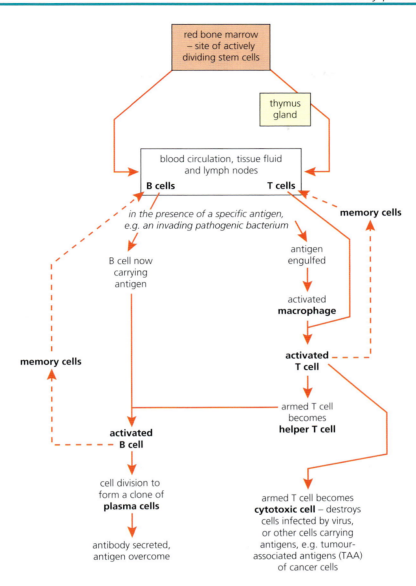

Memory cells and immunity

Immunity is the ability of the body to resist an infection by a pathogen. Long-lived and specific immunity is the result of the presence of certain memory cells, retained after a previous infection by that pathogen. Memory cells are specifically activated B cells. They are long-lived cells, in contrast to plasma cells and other activated B cells.

Memory cells make possible an early and effective response in the event of a reinfection of the body by the same antigen (Figure 11.6). This is the basis of natural immunity (see below).

2 **Identify** where antigens and antibodies may be found in the body.

3 Remind yourself of the appearance of the types of white blood cells observed in a blood smear preparation. See for example, the photomicrograph in Figure 6.6 (page 256). Compare this image with a prepared slide of a human blood smear, so that you are familiar with the appearance of phagocytic white blood cells and lymphocytes.

Draw and fully **annotate** some of the white blood cells you observe, making clear the differences between them.

■ **Figure 11.6**
Profile of antibody
production in
infection and
reinfection

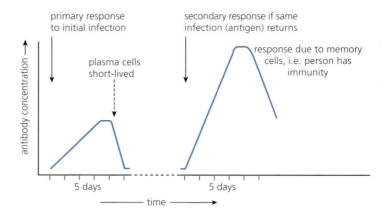

Memory cells are retained in lymph nodes. They allow a quick and specific response if the same antigen reappears.

■ **Figure 11.6**
Profile of antibody production in infection and reinfection

Transfusions of blood can bring about an immune reaction

Human blood cells carry antigens on the plasma membrane, of which the ABO system and the Rhesus system are most important as far blood transfusions are concerned. You may already know your own blood grouping; each of us carries a particular combination and this has to be determined before we can receive a transfusion of blood.

Blood groups demonstrate a special example of the antigen–antibody reaction. Looking at the detail of the **ABO system** in Figure 11.7, you will notice that people tend to have an antibody in the plasma against whichever antigen they lack. This is always present, even though the blood has not been in contact with the relevant antigen. So, for example, if a person of blood group A accidentally receives a transfusion of group B blood, then the anti-B antibodies in the recipient's plasma make the 'foreign' B cells clump together (the term is **agglutinate**). The clumped blood cells block smaller vessels and capillaries, which may be fatal. Consequently, you can see why blood of group O is so useful for transfusion purposes – it has neither A nor B antigens on the red cells.

Incidentally, if a small quantity of blood is given in a transfusion, the type of antibodies in the plasma received does not matter because of dilution by the plasma of the recipient's blood. But for a large transfusion, the match of antigens and antibodies needs to be perfect.

Remember, the inheritance of ABO blood groups is controlled by multiple alleles and is inherited according to Mendelian laws (page 155).

The **Rhesus system** is different again. Rhesus negative people (without the Rhesus antigen on their red cells) do not carry antibodies in their plasma. That is, unless they have been previously sensitized.

In practice, sensitization of a Rhesus negative woman may occur if she has carried a Rhesus positive foetus in the uterus (the Rhesus factor is also inherited). Late on in pregnancy there is a likelihood of some mixing of maternal and fetal bloods. As a result, the mother is sensitized by Rhesus positive antigens and she makes Rhesus antibodies. These would remain in her plasma and, should a second pregnancy occur, again with a Rhesus positive fetus, then antibodies would pass across the placenta from her plasma and destroy the red blood cells of the Rhesus positive fetus. In fact, treatment with anti-Rhesus antibodies immediately the first pregnancy ends prevents the problem.

Inflammation – the initial response to trauma

Inflammation is the initial, rapid and localized response that tissues make to damage, whether due to a cut, scratch, bruising or a deep wound. We are quickly aware that the site of a cut or knock has become swollen, warm and painful. Inflammation is triggered by inflammatory signalling molecules, principally **histamine**. This is released from mast cells (cells of the immune system packed with chemicals that trigger inflammation when released) and white blood cells, at the site of trauma. The volume of blood in the damaged area is increased, and white blood cells and plasma accumulate outside the enlarged capillaries. This increased blood flow removes any toxic products that are released by invading microorganisms and by damaged cells. Ultimately, tissue repair is initiated, also.

1 ABO system	Blood group A	Blood group B	Blood group AB	Blood group O
red blood cell surface	A antigens	B antigens	A + B antigens	neither
plasma	anti-B antibodies	anti-A antibodies	neither	both anti-A antibodies and anti-B antibodies
blood groups that may be used for transfusion	A, O	B, O	A, B, AB, O	O

Note: Blood group O is the universal donor blood group. Blood group AB is the universal recipient.

2 Rhesus system	Rhesus	Rhesus
Rhesus antigens on red cells	present (positive)	absent (negative)
Antibodies to Rhesus antigens in blood plasma	none	none (unless previously sensitized)

Note: Rhesus blood should only be given to a Rhesus negative person.

■ **Figure 11.7** Blood group and transfusion possibilities

Histamine and allergic symptoms

An allergy is an exaggerated response by the body to antigens. Many people suffer an allergy to pollen grains – they say they have 'hay fever'. In the presence of antigens on the surface of pollen grains, plasma cells release antigens that attach to mast cells, thereby triggering the release of histamine and other inflammatory molecules. These act on a variety of cells in the body and bringing about typical allergy symptoms. These include contraction of smooth muscle – causing breathing difficulties. Anti-histamine drugs counteract allergy symptoms by blocking histamine receptors around the body.

Vaccines and vaccination

Vaccines contain antigens that trigger immunity but do not cause disease. **Vaccination** is the deliberate administration of antigens that have been made harmless, after they are obtained from disease-causing organisms, in order to confer future immunity. The practice of vaccination has made important contributions to public health, for example, in the case of measles (page 452), smallpox (page 453) and many other diseases.

Vaccines are administered either by injection, by nasal spray or by mouth. Briefly and *without causing infection*, they cause the body's immune system to make antibodies against the disease, and then to **retain the appropriate memory cells**. Active artificial immunity is established in this way. The profile of the body's response in terms of antibody production, if it is re-exposed to the antigen, is normally exactly the same as if the immunity was acquired by overcoming an earlier infection.

Vaccines are manufactured from:

- dead or attenuated (weakened) bacteria
- purified polysaccharides from bacterial walls
- inactivated viruses
- toxoids (inactivated toxins)
- recombinant DNA produced by genetic engineering.

Vaccines are often produced using the immune system responses of other animals, too.

In communities and countries where vaccines are widely available and have been taken up by 85–90% of the relevant population, vaccination has reduced some previously common and dangerous diseases to very uncommon occurrences. As a result, in these places, the public sometimes becomes casual in its regard for the threat such diseases still represent.

In summary – types of immunity

The immune system provides protection to the body from the worst effects of many of the pathogens that may invade. Note that immunity may be acquired actively or passively, naturally or artificially:

- **Actively** – *naturally*, as when our body responds to invasion by a pathogen, or *artificially*, after injection of killed or weakened antigens in a vaccine – causing memory cells to be made.

- **Passively** – *naturally*, as when maternal antibodies enter the fetus through the placenta, or *artificially*, such as when ready-made antibodies are injected into the body.

The effectiveness of vaccination programmes – epidemiological data

Epidemiology is the study of the occurrence, distribution and control of diseases. A study of community disease patterns of this sort *may* be a sound source of evidence. However, bear in mind that, when the community-wide administration of a vaccine is closely followed by a decrease in the targeted disease, a causal relationship has *not* been proven. We cannot be certain it was the vaccine that prevented the infections because there are so many uncontrolled variables affecting a community and its health.

Tuberculosis (TB), a major worldwide public health problem of long standing, is caused by a rod-shaped bacterium, *Mycobacterium tuberculosis*. There is evidence that TB was present in some of the earliest human communities and it has persisted as a major threat to health where people live in crowded conditions. Today, in most developed countries, this disease is relatively rare, but globally almost 14 million people have active TB. In developing countries with a high incidence of this disease, it primarily affects young adults. In developed countries where TB had, until very recently, ceased to be a major health threat, the rising incidence is largely among people with HIV/AIDS who are immunocompromised (page 278), and also among people arriving from other countries.

A vaccine against tuberculosis, the Bacillus Calmette-Guérin (**BCG**) vaccine, is prepared from a strain of attenuated (weakened) live bovine tuberculosis bacillus. It is most often used to prevent the spread of TB among children.

Look at the pattern of deaths from TB in a developed country, 1900–2000, shown in Figure 11.8 A. *Examine the graph carefully and then answer question 4.*

In the prevention and control of **measles**, the first measles vaccines became available in 1963. It was then replaced by a superior version in 1968. Epidemiological evidence of the effect of the widespread use of these vaccines in one developed country is shown in Figure 11.8 B. *Examine this data carefully, and then answer question 5.*

■ **Figure 11.8**
The effectiveness
of vaccination
programmes –
epidemiological data

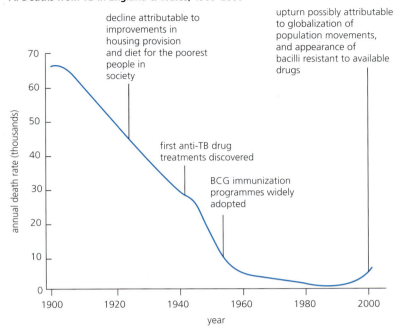

A. Deaths from TB in England & Wales, 1900–2000

decline attributable to
improvements in
housing provision
and diet for the poorest
people in
society

upturn possibly attributable
to globalization of
population movements,
and appearance of
bacilli resistant to available
drugs

first anti-TB drug
treatments discovered

BCG immunization
programmes widely
adopted

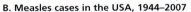

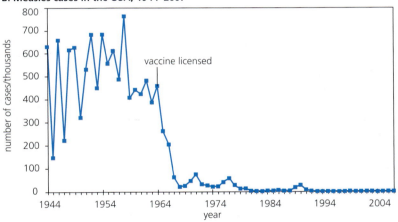

B. Measles cases in the USA, 1944–2007

vaccine licensed

4   Using the epidemiological evidence in Figure 11.8 A, **suggest** the actions needed to combat the spread
    of TB in a community *in order of priority,* paying particular attention to the importance to be placed on a
    vaccination programme.

5   **Analyse** the extent to which the epidemiological data in Figure 11.8 B supports the idea of vaccination
    against measles as an effective procedure for maintaining community health.

The origins of vaccination

It has long been known that people who recovered from plague or smallpox rarely contracted those diseases again. The possibility of a patient acquiring immunity from smallpox was known in the East in the seventeenth century, if not earlier. For example, an English ambassador's wife saw that vaccination was practised in Turkey early in the eighteenth century and had observed how successful it was there. She tried to introduce the practice of vaccination into Britain. Sadly, she had little success.

The first attempt in the West at immunization was made by **Edward Jenner** (1749–1823), a country doctor from Gloucestershire, UK. At the time, many people who got smallpox died. However, those who had earlier contracted cowpox never died. (Workers who handled cows typically caught cowpox at some stage.) Jenner saw the significance of the protection the patients had acquired. He extracted fluid from a cowpox pustule on an infected milkmaid, and injected it *into himself and into the arm of an eight-year-old boy* (Figure 11.9). The child got a mild cowpox infection but, when exposed to smallpox, remained healthy. Jenner named this technique **vaccination**, after the cowpox *vaccinia* (a virus). Of course, he did not understand the cause of smallpox; the chemical nature of viruses was not reported until 1935.

The French scientist **Louis Pasteur** (1822–95) discovered that cultures of chicken cholera bacterium, allowed to age for 2–3 months, produced only a mild infection of cholera when inoculated into chickens. The old cultures had become less pathogenic – today we say they had **attenuated**. Fresh and virulent strains of the chicken cholera bacterium failed to infect chickens that were previously exposed to the mild form of the disease.

Pasteur, one of the greatest experimentalists in microbiology, went on to have notable successes in immunization (Figure 11.9). However, he recognized the original contribution of Jenner to the discovery of immunity by using the name 'vaccine' for injections of the attenuated organisms that he developed to prevent chicken cholera disease, and (later) anthrax (due to a bacterium) and rabies (due to a virus).

■ **Figure 11.9**
Edward Jenner and Louis Pasteur at work

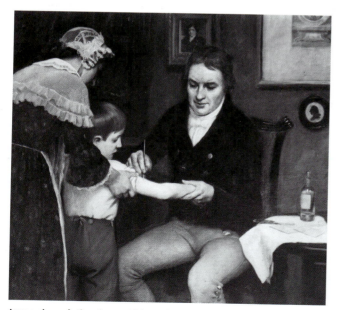

Jenner inoculating James Phipps, from a drawing by William Thompson, c. 1880 (*by courtesy of the Wellcome Trustees*)

Pasteur made a major contribution to our understanding of diseases of humans and other animals; working with dogs, he showed that an injection of the attenuated rabies microorganism can produce immunity to the disease

Nature of Science **Ethical implications of research**

■ **The ethical implications of Jenner's research into a vaccine for smallpox**

- See 'What is ethics?' and 'Making ethical decisions' on the accompanying website, in *Appendix 3: Defining ethics and making ethical decisions*.
- See 'An introduction to today's regulations for the testing and approval of drugs' on the accompanying website, in *Appendix 3; Defining ethics and making ethical decisions*.
- What improvements could be made (with the advantage of the hindsight that we have today) to the testing procedure that Dr Jenner devised?
- Why, do you imagine, did the boy's mother allow this experiment to take place? Do we need to make allowances for people who take a calculated risk (or who have exceptional foresight)?

Smallpox – an infectious disease eradicated by vaccination

Smallpox, a highly contagious disease, was once endemic throughout the world. It killed or disfigured all those who contracted it. Smallpox was caused by a DNA *variola* virus. The virus was stable, meaning it did not mutate or change its surface antigens, as others do.

Eventually, a suitable vaccine was identified – made from a harmless, but related, virus, *vaccinia*. This was used in a 'live' state and could be freeze dried for transport and storage. Consequently, the virus was relatively easy to handle and stable for long periods in tropical climates.

How eradication came about

Smallpox has been eradicated (the last case occurred in Somalia in 1977). The development of a vaccine played an important part in this achievement, which was the outcome of a determined World Health Organization (WHO) programme, begun in 1956. This involved careful surveillance of cases in isolated communities and within countries sometimes scarred by wars – altogether a most remarkable achievement. The reasons why smallpox was eradicated when so many other diseases continue are listed in Table 11.2.

■ **Table 11.2**
Why smallpox was eradicated

Patients with the disease were easily identified; they had obvious clinical features.
Transmission was by direct contact only.
On diagnosis, patients were isolated and all their contacts traced. All were vaccinated.
It had a short period of infectivity – of about 3 to 4 weeks.
Patients who recovered did not retain any virus in the body, so 'carriers' did not exist.
There were no animals that acted as a vectors or 'reservoirs' of the infection (and which, otherwise, would have been able to pass on the virus to other humans).
The virus was stable – it did not mutate or change its surface antigens. Minor changes in antigens (**antigenic shift**) may cause memory cells to fail to recognize a pathogen.
Last but not least – the issue was tackled through international cooperation by the WHO.

6 **Identify** the steps of plasma cell formation that are avoided in cases of reinfection, due to the existence of memory cells.

7 **State** what we mean by immunity.

Vaccines today

The deliberate administration of antigens that have been made harmless, after they are obtained from disease-causing organisms, in order to confer future immunity, is a very important contribution to public health. So successful has vaccination been – *where vaccines are widely available and the take up is by about 85–90% of the relevant population* – that, in many human communities, some formerly common and dangerous diseases are rarely seen. Unfortunately, as a result, the public has become casual about the threat such diseases still pose.

The recommended schedule of vaccinations for children brought up in one developed country can be accessed via the website www.doh.gov.uk

■ Monoclonal antibody production and uses

White cells in the blood provide our main defences against invasion of the body by harmful microorganisms. Antibodies are effective in the destruction of antigens within the body, but the plasma cells (the product of a B lymphocyte, Figure 11.3) that secrete them have an extremely short lifespan and cannot, themselves, divide. As a consequence, antibodies cannot be used outside the body.

   Monoclonal antibodies are the product of a recently invented process through which antibodies are made available in the long term, for applications in entirely new circumstances. A monoclonal antibody is a single antibody that is stable and that can be used over a period of time. Each specific antibody is made by one particular type of B cell. The problem of the normally brief existence of a plasma cell is overcome by fusing the specific lymphocyte with a cancer cell – which, unlike other body cells, goes on dividing indefinitely. The resulting hybrid cell, known as a **hybridoma cell**, divides to form a clone of cells which persists and which conveniently goes on secreting the antibody in significant quantities (Figure 11.10). Hybridoma cells are virtually immortal, provided they are kept in a suitable environment.

■ **Figure 11.10**
Formation of monoclonal antibodies

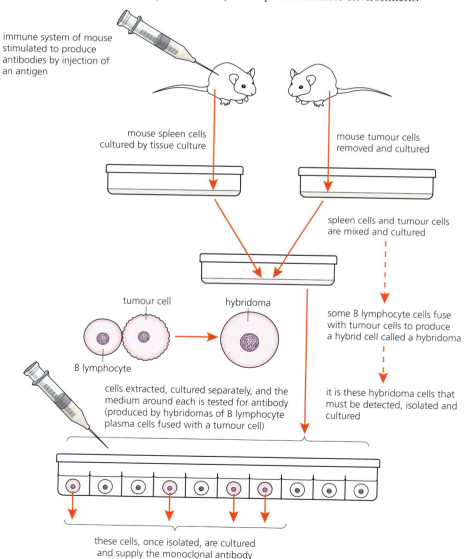

immune system of mouse stimulated to produce antibodies by injection of an antigen

mouse spleen cells cultured by tissue culture

mouse tumour cells removed and cultured

spleen cells and tumour cells are mixed and cultured

tumour cell

hybridoma

B lymphocyte

cells extracted, cultured separately, and the medium around each is tested for antibody (produced by hybridomas of B lymphocyte plasma cells fused with a tumour cell)

some B lymphocyte cells fuse with tumour cells to produce a hybrid cell called a hybridoma

it is these hybridoma cells that must be detected, isolated and cultured

these cells, once isolated, are cultured and supply the monoclonal antibody

8   Study the production of monoclonal antibodies, illustrated in Figure 11.10. **Construct** a concise list of the sequence of steps in their production.

The original monoclonal antibodies were developed from mouse cells, but some patients have developed an adverse reaction to antibodies made this way, since they are 'foreign' proteins to the patient's immune system. Genetically engineered antibodies compatible with the patient's immune system are sought, to avoid triggering the immune response.

Monoclonal antibodies are used in pregnancy testing kits

The use of monoclonal antibodies in medicine is already established, and is under further development in these and other areas. For example, in **diagnosis**, monoclonal antibodies are used in pregnancy testing.

A pregnant woman has a significant concentration of the hormone human chorionic gonadotrophin (HCG, page 494) in her urine, whereas a non-pregnant woman has a negligible amount. Monoclonal antibodies to HCG have been engineered to carry (become attached to) coloured granules, so that in a simple test kit the appearance of a coloured strip in one compartment provides immediate and visual confirmation of pregnancy. How this works is illustrated in Figure 11.11.

■ **Figure 11.11**
Detecting pregnancy
using monoclonal
antibodies

pregnancy testing kit

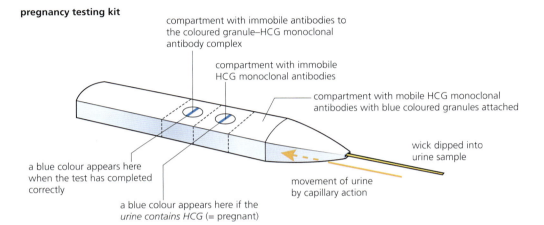

compartment with immobile antibodies to
the coloured granule–HCG monoclonal
antibody complex

compartment with immobile
HCG monoclonal antibodies

compartment with mobile HCG monoclonal
antibodies with blue coloured granules attached

wick dipped into
urine sample

movement of urine
by capillary action

a blue colour appears here
when the test has completed
correctly

a blue colour appears here if the
urine contains HCG (= pregnant)

how the positive test result is brought about

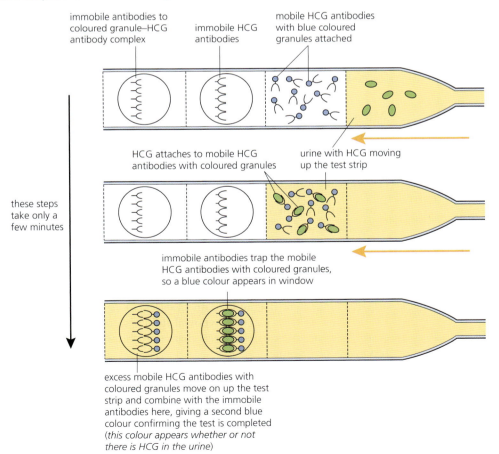

immobile antibodies to
coloured granule–HCG
antibody complex

immobile HCG
antibodies

mobile HCG antibodies
with blue coloured
granules attached

HCG attaches to mobile HCG
antibodies with coloured granules

urine with HCG moving
up the test strip

these steps
take only a
few minutes

immobile antibodies trap the mobile
HCG antibodies with coloured granules,
so a blue colour appears in window

excess mobile HCG antibodies with
coloured granules move on up the test
strip and combine with the immobile
antibodies here, giving a second blue
colour confirming the test is completed
(*this colour appears whether or not
there is HCG in the urine*)

11.2 Movement *– the roles of the musculoskeletal system are movement, support and protection*

Movement is a characteristic of all living things. It occurs within cells (for example, in cytoplasmic streaming), within organisms (such as in the pumping action of the heart) and as movements of whole organisms, known as **locomotion**. It is the issue of muscular locomotion in humans that is examined first.

■ Roles of the components of our locomotory system

Locomotion is the result of the interactions of nervous, muscular and skeletal systems. The component parts and their roles in locomotion are as follows:

- **Bones** support and partially protect the body parts. Also, they articulate with other bones at moveable **joints** and they provide anchorage for the muscles. In mammals, the skeleton consists of the axial skeleton (skull and vertebral column) and the appendicular skeleton (limb girdles and limbs).

- **Ligaments** hold bones together and form protective capsules around the moveable joints. Ligaments are made of fibres of strong but very slightly elastic connective tissue.

- **Muscles** cause movements by contraction. Skeletal muscle is one of three types of muscle in the mammal's body. Skeletal muscles occur in pairs, anchored to bones across joints. They are arranged so that when one of the pair contracts the other is stretched, a system known as **antagonistic pairing** (Figure 11.12). Contractions of skeletal muscle may either maintain the posture and position of the body or go on to bring about movement at joints.

- **Tendons** attach muscles to bones at their points of anchorage. They are cords of dense connective tissue.

- **Nerves** are bundles of many nerve fibres of individual nerve cells (neurons). They connect the central nervous system (brain and spinal cord) with other parts of the body, including the skeletal muscles. Nerve impulses are transmitted in a few milliseconds, travelling along individual nerve fibres to particular points in the body. Nervous control is, therefore, precise and specific, as well as quick. Nerve impulses stimulate muscles to contract and the nervous system as a whole coordinates movement.

Movements at joints

Moveable joints in the body are of different types, but they all permit controlled movements. They are known as **synovial joints**, because a thick viscous fluid, the synovial fluid, is secreted and retained in the joint for lubrication.

We will consider a hinge joint, as found in the human elbow.

The human elbow joint

Moveable joints are contained in fibrous **capsules** that are attached to the immediately surrounding bone. The capsule consists of tough connective tissue, but is flexible enough to permit movement. Also present at the joint, but outside the capsule, are ligaments that hold the bones together, preventing dislocations.

At the elbow, the humerus of the upper arm articulates with the radius and ulna of the lower arm at a hinge joint (Figure 11.13 and Table 11.3). At the joint, a fluid-filled space separates the articulating surfaces. Here, **synovial fluid** is secreted by the **synovial membrane**. The fluid nourishes the living cartilage layers that cover the articulating surfaces of the bones, as well as serving as a lubricant.

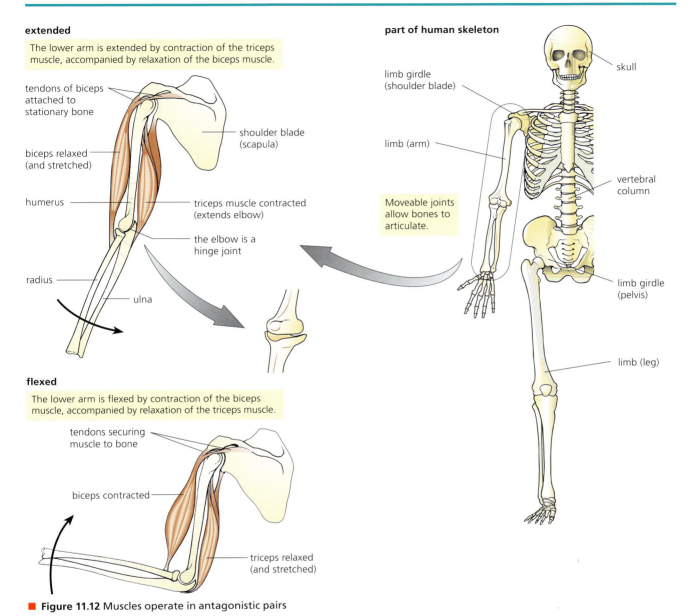

extended

The lower arm is extended by contraction of the triceps muscle, accompanied by relaxation of the biceps muscle.

tendons of biceps attached to stationary bone

biceps relaxed (and stretched)

humerus

triceps muscle contracted (extends elbow)

shoulder blade (scapula)

the elbow is a hinge joint

radius

ulna

flexed

The lower arm is flexed by contraction of the biceps muscle, accompanied by relaxation of the triceps muscle.

tendons securing muscle to bone

biceps contracted

triceps relaxed (and stretched)

part of human skeleton

limb girdle (shoulder blade)

limb (arm)

Moveable joints allow bones to articulate.

skull

vertebral column

limb girdle (pelvis)

limb (leg)

■ **Figure 11.12** Muscles operate in antagonistic pairs

■ **Table 11.3** Components of the elbow joint (a synovial joint) and their functions

Components	Functions
humerus, radius and ulna	the bones of the skeleton, together with the muscles attached across joints, function as a system of levers to maintain body posture and bring about actions, typically movements
biceps muscle	anchored to shoulder blade and attached to radius, so contraction flexes the lower arm (and stretches triceps)
triceps muscle	anchored to shoulder blade and attached to ulna so contraction extends the lower arm (and stretches biceps)
ligaments	hold bones (humerus, radius and ulna) in correct positions at the joint (combats dislocation)
capsule	contains and protects the joint, without restricting movement
synovial membrane	secretes synovial fluid
synovial fluid	lubricates the joint, nourishes the cartilage and removes any (harmful) detritus from worn bone and cartilage surfaces

■ **Figure 11.13**
The hinge joint of the
elbow

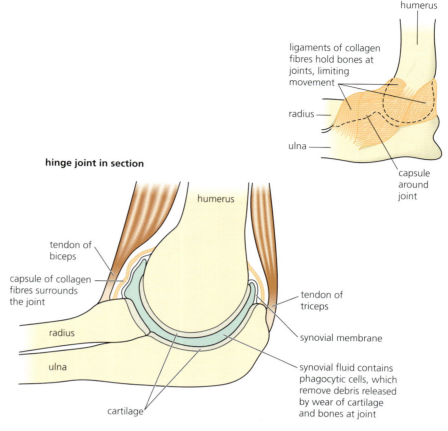

Draw and annotate a diagram of the human elbow, showing joint structure, named bones and antagonistic muscles *in situ*, by combining information from Figures 11.12 and 11.13.

Bones, joints and muscles as levers

We use machines to apply forces to objects in the world about us. A simple machine, commonly used, is a lever. The skeleton of an animal, together with the muscles attached across the joints, functions as a system of levers. Each joint acts as a pivot point or **fulcrum**. The force applied (when the muscle contracts) is called the **effort**. The load or force to be overcome is known as the **resistance**. The further away from the fulcrum the effort is applied, the greater the leverage; that is, the smaller the force that is required to raise the load (Figure 11.14).

Comparative movements at joints

You will have noticed that we have contrasting degrees of movement at our knees and hips (and at shoulder and elbow). This is because of a fundamental difference in the types of joints involved (Figure 11.15 and Table 11.4).

At the hip (and shoulder) is a **ball-and-socket joint**, which is also a synovial joint. In this type of joint, the ball-like surface of one bone (for example, the head of the humerus of the upper arm) fits into a cup-like depression on another bone (the shoulder blade). This type of joint permits movements in all three planes, a type of movement described as circumduction.

At the knee (and elbow), there is a **hinge joint**, which restricts movement to one plane. This is because of the shape of the articulating surfaces and also the position of the ligaments that hold the bones together. Movements at the elbow and knee are described as flexions and extensions.

■ **Figure 11.14**
The most common lever system in the body

This is the most common lever system in the body. The effort is greater than the load, but the distance moved by the effort (the muscle) is less than that moved by the load. Movement occurs with little shortening of muscle.

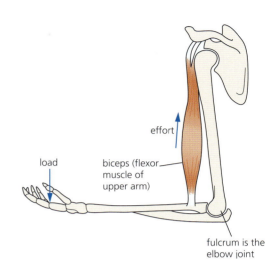

■ **Figure 11.15**
Movement at shoulder and elbow joints

Here, movements occur at joints where a cavity filled with fluid (synovial fluid) separates the articulating surfaces.

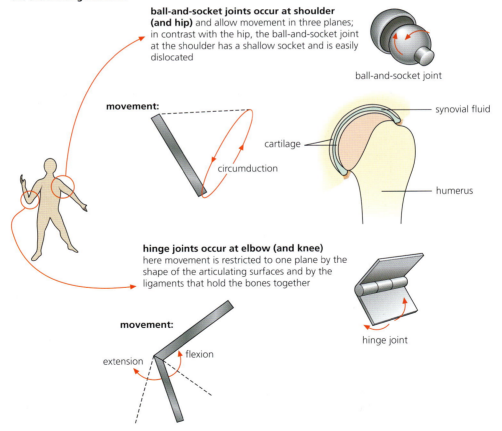

ball-and-socket joints occur at shoulder (and hip) and allow movement in three planes; in contrast with the hip, the ball-and-socket joint at the shoulder has a shallow socket and is easily dislocated

ball-and-socket joint

movement:

circumduction

synovial fluid

cartilage

humerus

hinge joints occur at elbow (and knee) here movement is restricted to one plane by the shape of the articulating surfaces and by the ligaments that hold the bones together

hinge joint

movement:

extension

flexion

■ **Table 11.4**
Movement of
shoulder and elbow
joints compared

Comparison	Shoulder joint	Elbow joint
Type of joint	synovial – ball-and-socket	synovial – hinge
Articulating bones	shoulder blade and humerus (head)	humerus, radius and ulna
Articulating surface(s)	between shoulder blade and humerus	between humerus and radius and ulna
Permitted movement	circumduction	flexion and extension

■ Antagonistic muscle action – in insect and mammal skeletons

You will remember that insects have bodies which are divided into a head, thorax and abdomen. The thorax has attached three pairs of jointed legs and two pairs of wings. Body and limbs are covered by a tough external skeleton, or **exoskeleton**, with flexible membrane between the body segments and in the joints of the limbs. The muscles for movement are attached to the inside of the skeleton. The insect's legs are a series of hollow cylinders, held together by joints. Across the joints are attached muscles in antagonistic pairs. Figure 11.16 compares how bones and exoskeleton provide anchorage for muscles and act as levers to bring about movement in insects and in mammals.

■ **Figure 11.16** Limb movement – in insects and mammals

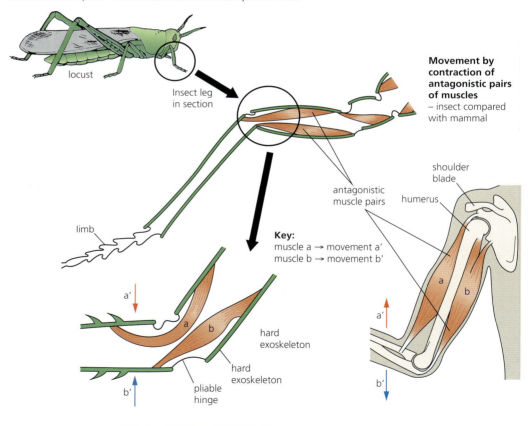

Insects are arthropods – with an external skeleton and jointed limbs.

locust

Insect leg in section

limb

Movement by contraction of antagonistic pairs of muscles
– insect compared with mammal

antagonistic muscle pairs

shoulder blade

humerus

a

b

a′

b′

Key:
muscle a → movement a′
muscle b → movement b′

a′

a

b

hard exoskeleton

hard exoskeleton

b′

pliable hinge

The controlled movement of a limb in any direction depends on the balance of opposing contraction of antagonistic pairs of muscles.

9 **Distinguish**
between the
following pairs:
endoskeleton and
exoskeleton; bone
and cartilage;
hinge joint and
ball-and-socket
joint; ligament and
tendon.

■ Skeletal muscle structure

Skeletal muscles are attached to the moveable parts of skeletons and their contraction brings about locomotion. The muscles are attached by **tendons** and work in antagonistic pairs (Figure 11.12).

Skeletal muscle consists of bundles of muscle fibres (Figure 11.17). The remarkable feature of a muscle fibre is the ability to shorten to half or even a third of the relaxed or resting length. Fibres appear striped under the light microscope, so skeletal muscle is also known as striated muscle. Actually, each fibre is, itself, composed of a mass of myofibrils, but we need the electron microscope to see this important detail.

■ **Figure 11.17**
The structure of skeletal muscle

skeletal muscle cut to show bundles of fibres

bundle of thousands of muscle fibres

tendon

connective tissue

photomicrograph of LS voluntary muscle fibre, HP (×1500)

alternating light and dark bands of muscle fibres (when viewed microscopically)

each muscle fibre contains several nuclei (i.e. a syncytium)

individual muscle fibre

each fibre consists of a mass of myofibrils

The ultrastructure of skeletal muscle

Skeletal muscle fibres are multinucleate and contain specialized endoplasmic reticulum. By electron microscopy, we can see that each muscle fibre consists of very many parallel **myofibrils**, within a plasma membrane known as the **sarcolemma**, together with cytoplasm. The cytoplasm contains **mitochondria** packed between the myofibrils. The sarcolemma is folded to form a system of transverse tubular endoplasmic reticulum, known as **sarcoplasmic reticulum**. This is arranged as a network around individual myofibrils. The arrangement of myofibrils, sarcolemma and mitochondria, surrounded by the sarcoplasmic membrane, is shown in Figure 11.18.

The striped appearance of skeletal muscle is due to an interlocking arrangement of two types of protein filaments, known respectively as **thick** and **thin filaments** – they make up the myofibrils. These protein filaments are aligned, giving the appearance of stripes (alternating **light and dark bands**). This is shown in the more highly magnified electron micrograph and interpretive drawing in Figure 11.19.

electron micrograph of TS through part of a muscle fibre, HP (x36 000)

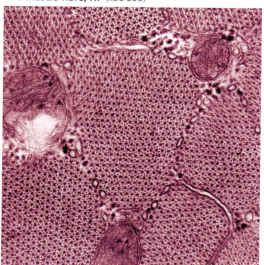

stereogram of part of a single muscle fibre

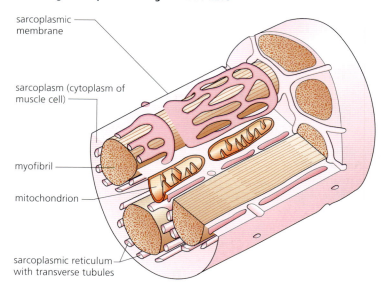

sarcoplasmic membrane

sarcoplasm (cytoplasm of muscle cell)

myofibril

mitochondrion

sarcoplasmic reticulum with transverse tubules

■ Figure 11.18 The ultrastructure of a muscle fibre

10 Explain the relationship to a muscle of:
a a muscle fibre
b a myofibril
c a myosin filament.

The **thick filaments** are made of a protein called **myosin**. They are about 15 nm in diameter. The longer **thin filaments** are made of another protein, **actin**. Thin filaments are about 7 nm in diameter and are held together by transverse bands, known as **Z lines**. Each repeating unit of the myofibril is, for convenience of description, referred to as a **sarcomere**. So we can think of a myofibril as consisting of a series of sarcomeres, attached end to end.

■ Figure 11.19
The ultrastructure of a myofibril

electron micrograph of an individual sarcomere

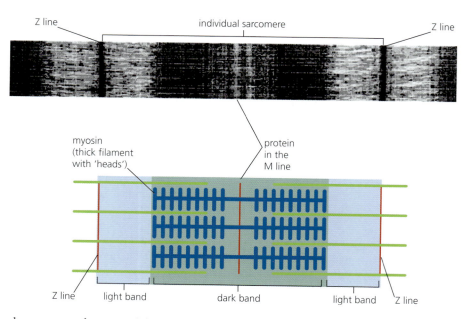

Z line

individual sarcomere

Z line

myosin (thick filament with 'heads')

protein in the M line

Z line   light band   dark band   light band   Z line

Draw and annotate a diagram of show the structure of a sarcomere; show Z lines, actin filaments, myosin filaments with 'heads', and the light and dark bands.

| **Nature of Science** | ### Skeletal muscle contracts by sliding of the filaments |

When skeletal muscle contracts, the **actin and myosin filaments slide past each other**, in response to nervous stimulation, causing shortening of the sarcomeres (Figure 11.20). This occurs in a series of steps, sometimes described as a ratchet mechanism. A great deal of **ATP** is used in the contraction process.

Shortening is possible because the thick filaments are composed of many myosin molecules, each with a **bulbous head** that protrudes from the length of the myosin filament. Along the actin filament is a complementary series of binding sites into which the bulbous heads fit. However, in muscle fibres at rest, the binding sites carry **blocking molecules** (a protein called **tropomyosin**); so, binding and contraction are not possible at rest.

Calcium ions play a critical part in the muscle fibre contraction mechanism, together with the proteins **tropomyosin** and **troponin**. The contraction of a sarcomere is best described in the following four steps.

■ **Figure 11.20**
Muscle contraction of a single sarcomere

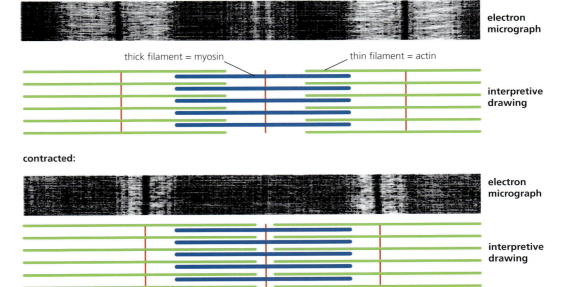

1 The myofibril is stimulated to contract by the arrival of an **action potential** (Figure 11.21). This triggers release of calcium ions from the sarcoplasmic reticulum, to surround the actin molecules. Calcium ions now react with an additional protein present (**troponin**) which, when so activated, triggers the removal of the blocking molecule, tropomyosin. The binding sites are now **exposed**.

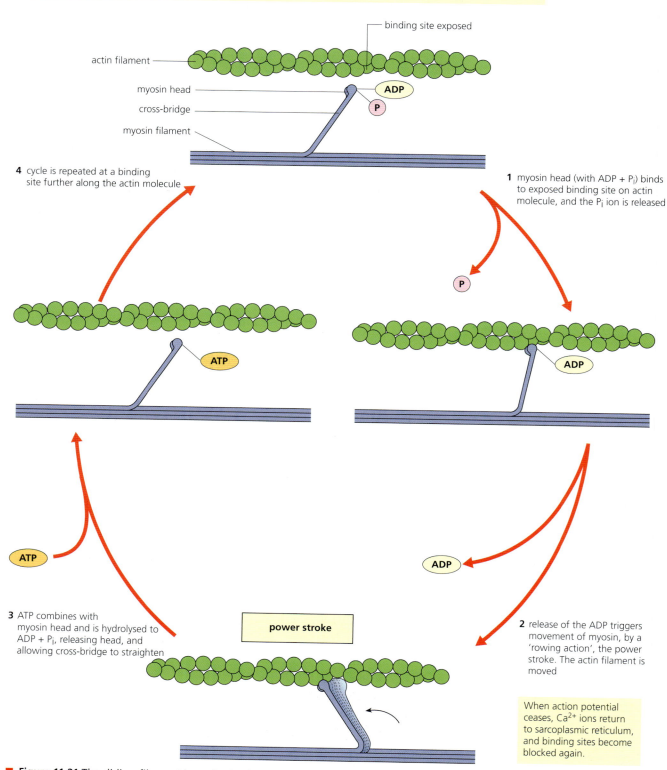

Arrival of action potential at myofibril releases Ca²⁺ ions from sarcoplasmic reticulum.

Ca²⁺ ions react with a protein (troponin), activating it. Activated troponin reacts with tropomyosin at the binding sites on the actin molecules, thereby exposing the binding sites.

Each myosin molecule has a 'head' that reacts with ATP → ADP + Pᵢ which remain bound.

binding site exposed

actin filament

myosin head — ADP
cross-bridge — P
myosin filament

4 cycle is repeated at a binding site further along the actin molecule

1 myosin head (with ADP + Pᵢ) binds to exposed binding site on actin molecule, and the Pᵢ ion is released

P

ADP

ATP

ATP

ADP

3 ATP combines with myosin head and is hydrolysed to ADP + Pᵢ, releasing head, and allowing cross-bridge to straighten

power stroke

2 release of the ADP triggers movement of myosin, by a 'rowing action', the power stroke. The actin filament is moved

When action potential ceases, Ca²⁺ ions return to sarcoplasmic reticulum, and binding sites become blocked again.

■ **Figure 11.21** The sliding-filament hypothesis of muscle contraction

2 Each bulbous head to which ADP and P$_i$ are attached (called a **charged bulbous head**) reacts with a binding site on the actin molecule beside it. The phosphate group (P$_i$) is shed at this moment.

3 The ADP molecule is then released from the bulbous head and this is the trigger for the **rowing movement** of the head, which tilts by an angle of about 45°, pushing the actin filament along. At this step, the **power stroke**, the myofibril has been shortened (**contraction**).

4 Finally, a fresh molecule of **ATP binds** to the bulbous head. The protein of the bulbous head includes the enzyme ATPase, which catalyses the hydrolysis of ATP. When this reaction occurs, the ADP and inorganic phosphate (P$_i$) formed remain attached, and the **bulbous head is now 'charged'** again. The charged head detaches from the binding site and straightens.

Nature of Science

Role of fluorescent dyes in the investigation of muscle contraction

The cyclic interactions in muscle contraction and their dependence on ATP was demonstrated by attaching fluorescent dye to the myosin molecules, prior to causing them to 'row' along the actin filament. When electromagnetic radiation of a particular wavelength illuminated this contracting tissue, the cyclic movement of the myosin heads was detectable. The velocity of movement observed could be correlated with **changing ATP concentrations**.

Muscles, controlled movements and posture

Muscles are involved in maintaining body posture and in subtle and delicate movements, as well as in vigorous or even violent actions. Consequently, nervous control of muscle contraction may cause relaxed muscle to contract slightly, moderately or fully, depending on the occasion. In these differing states of contraction, the overall lengths of the sarcomeres are changed accordingly. These relative changes are illustrated diagrammatically in a single sarcomere in Figure 11.22. Below them is a representation of part of a myofibril, seen at a particular stage of contraction. This representation is not diagrammatic, but is based on an interpretation of an electron micrograph. Now see question 11 (which is concerned with analysing the state of contraction of myofibrils) and question 12 (which involves accurate measurements of the length of sarcomeres).

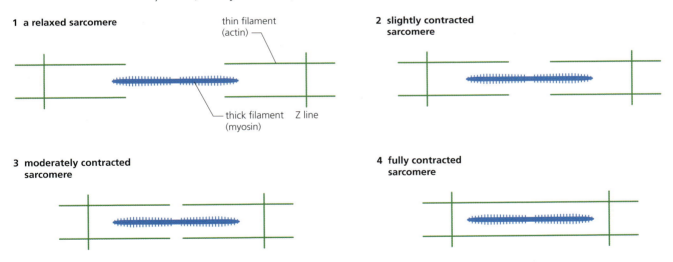

1 a relaxed sarcomere

thin filament (actin)

thick filament (myosin)   Z line

2 slightly contracted sarcomere

3 moderately contracted sarcomere

4 fully contracted sarcomere

sketch representing a myofibril at a particular stage in contraction, based on interpretation of a TEM of a sample of striated muscle

■ **Figure 11.22** Analysing states of contraction in striated muscle fibres

11 a Identify the approximate state of contraction illustrated in the sketch of the electron micrograph of a myofibril shown in Figure 11.22.

b Sketch a similar myofibril, fully contracted. Label your drawing (showing sarcomere, Z lines, light band and dark band).

12 Viewing a prepared stained slide of stripped skeletal muscle, **measure** the length of the sarcomeres, using an eye-piece graticule, calibrated as shown in Figures 1.8 and 1.9, pages 9 and 10.

11.3 The kidney and osmoregulation – *all animals excrete nitrogenous waste products and some animals also balance water and solute concentrations*

The chemical reactions of metabolism give by-products. Some of these would be toxic if they were allowed to accumulate in the organism. **Excretion is the removal from the body of the waste products of metabolism.** Excretion is a characteristic activity of all living things. Metabolites that are present in excessive concentrations are also excreted.

In mammals, excretion plays an important part in the process by which the internal environment is regulated to maintain more or less constant conditions (homeostasis).

Associated with excretion, and very much part of homeostasis, is the process of osmoregulation. Osmoregulation is the maintenance of a proper balance of water and dissolved substances in the organism.

■ Why osmoregulation is important in animals

All animals need to balance water uptake and loss from the body. We know that water enters and exits cells by osmosis (page 39). If water uptake into animal cells is excessive, the hydrostatic pressure that quickly develops stretches the plasma membrane to the point of bursting. The problem is illustrated by the fresh water protozoan, *Amoeba*, through the constant pumping action of its contractile vacuole (Figure 11.23). If the cytoplasm of the amoeba is anaesthetized, the contractile vacuole ceases pumping and the cell rapidly bursts.

■ **Figure 11.23**
Osmoregulation in fresh water *Amoeba*

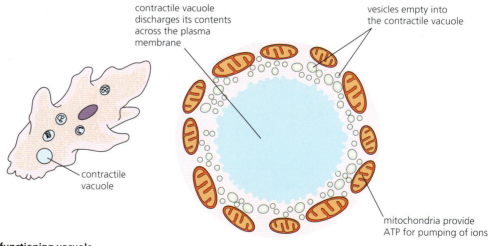

contractile vacuole discharges its contents across the plasma membrane

vesicles empty into the contractile vacuole

contractile vacuole

mitochondria provide ATP for pumping of ions

the functioning vacuole

1 water + ions from cytosol pass into vacuole

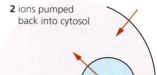
2 ions pumped back into cytosol

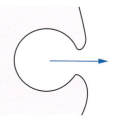
3 contractile vacuole contents discharged

Alternatively, if water loss is excessive, a cell will shrink and die, for water is an essential and major component of cytoplasm of all living organisms – it makes up about 90% of cell volume (Figure 1.46, page 41).

■ Osmoconformer or osmoregulator?

Some animals maintain the osmotic concentration (osmolarity) of their cells and body fluids at the same concentration as that of the environment. These are the **osmoconformers**. Many marine non-vertebrates osmoconform, maintaining their osmolarity at the same level as sea water. There is no tendency for water uptake (or loss) to occur from their cells and tissues – across their gills, for example. This is achieved by retaining in their cells and body fluids some of the dissolved ions from the environment.

Other animals, including humans, control their internal osmolarity independently of environmental conditions (they are **osmoregulators**). The unit of measurement of osmolarity is milliosmoles per litre ($mOsm\,dm^{-3}$). Sea water (roughly 3% solution of sodium chloride) has an omolarity of about $1000\,mOsm\,dm^{-3}$. Fresh water has a variable, but extremely low osmolarity, by comparison. The osmolarity of human blood is about $300\,mOsm\,dm^{-3}$. Our osmolarity is maintained at this level by regulation of the balance of dissolved substances and water in body fluids. The concentrations of inorganic ions, such as Na^+ and Cl^-, and of sugars and amino acids (non-electrolytes) are regulated in the blood and tissue fluids, along with the water content. The osmolarity of these liquids is maintained at the same level as that of the cell cytoplasm.

The kidneys are where any excess solutes and water are removed from the blood circulation. How the kidneys bring this about we will discuss next. The steps to osmoregulation by negative feedback are summarized in Figure 11.34, page 478.

Finally, we should remember that a major threat for terrestrial organisms, in general, is the danger of excessive water loss, leading to dehydration. This is because fresh water for uptake or drinking can sometimes be in short supply, and mechanisms for gaseous exchange involve continuous and significant loss of water vapour (Figure 11.24).

■ **Figure 11.24**
Water balance in humans – typical figures

Exhaled air is saturated with water vapour, an invisible component except in freezing weather or when breathed on to a very cold surface.

The water balance of the body:
- about 3 litres is taken in and lost daily
- about 10–15% of this total is lost from the lungs.

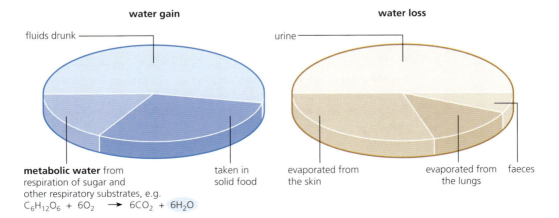

water gain

fluids drunk

metabolic water from respiration of sugar and other respiratory substrates, e.g.
$C_6H_{12}O_6 + 6O_2 \rightarrow 6CO_2 + 6H_2O$

taken in solid food

water loss

urine

evaporated from the skin

evaporated from the lungs

faeces

13 Explain why gaseous exchange in a mammal involves continuous loss of water vapour from the body.

■ The challenge of nitrogen excretion in animals

In animals, any excess protein in the diet cannot be stored. Consequently, after proteins have been hydrolysed to amino acids, excess amino acids are **deaminated** – by the removal of the amino group. Once this group is removed, a likely initial product is **ammonia**. This is extremely toxic to cells, if it is allowed to accumulate. So, the amino group may be converted into a much less dangerous nitrogenous excretory product. **Urea** is an example – a chemical compound synthesized from carbon dioxide and ammonia. In dilute solution, urea may be safely excreted from the body, as it is in mammals (Figure 11.25). How different animals respond to the excretion of nitrogenous waste, and to other demands on their excretory systems, may be related to structure, physiology, environment and evolutionary history. Examples of this are shown by study of excretion in:

- insects
- mammals that are adapted to contrasting environments
- other vertebrate groups.

We will return to this later, after the issue of human kidney function has been examined.

■ **Figure 11.25** Deamination

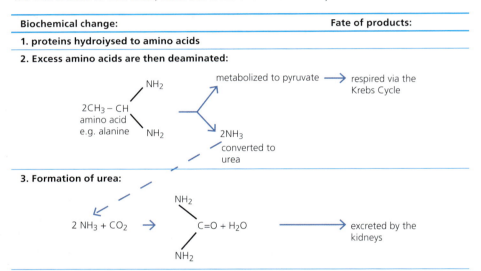

Biochemical change:	Fate of products:
1. proteins hydrolysed to amino acids	
2. Excess amino acids are then deaminated:	

3. Formation of urea:

■ The human kidney – an organ of excretion and osmoregulation

The role of our kidneys is to regulate the body's internal environment by constantly adjusting the composition of the blood. Waste products of metabolism are transported from the metabolizing cells by the blood circulation, removed from the blood in the kidneys and excreted in a solution called **urine**. At the same time, the concentrations of inorganic ions, such as Na^+ and Cl^-, and of water in the body are also regulated in the kidneys. The functioning of the kidney is another example of **homeostasis by negative feedback** (see Figure 6.52, page 301).

The position of the kidneys in humans is shown in Figure 11.26. Each kidney is served by a **renal artery** and drained by a **renal vein**. Urine from the kidney is carried to the bladder by the **ureter**, and (occasionally) from the bladder to the exterior by the **urethra**, when the bladder sphincter muscle is relaxed. Together these structures are known as the **urinary system**.

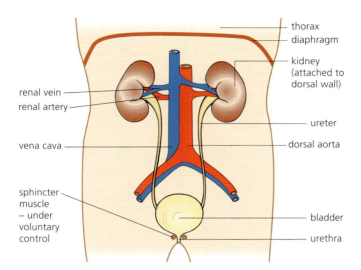

■ **Figure 11.26**
The human urinary
system

In section, a kidney can be seen to consist of an outer **cortex** and inner **medulla**, and these are made up of a million or more tiny tubules, called **nephrons**. The shape of a nephron and its arrangement in the kidney are shown in Figure 11.27.

Blood vessels are closely associated with each of the distinctly shaped regions of the nephrons. For example, the first part of the nephron is formed into a cup-shaped renal or **Bowman's capsule**, and the capillary network here is known as the **glomerulus**. These occur in the cortex. The **convoluted tubules** occur partly in the cortex and partly in the medulla, but notice that the extended **loops of Henle** and **collecting ducts** occur largely in the medulla.

Each region of the nephron has a specific role to play in the work of the kidney, and the capillary network serving the nephron plays a key part, too, as we shall now see.

Drawing and labelling as diagram of the human kidney

Construct your drawing of the human kidney, seen in longitudinal section (see Figure 11.27), so that the cortex region occupies no more than 20% of the width of the organ. Show the series of pyramid-like parts of the medulla that point towards the pelvis region, together with the renal artery (narrow) and vein (broad) which enter and exit just above the ureter. Exclude representation of individual nephrons on this scale of drawing.

The labels required are: cortex, medulla, pelvis, ureter, and renal artery and vein.

14 Distinguish
between excretion, egestion, osmoregulation and secretion by means of both definitions and examples.

The formation of urine

In humans, about 1.0–1.5 l of urine is formed each day, typically containing about 40–50 g of solutes, of which **urea** (about 30 g) and **sodium chloride** (up to 15 g) make up the bulk. The nephron produces urine in a continuous process, which we can conveniently divide into five steps to show how the blood composition is so precisely regulated.

Step 1: Ultrafiltration in the renal capsule

In the glomerulus, much of the water and many relatively small molecules present in the blood plasma, including useful ions, glucose and amino acids, are forced out of the capillaries, along with urea, into the lumen of the capsule (Figure 11.28). This fluid is called the **glomerular filtrate** and the process is described as **ultrafiltration**, because it is powered by the pressure of the blood. The **blood pressure** here is high enough for ultrafiltration because the input capillary (afferent arteriole) is significantly wider than the output capillary (efferent arteriole).

LS through kidney showing positions of nephrons in cortex and medulla

nephron with blood capillaries

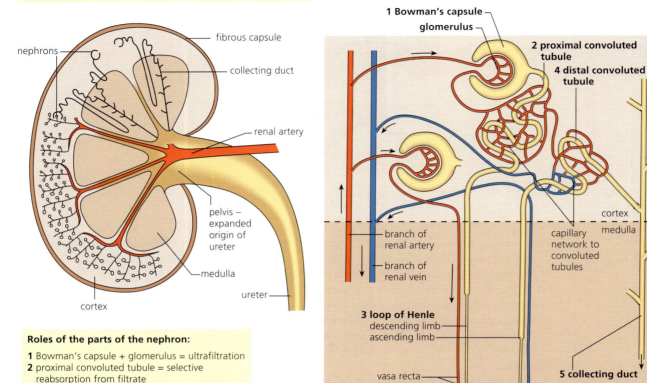

Roles of the parts of the nephron:

1 Bowman's capsule + glomerulus = ultrafiltration
2 proximal convoluted tubule = selective reabsorption from filtrate
3 loop of Henle = water conservation
4 distal convoluted tubule = pH adjustment and ion reabsorption
5 collecting duct = water reabsorption

photomicrograph of the cortex of the kidney in section, showing the tubules, renal capsules and capillary networks

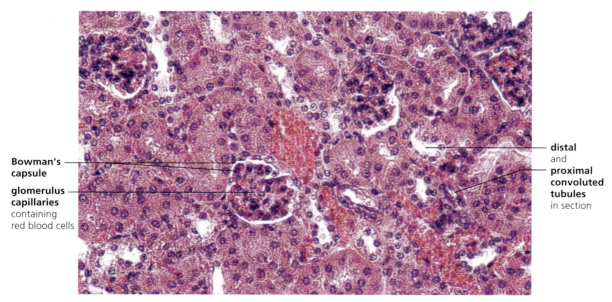

■ **Figure 11.27** The kidney and its nephrons – structure and roles

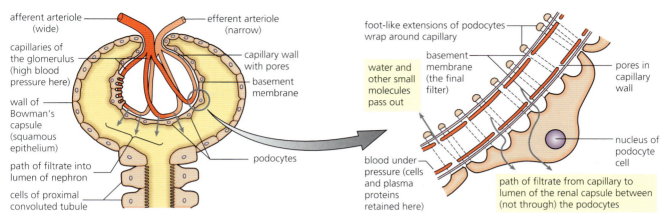

afferent arteriole (wide)

efferent arteriole (narrow)

capillaries of the glomerulus (high blood pressure here)

capillary wall with pores

basement membrane

wall of Bowman's capsule (squamous epithelium)

path of filtrate into lumen of nephron

cells of proximal convoluted tubule

podocytes

foot-like extensions of podocytes wrap around capillary

basement membrane (the final filter)

water and other small molecules pass out

pores in capillary wall

nucleus of podocyte cell

blood under pressure (cells and plasma proteins retained here)

path of filtrate from capillary to lumen of the renal capsule between (not through) the podocytes

false-colour scanning electron micrograph of podocytes (pale purple) with their extensions wrapped around the blood capillaries (pink/red) (× 3500)

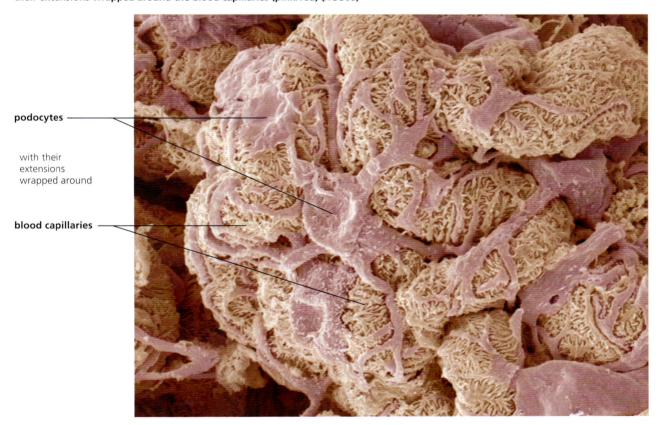

podocytes

with their extensions wrapped around

blood capillaries

■ **Figure 11.28** The site of ultrafiltration

15 State the source of energy for ultrafiltration in the glomerulus.

The barrier between the blood plasma and the lumen of the Bowman's capsule functions as a filter or 'sieve' through which ultrafiltration occurs. This sieve is made of two layers of cells (the endothelium of the capillaries of the glomerulus and the epithelium of the capsule wall), between which is a basement membrane.

You can see this arrangement in Figure 11.28.

Notice that the cells of the inner wall of the capsule are called **podocytes**, because they have foot-like extensions. These wrap around the capillaries of the glomerulus, leaving a network of slits between the extensions.

Similarly, the endothelium of the capillaries has **pores**. These are large enough for fluid to pass through, but not large enough for the passage of blood cells. This detail has only become known because of studies using the electron microscope; these filtration gaps are very small indeed.

Finally, there is the **basement membrane**, which is a layer that surrounds and supports the capillary walls. This structure consists of a mesh-work of glycoproteins that allows the filtrate to pass, but that retains almost all of the plasma proteins.

The fluid that has filtered through into the renal capsule is very similar to blood plasma, but it has a significant difference (Table 11.5). Not only are blood cells retained in plasma, but the majority of blood proteins and polypeptides also remain there.

■ Table 11.5
A comparison of chief components of the blood plasma and the glomerular filtrate (*mol dm^{-3}/**mg dm^{-3})

	Blood plasma	Filtrate
urea*	5	5
glucose*	5	5
sodium ions*	150	145
chloride ion*	110	115
proteins**	740	3–4

Step 2: Selective reabsorption in the proximal convoluted tubule

The proximal convoluted tubule is the longest section of the nephron and it is here that a large part of the filtrate is reabsorbed into the capillary network. The walls of the tubule are one cell thick and their cells are packed with mitochondria; we expect this, if **active transport** is a key part mechanism in reabsorption. The cell membranes of the cells of the tubule wall (in contact with the filtrate) all have a 'brush border' of microvilli. These enormously increase the surface area where reabsorption occurs (Figure 11.29). The individual mechanisms of transport are given in Table 11.6.

■ Table 11.6
Mechanisms of selective reabsorption in the proximal convoluted tubule

Component of filtrate	Mechanism of reabsorption
sugars, amino acids and other essential metabolites	sugars by active transport across the cell surface membrane – by the action of special carrier proteins in a process known as co-transport (Figure 11.30); a carrier protein uses the diffusion of hydrogen ions down their electrochemical gradient into the cell to drive the uptake of molecules, typically against their concentration gradient
	amino acids and other essential metabolites by active transport
ions (Na$^+$, Cl$^-$ and others)	by a combination of active transport, facilitated diffusion and some exchange of ions
urea	diffusion
proteins	pinocytosis
water	osmosis

16 The cells of the walls of the proximal convoluted tubule have a brush border. **Describe** what this means and **explain** how it helps in tubule function.

Step 3: Water conservation in the loop of Henle

The function of the loop of Henle is to enable the kidneys to conserve water. Since urea is expelled from the body in solution, some water loss in excretion is inevitable. There is a potential problem here. Water is a major component of the body and it can be a scarce resource for terrestrial organisms. It is important, therefore, that (when necessary) mammals are able to form urine which is more concentrated than the blood, thereby reducing water loss to a minimum. Human urine can be five times as concentrated as the blood, due to the action of the loop of Henle.

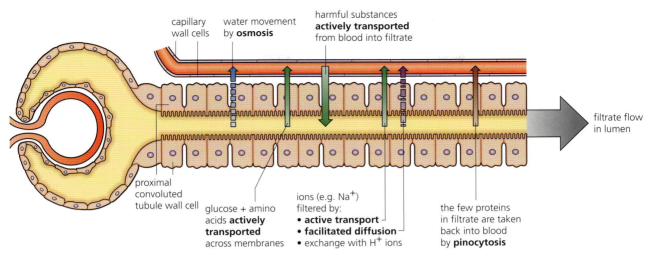

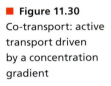 **Figure 11.29** Reabsorption in the proximal convoluted tubule

Figure 11.30
Co-transport: active
transport driven
by a concentration
gradient

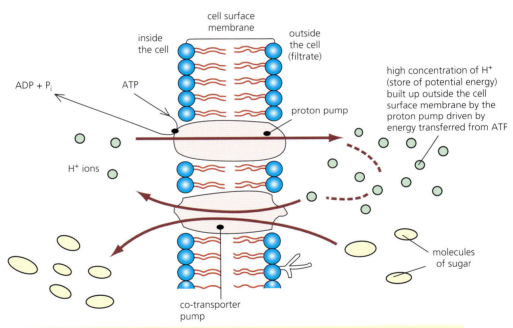

The electrochemical gradient in H^+ ions between the exterior and the interior of the cell drives the active transport of metabolites (e.g. sugars, amino acids) across the membrane as the H^+ flow down their electrochemical gradient via the co-transporter pump.

The structure of the loop of Henle with its **descending** and **ascending limbs**, together with a parallel blood supply, the **vasa recta**, is shown in Figure 11.31. The vasa recta is part of the same capillary network that surrounds a nephron.

The working loops of Henle and their capillary loops create and maintain an osmotic gradient in the **medulla** of the kidney. The gradient across the medulla is from a less concentrated salt solution near the cortex to the most concentrated salt solution at the tips of the pyramid of the medulla (Figure 11.27). The pyramid region of the medulla consists mostly of the collecting ducts. Thus, the loop of Henle maintains hypertonic conditions around the collecting ducts. The osmotic gradient allows water to be withdrawn from the collecting ducts if circumstances require it. How this occurs we shall discuss shortly.

First, there is the question of how the gradient itself is created by the loops of Henle. The gradient is brought about by a mechanism known as a counter-current multiplier.

Counter-current exchange

Here, the principles of counter-current exchange involve exchange between fluids flowing in opposite directions in two systems. The annotations in Figure 11.31 show how the counter-current mechanism works. When following these annotations, remember that the descending and ascending limbs lie close together in the kidney.

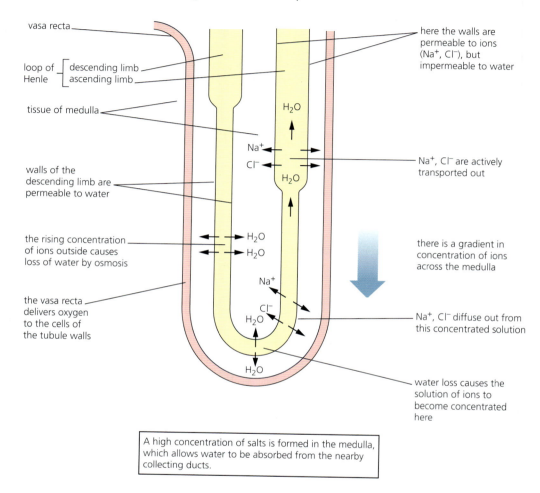

■ **Figure 11.31**
The functioning loop of Henle

vasa recta

loop of { descending limb
Henle { ascending limb

tissue of medulla

walls of the descending limb are permeable to water

the rising concentration of ions outside causes loss of water by osmosis

the vasa recta delivers oxygen to the cells of the tubule walls

here the walls are permeable to ions (Na$^+$, Cl$^-$), but impermeable to water

Na$^+$, Cl$^-$ are actively transported out

there is a gradient in concentration of ions across the medulla

Na$^+$, Cl$^-$ diffuse out from this concentrated solution

water loss causes the solution of ions to become concentrated here

H$_2$O
Na$^+$
Cl$^-$
H$_2$O
H$_2$O
H$_2$O
Na$^+$
Cl$^-$
H$_2$O
H$_2$O

A high concentration of salts is formed in the medulla, which allows water to be absorbed from the nearby collecting ducts.

Look first at the second *half of the loop*, the **ascending limb**. In the upper (thick-walled) part of the ascending limb, sodium and chloride ions are pumped out of the filtrate into the fluid between the cells of the medulla, called the interstitial fluid. The energy to pump these ions is transferred from ATP. In the lower (thin-walled) part of the ascending limb, sodium and chloride ions diffuse out into the interstitial fluid. This movement of sodium chloride out of the tubule helps maintain the osmolarity of the interstitial fluid in the medulla. All along the ascending limbs, the walls are unusual in being impermeable to water. So, water in the ascending limb is retained in the filtrate as salt is pumped out.

Opposite is the *first* half of the loop, the **descending limb**. This limb is fully permeable to water, but is of very low permeability to solutes generally. Here, water passes out into the interstitial fluid by osmosis, due to the salt concentration in the medulla.

Exchange in this counter-current multiplier is a dynamic process that occurs in the whole length of the loop. At each level in the loop, the salt concentration in the descending limb is slightly higher than the salt concentration in the adjacent ascending limb. As the filtrate flows, the concentrating effect is multiplied and so the fluid in and around the hairpin bend of the loops of Henle is saltiest.

The function of the vasa recta is to deliver oxygen to and remove carbon dioxide from the metabolically active cells of the loop of Henle. The vasa recta also absorbs water that has passed into the medulla at the collecting ducts. We discuss the working of these ducts in stage 5, below.

Step 4: Blood pH and ion concentration regulation in the distal convoluted tubule

The cells of the walls of the distal convoluted tubule are of the same structure as those of the proximal convoluted tubule, but their roles differ somewhat.

Here, the cells of the tubule walls adjust the composition of the blood – in particular, the **pH**. Also in the distal convoluted tubule, the **concentration of useful ions** is regulated. In particular, the concentration of potassium ions (K^+) is adjusted by secretion of any excess present in the plasma into the filtrate. Similarly, the concentration of sodium ions (Na^+) in the body is regulated by varying the amount of sodium chloride that is reabsorbed from the filtrate.

Step 5: Water reabsorption in the collecting ducts

When the intake of water exceeds the body's normal needs, the urine produced is copious and dilute. We notice this after drinking a lot of water. On the other hand, if we have taken in very little water, have been sweating heavily (part of our temperature regulation mechanism) or if we have eaten very salty food, perhaps, then a small volume of concentrated urine is formed.

Osmoregulation, the control of the water content of the blood (and therefore of the whole body), is a part of homeostasis – another example of regulation by negative feedback.

How exactly is this brought about?

The **hypothalamus**, part of the floor of the forebrain (Figure 11.32), controls many body functions. The composition of the blood is continuously monitored here, as it circulates through the capillary networks of the hypothalamus. Data is also received at the hypothalamus from sensory receptors located in certain organs in the body. All these inputs enable the hypothalamus to control accurately the activity of the pituitary gland.

The **pituitary gland** is situated below the hypothalamus, but is connected to it (Figure 11.32). The pituitary gland, as a whole, produces and releases hormones (it is part of our endocrine system) – in fact, it has been called the 'master' hormone gland. In the process of osmoregulation, it is the posterior part of the **pituitary** that stores and releases **antidiuretic hormone (ADH)** – among other hormones. (Other parts of the pituitary also secrete hormones, regulating a range of other body activities and functions.)

ADH is actually produced in the hypothalamus and stored in vesicles at the ends of neurosecretory cells in the posterior pituitary gland. When nerve impulses from the hypothalamus trigger release of ADH into the capillary networks in the posterior pituitary, ADH circulates in the bloodstream. However, the targets of this hormone are the walls of the collecting ducts of the kidney tubules.

When the water content of the blood is low, **antidiuretic hormone (ADH)** is secreted from the posterior pituitary gland. When the water content of the blood is high, little or no ADH is secreted.

How does ADH change the permeability of the walls of the collecting ducts?

The cell surface membranes of the cells that form the walls of the collecting ducts contain a high proportion of channel proteins (**aquaporin proteins**); each molecule is capable of forming an open pore running down its centre. You can see the structure of a fluid mosaic membrane and its channel proteins in Figure 1.34, page 31, and Figure 1.41, page 37.

When there is an excess of ADH in the blood circulating past the kidney tubules, this hormone binds to receptor molecules in the collecting-duct membrane, causing the protein channels in the membranes to open. As a result, much water diffuses out into the medulla and very little diffuses from the medulla into the collecting ducts (Figure 11.33).

Can you explain why? You may need to go back to step 3 above.

The water entering the medulla is taken up and redistributed in the body by the blood circulation. Only a small amount of very concentrated urine is formed. Meanwhile, the action of the liver continually removes and inactivates ADH. This means that the presence of freshly released ADH has a regulatory effect.

■ **Figure 11.32**
Hypothalamus and
pituitary gland, and
the release of ADH

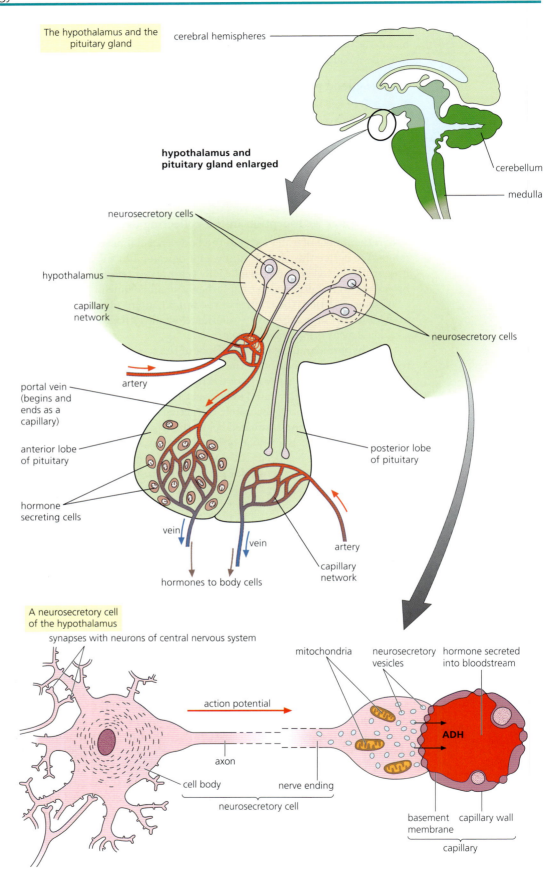

The hypothalamus and the
pituitary gland

cerebral hemispheres

**hypothalamus and
pituitary gland enlarged**

cerebellum

medulla

neurosecretory cells

hypothalamus

capillary
network

neurosecretory cells

portal vein
(begins and
ends as a
capillary)

artery

anterior lobe
of pituitary

posterior lobe
of pituitary

hormone
secreting cells

vein

vein

artery

capillary
network

hormones to body cells

A neurosecretory cell
of the hypothalamus

synapses with neurons of central nervous system

mitochondria

neurosecretory
vesicles

hormone secreted
into bloodstream

action potential

ADH

axon

cell body

nerve ending

neurosecretory cell

basement
membrane

capillary wall

capillary

When we have:
• drunk a lot of water
the hypothalamus detects this and stops the posterior pituitary gland secreting ADH.

When we have:
• taken in little water
• sweated excessively
• eaten salty food
the hypothalamus detects this and directs the posterior pituitary gland to secrete ADH.

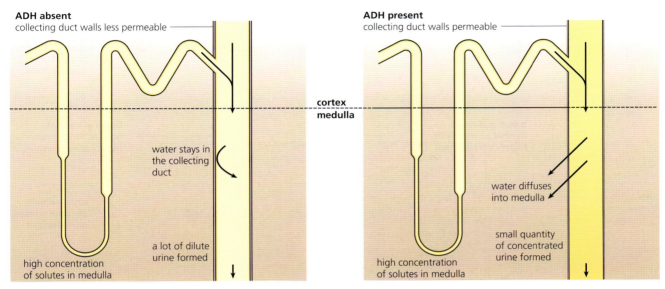

ADH absent
collecting duct walls less permeable

water stays in the collecting duct

high concentration of solutes in medulla

a lot of dilute urine formed

cortex
medulla

ADH present
collecting duct walls permeable

water diffuses into medulla

high concentration of solutes in medulla

small quantity of concentrated urine formed

■ **Figure 11.33** Water reabsorption in the collecting ducts

When ADH is absent from the blood circulating past the kidney tubules, the protein channels in the collecting-duct plasma membranes are closed. The amount of water that is retained by the medulla tissue is now minimal. The urine becomes copious and dilute.

A summary of osmoregulation by the kidneys is shown in Figure 11.34.

Annotation of a diagram of the nephron

You are now in a position to construct a diagram of an individual nephron, extending from the Bowman's capsule to the collecting duct, with its associated blood supply.

Use the knowledge you have gained to annotate the following regions and structures, highlighting the features of each and recording its function in the production of urine: Bowman's capsule; proximal convoluted tubule; loop of Henle; distal convoluted tubule; collecting duct. Use the diagram of the nephron in Figure 11.27, page 470, to help you.

Include also the blood vessels: afferent arteriole; glomerulus; efferent arteriole; capillary bed surrounding the convoluted tubules; vasa recta; venules.

17 Predict in what conditions in the body ADH is released.

■ Differences in composition of blood plasma, glomerular filtrate and urine

The composition of urine that is excreted from the body is variable – depending on diet (including salt intake and amount of protein consumed), water intake, degree of physical activity, environmental conditions, water loss by other routes (particularly in sweat) and our state of health.

On the other hand, the composition of the blood is held more or less constant. This is due to the efficiency of the homeostatic mechanisms of the body, particular those operating at the kidney tubules.

■ **Figure 11.34**
Homeostasis by
osmoregulation
in the kidneys – a
summary

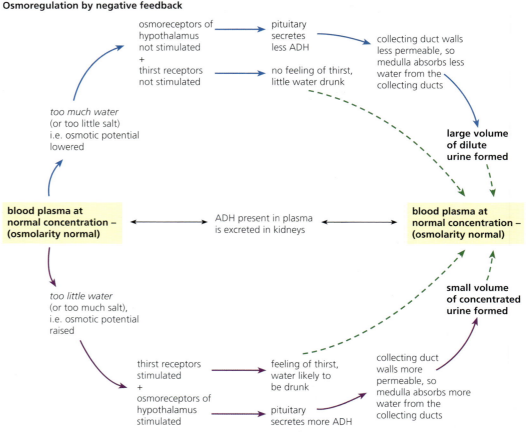

Osmoregulation by negative feedback

Similarly, the composition of the filtrate passing into the renal capsule is remarkably constant. This is the result of ultrafiltration in the glomerulus, the pressure of the blood, and the sizes of the blood proteins and polypeptides dissolved in the plasma – they are mostly too large to be filtered.

In Table 11.7, we have evidence of the power of the ultrafiltration mechanism and the scale of selective reabsorption. And we see the largest divergence in concentration when the composition of the filtrate is compared with composition of urine, formed in one day.

■ **Table 11.7**
Composition of
blood, glomerular
filtrate and urine

Substance	Total amount in the plasma	Filtered (per day)	Reabsorbed (per day)	Urine (excreted per day)
water	3 litres	180 litres	178–179 litres	1–2 litres
glucose	2.7 g	162.0 g	162.0 g	0.0 g
urea	0.9 g	54.0 g	27.0 g	27.0 g*
proteins and polypeptides	200.0 g	2.0 g	1.9 g	0.1 g
Na$^+$ ions	9.7 g	579.0 g	575.0 g	4.0 g
Cl$^-$ ions	10.7 g	640.0 g	633.7 g	6.3 g
*some is also secreted by the cells of the tubule into the urine				

18 The composition
of the blood in
the renal veins
leaving the kidneys
differs significantly
from that in the
renal arteries.
Identify and **list**
the differences you
would anticipate.

■ Kidney failure

Kidney failure may be caused by bacterial infection, by external mechanical damage or by high blood pressure. In the event of renal failure, urea, water and sodium ions start to accumulate in the blood. In mild cases, regulation of diet (particularly of fluids, salt and proteins consumed) may be sufficient to minimize the task of the remaining kidney tubules, so the body copes.

Detection of kidney failure, or other medical conditions

Samples of urine may be tested for the presence of:
- abnormal components, such as blood cells and proteins
- drugs, in anti-doping investigations
- glucose, in the case of suspected diabetes.

Treatment of kidney failure

In cases where more than 50% of kidney function has been lost, **hemodialysis** may be required every few days, in addition to a strict prescribed diet. In dialysis, the blood circulation is connected to a dialysis machine as shown in Figure 11.35. Blood is repeatedly circulated outside the body for 6–10 hours, through a fine tube of cellophane (a partially permeable membrane). This is bathed in dialysate, a fluid of equal solute potential and similar composition to that of blood leaving a healthy kidney. This prevents net outward diffusion of the useful components of blood (mainly water, ions, sugars and amino acids) but allows diffusion of urea and other toxic substances outwards.

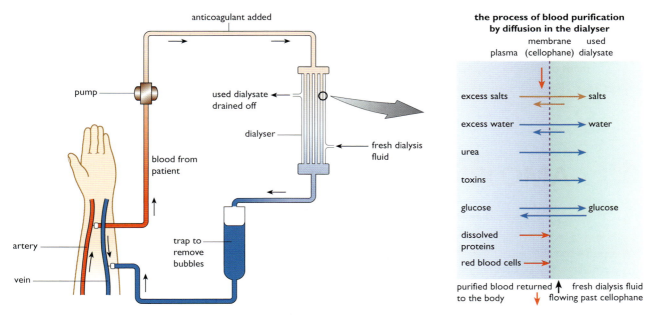

■ **Figure 11.35** The principle of dialysis

19 Explain precisely why it is that no kidney transplant is ever a perfect match, except in the case of an identical twin being the donor.

Ideally, acute renal failure is rectified by **kidney transplant** from a donor whose cell type is sufficiently compatible with that of the recipient. A kidney from a non-compatible donor would generate an immunological reaction, leading to rejection of the kidney. No match is 'perfect'; however, at transplantation the antibody-producing cells of the recipient are suppressed. Subsequently, drugs that suppress the body's response to 'foreign' proteins (the immune response, page 271) have to be administered permanently.

■ Mechanisms for excretion in other animals

Excretion in insects

Insects (phylum Arthropoda, page 234) are adapted to survive and prosper as terrestrial organisms. Their ability to conserve water in the process of excretion is an important factor in their success. Remember, within the insect's body is an open blood circulation system (in a hemocoel); their blood does not circulate in discrete blood vessels under relatively high pressure, as in vertebrates.

Another significant difference with insects is that their cells form uric acid as the nitrogenous excretory material. This acid, as its potassium salt, is removed from the blood at the Malpighian tubules. These closed tubules lie in the hemocoel and empty into the alimentary canal at the junction of mid and hind gut (Figure 11.36).

Look in the diagram at the position and role of the Malpighian tubules.

The upper part of the tubule secretes potassium urate into the lumen of the tubule. Carbon dioxide and water diffuse in, too. In the lower tubule, these contents react together forming uric acid, potassium hydrogencarbonate and water. Now, uric acid passes into the gut, and hydrogencarbonate and water are transported back into the blood. In the rectum, much water is withdrawn from the faeces and is transported back into the hemocoel; the uric acid becomes solid pellets that leave the body with the faeces.

Water loss from an insect's body is also reduced by the presence of a waxy cuticle over the external surface of the exoskeleton. A system of tubules (tracheae) pipe oxygen to all the body tissues; these tracheae have openings known as spiracles. Valves control movement of gases through the spiracles and further reduce loss of water vapour during gaseous exchange.

■ **Figure 11.36**
Excretion in insects

the position of Malpighian tubules in an insect

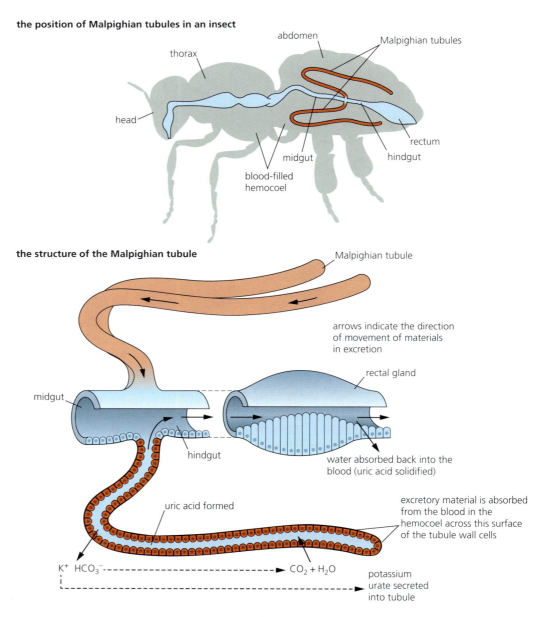

the structure of the Malpighian tubule

Excretion in mammals is adapted to contrasting environments

In mammals, we have seen that urea is expelled from the body in solution, so some water loss in excretion is inevitable. However, where necessary, mammals are able to form urine that is more concentrated than the blood plasma, maximizing the amount of water retained and enabling them to colonize dry land. Nephrons achieve this function due to the loop of Henle.

In fact, the lengths of both the loop of Henle and the collecting ducts, and the general thickness of the medulla region of the kidneys, increases progressively in mammals that are best adapted to drier habitats (Figure 11.37).

■ **Figure 11.37**
Water conservation in mammals – a comparative study

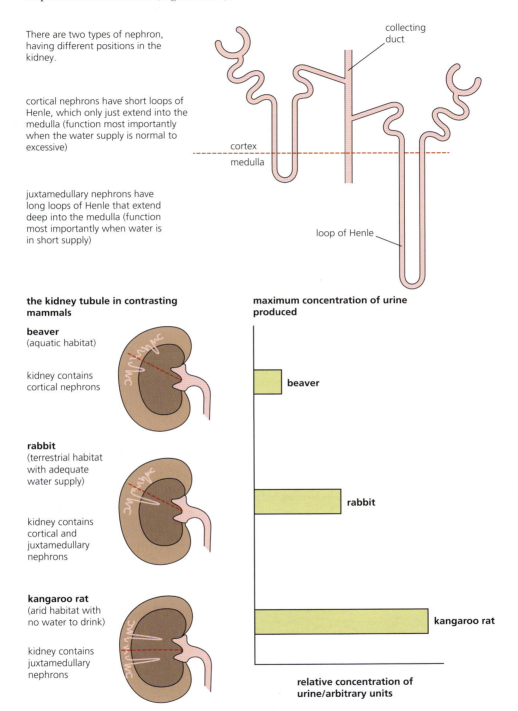

There are two types of nephron, having different positions in the kidney.

cortical nephrons have short loops of Henle, which only just extend into the medulla (function most importantly when the water supply is normal to excessive)

juxtamedullary nephrons have long loops of Henle that extend deep into the medulla (function most importantly when water is in short supply)

collecting duct

cortex
medulla

loop of Henle

the kidney tubule in contrasting mammals

beaver
(aquatic habitat)

kidney contains cortical nephrons

rabbit
(terrestrial habitat with adequate water supply)

kidney contains cortical and juxtamedullary nephrons

kangaroo rat
(arid habitat with no water to drink)

kidney contains juxtamedullary nephrons

maximum concentration of urine produced

beaver

rabbit

kangaroo rat

relative concentration of urine/arbitrary units

■ Nitrogen excretion – habitat and evolutionary history

The three common products of nitrogenous excretion in animals are ammonia, urea and uric acid. The excretion of each in relation to the quantity of water required for safe disposal is markedly different, as shown in Table 11.8. Excretory product can be correlated with different groups of animals (and their typical environments).

■ **Table 11.8**
Nitrogenous excretion – animal groups and habitats

Compound (structural formula)	Volume of water for safe disposal of 1 g	Animal groups	Typical habitat
ammonia	$500\,cm^3$	bony fish and aquatic non-vertebrates	fresh water
urea	$50\,cm^3$	most vertebrates with the exception of bony fish	terrestrial
uric acid	$1\,cm^3$	birds, reptiles and most terrestrial arthropods	dry or arid conditions

Nature of Science

Curiosity about particular phenomena

■ Investigation of water loss in desert animals

Animals of arid or desert regions clearly survive with little or no liquid water in their diets. This group of animals includes the kangaroo rat (*Dipodomys* species), which lives in hot dry deserts, but hides in a burrow during daylight. It is able to survive without access to drinking water.

Physiologists have investigated the metabolism, diet, and breathing and excretory losses of water in this, and other, animals. They noted that extremely concentrated urine is formed (see Figure 11.37) and no sweat is produced. Typical results from the measurements and estimates of water gain and water loss over a 28-day period are shown in Figure 11.38. This data establishes why it is that survival is possible for a well-adapted animal. Several species are so adapted. In fact, deserts have a remarkable fauna (and flora) that is only visible at certain times.

the water relations of the kangaroo rat, a desert-adapted mammal

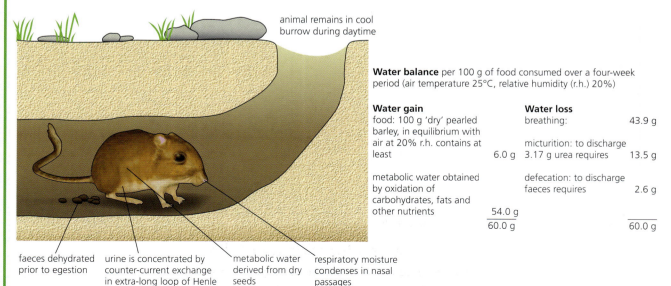

animal remains in cool burrow during daytime

Water balance per 100 g of food consumed over a four-week period (air temperature 25°C, relative humidity (r.h.) 20%)

Water gain		Water loss	
food: 100 g 'dry' pearled barley, in equilibrium with air at 20% r.h. contains at least	6.0 g	breathing:	43.9 g
metabolic water obtained by oxidation of carbohydrates, fats and other nutrients	54.0 g	micturition: to discharge 3.17 g urea requires	13.5 g
		defecation: to discharge faeces requires	2.6 g
	60.0 g		60.0 g

faeces dehydrated prior to egestion

urine is concentrated by counter-current exchange in extra-long loop of Henle

metabolic water derived from dry seeds

respiratory moisture condenses in nasal passages

■ **Figure 11.38** Water loss in a desert animal

11.4 Sexual reproduction *– sexual reproduction involves the development and fusion of haploid gametes*

20 Suggest what evidence there is that, in nature, only some individuals of any species succeed in reproducing.

In sexual reproduction, sex cells (**gametes**) fuse to form a zygote, which then grows into a new individual. The gametes are produced in paired glands called gonads – male gametes, or **sperms**, are formed in **testes** (Figure 6.60, page 307); female gametes, ova or **oocytes** (singular, ovum or **oocyte**) are formed in **ovaries** (Figure 6.61, page 308).

The process of gamete formation, known as **gametogenesis**, involves not only mitosis but also meiotic division, thereby halving the normal chromosome number. That is, gametes are haploid. **Fertilization** restores the diploid number of chromosomes.

The structure of the gonads and the steps of gamete formation are what we consider first.

■ Gametogenesis

In gametogenesis, many gametes are produced, although relatively few of them are ever used in reproduction. The processes of gamete formation in testes and ovaries have a common sequence of phases.

First, there is a **multiplication phase** in which the gamete mother cells divide by mitotic cell division (Figure 1.58, page 54). This division is then repeated to produce many cells with the potential to become gametes.

Secondly, each developing sex cell undergoes a **growth phase**.

Third and finally, comes the **maturation phase**. This involves meiosis and results in the formation of the haploid gametes. The products of meiosis I are secondary **spermatocytes** and secondary **oocytes**, and the products of meiosis II are **spermatids** and **ova**. The steps of meiosis are described on page 414, Figure 10.1.

These phases in sperm and ovum production are summarized in Figure 11.39; the differences between gametogenesis in testis and ovary are listed in Table 11.9 (page 489).

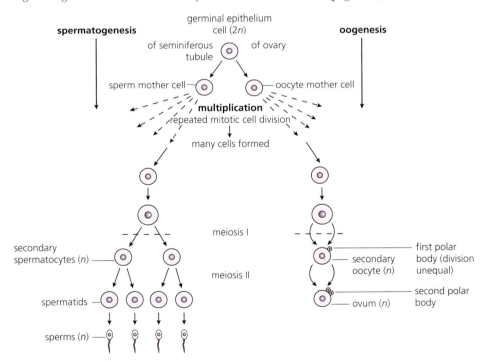

■ **Figure 11.39** The phases and changes during gametogenesis

■ The structure and functioning of the testis

In the human fetus, testes develop high on the posterior abdominal wall and migrate to the scrotum in about the seventh month of pregnancy. In the scrotum, the paired testes are held at a temperature 2–3°C below body temperature after birth. This lower temperature is eventually necessary for sperm production – testes that fail to migrate do not later produce sperms.

Spermatogenesis begins in the testes at puberty and continues throughout life. Each testis consists of many **seminiferous tubules**. These are lined by germinal epithelial cells which divide repeatedly. Tubules drain into a system of channels leading to the epididymis, a much coiled tube which leads to the **sperm duct**. Between the individual seminiferous tubules is connective tissue containing blood capillaries, together with groups of **interstitial cells**. These latter cells are hormone secreting (the testis is also an endocrine gland). Testes are suspended by a spermatic cord containing the sperm duct and blood vessels (Figure 11.40).

In the seminiferous tubules, the **germinal epithelial cells** are attached to the basement membrane, along with the **nutritive cells**. Cells from the subsequent steps of sperm production (spermatogonia, primary spermatocytes, secondary spermatocytes and spermatids) occur lodged in the surface of the Sertoli cells (nutritive cells) on which they are dependent until they mature into spermatozoa (sperms) (Figures 11.41 and 11.42).

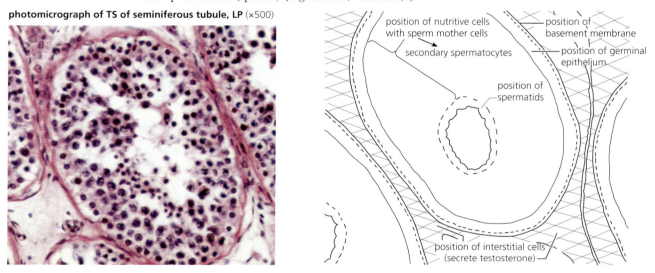

photomicrograph of TS of seminiferous tubule, LP (×500)

position of nutritive cells with sperm mother cells

position of basement membrane

secondary spermatocytes

position of germinal epithelium

position of spermatids

position of interstitial cells (secrete testosterone)

■ **Figure 11.40** Photomicrograph of testis tissue in section, and interpretive drawing

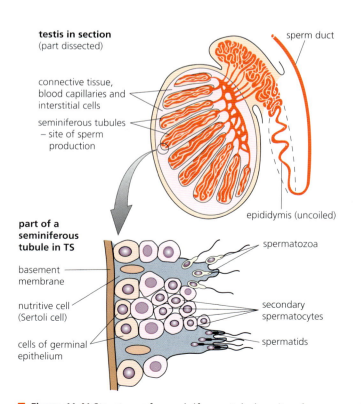

testis in section (part dissected)

sperm duct

connective tissue, blood capillaries and interstitial cells

seminiferous tubules – site of sperm production

epididymis (uncoiled)

part of a seminiferous tubule in TS

spermatozoa

basement membrane

nutritive cell (Sertoli cell)

secondary spermatocytes

cells of germinal epithelium

spermatids

■ **Figure 11.41** Structure of a seminiferous tubule – site of sperm production

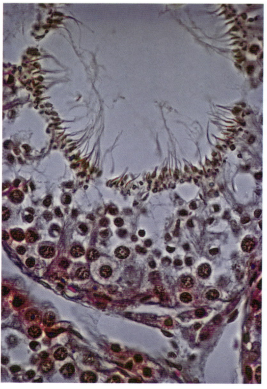

photomicrograph of TS of seminiferous tubule, HP (×1000)

You need to be able to draw a diagram of a section through part of a seminiferous tubule, showing the stages of gametogenesis. You must be able to annotate your diagram with details of the structures present and their functions.

The mature sperms, and the production of semen

The sperms are immobile when first formed. From the seminiferous tubules, they pass into the much coiled epididymis, where maturation is completed and storage occurs. During an ejaculation, the sperms are moved by waves of contraction in the muscular walls of the sperm ducts. Sperms are transported in a nutritive fluid that is secreted by glands, mainly the seminal vesicles and prostate gland. These glands add their secretions just at the point where the sperm ducts join with the urethra, below the base of the penis (Figure 6.60, page 307). As well as providing nutrients for the sperms, semen is a slightly alkaline fluid, the significance of which we will return to later. During an ejaculation, the sphincter muscle at the base of the bladder is closed.

21 List the structures you will annotate in your diagram of a mature sperm and **outline** the roles you will amplify.

■ **Figure 11.42**
The structure of a mature spermatozoon

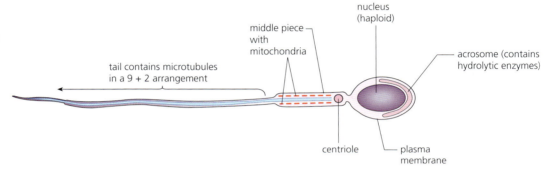

You need to be able to draw a diagram of a mature sperm and to annotate it with details of structures present and their functions.

The roles of hormones in spermatogenesis

The onset of puberty is triggered by a part of the brain called the hypothalamus. Here, production and secretion of a releasing hormone cause the nearby **pituitary gland** (the master endocrine gland) to produce and release into the blood circulation two hormones, known as **follicle-stimulating hormone (FSH)** and **luteinizing hormone (LH)**. These hormones are so named because their roles in sexual development in humans were first discovered in the female reproductive system. However, FSH and LH operate in males also (Figure 11.43).

So, at puberty, a hormone from the hypothalamus triggers the secretion of FSH and LH by the anterior lobe of the pituitary. In the male, the first effect of FSH is to initiate sperm production in the testes. LH stimulates the endocrine cells of the testes to secrete testosterone. Subsequently, testosterone and FSH together maintain continued sperm production and growth of the essential Sertoli cells that support sperms with nutrients as they grow and develop in the testes.

Subsequently, secretion of testosterone continues throughout life. Over-secretion of testosterone is regulated by **negative feedback control**, as an excessively high level of testosterone in the blood inhibits secretion of LH. Only when the concentration of LH in the blood has fallen significantly, will testosterone production recommence. (Similarly, over-activity of the nutritive cells inhibits secretion of FSH for a while.)

■ **Figure 11.43**
Hormone regulation
of sperm production

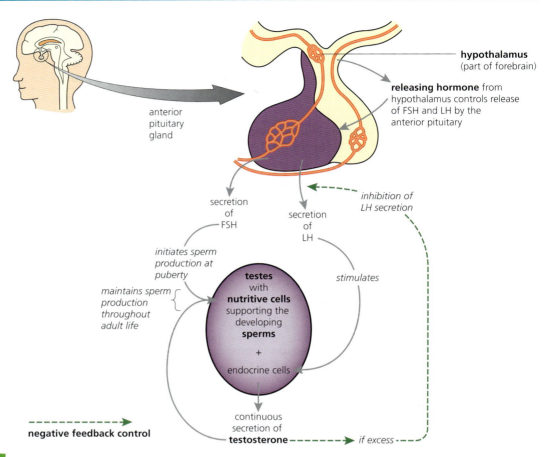

anterior
pituitary
gland

hypothalamus
(part of forebrain)

releasing hormone from
hypothalamus controls release
of FSH and LH by the
anterior pituitary

secretion
of
FSH

secretion
of
LH

*inhibition of
LH secretion*

*initiates sperm
production at
puberty*

stimulates

*maintains sperm
production
throughout
adult life*

testes
with
nutritive cells
supporting the
developing
sperms

+

endocrine cells

continuous
secretion of
testosterone

if excess

- - - - - - - - - ->
negative feedback control

Nature of Science

Assessing risks and benefits associated with scientific research

■ Risks to human male fertility from the female contraceptive pill

The oral contraceptive pill ('the pill') contains two hormones that are chemically very similar to estrogen and progesterone. The steps to contraception require the woman to take one pill at the same time each day, for 21 days. Then an inert pill (a placebo) is taken for seven days (or no pill at all), and during this time menstruation occurs. The effect of the pill is to:

- stop the ovaries releasing an egg each month (ovulation)
- thicken the mucus in the cervix, making it difficult for sperm to reach the egg
- make the lining of the uterus thinner, so it is less likely to accept a fertilized egg.

The pill works by depressing the release of FSH and LH from the pituitary gland (without completely stopping their release). This restricts the growth of follicles in the ovaries, and so a secondary oocyte does not grow and is not released.

Now, as a consequence of the following factors, humans are experiencing increased exposure to estrogens:

- the widespread use of the pill
- the way the kidneys constantly remove hormones from the blood and transfer them to the urine
- the discharge of treated sewage effluent into rivers
- rivers being the source of much of our drinking water.

At the same time as this increasing exposure, there is evidence of decreasing fertility in human males.

How may these be linked? Look again at Figure 11.43. Can you see a link between changing male fertility and environmental exposure to estrogens? What risk assessments were missed or inadequate in the policy of widespread availability of the pill? What steps in the original research may have been overlooked?

■ The structure and functioning of the ovaries

In the female, the ovaries are about 3 cm long and 1.5 cm thick. These paired structures are suspended by ligaments near the base of the abdominal cavity. As well as producing egg cells, the ovaries are also endocrine glands. They secrete the female sex hormones **estrogen** and **progesterone**. A pair of oviducts extend from the uterus and open as funnels close to the ovaries. The oviducts transport oocytes and are the site of fertilization. In the event of fertilization, development of the fetus will occur in the uterus.

The steps of oogenesis occur in the ovary. Ovulation, the process by which an egg is released to the oviduct, occurs at the secondary oocyte stage. Development of a secondary oocyte into an ovum is triggered in the oviduct, if fertilization occurs. Consequently, a thin section through a mature ovary, examined by light microscopy, shows the developing oocytes at differing stages (Figure 11.44).

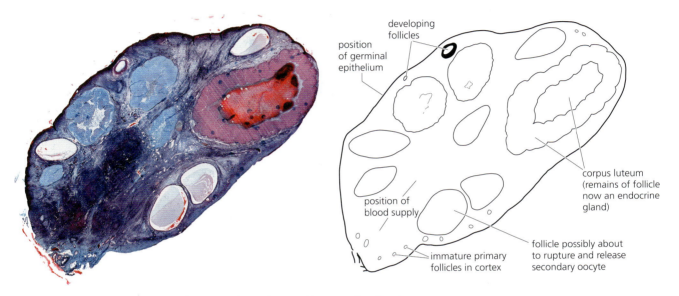

■ **Figure 11.44** Photomicrograph of an ovary in section, and interpretive drawing

The structure of the ovary and the steps of oogenesis

Oogenesis begins in the ovaries of the fetus before birth, but the final development of oocytes is only completed in adult life (Figure 11.45). The germinal epithelium, which lines the outer surface of the ovary, divides by mitotic cell division (Figure 1.58, page 54) to form numerous oogonia. These cells migrate into the connective tissue of the ovary, where they grow and enlarge to form oocytes. Each oocyte becomes surrounded by layers of follicle cells, and the whole structure is called a **primary follicle**.

By mid-pregnancy, production of oogonia in the fetus ceases – by this stage there are several million in each ovary. Very many degenerate, a process that continues throughout life. At the onset of puberty, the number of primary oocytes remaining is about 250 000. Less than 1% of these follicles will complete their development; the remainder never become secondary oocytes or ova.

Between puberty, at about 11 years, and the cessation of ovulation at menopause, typically at about 55 years of age, primary follicles begin to develop further. Several start growth each month, but usually only one matures. Development involves progressive enlargement and, at the same time, the follicles move to the outer part of the ovary. The primary follicle then undergoes **meiosis I** (pages 413–416), but the cytoplasmic division that follows is unequal, forming a tiny polar body and a **secondary oocyte** (Figure 11.46). The second meiotic division, **meiosis II** (pages 416–417), then begins, but it does not go to completion. In this condition the **egg cell** (it is still a secondary oocyte) is released from the ovary (ovulation), by rupture of the follicle wall (Figure 11.45).

summary of changes from oogonium to ovum
– steps in the growth and maturation phases of gametogenesis in the ovary

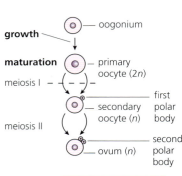

growth — oogonium

maturation — primary oocyte (2n)

meiosis I

secondary oocyte (n) — first polar body

meiosis II

ovum (n) — second polar body

Secondary oocyte **begins** meiosis II but this does not complete until sperm nucleus penetrates cytoplasm of oocyte.

diagrammatic representation of the sequence of events in the formation of a secondary oocyte for release and the subsequent changes in the ovary

each menstrual cycle a few primary follicles start to develop – **meiosis I** is completed, forming a secondary oocyte and the first polar body (unequal division of cytoplasm) which fails to develop further

primary follicle (**oocyte** surrounded by a layer of **follicle cells**)

oogonia grow to become oocytes

germinal epithelium – oogonia formed by **mitosis** (in uterus)

corpus luteum degenerates quickly if fertilization does not occur

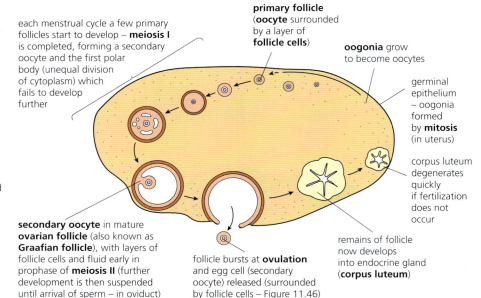

secondary oocyte in mature **ovarian follicle** (also known as **Graafian follicle**), with layers of follicle cells and fluid early in prophase of **meiosis II** (further development is then suspended until arrival of sperm – in oviduct)

follicle bursts at **ovulation** and egg cell (secondary oocyte) released (surrounded by follicle cells – Figure 11.46)

remains of follicle now develops into endocrine gland (**corpus luteum**)

■ Figure 11.45 The ovary, and stages in oogenesis

■ Figure 11.46 The structure of a mature secondary oocyte

22 List the structures you will annotate in your secondary oocyte diagram and **outline** the roles you will amplify.

23 Distinguish between the roles of FSH in sperm and ovum production.

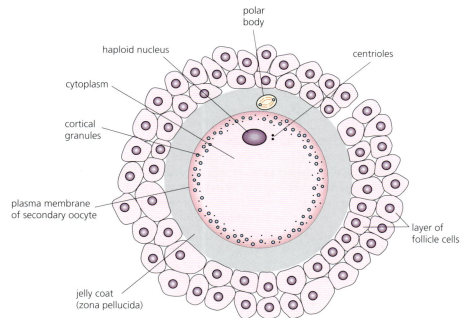

polar body

haploid nucleus

centrioles

cytoplasm

cortical granules

plasma membrane of secondary oocyte

layer of follicle cells

jelly coat (zona pellucida)

You need to be able to draw a diagram of a section through an ovary to show the stages of gametogenesis, and to annotate it with details of the structures present and their functions. You also need to be able to draw a diagram of a mature secondary oocyte, and to annotate it with details of structures present and their functions.

Ovulation occurs from one of the two ovaries about once every 28 days. Meanwhile, the remains of the primary follicle immediately develops into the yellow body, the **corpus luteum**. This is an additional but temporary endocrine gland, with a role to play if fertilization occurs (see below).

Gametogenesis in testis and ovary are compared in Table 11.9.

■ **Table 11.9**
Spermatogenesis and
oogenesis compared

Spermatogenesis	Oogenesis
Spermatogonia are formed from the time of puberty, throughout adult life.	Oogonia are formed in the embryonic ovaries, long before birth.
All spermatogonia develop into sperms, nurtured by the nutritive cells of the seminiferous tubules of the testes.	Oogonia become surrounded by follicle cells, forming tiny primary follicles, and remain dormant within ovary cortex. Most fail to develop further – they degenerate.
Millions of sperms are formed daily.	From puberty, a few primary oocytes undergo meiosis I to become secondary oocytes each month. Only one of these secondary oocytes, surrounded by a much enlarged follicle, forms a Graafian follicle – the others degenerate.
Four sperms are formed from each spermatogonium.	One ovum is formed from each oogonium (the polar bodies degenerate, too).
Sperms are released from the body by ejaculation.	The Graafian follicle releases a secondary oocyte into the oviduct at ovulation.
Meiosis I and II go to completion during sperm production.	Meiosis II reaches prophase and then stops until a male nucleus enters the secondary oocyte, triggering completion of meiosis II.
Sperms are small, mobile gametes.	A fertilized ovum is non-motile and becomes lodged in the endometrium of the uterus, where cell divisions lead to embryo formation.

■ Fertilization

Internal versus external fertilization

Motile male gametes, such as the sperms shown in Figure 11.42, require a watery medium in which to move. In aquatic animals, such as fish and amphibians, the male and female gametes are shed into the water and **fertilization occurs externally**.

In organisms that have colonized the land, such as mammals and birds (and the flowering plants), **internal fertilization** in an environment suitable for transport of the male gamete is necessary.

Fertilization in mammals

Internal fertilization in most terrestrial animals involves the male gametes being introduced into the female's reproductive organs during sexual intercourse. The erect penis is placed in the vagina. In mammals, internal fertilization occurs in the upper part of the oviduct. The sperms are introduced and semen may be ejaculated (3–5 cm^3 in humans) close to the cervix. Typically, more than one hundred million sperms are deposited. The pH of the vagina is quite acid, but the alkaline secretion of the prostate gland, a component of the semen, helps to neutralize the acidity and provides an environment in which sperms can survive.

Waves of contractions in the muscular walls of the uterus and the oviducts assist in drawing semen from the cervix to the site of fertilization. In this way, a few thousand of the sperms reach the upper uterus and swim up the oviducts.

How polyspermy is prevented

One or more of the few sperms that reach a secondary oocyte pass between the follicle cells surrounding the oocyte. The entry of more than one sperm into the oocyte is known as **polyspermy**. Fertilization involves a mechanism that *prevents* polyspermy.

Look at the steps to fertilization in Figure 11.47, now.

First, the coat that surrounds the oocyte, which is made of glycoprotein and is called the zona pellucida, has to be crossed. At the tip of the sperm is a membrane-bound sac of hydrolytic enzymes, called the **acrosome**. In contact with the zona pellucida, these enzymes are released and digest a pathway for the sperm to the oocyte membrane – a process known as the **acrosome reaction**.

Now, the plasma membrane around the head of a sperm fuses with the plasma membrane of the oocyte. The male nucleus enters the oocyte. As this happens, granules in the outer cytoplasm of the oocyte release their contents outside the oocyte by exocytosis. This is known as the **cortical reaction**. The result is that the oocyte plasma membrane cannot be crossed by another sperm. In this way, the possibility of fusion of more than one male nucleus with the oocyte nucleus is prevented.

As the sperm nucleus enters the oocyte, completion of meiosis II is triggered and the second polar body is released. The male and female haploid nuclei come together to form the diploid nucleus of the zygote. Fertilization is completed.

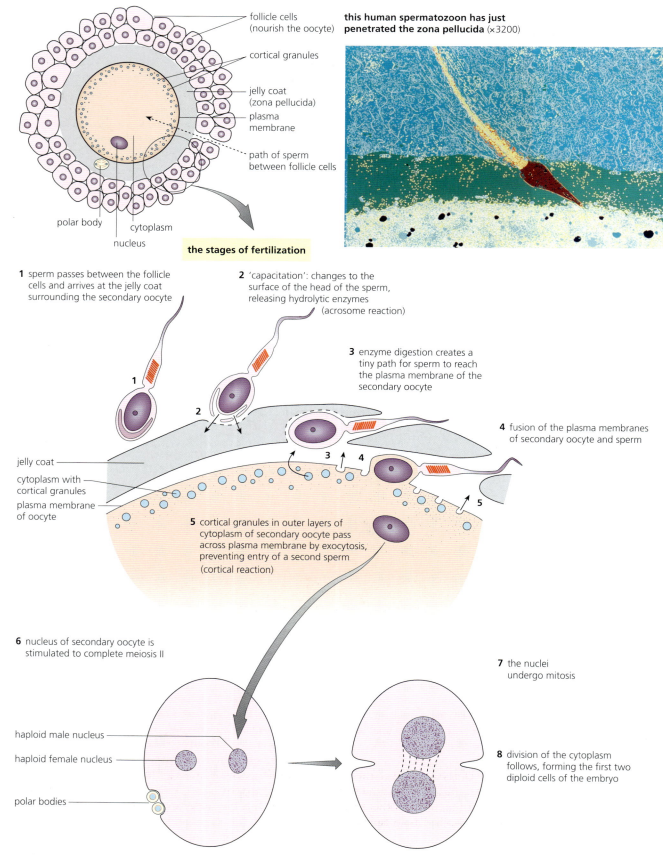

follicle cells
(nourish the oocyte)

cortical granules

jelly coat
(zona pellucida)

plasma
membrane

path of sperm
between follicle cells

polar body

cytoplasm

nucleus

this human spermatozoon has just
penetrated the zona pellucida (×3200)

the stages of fertilization

1 sperm passes between the follicle
cells and arrives at the jelly coat
surrounding the secondary oocyte

2 'capacitation': changes to the
surface of the head of the sperm,
releasing hydrolytic enzymes
(acrosome reaction)

3 enzyme digestion creates a
tiny path for sperm to reach
the plasma membrane of the
secondary oocyte

4 fusion of the plasma membranes
of secondary oocyte and sperm

jelly coat

cytoplasm with
cortical granules

plasma membrane
of oocyte

5 cortical granules in outer layers of
cytoplasm of secondary oocyte pass
across plasma membrane by exocytosis,
preventing entry of a second sperm
(cortical reaction)

6 nucleus of secondary oocyte is
stimulated to complete meiosis II

7 the nuclei
undergo mitosis

haploid male nucleus

haploid female nucleus

polar bodies

8 division of the cytoplasm
follows, forming the first two
diploid cells of the embryo

■ **Figure 11.47** Fertilization of a human secondary oocyte

Early development and implantation

Fertilization occurs in the upper oviduct. As the zygote is transported down the oviduct by ciliary action, mitosis and cell division commence. The process of the division of the zygote into a mass of daughter cells is known as **cleavage**. This is the first stage in the growth and development of a new individual. The embryo does not increase in mass at this stage. By the time the embryo has reached the uterus, it is a solid ball of tiny cells called **blastomeres**, the whole no larger than the fertilized egg cell from which it has been formed. Division continues and the blastomeres organize themselves into a fluid-filled ball, the blastocyst (Figure 11.48).

In humans, by day 7, the blastocyst consists of about 100 cells. It now starts to become embedded in the endometrium, a process known as **implantation**. Implantation takes from day 7 to day 14, approximately. At this stage, some of the blastomeres appear grouped as the **inner cell mass** and these cells will eventually become the fetus. Once implanted, the embryo starts to receive nutrients directly from the endometrium of the uterus wall (Figure 11.50).

■ **Figure 11.48**
The site of fertilization and early stages of development

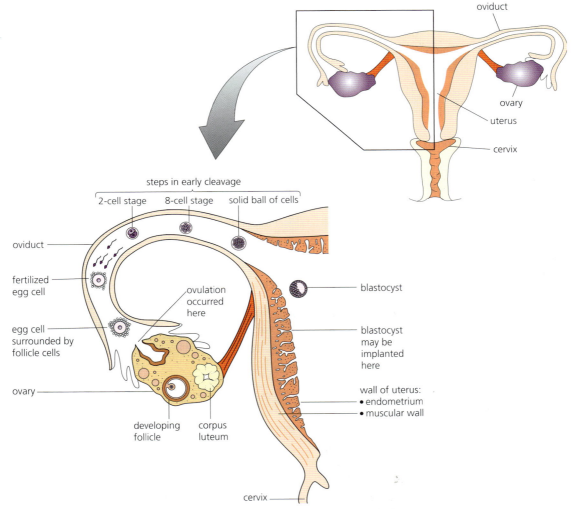

Gestation

The period of development in the mother's body, lasting from conception to birth, is known as **gestation** (taking 38 weeks in humans). The rate of growth and development during gestation is much greater than in any other stage of life. In the first two months of gestation, the developing offspring is described as an **embryo**.

Comparative aspects of gestation and development of young at birth

Data on the length of gestation in various mammals is given in Table 11.10. *When this data is plotted, to what extent is there a positive correlation between adult size and the length of gestation evident? Does the human fit into this relationship?*

■ **Table 11.10** Size of animal and their gestation period

	Length (feet)	Gestation (days)
rabbit	1	32
large squirrel	3	40
polar bear	8	225
horse	11	336
elephant	22	645

The extent of development at birth is a complex factor to quantify. For example, there is an ecological dimension and two strategies for survival can be identified (alticial and precocial, Table 11.11). In fact, these are two extremes – most organisms fit somewhere in between. The human is a case in point – not all human communities produce infants only very occasionally.

■ **Table 11.11** State of young at birth – alternative scenarios

'Alticial' offspring	'Precocial' offspring
e.g. mice	e.g. gorilla
born helpless, after a short gestation	alert at birth (eyes open) after a longer gestation
brain under-developed at birth – subsequently grows to 7.5 times its birth size	brain able to control the limbs immediately – subsequently grows to 2.5 times the birth size
small bodied, small brained and fast feeding	large bodied, big brained and slow feeding
tend to produce a large number of offspring at a rapid rate, but investing little in each infant (known as ***r*-species** to ecologists)	tend to produce few or one infant only occasionally, but investing heavily in each infant (known as ***K*-species** to ecologists)

■ From zygote to embryo to fetus in humans

From early in the development of the embryo, this tiny and delicate structure is contained, supported and protected by a membranous, fluid-filled sac. It is the outer layers of the tissues of the embryo that grow and give rise to the membranes and that also form the **placenta** (see below). By the end of two months' development, the beginnings of the principal adult organs can be detected within the embryo and the placenta is operational. During the rest of gestation, the developing offspring is called a **fetus**.

■ **Figure 11.49** Human embryo at the six-week stage

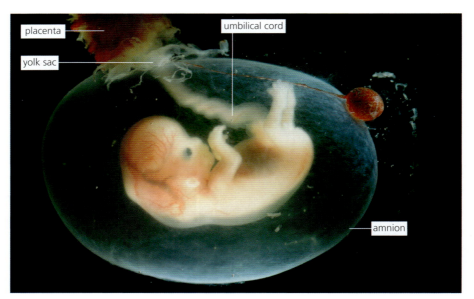

The placenta – structure and function

The **placenta** is a disc-shaped structure composed of maternal (endometrial) and fetal membrane tissues. Here the maternal and fetal blood circulations are brought very close together over a huge surface area, but they do not mix. Placenta and fetus are connected by arteries and a vein in the umbilical cord (Figure 11.50).

Exchange in the placenta is by diffusion and active transport. Movements across the placenta involve:

- **respiratory gases**, which are exchanged; oxygen diffuses across the placenta from the maternal hemoglobin to the fetal hemoglobin, and carbon dioxide diffuses in the opposite direction
- **water**, which crosses the placenta by osmosis; **glucose**, which crosses by facilitated diffusion; and **ions** and **amino acids**, which are transported actively
- **excretory products**, including urea, leaving the fetus
- **antibodies** present in the mother's blood, which freely cross the placenta, so the fetus is initially protected from the same diseases as the mother (**passive immunity**, page 450).

The placenta is a barrier to bacteria, although some viruses can cross it.

■ **Figure 11.50**
The placenta – site of exchange between maternal and fetal circulations

24 Explain why it is so important that the blood of mother and offspring do not mix together in the placenta.

25 List the structural features of the placenta that contribute to efficient exchange and **explain** why each is important.

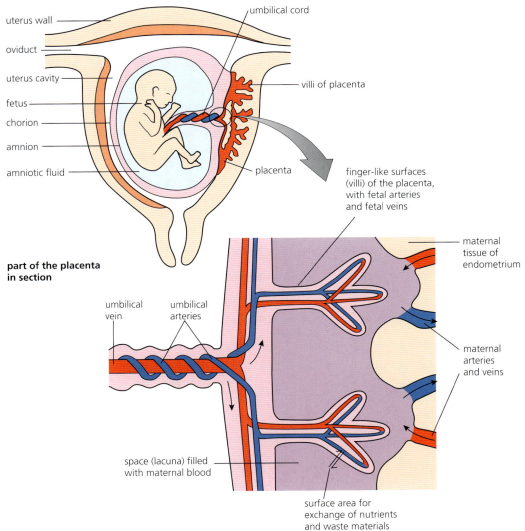

human fetus with placenta, about 10 weeks

uterus wall
oviduct
uterus cavity
fetus
chorion
amnion
amniotic fluid
umbilical cord
villi of placenta
placenta

finger-like surfaces (villi) of the placenta, with fetal arteries and fetal veins

part of the placenta in section

umbilical vein
umbilical arteries
space (lacuna) filled with maternal blood

maternal tissue of endometrium
maternal arteries and veins
surface area for exchange of nutrients and waste materials

The placenta as endocrine gland

The placenta is also an endocrine gland, initially producing an additional sex hormone known as **human chorionic gonadotrophin (HCG)**. HCG appears in the urine from about seven days after conception. We have already noted that it is the presence of HCG in a sample of urine that is detected using monoclonal antibodies in a pregnancy-testing kit (Figure 11.11, page 455).

HCG is initially secreted by the cells of the blastocyst, but later it comes entirely from the placenta. The role of HCG is to maintain the corpus luteum as an endocrine gland (secreting progesterone) for the first 16 weeks of pregnancy.

When the corpus luteum eventually does break down, the placenta itself secretes estrogen and progesterone (Figure 11.51). Without maintenance of these hormone levels, conditions favourable to a fetus are not maintained in the uterus and a spontaneous abortion results.

■ The process of birth and its hormonal control

Immediately before birth, the level of **progesterone** declines sharply. As a result, progesterone-driven inhibition of contraction of the muscle of the uterus wall is removed.

At the same time, the posterior pituitary begins to release a hormone, **oxytocin** (Figure 11.51). This relaxes the elastic fibres that join the bones of the pelvic girdle, especially at the front, and thus aids dilation of the cervix for the head (the widest part of the offspring) to

■ **Figure 11.51**
Blood levels of sex hormones during gestation

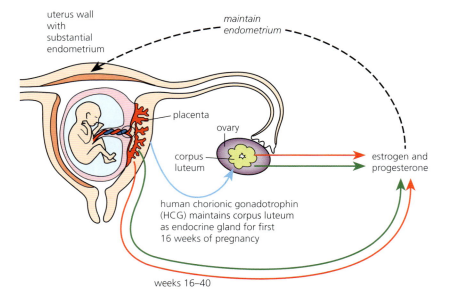

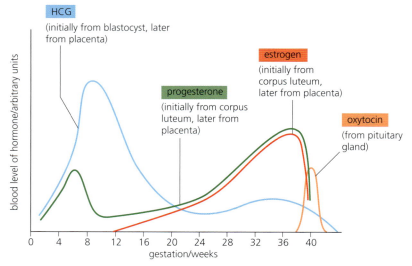

pass through. Oxytocin also stimulates rhythmic contractions of the muscles of the uterus wall. Subsequently, control of contractions during birth occurs via a **positive feedback loop** (Figure 11.52). The resulting powerful, intermittent waves of contraction of the muscles of the uterus wall start at the top of the uterus and move towards the cervix. Progressively during this process (known as labour), the rate and strength of the contractions increase, until they expel the offspring.

Finally, less powerful uterine contractions separate the placenta from the endometrium, and cause the discharge of the placenta and remains of the umbilicus as the afterbirth.

■ **Figure 11.52** The positive feedback loop in the control of labour

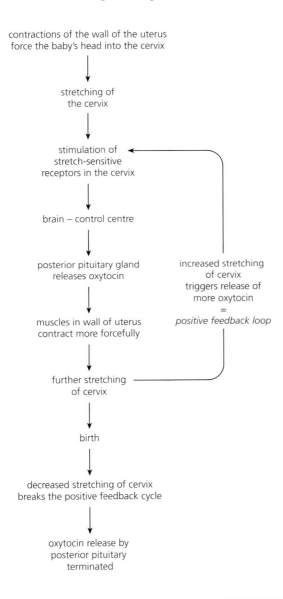

26 **Suggest** why the immediate production of HCG by the embryo while it still consists of relatively few cells is significant in a successful outcome to gestation.

27 **Distinguish** between negative and positive feedback processes.

■ *Examination questions – a selection*

Questions 1–5 are taken from IB Diploma biology papers.

Q1 Which of the following events form the basis of immunity upon which the principle of vaccination is based?

	Clonal selection	Production of memory cell	Production of monoclonal antibodies	Challenge and response
A	No	Yes	Yes	Yes
B	No	Yes	No	Yes
C	Yes	Yes	Yes	Yes
D	Yes	Yes	No	Yes

Higher Level Paper 1, Time Zone 2, May 13, Q38

Q2 The images below show muscle tissue.

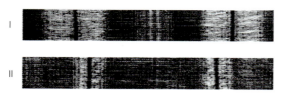

Which image shows contracted muscle tissue?
A I, because the dark band is narrower.
B II, because the Z lines are closer together.
C II, because there is less overlap between actin and myosin.
D I, because the dark bands are darker.

Higher Level Paper 1, Time Zone 2, May 13, Q27

Q3 Which cells activate helper T-cells by antigen presentation?
A B-cells
B Bacteria
C Macrophages
D Plasma cells

Higher Level Paper 1, Time Zone 0, Nov 10, Q 37

Q4 The diagram below shows a small portion of the tissue in a transverse section of a testis.

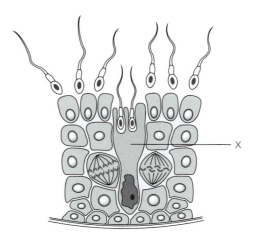

a Identify the structure labelled with an X. (1)
b Outline the function of this cell. (1)

Higher Level Paper 2, Time Zone 1, May 12 Q4bi and 4bii

Q5 Blood is a liquid tissue containing glucose, urea, plasma proteins and other components.
List the other components of blood. (5)

Higher Level Paper 2, Time Zone 1, May 10, Q4a

Questions 6–10 cover other syllabus issues in this chapter.

Q6 a Distinguish between passive and active immunity (4)
b Outline the sequence of steps by which the human body may acquire naturally active immunity to a viral infection. (6)

Q7 Muscles contract when the fibres shorten. This shortening is brought about when myosin and actin filaments of the sarcomeres slide past each other. Explain by means of annotated drawings how ATP and calcium ions (Ca^{2+}) enable the myosin and actin filaments to interact to shorten a muscle. (10)

Q8 **a** Outline the principle of a countercurrent
mechanism. (4)
    b Describe the roles of:
      i the collecting tube
      ii the descending and ascending limbs of the
loop in water conservation by the loop of
Henle. (8)
    c Distinguish the different types of nitrogenous
waste in animals. (3)

Q9 **a** Identify the stage in gestation at which the
placenta forms, and the tissues involved. (4)
    b Explain why is it essential that maternal and
fetal blood circulations do not mix in the
placenta. (4)
    c Describe what substances are required by
the fetus and how each is transferred to the
fetal blood at the placenta. (6)
    d Outline the additional role of the placenta
as an endocrine gland. (3)

Q10 State the differences in composition of blood
proteins, glucose and urea found in:
    a blood before entering the glomerulus
    b glomerular filtrate in the Bowman's capsule
    c filtrate inside the descending limb of the loop
of Henle. (6)

Answers to self-assessment questions in Chapters 1–11

1 Cell biology

page 2

1   The processes characteristic of living things are:

- transfer of energy (respiration)
- feeding or nutrition
- metabolism
- excretion
- movement and locomotion
- responsiveness or sensitivity
- reproduction
- growth and development

2   a $1\,mm = 1000\,\mu m$

Since $^{1000}/_{100} = 10$, 10 cells of $100\,\mu m$ will fit along a $1\,mm$ line.

   b   The drawing of *E. coli* is $64\,mm$ in length (actual length is $2.0\,\mu m$).

Magnification = size of image ÷ size of specimen

$M = 64 \times 1000 \div 2\,cm = \times 32\,000$

page 11

3   Size of image is $115\,mm$ approx. Length of scale bar corresponding to $0.10\,mm$ is $25\,mm$.

Observed length = $115 \div 25 = 4.6 \times 0.1\,mm = 0.46\,mm$ or approximately $460\,\mu m$

4   With ×6 eyepiece and ×10 objective lens magnification = $6 \times 10 = \times 60$

page 12

5   a

Dimensions/ mm	Surface area/ mm²	Volume/ mm³	SA:V ratio
1 × 1 × 1	6	1	6:1 = 6.0
2 × 2 × 2	24	8	24:8 = 3.0
4 × 4 × 4	96	64	96:64 = 1.5
6 × 6 × 6	216	216	216:216 = 1.0

   b   Small cells and organisms have a large surface area to volume ratio (that is, the surface area available for diffusion). As the cell increases in size (volume), the surface area to volume ratio (SA:V) decreases.

   c   As cell size increases, the efficiency of diffusion for the removal of waste products decreases. The rate of production of heat and waste products and the rate of oxygen consumption is a function of volume. The rate of exchange of resources and waste products is a function of surface area.

page 17

6   Important points to consider about use of ES cells.

In support	In opposition
Reduced suffering of patients with wide variety of conditions	Production of ES cells involves the death of early-stage embryos
ES cells retain the ability to divide and differentiate into many tissue types	There is a danger of rejection and the immune system must be suppressed, leading to the appearance of new diseases
Replacement of diseased or dysfunctional cells or tissue is possible	Cells can overgrow and develop into cancerous tumours

page 19

7   The resolution is the microscope's ability to distinguish between two small objects which are very close together and it is determined by the wavelength of light (or electron beam). Magnification is the number of times larger an image is than the specimen object. The magnification obtained with a compound microscope depends on which lenses are used.

page 23

8   A stained nucleus can be seen with a light microscope due to the microscope's resolution. This resolution is not sufficient to distinguish ribosomes due to their smaller size.

page 26

9   Electron microscopes have allowed scientists to examine cell ultrastructure, such as ribosomes and other components, since they have very high resolution and magnification.

The electron microscope has powers of magnification and resolution that are greater than those of an optical microscope. The wavelength of visible light is about $500\,nm$, whereas that of the beam of electrons used is $0.005\,nm$. At best the light microscope can distinguish two points which are $200\,nm$ ($0.2\,\mu m$) apart, whereas the transmission electron microscope can resolve points $1\,nm$ apart when used on biological specimens. Given the sizes of cells and of the organelles they contain, it requires the magnification and resolution achieved in transmission electron microscopy to observe cell ultrastructure – in suitably prepared specimens.

10

Common organelles	Principal role
Nucleus with nuclear envelope	Cell management
Mitochondria	Aerobic stages of respiration
Endoplasmic reticulum	Synthesis of substances
Ribosomes	Protein synthesis

Only present in plant cells: chloroplast, large permanent vacuole.

11 Follow the guidance provided on pages 8 and 20. All of the following structures must be included for the palisade mesophyll cell:

- cell wall with some thickness
- plasma membrane, shown as a single line
- large vacuole – large circular structure located in the middle of the cell
- chloroplasts – circular organelles surrounding the large vacuole (not larger than half the nucleus)
- nucleus – spherical structure larger than chloroplasts
- starch granule – circular bolded dot

page 27

12 This cell has the range of organelles typical of an animal cell. The unique feature is that the cell surface membrane on the side of the cell facing the lumen is in the form of a 'brush border' – vast numbers of microvilli. These greatly increase the surface area for selective absorption of metabolites into the cell from the fluid present in the lumen.

page 29

13 Length of image is approximately 50 mm. From the scale bar, 1 μm = 20 mm. Therefore, actual length of *E. coli* is 50 ÷ 20 μm = 2.5 μm.

M = 50 × 1000 ÷ 2.5

M = × 22 000 approx.

page 30

14 Prokaryotic chromosomes are not associated with histone protein 'naked' and not enclosed in a nucleus. In eukaryotes there are several linear chromosomes (an even number), associated with histone protein and enclosed by a nuclear membrane.

15 a A **cell wall** is totally permeable, made of cellulose, and does not change shape easily, whereas the **plasma membrane** is made of phospholipids associated with integral and peripheral proteins; it is flexible.

b The **nucleus** is an organelle with a double membrane that contains chromatin; the **nucleoid** is the region where the circular prokaryotic DNA is located.

c See page 26, Figure 1.31 on page 28 and Figure 1.32 on page 29.

d See pages 21–23, Figure 1.22 on page 22 and Figure 1.24 on page 23.

page 31

16
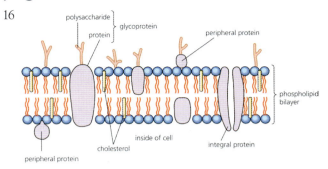

page 33

17 A **lipid bilayer** is made of two layers of phospholipids, whereas a **double membrane** of an organelle (like the one present in a mitochondrion) is composed of *two bilayers* of phospholipids.

page 37

18 a

Dimensions/mm	SA:V ratio
10 × 10 × 10	600/1000 = 0.6
5 × 5 × 5	150/125 = 1.2
4 × 4 × 4	96/64 = 1.5
2.5 × 2.5 × 2.5	37.5/15.6 = 2.4

b
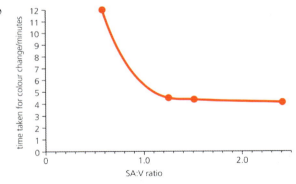

c Hydrogen ions enter the gelatine blocks by diffusion and, there, cause the observed colour change. Relatively large blocks (10 mm cubes) have a low SA:V ratio – meaning that little of the interior matter is close to the external environment, and the diffusion path is a long one. Here, colour change is slowest. The reverse is true of the smallest gelatine blocks (2.5 mm cubes).

page 38

19 The difference is the permeability of the membrane traversed; in facilitated diffusion this is due to the properties of the substance that passes – it triggers the opening of pores through which diffusion occurs.

	Simple diffusion	Facilitated diffusion
Specificity	not specific	specific
Passage directly through phospholipid membrane	yes	no
Globular protein channels	not required	required

page 39

20 The concentrated solution of glucose (in which the concentration of free water molecules is low) will show a net gain of water molecules at the expense of the dilute glucose solution.

21 The presence of water is required to germinate the fungal spore. Hydration or water uptake is an active process that requires a change in permeability of the spore wall. Many spores require an external or exogenous source of sugar, having no internal reserves, and will take this up from the jam.

page 43

22 This gives validity and reliability to the data, reducing the possibility of errors that might occur when working with one single strip.

page 45

23 Uptake of ions is by active transport involving metabolic energy (ATP) and protein pump molecules located in the plasma membranes of cells. As with all aspects of metabolism, this is a temperature-sensitive process and occurs more speedily at 25°C than at 5°C.

Individual ions are pumped across by specific, dedicated protein molecules. Because there are many more sodium ion pumps than chloride ions pumps, more of the former ion is absorbed.

page 46

24 a Proteins act as channels for facilitated diffusion, as pumps, enzymes and hormone-receptor sites, and they have both hydrophilic and hydrophobic regions based on the charges of their amino acids. Lipids are hydrophobic molecules with long mono- or polyunsaturated chains of carbohydrates.

b Active transport is a type of selective transport that moves solutes against their concentration gradient with the expenditure of energy, whereas bulk transport involves the movement of vesicles in a process known as cytosis; it also requires energy.

c Endocytosis is the uptake of fluid or tiny particles across the plasma membrane. Exocytosis is exporting across the plasma membrane.

page 49

25 • Hydrolysis and condensation reactions between molecules, mediated by enzymes.
  • Synthesis of nucleic acids by RNA polymerase-type enzymes.
  • Metabolic processes to transfer energy from compounds and from light as a source of energy, controlled by enzymes.
  • Lytic enzymes would have allowed primitive cells to digest other cells or organic compounds.

page 50

26 Fossils located in lower strata in a sedimentary rock were deposited longest ago. This is why we would expect a greater deviation from modern structures.

page 52

27 It is possible to see:

  • the nuclear membrane and its pores with the complex pore proteins
  • the nucleolus (or nucleoli)
  • the chromatin.

page 56

28 During mitosis the two chromatids of each chromosome must be separated. The DNA molecule in a chromosome is incredibly long – typically 50 000 μm in length. By supercoiling, the DNA of each chromatid is reduced to a manageable length for separation.

page 58

29 a

Stages of mitosis	a % of dividing cells at each stage of mitosis*	b Time
prophase	70	70/100 × 60 = 42 min
metaphase	10	10/100 × 60 = 6 min
anaphase	5	5/100 × 60 = 3 min
telophase	15	15/100 × 60 = 9 min

*Your pie chart should comprise sectors with the following angles at the centre: prophase 252°, metaphase 36°, anaphase 18° and telophase 54°.

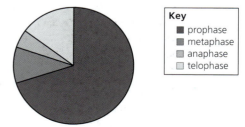

Key
■ prophase
■ metaphase
■ anaphase
□ telophase

page 59

30 • Ionizing radiations: these include X-rays that may trigger the formation of ions inside the nucleus, leading to DNA damage.
• Chemicals: accumulation of chemicals, for example asbestos, in some organs may produce DNA damage after years of continued exposure.
• Non-ionizing radiation, including UV light.

31 Cancer cells divide uncontrollably, forming a tumour. The cells of a malignant tumour invade and damage surrounding organs. Cancer cells may detach and be transported by the blood circulation (metastasize) to distant sites in the body. Normal cells do not behave in this way.

The switch to uncontrolled growth of cancer cells is caused by the accumulation of harmful genetic changes (mutations) in a range of genetically controlled mechanisms that regulate the cell cycle.

The cell cycle is a regulatory loop. It ensures that DNA is faithfully copied and that the replicated chromosomes move to the new (daughter) cells. Further, cell division is restricted – only a selected range of cells can divide again. Normally, if any cell is severely damaged, or grows abnormally, or becomes redundant in the further development of an organ, it undergoes programmed cell death. Thus, many controls prevent cancerous cell behaviour in normal cells.

So cancer cells arise as a result of the coincidence of mutations in several normal growth and behaviour genes. Once some cancers are established, malignant tumour cells typically release a specific protein that dictates the formation of a network of blood vessels supplying the tumour. This occurs at the expense of the supply of nutrients to surrounding healthy cells, which enhances abnormal cell division and growth, and leads to further metastasis.

2 Molecular biology

page 63

1 An **atom** is the basic unit of a chemical element that, when bound together with another atom, forms a molecule. A **molecule** represents the smallest unit of a chemical compound taking part in a chemical reaction. An **ion** is an atom or molecule with a net electric charge.

2 Carbon (C) exists in the atmosphere as CO_2, and when it enters the oceans it forms carbonic acid and bicarbonate ions by dissociation reactions. It also could be present as carbonate ions and dissolved CO_2.

page 67

3 Significant properties of carbon:

• It can form up to four covalent bonds.
• The covalent bonds between carbon and other elements are stable and strong.
• Carbon atoms are able to react with other carbon atoms to form extended and stable chains.
• When other atoms and molecules attach to carbon atoms, asymmetric molecules result, creating a huge variety of compounds.
• Carbon atoms can form double and triple bonds, giving unsaturated compounds and leading to an even wider variety of molecules.

page 69

4 **Ionic bonding** is the electrostatic attraction between oppositely charged ions/particles. In solid sodium chloride (crystals of the salt), ionic bonds hold the ions together in a regular arrangement (known as a crystal lattice). In solution, however, water molecules surround the sodium and chloride ions, causing them to be separated and dispersed.

Covalent bonding involves the sharing of electron pairs between atoms. Covalent bonds are the strongest bonds occurring in biological molecules. This means they need the greatest input of energy to break them. So, covalent bonds provide great stability to biological molecules, many of which are very large and elongated. Bonding of this kind is common in non-metal elements such as hydrogen, nitrogen, carbon and oxygen.

page 73

5 In a solution of glucose, water is the solvent and sucrose is the solute.

6 Significance of water as a coolant:

• Water can be evaporated at room temperature (below boiling point) by breaking the hydrogen bonds between water molecules. Energy is required for this, leading to the cooling effect of water when it evaporates.

Significance of water as transport medium:

• Water molecules stick together with hydrogen bonds, allowing plants to transport water through their xylem from the roots up into the leaves for photosynthesis. The cohesion and adhesion properties of water, due to strong forces between the water molecules, create columns of water inside plants.
• Due to the high heat capacity of water (large amounts of energy are required to break the hydrogen bonds), blood can transport heat inside blood vessels.
• Water acts as a solvent for many substances; it is used as a transport medium for minerals and respiratory gases.

page 74

7   The design could include the use of solar collectors and distilling mechanisms. The use of a solar still may simplify the method of collecting water and separating the salts dissolved in it (desalination). The surface area and power collectors could also be considered.

page 76

8   The presence of sugar in the cytoplasm or vacuole of cells has a powerful osmotic effect (See 'Osmosis – a special case of diffusion', p39), whereas if glucose molecules are condensed to starch (or glycogen, in animal cells), these carbohydrate reserves have no effect on cell water relations. Meanwhile, starch and glycogen may be speedily hydrolysed to sugar if and when the need arises.

page 79

9   Cellulose is made of straight and uncoiled fibres. It is tough and durable, with a flat and rigid structure, due to linkages between fibres. Since each cellulose molecule is flat it can stack with other cellulose molecules, creating hydrogen-bonded fibrils which are very strong.

page 81

10 A monomer is a molecule than can be bonded to many other similar molecules to form a polymer. Examples are galactose, found in milk, and fructose, found in fruits.

A polymer consists of a large number of similar units (monomers) bonded together. Examples are glycogen in animals, and starch and cellulose in plants.

page 82

11 See 'Condensation and hydrolysis reactions', pages 77–78.

page 89

12 The hydrophobic properties of a lipid are combined with the hydrophilic properties of an ionized phosphate group, resulting in a 'head and tail' molecule.

page 92

13

Fibrous proteins	Globular proteins
spider silk and collagen	RuBisCo, insulin, immunoglobulins and rhodopsin

14 In a polypeptide of only five amino acids there can be 20^5 different types, or 3 200 000.

So in a polypeptide of 25 animo acids, 20^{25} different types, and in one of 50, 20^{50}.

page 96

15 Their three-dimensional shape allows enzymes to create a unique site of reaction called the active site. Also it allows enzymes to work in a watery environment since they can be soluble in water.

page 98

16 Pre-incubation is required to ensure that when the reactants are mixed, the reaction occurs at the known, pre-selected temperature.

page 99

17 **pH** is a measure of acidity or alkalinity of a solution. A **buffer solution** acts to resist pH change when diluted, or if a little acid or alkali are added. pH is very important in enzyme experiments because pH affects the shape of enzymes, almost all of which are proteins. (See 'pH and Buffers' in Appendix 1: Background chemistry for biologists.)

page 101

18 a   Controlled variables: concentration of and source of the catalase solution; volume and concentration of hydrogen peroxide solution; temperature of the reaction; pH of the solutions.

    b   The independent variable is the time for the reaction to take place, recorded in seconds.

    c   The dependent variable is the volume of gas (O_2) released and collected into the measuring cylinder. This is recorded in cm^3.

    d   Possible source of error: leaking of the gas from the system when not sealed properly.

    e   This could be detected by observing low volumes of the gas at the beginning of the reaction and can be avoided by reducing the volume of hydrogen peroxide used, in order to reduce the pressure exerted by the O_2 released during the reaction.

19 a   If substrate is present in excess, increasing the concentration further will have no effect on the rate of reaction.

    b   The rate of reaction will increase.

page 104

20 A catalyst is a substance that alters the rate of a chemical reaction, but remains unchanged at the end.

An **inorganic catalyst** is typically a metal such as platinum in a finely divided state, e.g. as platinised mineral wool. Inorganic catalysts are able to withstand high temperatures, high pressures and extremes of pH, if necessary.

Enzymes are typically made of protein and so are easily denatured by high temperature, for example. They are extremely specific – most are specific to one type of substrate molecule for example.

page 105

21 A base found in inorganic chemistry is an acceptor of hydrogen ions from other compounds and has the ability to neutralize acidic solutions. These compounds are commonly inorganic, whereas the nitrogenous base is organic. The carbon 'backbone' of nitrogenous bases are ring molecules containing two or more nitrogen atoms, covalently bonded together.

page 110

22 After three generations 25% will be intermediate light and 75% will be light.

page 113

23 serine – aspartic acid – lysine – stop (Ser-Asp–Lys–STOP)

This sequence might be located at the end of the gene.

page 114

24 a   Gly – Asn – Pro – Phe – Val – Thr – His – Cys

b   CCA–TTA–GGA–AAA–CAA–TGA–GTA–ACA

c   Triplet codes are in DNA. Codons are in mRNA which could be found in both nucleus and cytoplasm. Anticodons are part of the tRNA molecule – which could be found inside the nucleus and in the cyctoplasm, but which are functionally active in the cytoplasm.

page 117

25 ATP is:

- a substance that moves easily within cells and organisms – by facilitated diffusion
- a very reactive molecule, able to take part in many steps of cellular respiration and in many reactions of metabolism
- an immediate source of energy, able to transfer energy in relatively small amounts, sufficient to drive individual reactions.

page 119

26 Glycolysis, which occurs in the cytosol, does not require oxygen, whereas the Krebs cycle and oxidative phosphorylation occurring inside the mitochondrion require oxygen.

27 Lactate and a limited amount of ATP.

page 121

28 In the respirometer, the far side of the U-tube manometer is the control tube (A). Here, conditions are identical to those in the respirometer tube, but in the former, no living material is present. However, any change in external temperature or pressure is equally experienced by both tubes, and their effects on the level of manometric fluid are equal and opposite so they cancel out.

29 Soda lime removes the carbon dioxide produced by the respiring maggots and the change in volume is due to the oxygen used inside the system. When water is used instead of the soda lime, the CO_2 released by the animals replaces the used volume of O_2.

Use of soda lime allows determination of the oxygen uptake by the respiring animals.

30 Glucose in mammals comes from ingested and digested organic compounds, whereas flowering plants use the glucose created during photosynthesis.

page 122

31 It reflects a green colour.

32 If a naked flame (Bunsen burner) is used to heat the water (Step 1), then the flame must be switched off before the propanone (acetone) is poured out – to avoid fire.

Care is needed when fresh leaves are immersed in boiling water using blunt forceps – to avoid scalding accident.

The contents of the centrifuge tubes to be placed in the centrifuge (Step 4) must be balanced in contents, so that they are of equal mass – to prevent damage to the centrifuge or to the tubes during centrifugation.

Pigment solution should be labelled with the contents and particularly the solvent, and stored in dark conditions – so that they are not confused with any other materials, and so that solvent vapour is not accidentally exposed to a naked flame.

33 **Chromatography** is the technique used to separate components of mixtures. It is an ideal technique for separating biologically active molecules since biochemists are often able to obtain only very small amounts.

page 124

34 a   Carbon is obtained from CO_2 in the air entering through the stomata in the leaves.

b   Hydrogen is obtained from water when it is broken down by light in a process known as photolysis.

page 126

35 a A thermometer is required as an increase in temperature will increase the amount of oxygen produced by the plant. It must be positioned near the plant, inside the test tube with the *Elodea*.

b It is inverted to allow immediate release of oxygen from the air spaces between the cells of the plant stem.

page 127

36 As the light intensity increases, the rate of photosynthesis rises – the rate of that rise is positively correlated with the increasing light intensity. However, at much higher light intensities the rate of photosynthesis reaches a plateau – now there is no increase in rate with rising light intensity.

37 Sugars formed in the leaf in the light may be stored there as starch. In darkness, the starch may be hydrolysed to sugars and these are then translocated to other sites of starch storage. The sugar may instead be oxidised immediately to carbon dioxide and water in respiration.

3 Genetics

page 135

1 Sickle-cell anemia is a genetic mutation caused by substitution of one of the nucleotides in the DNA, leading to a change in amino acid number six of the hemoglobin molecule. This supports the concept of a gene as a linear sequence of bases, as the nucleotide change affects the sequence of amino acid residues joined in translation.

page 141

2 The flow chart must include the sub-phases of interphase (first phase of growth, synthesis of DNA, second phase of growth) and the phases of mitosis and cytokinesis.

page 142

3 In order to maintain the number of chromosomes in all cells and to allow the correct functioning of the daughter cells formed after division.

page 145

4 Non-chromosomal birth defects increase as maternal age increases after 34 years. The increase in these birth defects is more pronounced after 39 years.

Any birth defects decrease when maternal age increases from 25 years to 34 and then there is a small increase until 39 years. Finally, there is a sudden increase from 39 years.

This suggests that there is a positive correlation of birth defects with maternal ages greater than 39 years.

page 147

5 This shows the karyotype of a male with non-disjunction of chromosome pair 21 (resulting from non-disjunction).

page 152

6 a The predicted ratio was 3:1.

b Actual ratios were:

- position of flowers: 3.145:1
- colour of seed coat: 3.147:1
- colour of cotyledons: 3.009:1

c A prediction of the likely outcome of a breeding experiment represents the probable results, provided that:

- fertilization is random
- there are equal opportunities for survival among the offspring
- large numbers of offspring are produced.

What is actually observed in a breeding experiment may not necessarily agree with the prediction. For example, there is a chance in this particular cross that:

- more pollen grains of one genetic constitution may fuse with egg cells than another
- more developing seeds of one type are predated and destroyed by insect larvae of species attacking the plant (so fewer zygotes of one type complete development)
- the cross produces too few progeny in total.

7 A person with sickle cell anemia is homozygous recessive, but someone with sickle cell trait is heterozygous.

8 Below is a table showing how the Law of Segregation relates to meiosis.

Segregation law	Meiosis
Applies to the production of gametes	Is the type of division that produces gamete cells
The two alleles of a gene separate	During anaphase I, homologous chromosomes separate and during anaphase II sister chromatids separate
Each one of the individual offspring receives one trait from each of their parents	Alleles separate during anaphase I and II, forming daughter cells containing only one version of the gene (allele)

page 154

9 The layout of your monohybrid cross will be as in Figure 3.19 (page 154), but the parental generation (P) will have genotypes (if you have chosen C as the allele for coat colour) of

$C^R C^R \times C^W C^W$

where C^R represents the allele for red coat, and C^W represents the allele for white coat. The gametes the parental generation produce will be: C^R and C^W

The offspring (F_1) will have genotype $C^R C^W$ and the phenotype will be 'roan'.

In a sibling cross of the F_2 generation the gametes of both siblings will be:

$\frac{1}{2}C^R + \frac{1}{2}C^W$

From an appropriate Punnett grid the offspring (F_2) to be expected and the proportions are:

genotypes	$C^R C^R$	$C^R C^W$	$C^W C^W$
ratio	1	2	2
phenotypes	red	roan	white

page 155

10 The **Jones** were groups A and B, giving the following possible outcomes:

Parental genotype	Parental genotype	Possible blood group of offspring
AA	BB	AB
AO	BO	A, B, AB or O
AA	BO	A or AB
AO	BB	B or AB

So the **Jones** could be the parents of *any* of the four children, but are the only parents who could have a child with AB blood group.

The **Lees** were B and O, giving the following possible outcomes:

Parental genotype	Parental genotype	Possible blood group of offspring
BB	OO	B
BO	OO	B or O

So the Lees might be the parents of *either* the B group or the O group child.

The **Gerbers** were both blood group O, so their offspring must also be blood group O. (Therefore the Lees were the parents of the B group child.)

The **Santiagos** were AB and O, giving the following possible outcomes:

Parental genotype	Parental genotype	Possible blood group of offspring
AB	OO	A or B

So the Santiagos were the parents of the A group child (since the Lees were the parents of the B group child).

page 156

11 a   Liz and Diana

    b   (i) David and Anne (ii) James

c   Eight: Richard, Judith, Anne, Charles, Sophie, Chris, Sarah and Gail

d   James and William, Arthur and Diana, etc.

page 157

12 The Punnett grid for the cross between a normal-handed individual (nn) and someone with brachydactylous hands (Nn) is shown here.

	N	**n**
n	Nn	Nn
n	nn	nn

The probability of having an offspring with brachydactylous hands is equal to 50%.

page 159

13 The genotype of a female with colour blindness can be represented as $X^b X^b$. Males cannot be carriers as only one X chromosome is present – the recessive allele is always expressed; the heterozygous condition (carrier) does not apply for sex-linked inheritance in males.

page 160

14 Males have only one copy of the X chromosome which is inherited from the mother. When the recessive allele for hemophilia is present on this X chromosome, males express the disease. Since the heterozygous condition does not exist in males, there are no male carriers. Females may be carriers but are rarely affected by the disease as they must receive two recessive alleles – one from each parent.

page 165

15 Reverse transcriptase (RT) generates complementary DNA from a molecule of RNA. HIV is a retrovirus that uses (RT) to convert its single stranded RNA into double stranded DNA (cDNA), in order to integrate it into the host genome. This generates a long term infection that is difficult to eliminate.

page 166

16 Gel electrophoresis uses a gel of agar within a chamber where a current is applied. A DNA sample is digested or cleaved using restriction enzymes; a dyeing solution is added and then the samples are inserted into the gel. The fragments separate based on their negative charge and size, creating visible bands – the smaller fragments are able to go faster and further than larger fragments. Fragments have to move through the porous structure of the gel towards the positive pole.

The buffer maintains the pH of the solution and allows the electrical current to flow between the negative and positive poles.

The power supply provides a continual electrical current, allowing the negatively charged DNA fragments to move towards the positive pole.

page 168

17

	DNA	RNA
Number of strands	2/double	1/single
Ribose	deoxyribose	ribose
Bases	adenine, thymine, cytosine and guanine	adenine, uracil, cytosine and guanine
Location	nucleus	nucleus and cytoplasm
Double helix	yes	no, only tRNA

page 169

18 a Cleaving DNA by a restriction enzyme at the recognition sequence (palindrome section) creates protruding 5′ and 3′ endings.

b The recognition sequence has complementary bases that allow other sticky ends with corresponding bases to join on to it.

page 171

19 The genes of prokaryotes are more easily modified than those of eukaryotes because:

- plasmids, the most useful vehicle for moving genes, do not occur in eukaryotes (except in yeasts) and, if introduced, may not survive and be replicated there
- eukaryotes are diploid organisms, so two forms (alleles) for every gene must be engineered into the nucleus. Prokaryotes have a single, circular 'chromosome', so only one of a gene has to be engineered into their chromosome
- transcription of eukaryotic DNA to mRNA is more complex than in prokaryotes, where it involves removal of short lengths of 'non-informative' DNA sequences – the introns
- machinery for triggering gene expression in eukaryotes is more complex.

page 173

20 a The genotype is the alleles of an organism and the genome is the whole of the genetic information of an organism.

b Restriction endonucleases cut, cleave or digest the DNA at specific recognition sites; ligase catalyses the joining of DNA strands.

c A bacterial chromosome is a circular chromosome and a plasmid is circular extra-chromosomal DNA material.

page 174

21 A fertilized egg cell is a diploid cell that will divide by mitosis during the development of new tissue cells. If a gene is inserted into this cell, all of the cells of the organism that develop from the embryo will have the gene – including the germinal cells, allowing the individual to pass on this property to its progeny.

22 All cells in the gall tissue are infected with *Agrobacterium* which means that the Ti plasmid has been integrated into each cell's genome.

page 175

23 Cereals will require less nitrogen-containing fertilizer, and they form the bulk of human food intake.

24 a Criticisms could include:

- Unknown secondary effects in humans, such as allergies.
- Passing antibiotic resistance genes to consumers or other organisms.
- Reducing biodiversity of crops.
- Reducing variability in the species of crops cultivated, leading to new pests that require more attention or new modifications.
- Cross-pollination of species, leading to super-weeds containing an insecticide gene.

b Arguments in support:

- Allergenic effects are restricted to certain groups of the human population and, in most cases, are similar to the natural allergies humans have for a specific product or food.
- No need to use antibiotic resistance genes in plasmids; fluorescent markers can be used instead.
- Biodiversity is preserved, since only certain areas are cultivated with the GM crops.
- Variability of the species is not reduced, but improved, since new features are added to the species.
- Cross-pollination could occur, but there is not much evidence of this happening with pollen coming from regular crops.
- GM crops increase yield and help to improve the quality of the products.

page 178

25 Identical twins are monozygotic, meaning that they come from the same dividing cell, so they have the same or very similar DNA. They cannot be distinguished by DNA fingerprinting. Non-identical twins come from two different egg cells, each one with a different combination of alleles, allowing fingerprinting techniques to find differences between them.

4 Ecology

page 185

1 a ecosystem b population c abiotic factor

   d community e biomass f habitat

   g abiotic factor

page 187

2 1 a autotrophic organism producing organic compounds by means of photosynthesis
                                                producer/autotroph

      b heterotroph organism   Go to 2

   2 a organism consuming grass or herbs only
                                        herbivore/primary consumer

      b organism that ingests other animals   go to 3

   3 a organism consuming organic matter that is living or recently killed   go to 4

      b organism that feeds on decaying matter   go to 5

   4 a organisms feeding on primary consumers
                                        secondary consumer

      b organisms feeding on secondary consumer
                                        tertiary consumer

   5 a organisms living on dead organic matter, secreting digestive enzymes   saprotroph

      b organisms that ingests dead organic matter
                                        detritivore or decomposer

page 191

3 Observed results table:

	Bell heather present	Bell heather absent	Sum
Bilberry present	12	55	67
Bilberry absent	88	45	133
Sum	100	100	200

Expected frequency table or contingency table:

	Bell heather present	Bell heather absent	Sum
Bilberry present	(67 × 100)/200 = 33.5	(67 × 100)/200 = 33.5	67
Bilberry absent	(133 × 100)/200 = 66.5	(133 × 100)/200 = 66.5	133
Sum	100	100	200

$$\chi^2 = \sum \frac{(O-E)^2}{E}$$
$$= \frac{(12-33.5)^2}{33.5} + \frac{(55-33.5)^2}{33.5} + \frac{(88-66.5)^2}{66.5} + \frac{(45-66.5)^2}{66.5}$$
$$= 13.79 + 13.79 + 6.95 + 6.95$$
$$= 41.48$$

Degrees of freedom = $(2-1) \times (2-1) = 1$

Critical value for a p-value of 0.05 is 3.34.

The value of χ^2 calculated is clearly larger than the critical value of 3.34 for one degree of freedom.

The results show that the distributions of the two species are not in independent of each other; the distributions of the two species are associated.

4 The answer depends on the location, but here are some suggestions:

- deforestation – can be addressed by reforesting zones using local species of trees
- reduction of natural habitats – counter by protecting regions where endemic species live
- urban development – restrict new building to urban areas
- air pollution – reduce pollutants in the air by imposing restrictions on traffic, making strong policies to force industries to use filters or to avoid burning of fuels

page 193

5 Saprotrophs recycle nutrients into the food chain, ultimately making them available to animals again.

page 195

6 Aquatic food chain A: phytoplankton → zooplankton → common mussel → dog whelk

   Aquatic food chain B: seaweeds → grey mullet → pollack → seal

   Trophic level of each organism:

- Seaweeds and phytoplankton: producers
- Zooplankton and grey mullet: primary consumers.
- Secondary consumers: common mussel, pollack
- Tertiary consumers: dog whelk and seal.

7 Sunlight.

page 197

8 Trophic levels of humans depend on what they feed on:

- when eating vegetables, cereals and fruits only: primary consumers
- when eating meat from herbivores: secondary consumers
- when eating other carnivore animals: tertiary or quaternary consumers

page 199

9 Energy is lost at each trophic level. Only 10–20% of the energy made available by producers passes to primary consumers. This is because plants use energy for their own metabolic processes such as creating ATP. Consumers use energy in movement; the energy dissipates and is lost as heat into the surrounding environment. Also, not all the organic matter is consumed.

The total amount of energy available to top consumers is small; a large number of producers are required to provide tertiary and quaternary consumers with enough energy.

page 200

10 From primary to secondary consumers: 11.4%

11 a Energy taken in: $3050 \, kJ \, m^{-2} \, yr^{-1}$

Total energy wasted: $2925 \, kJ \, m^{-2} \, yr^{-1}$

Energy value of the new biomass: $125 \, kJ \, m^{-2} \, yr^{-1}$

Percentage of energy in biomass of the cow: 4.09%

b Energy transfer between primary and secondary consumer: 4.09% (same as the biomass)

page 202

12 • Differences in the rates of flux between atmosphere and land biota: carbon is fixed by photosynthesis, giving the greatest pool of carbon in land biota. Flux rate due to respiration takes back carbon stored in land biota into the atmosphere pool, and this flux is enhanced by the burning of materials, thus making the atmosphere pool greater than the land biota. Also, the atmosphere receives a flux of carbon from other pools, such as soil, surface ocean, the mantle and with the burning of fossil fuels from the pool of 'sedimentary rocks'.
- Differences in rates of flux between deep ocean and sedimentary rocks: deep oceans store a total of 38 000 GT and sedimentary rocks 1 000 000 GT, meaning that the sedimentary rock pool is much greater in storage of carbon than deep oceans.

13 The largest amount of carbon stored is in the sedimentary rocks and in the Earth's crust (mantle). Total percentage is: 95.6%.

page 203

14 a as CO_2

b hydrogencarbonate ions

c carbonates (e.g. Ca_2CO_3)

page 205

15 The atmospheric carbon dioxide varies between high and low values within each 12-month period of the graph due to photosynthesis on lands in the northern hemisphere – the peaks are winter levels, and the troughs are summer levels of atmospheric CO_2.

page 207

16 Plants absorb CO_2 into their tissues, thus reducing atmospheric concentrations of this gas. Destruction of rainforest will reduce the number of photosynthetic organisms available to fix CO_2 into organic molecules, leading to an increase of this gas in the atmosphere.

5 Evolution and biodiversity
page 212

1 Fossils form when the skeleton is left after the soft parts of a body have decayed and when the bones become covered with sediment. When more sediment is added, this creates enough pressure to harden them and they may become impregnated with silica or carbonate ions.

page 213

2 The breeding of domesticated plants and animals has created varieties with little external resemblance to their wild ancestors. Darwin bred pigeons, and noted there were more than a dozen distinctive varieties of pigeon, all of which were descended from the rock dove. Darwin argued that if so much change can be induced in so few generations, then species must be able to evolve into other species by the gradual accumulation of minute changes, as environmental conditions alter and natural selection operates.

3 A species is a group of organisms that can interbreed and produce fertile offspring. All these dog types can breed and their offspring will be fertile.

page 219

4   Natural selection results in gradual change in populations due to the possession of more favourable traits by some individuals in the face of environmental change. These individuals are more likely to survive, breed, and produce offspring than others. In artificial selection it is humans that select individuals for breeding with the more favourable characteristic in each generation, bringing about quick change.

page 221

5   Modern genetics can identify relationships between species of organisms, based on molecular evidence, that confirm and support the theory of natural evolution. See 'Neo-Darwinism', page 220.

page 222

6   a   Exposure of pathogenic bacteria to sub-lethal doses of antibiotic may increase the chances of resistance developing in that population of pathogens.

b   By varying the antibiotics used, there is increased likelihood of killing all the pathogens in a population, including any now resistant to the previous antibiotics used. This approach works until multiple-resistance strains have evolved, such as in strains of *Clostridium difficile* and *Staphylococcus aureus*.

page 223

7   The idea that the Earth was much older than previously thought gave a new and realistic timescale for the process of heritable change known as evolution.

8   Impact events, sustained and significant global warming and cooling that also gave rise to sea level rises; plate tectonics and volcanic activity with flood events, leading to release of sulfur gases into the atmosphere, killing photosynthetic organisms; anoxic events of the ocean, amongst others.

page 225

9   The international naming system is used to avoid confusion among scientists around the world when naming organisms.

page 228

10   The domain for both organisms in the figure is the Eukarya domain. The flow diagrams will be modified by adding an upper level from which both kingdoms come.

page 238

11   The design can include a dichotomous key of plants and animals common in gardening. Consider the location of the gardener, since the environment affects the distribution and presence of plants and animals.

6 Human physiology

page 248

1   Parasitic nutrition and saprotrophic nutrition.

page 249

2

Structure	Associated gland	Function
mouth and esophagus*	salivary glands	break up of food by teeth, lubrication of food and throat by saliva
stomach*	wall of stomach	beginning of protein digestion – secretion of gastric juice
		mechanical digestion by churning action of stomach wall
small intestine*	wall of small intestine	digestion of carbohydrates, lipids and proteins, movement and mixing of food by peristalsis
		site of absorption of digested food
pancreas	pancreas	secretion of enzymes that complete digestion (also an endocrine gland)
liver	liver	produces bile (also processes the products of digestion after absorption
gall bladder	none	stores bile and releases it into the first portion of the small intestine
large intestine*	goblet cells in the wall of the intestine	water and mineral salts absorbed

*These regions also produce copious quantities of mucus.

page 251

3   a   Digestion reduces the size of macromolecules by hydrolysing them in order to allow the absorption of small molecules through the intestinal wall.

b   Water and foods such as fruit juices that contain negligible amounts of proteins and carbohydrates. Vitamins and minerals are also absorbed without the need for digestion.

c   Cellulose

4   The type of reaction is hydrolysis, which requires water to break down the peptide bonds between amino acids and the bonds between sugars in starch.

page 253

5   The mechanisms included in the drawing are:

- Diffusion of minerals and facilitated transport of glucose with sodium at the same time.
- Amino acids actively enter into the intestinal cells and are transported with sodium and glucose, but energy is required for this process. After this both glucose and amino acids are moved by facilitated transport into the hepatic portal vein.
- Fatty acids and glycerol are integrated with proteins, phospholipids and cholesterol to be transported inside the lacteals as a lipoprotein called chylomicron.

page 254

6   Absorption: uptake into the body (blood circulation or lacteals) of the useful products of digestion, from the gut lumen. Assimilation: uptake of nutrients into cells and tissues.

page 255

7   For example, it means that blood cannot be directed to a respiratory surface immediately before or after servicing tissues that are metabolically active (and so have a high rate of respiration). Rather, the blood circulates randomly around the blood spaces and blood vessels.

page 256

8   Phagocytic leucocytes function anywhere in the body that an infection occurs, but in addition to their presence in the blood plasma, lymph and lymph glands, they are always present in the airways and alveoli of the lungs, and in the liver, lining the rows of liver cells (hepatocytes) past which the blood flows.

page 258

9   Pulmonary system: the heart sends deoxygenated blood into the pulmonary artery to the lungs. Oxygenated blood enters the heart through the pulmonary vein from the lungs.

Systemic circuit: the heart sends oxygenated blood into the aorta (major artery) and receives deoxygenated blood from the body through the vena cava.

page 262

10   These non-elastic strands keep the heart-valve flaps pointing in the direction of the blood flow. They stop the valves turning inside out when the pressure rises due to ventricular contraction.

page 263

11   a   Pressure in the aorta is always significantly higher than that in the atria because blood is pumped under high pressure into the aorta, and during diastole and atrial systole the semilunar valves prevent backflow from the aorta. Meanwhile, blood enters the atria under low pressure from the veins, and the pumping action of the atria is slight compared to that of the ventricle which generates our 'pulse'.

b   Pressure falls abruptly in the atrium once ventricular systole is underway as atrial diastole begins then.

c   The semilunar valve in the aorta does not open immediately but opens as soon as the pressure in the ventricles exceeds that of the pressure in the aorta, so preventing backflow.

d   When ventricular diastole commences, the bicuspid valve will open when pressure in the ventricles falls below that in the atria, so preventing backflow.

e   About 50% of the cardiac cycle is given over to diastole – the resting phase in each heart beat. The heart beats throughout life and takes limited rest at these moments.

page 265

12   a   The velocity and pressure are higher when the blood is in the aorta, following contraction of heart muscle, compared to when it is in the veins.

b   Artery walls have elastic muscular fibres that help to maintain the blood pressure. Also they have a thick outer layer of collagen to withstand the pressure.

c   The branching of the capillaries reduces the pressure, and their porous membranes allow liquid to leak out, forming the tissue fluid that bathes cells. This facilitates the exchange of materials between blood and cells.

page 267

13   Released during a stressful situation, such as when feeling fear or anger, in order to prepare the body for 'fight or flight' actions.

page 270

14   A bacterial cell performs all the function of life, so it has a complex metabolism. Enclosed within a cell wall and a plasma membrane, it contains cytoplasm, a nucleoid region, ribosomes and mesosome among other important

structures. Also, there are a flagellum and pili for locomotion and for attachment to other bacteria.

A virus is an infectious agent, with a limited amount of resources enclosed in a capsid made of proteins. This capsid contains a nucleic acid and RNA viruses also contain a retroviral enzyme. Viruses depend on the enzymes and cellular machinery of the host cell, making them all parasitic.

page 271

15 Mucus protects the tissue from drying out and acts as a surfactant. It also helps by trapping dust and bacteria, reducing the chances of an infection.

page 273

16 Damaged cells produce a signal that causes platelets to collect at the site of damage and **release clotting factors**. Clotting factors **activate thrombin** which then stimulates the conversion of fibrinogen into **fibrin**.

page 274

17 • They tested their penicillin on human patients after very few tests on animals
   • They then tested humans (singly, or very small groups) already ill with acute and life-threatening bacterial infections, without any prior investigation of the effect of their penicillin on healthy people
   • They worked with small groups, and with minimal quantities of their preparation
   • Their penicillin preparation was impure so there could have been side effects from the impurities.

page 275

18 Antibiotics either destroy the mechanisms that produce the cell wall in bacteria or inhibit some of the transcriptional processes that create enzymes inside the bacteria. Viruses do not have a cell wall and do not have any molecular pathways to produce enzymes (using host cell facilities for this).

19 Intensive agriculture leads to the excessive use of antibiotics in farm animals, in order to control bacterial diseases that spread easily among groups of enclosed or closely spaced animals. The dangers are related to antibiotic resistance, as some super-resistant bacteria may be created by this intensive and extensive use of the antibiotics.

page 276

20 Commercial challenges: strict regulations for creating new medicines and limited financial reward are two of the problems that pharmaceutical companies face. New antibiotics are not used extensively and, if they are used, the treatment lasts for two weeks. Compare this with the financial gain from developing a medicine that is taken every day indefinitely.

Medical challenges: some Gram-negative bacteria have a double cell wall that limits the entrance of the antibiotic.

21 Antigens are found on the surface of bacterial cells that invade the body. They are also found on the surface of red blood cells, platelets and white blood cells. Antibodies are produced by our own cells in the circulatory system in response to non-self antigens. Antibodies exist in the blood and lymph, and may be carried in the plasma solution anywhere that blood 'leaks out' to – including sites of invasions.

page 278

22 In order to produce antibodies, activation of lymphocyte genes is required. Transcription of mRNA, followed by translation, creates the antibodies. Translation happens in the RER and the antibody proteins are secreted into vesicles. After this, the Golgi apparatus modifies the antibodies, making them functionally active. Finally, they are secreted out of cells by exocytosis of vesicles.

page 280

23 AIDS patients have compromised immune systems, due to the HIV viral infection, and are not able to keep common infectious diseases under control, such that these common diseases contribute to the depletion of the already compromised immune system.

page 281

24

Characteristic	How it influences diffusion
a large, thin surface area	the greater the surface area and the shorter the distance that gases (O_2 and CO_2) have to diffuse, the quicker gas exchange occurs
a ventilation mechanism that moves air (or water) over the respiratory surface	the higher the concentration of oxygen on the 'supply side' of the respiratory membrane, the quicker gas can diffuse across the surface
a blood circulation that speeds up the removal of dissolved oxygen, with a respiratory pigment that increases the gas-carrying capacity of the blood	the quicker oxygen is picked up and transported away from the gas exchange surface, the greater the rate of diffusion

The relationship of these factors is summarized by **Fick's Law of Diffusion**:

$$\frac{\text{surface area} \times \text{difference in concentration}}{\text{length of diffusion pathway}}$$

page 282

25 Inspiration		Expiration
relax	**internal intercostal muscles**	contract
contract	**external intercostal muscles**	relax
contract	**diaphragm**	relax
relax	**abdominal wall muscles**	contract
move upwards and outwards	**effects on rib cage**	move downwards and inwards

page 285

26 Carbon dioxide is an acidic gas which, if it accumulates in the blood, alters the pH of the plasma solution, leading to respiratory acidosis. This imbalance causes increase in heartbeat, high blood pressure, swellings, difficulty breathing, discoloration of the skin (turning blue); coma or even death may result due to the lack of oxygen.

page 286

27 Gaseous exchange involves the action of the respiratory membrane (in the alveoli) only; cellular respiration happens in the cell cytoplasm and the mitochondria, where energy is released from organic compounds.

page 289

28 In both men and women the habit of smoking was declining. In the case of men, this followed a previous prolonged period when heavy smoking was extremely prevalent, followed by a decline in smoking since 1950. The incidence of male lung cancer was positively correlated with this decline. In the case of women, smoking had been a relatively recently adopted habit, and although now declining, the incidence of smoking had been on the increase up until 1970. The resulting lung cancer was still on the increase, for there is a significant delay in the onset of cancer after smoking is taken up.

page 294

29 a The resting potential uses energy in the form of ATP for the active transport pumps (Na^+/K^+ pumps).

b An action potential is due to influx of Na^+ ions and the efflux of K^+ ions through voltage-gated channels that charge the membrane.

page 296

30 See 'Speed of conduction of the action potential', page 295 and Figure 6.46 on page 296.

31 a Stimulations of the nerve fibre are insufficient to generate an action potential.

b (I) Sodium channels open and sodium ions rush in.

(II) Interior of the axon becomes increasingly more positive with respect to the outside.

(III) Interior of the axon starts to become less positive again.

(IV) Slight oversheet (more negative) than the resting potential (= hyperpolarization) before sodium/potassium ion pump, with facilitated diffusion re-establishing the resting potential.

page 298

32 a The Golgi apparatus sends vesicles to the presynaptic membrane containing neurotransmitters.

b The mitochondria provide ATP as a source of energy.

page 302

33 Mitochondria, because liver cells are metabolically very active, and RER, because the liver is a secretory organ that produces many useful substances.

page 304

34 Endergonic reactions require an input of energy (ATP), because the products have more potential energy than the reactants.

page 312

35 See 'Hormones in the control of reproduction', pages 310–312; numbered points 1–5 and Figure 6.64.

page 313

36 *In vivo* describes an experiment that takes place within a living organism. *In vitro* literally means 'in glass' and describes an experiment that is performed in a test tube or a Petri dish, often using cell extracts.

7 Nucleic acids

page 317

1 See 'Nucleotides', page 105–106, and Figures 2.47, and 2.48. In summary:

nitrogenous base + pentose sugar → nucleoside
nucleoside + phosphoric acid → nucleotide
many nucleotides condensed together → nucleic acid

page 318

2 a Supercoiling of chromatin during metaphase helps to ensure correct segregation of chromosomes at cell division.

b The attraction between histones and DNA facilitates supercoiling.

page 319

3 Hershey and Chase would have expected that only the bacteria infected with the virus labelled with ^{35}S (an element in proteins but not in nucleic acids) would have produced radioactive virus. Remember, they were not aware that the protein coat of the virus remained outside of the host cell at the time of infection.

page 320

4 A hydrogen bond is an electrostatic attraction between the positively charged region of one molecule and the negatively charged region of a neighbouring one; the attraction gives rise to a relatively weak bond (compared to a covalent bond). Collectively, the H-bonds of DNA hold the polynucleotide strands together.

page 325

5 The strands of the molecule of DNA that run parallel to each other but with opposite 3′ to 5′ alignments.

6 a Replication can only proceed by adding nucleotides to the 3′, so one strand of a DNA molecule can be replicated as a single long molecule. The other strand is replicated in short portions that must be joined together.

b In transcription only the coding or antisense strand of DNA is transcribed, creating an mRNA molecule identical in sequence to the sense strand in the DNA.

7 Transcription: first part of gene expression in which a particular section of a molecule of DNA (gene) is copied into mRNA by the enzyme RNA polymerase.

Translation: the process of reading a molecule of mRNA by a ribosome during the synthesis of a polypeptide.

Antisense (coding) strand: this is the strand of DNA that is transcribed by the enzyme RNA polymerase.

Sense strand: the segment of double-stranded DNA running from 5′ to 3′, and complementary to the antisense strand. It has the same base sequence as the mRNA, but with thymine instead of uracil.

page 327

8 After the DNA double helix has been unwound and the coding strand exposed, RNA polymerase recognizes and binds to the promoter region. This enzyme now draws on the pool of free nucleotides to complement the DNA sequence (A with U, C with G), and catalyses the reaction that attaches them in a chain. mRNA is formed.

page 334

9 a mRNA

b tRNA

10 See 'Peptide linkages' and Figure 2.31, page 90.

page 336

11 Inside the nucleus there are enzymatic proteins that participate in the processes of DNA duplication and transcription, such as DNA and RNA polymerases, helicases, topoisomerases and single strand DNA binding proteins. Also there are histone proteins that help to organize the DNA and participate in the control of transcriptional processes. Other proteins regulate the exit of molecules from the nuclear envelope.

page 341

12 See 'The structure of proteins' and Figure 2.32, page 91.

page 342

13 • They act as receptors for hormones (hormone-binding sites), allowing cells to receive a hormonal signal which triggers a set of intracellular reactions.
• Electron carriers are arranged in chain of transporters. These move electrons from one carrier to another, producing energy.
• Pumps for active transport, using energy from ATP, are able to move substrates against their concentration gradient.

8 Metabolism, cell respiration and photosynthesis

page 346

1 See Figure 8.1, page 345.

page 348

2 See 'Enzyme specificity and the induced-fit model' and Figures 2.35 and 2.36, pages 94–5.

page 350

3 a

rate of reaction of catalase with of H_2O_2 in absence and presence of heavy metal ions

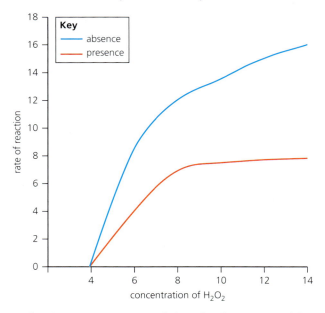

b A non-competitive inhibitor binds to a part of the enzyme other than the active site. As the inhibitor does not compete with a substrate molecule for the active site, an excess of substrate does not overcome the inhibition. From the graph we see that an excess of the substrate H_2O_2 has not overcome the inhibition, so copper ions are a non-competitive inhibitor of this enzyme. This is supported by the data in the graph where the rate of reaction levels off even with the increasing concentration of H_2O_2.

page 354

4 NADH, ATP and pyruvate.

page 356

5 a Dehydrogenases: these enzymes oxidize a substrate by a reduction reaction, transferring hydrogen ions (hydrides) to an electron acceptor such as NAD^+.

b Decarboxylases: enzymes that add or remove a carboxyl group from organic compounds, usually released as CO_2.

6 In the absence of oxygen, reduced NAD ($NADH_2$) accumulates and oxidized NAD reserves are used up. In the absence of NAD^+, pyruvate production by glycolysis slows and stops, so subsequent steps in respiration stop too.

page 357

7 See 'Phosphorylation by chemiosmosis' and Figure 8.12, pages 357–8.

page 358

8 From the scale bar: 15 mm represents 1 μm. The length of the mitochondrion is approximately 64 mm, so the actual length of the mitochondrion = 64/15 = 4.3 μm. The actual width of the mitochondrion = 21/15 = 1.4 μm.

page 359

9 a The substrate is the molecule that is the starting point for a biochemical reaction; it forms a complex with a specific enzyme. An intermediate is a metabolite formed as a component of a metabolic pathway.

b Glycolysis is the breakdown of glucose in the cytoplasm, producing a small yield of ATP. The Krebs cycle happens inside the mitochondrion and it results in the reduction of electron carriers; it produces one molecule of ATP per molecule of acetyl entering the mitochondrion.

c Oxidation is the loss of electrons and reduction is the gain of electrons.

page 361

10

Components	Photosystem I	Photosystem II
Pigment molecules (proteins)	Receive the photon of light and pass it to other pigment molecules, until it reaches the reaction centre. These are called pigmented antennae.	
Reaction centre	An enzyme that uses light of a wavelength 700 nm to reduce molecules.	An enzyme that uses light of a wavelength 680 nm to reduce molecules.
Accessory pigments	Constitute part of the pigmented antennae and help in moving photons of light.	
Electron carrier molecules	Ferredoxin and ferredoxin-NADP reductase transfer electrons to reduce NADP into NADPH.	Plastoquinone, cytochrome and plastocyanin proteins help the excited electron to move hydrogen ions from the stroma into the thylakoid space.
Oxygen-evolving complex	NA	With light, performs the photolysis of water to replace the electron excited by the photon of light. Produces hydrogen ions and oxygen.
ATP synthase	Located in the thylakoid membrane and forms ATP from ADP and Pi.	

page 363

11 Electrons displaced from the reaction centre of photosystem II in an excited state are first passed to the reaction centre of photosystem I. Here they are again

raised to an excited state and this time they are passed to oxidized NADP (NADP⁺) to form reduced NADP (NADP + H⁺).

12 They are formed there as they will be used during the Calvin cycle. This uses enzymes that are located in the stroma of the chloroplasts.

13 See 'The transfer of light energy' and Figure 8.18, pages 362–3.

page 364

14 An **isotonic** solution is of the same osmotic concentration as the organelles, therefore they are not disrupted by excess movement of water into or out of their delicate structure. As a **buffer** solution, this medium maintains the pH at that of the cell contents, thereby ensuring enzyme action continues.

The presence of **ice** keeps the solution and organelles at or just above 0°C, thereby preventing unwanted biochemical reactions whilst the extracted organelles are held before investigation.

page 367

15 *Chlorella* cells in suspension function biochemically like mesophyll cells but can be cultured in suspension, sampled without interference of the biochemistry of other cells, and supplied directly and quickly with light, intermediates and inhibitors (if required) in such a way that all the cells are treated identically. Gaseous exchange and diffusion occur without the complexity (and delays) of the intact leaf with its air spaces and stomata. In the lollipop chamber the environment of the cells (light intensity, etc) was more nearly identical for the whole chamber than it would be in a spherical flask, for example.

page 370

16 a   The light-dependent reaction happens in the presence of light, whereas the light-independent reaction does not require light to occur; it uses the products from the light-dependent one. The light-dependent reactions occur in the thylakoids in the presence of light and the light-independent reactions in the stoma in the absence of light.

b   Photolysis uses light to break down water, producing hydrogen ions and oxygen, and photophosphorylation uses light energy to phosphorylate ADP to form ATP.

17 The starting materials (carbon dioxide and water) are in an oxidized (low energy) state. The end product of photosynthesis – glucose – is in a reduced (higher energy) state.

18 The drawing must include: outer and inner membranes, stroma, grana (singular granum) and thylakoid. The grana increase surface area for more reactions; the thylakoid space is reduced to allow for the accumulation of ions; the stroma contains enzymes that are used in the Calvin cycle.

19 Photoactivation of photosystem I and production of ATP.

9 Plant biology

page 375

1   The features of root hairs that facilitate absorption from the soil are:

- they greatly increase the surface area of the root
- they grow in close contact with the film of soil water that occurs around mineral particles where the essential resources (water and soluble ions) occur
- their walls are of cellulose and are in intimate contact with cells of the cortex in a region that is permeable to water.

2   The symplast pathway of water movement is through the cytoplasm (by diffusion), including through the cytoplasmic connections between cells.

The apoplast pathway of water movement is the mass flow through the space in the wall cellulose (about 50% of the wall volume).

page 378

3   Plant growth is dependent on a supply of chemically combined nitrogen (for amino acid and protein synthesis) of which nitrates are typically the most readily available. However, nitrates are also taken up by microorganisms and may be released in the soil at times other than when plant demand is at its peak. They are also very soluble and are easily leached away into ground water in heavy rain. By taking up nitrate whenever it becomes available (and storing it in cells), plants can maintain growth at peak times.

4   In waterlogged soil there are no air spaces and the dissolved oxygen is reduced. Roots need oxygen to respire and perform other metabolic activities.

page 381

5   The girth of the tree trunk shrinks during the daytime when transpiration occurs at a greater rate and increases at night when transpiration is reduced. In the day, reduced water in the xylem vessels causes shrinkage in the overall girth of the tree, demonstrating the cohesive and adhesive properties of water molecules and the tension between water molecules and xylem walls.

6   The Casparian strip is made of a dense material that restricts the passive movement of water through the apoplast pathway, forcing it to go into the symplast pathway.

page 382

7   Stomata work by turgor pressure, opening when water is absorbed by the guard cells from the surrounding epidermal cells. Fully turgid guard cells push into the epidermal cell besides them because of the way cellulose is laid down in the guard cell walls (thinly in the lateral walls, greatly thickened in dorsal and ventral walls – see Figure 9.10, page 382).

page 383

8   Turgor in the guard cells is responsible for the opening of the stomatal pore. In very dry conditions there is insufficient supplies of water to the cells for the pores to reach maximum opening or to sustain open pores for long in the face of continuing water loss by transpiration.

page 387

9   This is because water is required to maintain the cells' turgor and, in non woody tissues, allow the cells to enlarge and grow – when the cell walls are expanding water (turgor pressure) is required to maintain this enlarging process. Also it is important that water is required for transport of minerals and sugars inside the plant.

10 In the veins and vascular bundles of the leaf, the water-conducting tissue (xylem) is on the upper side (with phloem below – because of their relative positions in stem vascular bundles with which they connect directly). So looking at the features of marram grass leaf, we can see the outer epidermis is strictly the lower epidermis.

11 Soil water, though present, is unavailable to the plant. This may be because it is frozen, for example, or because the external water has a lower water potential than that of water in the plant (apoplast and symplast), due to excess dissolved salts, for example.

page 391

12 a   It is generated during the Calvin cycle in photosynthesis in leaves supplied with radioactive carbon dioxide in the light.

    b   Catkins (flowers) produced during spring and also the stem.

    c   The distance is determined by the exact location that an individual aphid inserts its proboscis into the phloem.

    d   Mean = 560 mm

e

Experiment	Rate/mm hr^{-1}
1	242.9
2	260.0
3	300.0
4	308.7
5	300.0

f   282.3 mm hr^{-1}

g   Slowest is experiment 1; fastest is experiment 4.

h   As these experiments involved different stems (and separate experiments), it is possible (i) that temperature differences accounted for different rates of translocation, and (ii) the differences in the conditions in the particular source and sink tissues that drive mass flow were responsible.

page 392

13 The companion cells provide all the metabolic functions for the sieve tubes, as well as providing for their own needs, so the ATP demand is very high. Energy is also needed to actively load the sieve tube with the sap containing solutes coming from the source cells.

page 394

14 a   See the annotations to Figure 9.18, page 389.

    b   Sample from near the leaves will contain a higher concentration of sucrose than the one taken from the base of the stem. Sugars made by photosynthesis in the leaves will be loaded in to the sieve tubes high up in the tree. The contents are moved down to the sink organs, where there will be a lower concentration of these sugars.

15 a   Here sugar (delivered by translocation) is converted to starch, lowering the solute potential.

    b   Here, if photosynthesis in the light produces sugar faster than it can be translocated to sink areas in the plant, the solute potential will be high.

16 Transpiration is the loss of water vapour through the stomata in the leaves of plants, whereas translocation is the movement of solutes, such as organic compounds, from source organs such as leaves to sink organs such as roots and fruits.

page 400

17 Light is required for the manufacture of sugar by photosynthesis; also light inhibits extension growth in stems, triggers a positive phototrophic response in stems, promotes expansion of leaf blades, is required for the synthesis of chlorophyll, triggers the switch from vegetative growth to flowering in many plants

(depending upon the length of the light and dark periods within 24 hour cycles) and is required to promote germination in some seeds. Stomatal pores widen in response to increasing light intensity.

page 404

18 Pollination is the arrival of pollen grains from the pollen sacs in the anthers, into the stigma of the carpel. Fertilization refers to the fusion of the gamete cells inside the ovary of the flower.

page 406

19 a This includes plums, peaches and avocados.

   b This includes pea pods and runner beans.

   c This includes gooseberries, cucumbers and tomatoes.

10 Genetics and evolution

page 417

1 Mitosis is the division of ($2n$) somatic cells, producing two genetically identical ($2n$) nuclei, and meiosis is the division of a ($2n$) somatic cell, to produce four (n) gamete cells, each with half of the genetic information.

Mitosis is used to replace damaged and old tissue cells, as well as for growth and embryonic development. Meiosis happens after gonads are mature and only produces gamete cells. See 'Meiosis' and Figure 3.8, page 142.

page 420

2 From the information available in Figure 10.6:

   a the heterozygotes were the F_1 progeny – round and yellow

   b the recombinants (due to reassortment) among the F_2 progeny were 'round and green' and 'wrinkled and yellow'.

page 422

3 Their position is on separate chromosomes. This is important as Mendel was able to observe independent segregation of these traits, as reflected in the law of independent assortment. If these genes were located on the same chromosome they would be linked and the proportions of the progeny would have been different from the ones Mendel found.

page 424

4 A mutant is an organism or genetic feature resulting from a change in the sequence of nucleotides in the DNA, or from a change in the chromosomes.

page 425

5 F_1

	Normal wing and normal body (BBVV)
Vestigial wing and ebony body (bbvv)	All heterozygous, normal wings and normal body (BbVv)

F_2

Genotypes	BBVV 1/16	BBvv 1/16	bbVV 1/16	Bbvv 1/16
	BBVv 2/16	Bbvv 2/16	bbVv 2/16	
	BbVV 2/16			
	BbVv 4/16			
Phenotypes	parental phenotype — normal wing and normal body	normal wing and ebony body	vestigial wing and normal body	parental phenotype — vestigial wing and ebony body
ratio	9	3	3	1

page 427

6 Expected results are equal to ¼ = 50 pea plants

Calculating χ^2

	Predicted	O	E	O – E	$(O – E)^2$	$(O – E)^2/E$
Tall, axial	1	55	50	5	25	0.5
Tall, terminal	1	51	50	1	1	0.02
Dwarf, axial	1	49	50	1	1	0.02
Dwarf, terminal	1	53	50	3	9	0.18
Total		208	$\Sigma(\chi^2)$			0.72

Degrees of freedom = 4 − 1 = 3

Critical value of χ^2 = 7.81

The difference observed is due to chance.

7 The recombinants will be

- $\dfrac{Tb}{tB}$

- $\dfrac{tB}{tB}$

page 429

8 The phenotype ratio was 1 white : 4 light : 6 medium : 4 dark : 1 black

page 431

9 The following factors affect the composition of a gene pool:

- Whole population level: founder effect; random genetic drift, and emigration and immigration
- Members of a population: mutations and selective predation. See Figure 10.16, page 432.

page 434

10 A species is a group of organisms that can interbreed and produce fertile offspring; a variety is a rank below species and it consists of members who differ from others of the same species in minor but heritable ways. Examples should be of garden plants or vegetables, of common varieties.

11 A **mutation** is a change in the sequence of bases in DNA that results in an altered polypeptide and has the potential to change the characteristics of an organism or an individual cell as a result of alterations in, or non-production of, proteins specified by the mutated DNA. Mutations occurring in body cells of multicellular organisms, i.e. somatic mutations, are only passed on to the immediate descendents of those cells, and disappear when the organism dies.

However, mutations occurring in ovaries or testes (or anthers or embryo sac of flowering plants) – **germ line mutations** – are mutations that may be passed to the offspring and persist from generation to generation.

As a result of this source of genetic variation, the individual offspring of parents may show variations in their characteristics. Some of these changes may confer an advantage. Favourable characteristics expressed in the phenotypes of some of the offspring may make them better able to survive and reproduce in a particular environment. Others will be less able to compete successfully, survive and reproduce. This is the principle of **natural selection**.

page 438

12 The frequency for indigenous Australians is 0.49. The frequency for indigenous North Americans is 0.04.

11 Animal physiology

page 444

1 The function of the thymus gland is to receive immature T cells (produced in the bone marrow) to select only those that (i) attack antigens of foreign cells, and (ii) exclude any T cells that also attack the body's own cells. The remaining T cells are allowed to mature and pass out of the thymus gland. These cells are able to protect the body against invasion, normally without danger of autoimmune diseases being triggered by the immune system.

page 447

2 Antigens are found on the surface of bacterial cells, that invade the body. They are also found on the surface of our red blood cells. Antibodies are produced by our own cells in the circulatory system in response to non-self antigens. Antibodies exist in the blood and lymph, and may be carried in the plasma solution anywhere that blood 'leaks out' to – including sites of invasions.

3 The drawing must demonstrate the student's ability to differentiate between the polynucleated cells and the mononucleated ones, as well as differentiating between the granular and the agranular types.

page 451

4 The first priority in combating TB is to make sure that the quality of peoples' housing ensures that they are not exposed to conditions favouring the transmission of TB and also that their diet and lifestyle choices enable them to be healthy. The first group to receive the vaccine is children. Then people at high risk, such as those living in poverty or in economically depressed zones where access to a hospital is not possible. After this, treat older people and patients with respiratory problems or with immunodeficiency syndromes. Finally, apply the vaccination programme to people who are located near reported areas of infection and then start extending the vaccination programme to the rest of the population.

5 The data supports the use of vaccines, as it is shows that there was a decrease in the number of cases after the vaccination programme started. This reduction was then maintained.

page 453

6 See 'Memory cells and immunity', page 447, and Figure 11.3 on page 445.

7 Immunity is resistance to the onset of disease after infection by harmful microorganisms or internal parasites.

page 454

8 • Generate antibody-producing cells by immunizing a mouse against the antigen of interest.
• Perform a blood test to determine the presence of the desired antibody.
• Remove the spleen from the mouse and culture the B cells with myeloma (cancer) cells that divide indefinitely.
• Allow the cells to produce hybridoma cells in the culture and then eliminate the B cells which did not fuse with the myeloma cells.
• Allow each hybridoma cell to divide, creating a clonal culture, where by all cells are clones from the first one.
• Check, after some weeks, that the desired antibody is produced by the hybridoma cloned cells.
• Clones that produce the desired antibody are mass cultured and frozen for future use.

page 460

9 The **exoskeleton** is the external skeleton that supports and protects an animal; it is made of chitin in insects and deposits of calcium carbonate in shelled animals. An **endoskeleton** is an internal support structure in animals, such the skeleton in humans.

A **bone** is a rigid organ, made of dense connective tissue, that provides support to organs and produces blood cells and blood components. **Cartilage** is a flexible connective tissue that cushions bones at joints, in order to reduce friction between the endings of the bones. It is also found in other parts in the body, such as the outside of the ear and in the larynx.

Hinge joints allow motion in one plane as the bones are aligned in one plane, whereas **ball-and-socket** joints offer a freedom of motion as there is a ball-shaped surface bone that moves inside the depression of another bone. Both joints are synovial.

A **ligament** is a fibrous connective tissue that connects bones to other bones, and also helps to support internal organs. A **tendon** is also a fibrous connective tissue, but one that connects muscles to bones and transmits the mechanical force from muscles to bones, being able to withstand the tension.

page 462

10 a Muscle fibre: these are long, multinucleated cells that form the basic unit of a muscle.

b Myofibril: these tubular structures are contained by the muscle fibre. They are composed of actin and myosin proteins, organized into thin and thick filaments, and are responsible for the muscle contraction.

c Myosin filament: this is a motor protein that uses ATP for muscle contraction, generating a force in skeletal muscle by means of a power stroke mechanism.

page 466

11 a This is a relaxed sarcomere. There is an M line, and thick (myosin) and thin (actin) filaments are visible in the centre of the sarcomere. (The H zone is visible.)

b This must show a shorter sarcomere, but with the same length of dark band (thick myosin). The Z lines from both sides must be closer than in the original picture.

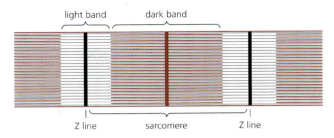

12 A relaxed sarcomere typically has a length of about 3.5 μm.

page 468

13 To be an efficient surface for the absorption of oxygen gas, the surface of the lungs has to be moist. Consequently, the process of gaseous exchange also results in continuous loss of water vapour from the body. See Figure 11.24, page 467.

page 469

14 • **Excretion** is the process by which waste products from metabolic activities are moved outside the body (for example, during urination), whereas **egestion** is the discharge of undigested material during defecation.

• **Osmoregulation** is the regulation of the osmotic potential body fluids by controlling the amounts of water and of salts present in the blood and tissue fluid. See 'Step 5: Water reabsorption in the collecting ducts', page 475.
Secretion is the production of useful substances outside the cell for a particular function in an organ. For example, gland cells secrete hormones into the bloodstream.

page 471

15 The source of energy for ultrafiltration in the glomerulus is respiration of heart muscle that 'fuels' contractions of the ventricles and thus creates blood pressure in the glomerulus.

page 472

16 See 'Step 2: Selective reabsorption in the proximal convoluted tubule', page 472.

page 477

17 ADH is released as part of the osmoregulation mechanism when the level of water in the body is low. This situation leads to an increased urine concentration, which in turn decreases water excretion. See Figure 11.34, page 478.

page 478

18 Same concentration of both blood cells and plasma proteins; same or almost the same concentration of glucose; urea concentration will be different with $30\,mg\,100\,ml^{-1}$ of urea in the renal artery and a lower amount of this waste product in the renal vein. See Table 11.7, page 478.

page 479

19 This is due to the nature of the kidney as a highly vascular tissue that contains endothelial cells with antigens belonging to the major histocompatibility complex or MHC. This can induce an immune response, leading to a critical rejection of the transplanted kidney. The MHC will be the same in identical twins.

page 483

20 The numbers of individuals of most populations remain more or less constant despite the fact that usually many more young are produced than reach adulthood and survive to breed themselves.

page 485

21 Structures and functions of a mature sperm cell:

- Tail: motility.
- Middle piece: contains many mitochondria to provide the required energy for the movement of the tail.
- Head: surrounded by the plasma membrane, containing a haploid nucleus.
- Acrosomal cap or acrosome (at the top of the head): contains enzymes for digesting the zona pellucida (jelly coat) in the oocyte.

page 488

22 Structures and functions of a mature secondary oocyte:

- Follicle cells: nourish and protect the oocyte.
- Zona pellucida (jelly coat): allows binding of sperm cells, prevents polyspermia (several sperm haploid nuclei entering), also prevents premature implantation of the embryo (ectopic pregnancy). This is also the place where the polar body is found.
- Plasma membrane: contains the haploid nucleus and contains microvilli to absorb nutrients from follicular cells.
- Cortical granules: during fertilization prevent polyspermia.
- Cytoplasm: place for metabolic reactions and where all organelles are located.
- Nucleus: haploid nucleus with half of the genetic information.

23 FSH in males acts on the Sertoli cells to stimulate sperm production. In females, it stimulates the growth of ovarian follicles.

page 493

24 The fetus is 'foreign' tissue to the mother (for example, they may not be of the same blood group) and carries antigens foreign to her. Ensuring the separation of the maternal and fetal blood supplies avoids an immunological response of the mother against the presence of the embryonic tissue. Such a response could lead to the end of pregnancy.

25 It has finger-like projections (villi), providing a large surface area for exchange of substances between the embryo's blood and the mother's blood. It has spaces around the villi (lacuna) that contain maternal blood to pass nutrients and oxygen to the embryo's blood.

page 495

26 The chief effect of HCG is to maintain the corpus luteum as an endrocrine gland, at least for the first 16 weeks of the pregnancy. The maintenance of the corpus luteum means that the endometrium persists, to the benefit of the implanted embryo, and menstruation is prevented. (The premature demise of the corpus luteum, leading to degeneration of the endometrium during early pregnancy, is a possible cause of a miscarriage.)

27 In **negative feedback** the effect of a deviation from the normal or set condition is to create a tendency to eliminate the deviation. Negative feedback is a part of almost all control systems in living things. The effect of negative feedback is to reduce further corrective action of the control system once the set-point value is reached.

In **positive feedback** the effect of a deviation from the normal or set condition is to create a tendency to reinforce the deviation. Positive feedback intensifies the corrective action taken by a control system, so leading to a vicious circle. Imagine a car in which the driver's seat was set on rollers (not secured to the floor), being driven at speed. The slightest application of the foot brake causes the driver to slide and to press harder on the brake as the car starts to slow, with an extreme outcome.

Biological examples of positive feedback are rare, but one can be identified at the synapse. When a wave of depolarization (a nerve impulse) takes effect in the post-synaptic membrane, the entry of sodium ions triggers the entry of further sodium ions at a greater rate. The depolarized state is established, and the impulse moves along the post-synaptic membrane.

Glossary

A

abiotic factor a non-biological factor (e.g. temperature) that is part of the environment of an organism

abscisic acid a plant growth substance tending to inhibit growth

absorption spectrum range of a pigment's ability to absorb various wavelengths of light

acetylcholine a neurotransmitter, liberated at synapses in the CNS

acid rain the cocktail of chemical pollutants that may occur in the atmosphere

action potential rapid change (depolarization) in membrane potential of an excitable cell (e.g. a neuron)

action spectrum range of wavelengths of light within which a process like photosynthesis takes place

activation energy energy required by a substrate molecule before it can undergo a chemical change

active site region of enzyme molecule where substrate molecule binds

active transport movement of substances across a membrane involving a carrier protein and energy from respiration

adenine a purine organic base, found in the coenzymes ATP and NADP, and in nucleic acids (DNA and RNA) in which it pairs with thymine

adenosine diphosphate (ADP) a nucleotide, present in every living cell, made of adenosine and two phosphate groups linked in series, and important in energy transfer reactions of metabolism

adenosine triphosphate (ATP) a nucleotide, present in every living cell, formed in photosynthesis and respiration from ADP and P_i, and functioning in metabolism as a common intermediate between energy-requiring and energy-yielding reactions

adrenaline (also called epinephrine) a hormone secreted by the adrenal medulla (and a neurotransmitter secreted by nerve endings of the sympathetic nervous system), having many effects, including speeding of heartbeat, and the breakdown of glycogen to glucose in muscle and liver

aerobic respiration respiration requiring oxygen, involving oxidation of glucose to carbon dioxide and water

alimentary canal the gut; a tube running from mouth to anus in vertebrates, where complex food substances are digested and the products of digestion selectively absorbed into the body

allele an alternative form of a gene, occupying a specific locus on a chromosome

allele frequency the commonness of the occurrence of any particular allele in a population

alpha cell (pancreas) glucagon-secreting cell of the islets of Langerhans in the pancreas

alveolus air sac in the lung

amino acid building block of proteins, of general formula $R.CH(NH_2).COOH$

amphipathic having hydrophobic and hydrophilic parts on the same molecule

anabolism the building up of complex molecules from smaller ones

anaerobic respiration respiration in the absence of oxygen, involving breakdown of glucose to lactic acid or ethanol

analogous structure similar in structure but of different evolutionary origin

analyse break down in order to bring out the essential elements or structure

anion negatively charged ion

annotate add brief notes to a diagram or graph

anther part of the stamen in flowers, consisting of pollen sacs enclosed in walls that eventually split open, releasing pollen

antibiotics organic compounds produced by some microorganisms which selectively inhibit or kill other microorganisms

antibody a protein produced by blood plasma cells derived from B lymphocytes when in the presence of a specific antigen, which then binds with the antigen, aiding its destruction

anticodon three consecutive bases in tRNA, complementary to a codon on RNA

antidiuretic hormone (ADH) hormone secreted by the pituitary gland that controls the permeability of the walls of the collecting ducts of the kidney

antigen a substance capable of binding specifically to an antibody

apoplast collective name for the cell walls of a tissue or plant

aquaporin a water channel pore (protein) in a membrane

aqueous humour fluid between lens and cornea of the eye

arteriole a very small artery

artificial classification classifying organisms on the basis of few, self-evident features

artificial selection selection in breeding exercises, carried out deliberately, by humans

asexual reproduction reproduction not involving gametes and fertilization

assimilation uptake of nutrients into cells and tissues

atherosclerosis deposition of plaque (cholesterol derivative) on inner wall of blood vessels

atrioventricular node mass of tissue in the wall of the right atrium, functionally part of the pacemaker mechanism

atrioventricular valve tricuspid or bicuspid valve

atrium (plural, **atria**) one of the two upper chambers of the mammalian four-chambered heart

autolysis self-digestion

autonomic the involuntary nervous system

autotrophic (organism) self-feeding – able to make its own elaborated foods from simpler substances

auxin plant growth substance, indoleacetic acid

axon fibre carrying impulses away from the cell body of a neuron

B

bacillus a rod-shaped bacterium

bacteriophage a virus that parasitizes bacteria (also known as a phage)

baroreceptor a sensory receptor responding to stretch, in the walls of blood vessels

basement membrane the thin fibrous layer separating an epithelium from underlying tissues

beta cell (pancreas) insulin-secreting cells of the islets of Langerhans in the pancreas

bicuspid valve valve between atrium and ventricle on the left side of the mammalian heart

bile an alkaline secretion of liver cells which collects in the gall bladder in humans, and which is discharged into the duodenum periodically

binary fission when a cell divides into two daughter cells, typically in reproduction of prokaryotes

binomial system double names for organisms, in Latin, the generic preceding the specific name

bioinformatics interdisciplinary science of storage, retrieval, organization and analysis of biological data

biomass total mass of living organisms in a given area (e.g. a quadrat)

biome a major life-zone over an area of the Earth, characterized by the dominant plant life present

bioremediation waste management techniques that involve the use of organisms to neutralize pollutants at contaminated sites

biosphere the inhabited part of the Earth

biotechnology the industrial and commercial applications of biology, particularly of microorganisms, enzymology and genetic engineering

biotic factor the influence of living things on the environment of other living things

bivalent a pair of duplicated chromosomes, held together by chiasmata during meiosis

blastocyst embryo as hollow ball of cells, at the stage of implantation

blind spot region of the retina where the optic nerve leaves

body mass index (BMI) body mass in $kg/(height in m)^2$

bone marrow tissue special connective tissue filling the cavity of certain bones

boreal forest northern coniferous forests (example of a biome)

brain the coordinating centre of the nervous system

breed (animal) the animal equivalent of a plant variety

bronchiole small terminal branch of a bronchus

bronchus a tube connecting the trachea with the lungs

brush border tiny, finger-like projections (microvilli) on the surface of epithelial cells of the small intestine

buffer a solution which minimizes change in pH when acid or alkali are added

bundle of His bundles of long muscle fibres that transmit myogenic excitation throughout the ventricle walls

C

C_3 **pathway** the light-independent reaction in photosynthesis, producing as its first product a 3-carbon compound, glycerate 3-phosphate

calculate obtain a numerical answer showing the relevant stages in the working (unless instructed not to do so)

Calvin cycle a cycle of reactions in the stroma of the chloroplast by which some of the product of the dark reaction is reformed as the acceptor molecule for carbon dioxide (ribulose biphosphate)

carcinogen any substance or radiation directly involved in causing cancer

cardiac cycle the stages of the heartbeat, by which the atrial and then the ventricular walls alternately contract (systole) and relax (diastole)

carnivore flesh-eating animal

carrier an individual that has one copy of a recessive allele that causes a genetic disease in individuals that are homozygous for this allele

carrier protein one of the types of protein in plasma membranes, responsible for active transport across the membranes

cartilage firm but plastic skeletal material (e.g. cartilage over bones at joints)

Casparian strip band of cells with impervious walls, found in plant roots

catabolism the breaking down of complex molecules in the biochemistry of cells

catalyst a substance that alters the rate of a chemical reaction, but remains unchanged at the end

cellular respiration controlled release (transfer) of energy from organic compounds in cells to form ATP

cellulase enzyme capable of hydrolysing cellulose

cellulose an unbranched polymer of 2000–3000 glucose residues, the major ingredient of most plant walls

central dogma the idea that transfer of genetic information from DNA of the chromosome to mRNA to protein (amino acid sequence) is irreversible

centromere constriction of the chromosome, the region that becomes attached to the spindle fibres in division

centrosome organelle situated near the nucleus in animal cells, involved in the formation of the spindle prior to nuclear division

cephalization development of a head at the anterior of an animal

cerebellum part of hindbrain, concerned with muscle tone, posture and movement

cerebral cortex superficial layer of grey matter on extension of forebrain, much enlarged in humans and other apes

cerebral hemispheres (cerebrum) the bulk of the human brain, formed during development by the outgrowth of part of the forebrain, consisting of densely packed neurons and myelinated nerve fibres

chemiosmosis movement of ions down an electrochemical gradient, linked to the generation of ATP

chemoautotroph an organism that uses energy from chemical reactions to generate ATP and produce organic compounds from inorganic substances

chemoheterotroph an organism that uses energy from chemical reactions to generate ATP and obtains organic compounds from other organisms

chemoreceptor a sense organ receiving chemical stimuli

chemosynthesis use of chemical energy from oxidation of inorganic compounds to synthesize organic compounds, typically from carbon dioxide and water

chiasma (plural, **chiasmata**) site of crossing over (exchange) of segments of DNA between homologous chromosomes

chlorophyll the main photosynthetic pigment of green plants, occurs in the grana membranes (thylakoid membranes) of the chloroplasts

chloroplast organelle that is the site of photosynthesis and contains chlorophyll

cholesterol a lipid of animal plasma membranes; a precursor of the steroid hormones, in humans, formed in the liver and transported in the blood as lipoprotein

choroid layer of blood vessels lying below the retina

chromatid one of two copies of a chromosome after it has replicated

chromatin a nuclear protein material in the nucleus of eukaryotic cells at interphase; forms into chromosomes during mitosis and meiosis

chromosome visible in appropriately stained cells at nuclear division, each chromosome consists of a long thread of DNA packaged with protein; chromosomes replicate prior to division, into chromatids. Contents of nucleus appears as granular chromatin between divisions

chyme partly digested food as it leaves the stomach

cilium (plural, **cilia**) motile, hair-like outgrowth from surface of certain eukaryotic cells

citric acid cycle see *Krebs cycle*

clade the branch of a phylogenetic tree containing the set of all organisms descended from a particular common ancestor which is not an ancestor of any non-member of the group

cladistics method of classifying living organisms that makes use of lines of descent only (rather than phenotypic similarities)

climax community the mature (stable) stage of a succession of communities

clone a group of genetically identical individuals (or cells)

CNS see *nervous system*

codominant alleles pairs of alleles that both affect the phenotype when present in a heterozygous state

codon three consecutive bases in DNA (or RNA) which specify an amino acid

coleoptile protective sheath around emerging leaves of germinating grass seeds

colon part of the gut, preceding the rectum

colostrum first milk secreted by the mother, after birth of young

commensalism a mutually beneficial association between two organism of different species

comment give a judgment based on a given statement or result of a calculation

community a group of populations of organisms living and interacting with each other in a habitat

compare give an account of similarities between two (or more) items or situations, referring to both (all) of them throughout

compare and contrast give an account of similarities and differences between two (or more) items or situations, referring to both (all) of them throughout

compensation point the point where respiration and photosynthesis are balanced

condensation reaction formation of larger molecules involving the removal of water from smaller component molecules

cone (retinal cell) a light-sensitive cell in the retina, responsible for colour vision

conjugate protein protein combined with a non-protein part

connective tissue tissues that support and bind tissues together

conservation applying the principles of ecology to manage the environment

contractile vacuole a small vesicle in the cytoplasm of many fresh water protozoa that expels excess water

construct display information in a diagrammatic or logical form

cornea transparent covering at the front of the eye

corpus luteum glandular mass that develops from an ovarian follicle in mammals, after the ovum is discharged

cotyledon the first leaf (leaves) of a seed plant, found in the embryo

covalent bond bond between atoms in which electrons are shared

cristae folds in the inner membrane of mitochondria

crossing over exchange of genetic material between homologous chromosomes during meiosis

crypt of Lieberkühn endocrine cells within the pancreas

cuticle layer of waxy material on outer wall of epidermis

cyanobacteria photosynthetic prokaryotes

cytokinesis division of cytoplasm after nucleus has divided into two

cytology study of cell structure

cytoplasm living part of the cell bound by the plasma membrane, excluding the nucleus

cytosol what remains of cytoplasm when the organelles have been removed

D

data recorded products of observations and measurements
qualitative data observations not involving measurements
quantitative data precise observations involving measurements

deamination the removal of NH_2 from an amino acid

decomposer organisms (typically microorganisms) that feed on dead plant and animal material, causing matter to be recycled by other living things

deduce reach a conclusion from the information given

define give the precise meaning of a word, phrase, concept or physical quantity

degenerate code the triplet code contains more codons than there are amino acids to be coded, so most amino acids are coded for by more than one codon

denaturation a structural change in a protein that results in a loss (usually permanent) of its biological properties

dendrite a fine fibrous process on a neuron that receives impulses from other neurons

depolarization (of axon) a temporary and local reversal of the resting potential difference of the membrane that occurs when an impulse is transmitted along the axon

describe give a detailed account

desertification the conversion of marginal cultivated land into desert, caused by climate change or by over-grazing or inferior cultivation

design produce a plan, simulation or model

determine obtain the only possible answer

detrital chain a food chain based on dead plant matter

detritivore an organism that feeds on detritus (dead organic matter)

dialysis separation of large and small molecules in solution by the inability of the former to pass through a selectively permeable membrane

diaphragm a sheet of tissues, largely muscle, separating thorax from abdomen in mammals

diastole relaxation phase in the cardiac cycle

dichotomous key one in which a group of organisms is progressively divided into two groups of smaller size

dicotyledon class of Angiospermophyta having an embryo with two seed leaves (cotyledons)

diffusion passive movement of particles from a region of high concentration to a region of low concentration

dihybrid cross one in which the inheritance of two pairs of contrasting characters (controlled by genes on separate chromosomes) is observed

diploid condition organisms whose cells have nuclei containing two sets of chromosomes

disaccharide a sugar that is a condensation product of two monosaccharides (e.g. maltose)

discuss offer a considered and balanced review that includes a range of arguments, factors or hypotheses. Opinions or conclusions should be presented clearly and supported by appropriate evidence

distinguish make clear the differences between two or more concepts or items

disulfide bond S—S bond between two S-containing amino acid residues in a polypeptide or protein chain

diuresis increased secretion of urine

division of labour the carrying out of specialized functions by different types of cell in a multicellular organism

DNA a form of nucleic acid found in the nucleus, consisting of two complementary chains of deoxyribonucleotide subunits, and containing the bases adenine, thymine, guanine and cytosine

dominant allele an allele that has the same effect on the phenotype whether it is present in the homozygous or heterozygous state

double bond a covalent bond involving the sharing of two pairs of electrons (rather than one)

double circulation in which the blood passes twice through the heart (pulmonary circulation, then systemic circulation) in any one complete circuit of the body

double fertilization a feature of flowering plants in which two male nuclei enter the embryo sac, and one fuses with the egg cell and one with the endosperm nucleus

draw represent by means of a labelled, accurate diagram or graph, using a pencil. A ruler (straight edge) should be used for straight lines. Diagrams should be drawn to scale. Graphs should have points correctly plotted (if appropriate) and joined in a straight line or smooth curve

duodenum the first part of the intestine after the stomach

E

ecology the study of relationships between living organisms and between organisms and their environment – a community and its abiotic environment

ecosystem a natural unit of living (biotic) components and non-living (abiotic) components (e.g. temperate deciduous forest)

edaphic factor factor influenced by the soil

effector an organ or cell that responds to a stimulus by doing something (e.g. a muscle contracting, a gland secreting)

egestion disposal of waste from the body (e.g. defecation)

egg cell an alternative names for an ovum

electron microscope (EM) microscope in which a beam of electrons replaces light, and the powers of magnification and resolution are correspondingly much greater

electron-transport system carriers that transfer electrons along a redox chain, permitting ATP to be synthesized in the process

embolism a blood clot blocking a blood vessel

embryo the earliest stages in development of a new animal or plant, from a fertilized ovum, entirely dependent on nutrients supplied by the parent

embryo sac occurs in the ovule of flowering plants, and contains the egg cell and endosperm nucleus

emulsify to break fats and oils into very tiny droplets

endemic species restricted to a particular region

endergonic reaction metabolic reaction requiring energy input

endocrine glands the hormone-producing glands that release secretions directly into the body fluids

endocytosis uptake of fluid or tiny particles into vacuoles in the cytoplasm, carried out at the plasma membrane

endoplasmic reticulum system of branching membranes in the cytoplasm of eukaryotic cells, existing as rough ER (with ribosomes) or as smooth ER (without ribosomes)

endoskeleton an internal skeleton system

endosperm the stored food reserves within the seeds of flowering plants

endothelium a single layer of cells lining blood vessels and other fluid-filled cavities

endothermic generation of body heat metabolically

enzyme mainly proteins (a very few are RNA) that function as biological catalysts

epidemiology the study of the occurrence, distribution and control of disease

epidermis outer layer(s) of cells

epigenetics study of heritable changes in gene activity not caused by changes in DNA

epiglottis flap of cartilage that closes off the trachea when food is swallowed

epiphyte plant living on the surface of other plants

epithelium sheet of cells bound strongly together, covering internal or external surfaces of multicellular organisms

epitope part of an antigen that is recognized by the immune system

erythrocyte red blood cell

estimate obtain an approximate value

estrous period of fertility (immediately after ovulation) during the estrous cycle

estrous cycle reproductive cycle in female mammal in the absence of pregnancy

etiolation the condition of plants when grown in the dark

eukaryotic (cells) cells with a 'good nucleus' (e.g. animal, plant, fungi and protoctista cells)

evaluate make an appraisal by weighing up the strengths and limitations

evolution cumulative change in the heritable characteristics of a population

ex situ not in its original or natural position or habitat

excretion removal from the body of the waste products of metabolic pathways

exergonic reaction metabolic reaction releasing energy

exocrine gland gland whose secretion is released via a duct

exocytosis secretion of liquids and suspensions of very fine particles across the membrane of eukaryotic cells

exoskeleton skeleton secreted external to the epidermis of the body

exothermic chemical reaction that releases energy as heat (an endothermic reaction requires heat energy)

expiratory emitting air during breathing

explain give a detailed account including reasons or causes

extensor muscle a muscle that extends or straightens a limb

F

F_1 generation first filial generation – arises by crossing parents (P) and, when selfed or crossed via sibling crosses, produces the F_2 generation

facilitated diffusion diffusion across a membrane facilitated by molecules in the membrane (without the expenditure of metabolic energy)

fermentation anaerobic breakdown of glucose, with end-products ethanol and carbon dioxide or lactic acid

fertilization the fusion of male and female gametes to form a zygote

fetus a mammalian embryo when it becomes recognizable (e.g. the human embryo from 7 weeks after fertilization)

field layer the layer of herbaceous plants in a forest or wood

filter-feeding feeding on tiny organisms which are strained from the surrounding medium

fimbria (singular, **fimbrium**) thin, short filaments protruding from some bacteria, involved in attachment

flaccid state of a tissue with insufficient water, as in wilting leaves

flagellum (plural, **flagella**) a long thin structure, occurring singly or in groups on some cells and tissues, and used to propel unicellular organisms, and to move liquids past anchored cells (flagella of prokaryotes and eukaryotes are of different internal structure)

flexor muscle a muscle that on contraction bends a limb (or part of a limb)

flower develops from the tip of a shoot, with outer parts (e.g. sepals, petals) surrounding the male and female reproductive organs

fluid mosaic model the accepted view of the structure of the plasma membrane, comprising a phospholipid bilayer with proteins embedded but free to move about

food chain a sequence of organisms within a habitat in which each is the food of the next, starting with a producer, which is photosynthetic

food web interconnected food chains

founder effect genetic differences that develop between an original breeding population and a small isolated interbreeding group of these organisms

fovea point on a retina of greatest acuity of vision

free energy part of the potential chemical energy in molecules that is available to do useful work when the molecules are broken

freeze etching preparation of specimens for electron microscope examination by freezing, fracturing along natural structural lines and preparing a replica

frequency commonness of an occurrence

fruit forms from the ovary after fertilization, as the ovules develop into seeds

functional group the chemically active part of a member of a series of organic molecules

fungus heterotrophic, non-motile, multicellular (usually) eukaryotic organism with 'plant' body – a mycelium of hyphae with cell walls of chitin; the fungi constitute a separate kingdom

G

gall bladder sac beside the liver that stores bile, present in some mammals (e.g. humans)

gamete sex cell (e.g. ovum, sperm)

ganglion part of a nervous system, consisting of nerve cell bodies

gaseous exchange exchange of respiratory gases (oxygen, carbon dioxide) between cells/organism and the environment

gastric relating to the stomach

gene a heritable factor that controls a specific characteristic

gene mutation change in the chemical structure (base sequence) of a gene resulting in change in the characteristics of an organism or individual cell

gene pool all the genes (and their alleles) present in a breeding population

gene probe an artificially prepared sequence of DNA made radioactive with ^{14}C, coding for a particular amino acid residue sequence

gene therapy various mechanisms by which corrected copies of genes are introduced into a patient with a genetic disease

generator potential localized depolarization of a membrane of a sensory cell

genetic code the order of bases in DNA (of a chromosome) that determines the sequence of amino acids in a protein

genetic counselling genetic advice to potential parents on the risks of having children with an inherited disease

genetic engineering change to the genetic constitution of individuals or populations by artificial selection

genome the whole of the genetic information of an organism

genotype the genetic constitution of an organism – the alleles of an organism

genus a group of similar and closely related species

germination the resumption of growth by an embryonic plant in seed or fruit, at the expense of stored food

gland cells or tissues adapted for secretion

global warming the hypothesis that the world climate is warming due to rising levels of atmospheric carbon dioxide, a greenhouse gas

glomerulus network of capillaries which are surrounded by the renal capsule

glycocalyx long carbohydrate molecules attached to membrane proteins and membrane lipids

glycogen a much-branched polymer of glucose, the storage carbohydrate of many animals

glycogenesis the synthesis of glycogen from glucose (the reverse is glycogenolysis)

glycolysis the first stage of tissue respiration in which glucose is broken down to pyruvic acid, without use of oxygen

glycoprotein membrane protein with a glycocalyx attached

glycosidic bond a type of chemical linkage between monosaccharide residues in polysaccharides

goblet cell mucus-secreting cell of an epithelium

Golgi apparatus a stack of flattened membranes in the cytoplasm, the site of synthesis of biochemicals

gonad an organ in which gametes are formed

gonadotrophic hormone follicle-stimulating hormone (FSH) and luteinizing hormone (LH), secreted by the anterior pituitary, which stimulate gonad function

granum (plural, **grana**) stacked discs of membranes found within the chloroplast, containing the photosynthetic pigments, and the site of the light-dependent reaction of photosynthesis

grey matter regions of the brain and spinal cord consisting largely of nerve cell bodies

growth more or less irreversible increase in size and amount of dry matter

gut the alimentary canal

H

habitat the locality or surroundings in which an organism normally lives or the location of a living organism

halophyte a plant adapted to survive at abnormally high salt levels (e.g. seashore or salt marsh plant)

haploid (cells) cells having one set of chromosomes, the basic set

heart rate number of contractions of the heart per minute

helicase an enzyme that unwinds the DNA double helix

hemoglobin a conjugated protein, found in red cells, effective at carrying oxygen from regions of high partial pressure (e.g. lungs) to regions of low partial pressure of oxygen (e.g. respiring tissues)

hepatic associated with the liver

herb layer layer of herbaceous plants (mainly perennials) growing in woodland

herbaceous non-woody

herbicide pesticide toxic to plants

herbivore an animal that feeds (holozoically) exclusively on plants

hermaphrodite organism with both male and female reproductive systems

heterotroph an organism incapable of synthesizing its own elaborated nutrients

heterozygous having two different alleles of a gene

hexose a monosaccharide containing six carbon atoms (e.g. glucose, fructose)

hibernation passing the unfavourable season in a resting state of sleep

histology the study of the structure of tissues

histone basic proteins (rich in the amino acids arginine and lysine) that form the scaffolding of chromosomes

holozoic ingesting complex food material and digesting it

homeostasis maintenance of a constant internal environment

homeotherm organism that maintains a constant body temperature

homologous chromosomes chromosomes in a diploid cell which contain the same sequence of genes, but are derived from different parents

homologous structures similar due to common ancestry

homozygous having two identical alleles of a gene

hormone a substance, formed by an endocrine gland and transported in the blood all over the body, but triggering a specific physiological response in one type of organ or tissue

host an organism in or on which a parasite spends all or part of its lifecycle

humus complex organic matter, the end-product of the breakdown of the remains of plants and animals, which covers the mineral particles of soil

hybrid an individual produced from a cross between two genetically unlike parents

hybridoma an artificially produced hybrid cell culture, used to produce monoclonal antibodies

hydrocarbon chain a linear arrangement of carbon atoms combined together and with hydrogen atoms, forming a hydrophobic tail to many large organic molecules

hydrogen bond a weak bond caused by electrostatic attraction between a positively charged part of one molecule and a negatively charged part of another

hydrolysis a reaction in which hydrogen and hydroxide ions from water are added to a large molecule causing it to split into smaller molecules

hydrophilic water loving

hydrophobic water hating

hydrophyte an aquatic plant

hydrosere a plant succession that originated from open water

hydrostatic pressure mechanical pressure exerted on or by liquid (e.g. water) also known as pressure potential

hyperglycemia excess glucose in the blood

hypertonic solution a more concentrated solution (one with a less negative water potential) than the cell solution

hypha the tubular filament 'plant' body of a fungus, which in certain species is divided by cross walls into either multicellular or unicellular compartments

hypoglycemia very low levels of blood glucose

hypothalamus part of floor of the rear of the forebrain, a control centre for the autonomic nervous system, and source of releasing factors for pituitary hormones

hypothesis a tentative (and testable) explanation of an observed phenomenon or event

hypotonic solution a less concentrated solution (one with a more negative water potential) than the cell solution

I

identify provide an answer from a number of possibilities

immunity resistance to the onset of a disease after infection by the causative agent

active immunity immunity due to the production of antibodies by the organism itself after the body's defence mechanisms have been stimulated by antigens

passive immunity immunity due to the acquisition of antibodies from another organism in which active immunity has been stimulated, including via the placenta, colostrum, or by injection of antibodies

immunization (e.g. inoculation/vaccination) the injection of a specific antigen, derived from a pathogen, to confer immunity against a disease

immunoglobin proteins synthesized by the B lymphocytes of the immune system

immunology study of the immune system

immunosuppressant a substance causing temporary suppression of the immune response

implantation embedding of the blastocyst (developed from the fertilized ovum) in the uterus wall

imprinting process occurring soon after birth, causing young birds to follow their mother

impulse see *action potential*

in situ in the original place (in the body or organism)

in vitro biological processes occurring in cell extracts (literally 'in glass')

in vivo biological process occurring in a living organism (literally 'in life')

inbreeding when gametes of closely related individuals fuse leading to progeny that is homozygous for some or many alleles

incubation period period between infection by a causative agent and the appearance of the symptoms of a disease

incus tiny, anvil-shaped bone, the middle ossicle of the middle ear in mammals

industrial melanism increasing proportion of a darkened (melanic) form of an organism, in place of the light-coloured form, associated with industrial pollution by soot

infectious disease disease capable of being transmitted from one organism to another

inhibitor (enzyme) a substance which slows or blocks enzyme action (a competitive inhibitor binds to the active site; a non-competitive inhibitor binds to another part of the enzyme)

inhibitory synapse synapse at which arrival of an impulse blocks forward transmissions of impulses in the post-synaptic membrane

innate behaviour behaviour that does not need to be learned

innervation nerve supply

inspiratory capacity amount of air that can be drawn into the lungs

intelligence the ability to learn by reasoning and to solve problems not yet experienced

interferon proteins formed by vertebrate cells in response to virus infections

intermediates metabolites formed as components of a metabolic pathway

interphase the period between nuclear divisions when the nucleus controls and directs the activity of the cell

interspecific competition competition between organisms of different species

intestine the gut

intracellular enzymes enzymes operating inside the cell

intraspecific competition competition between organisms of the same species

intron a non-coding nucleotide sequence of the DNA of chromosomes, present in eukaryotic chromosomes

invagination the intucking of a surface or wall

ion charged particle formed by the transfer of electron(s) from one atom to another

ionic bonding strong electrostatic attraction between oppositely charged ions

iris circular disc of tissue, in front of the lens of the eye, containing circular and radial muscles

irreversible inhibition inhibition by inhibitors that bind tightly and permanently to an enzyme, destroying its catalytic properties

islets of Langerhans groups of endocrine cells scattered through the pancreas

isomers chemical compounds of the same chemical formula but different structural formulae

isotonic being of the same osmotic concentration and therefore of the same water potential

isotopes different forms of an element, chemically identical but with slightly different physical properties, based on differences in atomic mass (due to different numbers of neutrons in the nucleus)

J

joule the SI unit of energy

K

karyogram the chromosomes of an organism in homologous pairs of decreasing length

karyotype the number and type of chromosomes present in an organism

keratin a fibrous protein found in horn, hair, nails and the upper layer of skin

keystone species species that have a key role in an ecosystem

kinesis random movements maintained by motile organisms until more favourable conditions are reached

kinetic energy energy in movement

kingdom the largest and most inclusive group in taxonomy

Krebs cycle part of tissue respiration

L

label add labels to a diagram

lactation secretion of milk in mammary glands

leaching washing out of soluble ions and nutrients by water drainage through soil

learnt behaviour in animals, behaviour that is consistently modified as a result of experiences

leucocyte white blood cell

lichens permanent, mutualistic associations between certain fungi and algae, forming organisms found encrusting walls, tree trunks and rocks

ligament strong fibrous cord or capsule of slightly elastic fibres, connecting movable bones

light-dependent step part of photosynthesis occurring in grana of the chloroplasts, in which water is split and ATP and $NADPH_2$ are regenerated

light-independent step part of photosynthesis occurring in the stroma of the chloroplasts and using the products of the light-dependent step to reduce carbon dioxide to carbohydrate

lignin complex chemical impregnating the cellulose of the walls of xylem vessels, fibres and tracheids, imparting great strength and rigidity

linkage group the genes carried on any one chromosome

lipid diverse group of organic chemicals essential to living things, insoluble in water but soluble in organic solvents such as ether and alcohol (e.g. lipid of the plasma membrane)

lipoprotein a complex of lipid and protein of various types which are classified according to density (e.g. LDL, HDL)

list give a sequence of brief answers with no explanation

liver lobule polygonal block of liver cells, a functional unit within the liver structure

locus the particular position on homologous chromosomes of a gene

loop of Henle loop of mammalian kidney tubule, passing from cortex to medulla and back, important in the process of concentration of urine

lumen internal space of a tube (e.g. gut, artery) or sac-shaped structure

lymph fluid derived from plasma of blood, bathing all tissue spaces and draining back into the lymphatic system

lymph node tiny glands in the lymphatic system, part of the body's defences against disease

lymphatic system network of fine capillaries throughout the body of vertebrates, which drain lymph and return it to the blood circulation

lymphocyte type of white blood cell

lysis breakdown, typically of cells

lysosome membrane-bound vesicles, common in the cytoplasm, containing digestive enzymes

M

macromolecule very large organic molecule – relative molecular mass 10 000+ (e.g. protein, nucleic acid or polysaccharide)

macronutrients ions required in relatively large amounts by organisms

macrophage phagocytic cells of the immune system found throughout the body

Malpighian body glomerulus and renal capsule of mammalian nephron

mandibles the lower jaw of vertebrates; in arthropods paired, biting mouthparts

matrix ground substance of connective tissue, and the innermost part of a mitochondrion

measure obtain a value for a quantity

mechanoreceptors a sensory receptor sensitive to mechanical stimulus

meiosis nuclear division with daughter cells containing half the number of chromosomes of the parent cell

melanic pigmented

menstrual cycle monthly cycle of ovulation and menstruation in human females

meristem plant tissue capable of giving rise to new cells and tissues

mesentery connective tissue holding body organs (e.g. gut) in position

mesophyll parenchyma cells containing chloroplasts

mesosome an invagination of the plasma membrane of a bacterium

metabolic pathway sequence of enzyme-catalysed biochemical reactions in cells and tissues

metabolic water water released within the body by oxidation, typically of dietary lipids

metabolism integrated network of all the biochemical reactions of life

metabolite a chemical substance involved in metabolism

metaphase stage in nuclear division (mitosis and meiosis) in which chromosomes become arranged at the equator of the spindle

metastasis spread of cancer from one organ to another within the body

methylation addition of a methyl group to a molecule by enzyme action

microarray a collection of microscopic DNA spots attached to a solid surface

microhabitat the environment immediately surrounding an organism, particularly applied to tiny organisms

micronutrient ions required in relatively small (trace) amounts by organisms

microtubule tiny, hollow protein tube in cytoplasm (e.g. a component of the spindle)

microvillus one of many tiny infoldings of the plasma membrane, making up a brush border

middle lamella a layer of pectins between the walls of adjacent cells

mitochondrion (plural, **mitochondria**) organelle in eukaryotic cells, site of Krebs cycle and the electron-transport pathway

mitosis nuclear division in which the daughter nuclei have the same number of chromosomes as the parent cell

mitral valve left atrioventricular valve

mode the most frequently occurring value in a distribution

monoclonal antibody antibody produced by a single clone of B lymphocytes; it consists of a population of identical antibody molecules

monocotyledon class of angiosperms having an embryo with a single cotyledon

monohybrid cross a cross (breeding experiment) involving one pair of contrasting characters exhibited by homozygous parents

monomer a molecule that chemically combines with other monomers to form a polymer

monosaccharide simple carbohydrate (all are reducing sugars)

morphology form and structure of an organism

motile capable of moving about

motor area area of the brain where muscular activity is coordinated

motor end plate the point of termination of an axon in a voluntary muscle fibre

motor neuron nerve cell that carries impulses away from the central nervous system to an effector (e.g. muscle, gland)

mRNA single-stranded ribonucleic acid formed by the process of transcription of the genetic code in the nucleus, that then moves to ribosomes in the cytoplasm

MRSA a bacterial infection that is resistant to a number of widely used antibiotics

mucilage mixture of various polysaccharides that become slippery when wet

mucosa the inner lining of the gut

mucus a watery solution of glycoprotein with protective and lubrication functions

muscle spindle sensory receptor in muscle, responding to stretch stimuli

mutagen an agent that causes mutation

mutant organism with altered genetic material (abruptly altered by a mutation)

mutation a change in the amount or the chemical structure (i.e. base sequence) of DNA of a chromosome

mutualism a case of symbiosis in which both organisms benefit from the association

mycelium a mass or network of hyphae

mycology the study of fungi

mycorrhiza a mutualistic association between plant roots and fungi, with the mycelium restricted to the exterior of the root and its cells (ectotrophic), or involving a closer association between hyphae and root cell contents (endotrophic)

myelin sheath an insulating sheath of axons of nerve fibres, formed by the wrapping around of Schwann cells

myelinated nerve fibre nerve fibre insulated by a lipid sheath formed from membranes of Schwann cells

myocardial infarction heart attack

myofibril contractile protein filament from which muscle is composed

myogenic originating in heart muscle cells themselves, as in generation of the basic heartbeat

N

natural classification organisms grouped by as many common features as possible, and therefore likely to reflect evolutionary relationships

nectary group of cells secreting nectar (dilute sugar solution) in a flower

nematocyst stinging cell of cnidarians (coelenterates) (e.g. *Hydra*)

Neolithic revolution the period of human development involving the first establishment of settled agriculture practices, and including the breeding and cultivation of crop plants and herd animals

nephron the functional unit of a vertebrate kidney

nerve bundle of many nerve fibres (axons), connecting the central nervous system with parts of the body

nerve cord in non-vertebrates, a bundle of nerve fibres and/or nerve ganglia running along the length of the body

nervous system organized system of neurons which generate and conduct impulses

  autonomic nervous system (ANS) the involuntary nervous system

  central nervous system (CNS) in vertebrates, the brain and spinal cord

  parasympathetic nervous system part of the involuntary nervous system, antagonistic in effect to the sympathetic nervous system

  peripheral nervous system (PNS) in vertebrates, neurons that convey sensory information to the CNS, and neurons that convey impulses to muscles and glands (effector organs)

  sympathetic nervous system part of the involuntary nervous system, antagonistic in effect to the parasympathetic nervous system

neuron nerve cell

neurotransmitter substance chemical released at the pre-synaptic membrane of an axon, on arrival of an action potential, which transmits the action potential across the synapse

neutrophil a type of white blood cell

niche both the habitat an organism occupies and the mode of nutrition employed

node of Ranvier junction in the myelin sheaths around a myelinated nerve fibre

noradrenaline (also called norepinephrine) neurotransmitter substance in the sympathetic nervous system

nuclear division first step in the division of a cell, when the contents of the nucleus are subdivided by mitosis or meiosis

nuclear membrane double membrane surrounding the eukaryotic nucleus

nuclear pores organized gaps in the nuclear membrane, exit points for mRNA

nucleic acid polynucleotide chain of one of two types, deoxyribonucleic acid (DNA) or ribonucleic acid (RNA)

nucleolus compact region of nucleus where RNA is synthesized

nucleoside organic base (adenine, guanine, cytosine, thymine) combined with a pentose sugar (ribose or deoxyribose)

nucleosome a sequence of DNA wound around eight histone protein cores – a repeating unit of eukaryotic chromatin

nucleotide phosphate ester of a nucleoside – an organic base combined with pentose sugar and phosphate (P_i)

nucleus largest organelle of eukaryotic cells; controls and directs the activity of the cell

nutrient a chemical substance found in foods that is used in the human body – any substance used or required by an organism as food

nutrition the process by which an organism acquires the matter and energy it requires from its environment

O

obesity condition of being seriously over-weight (BMI of 30+)

olfactory relating to the sense of smell

omnivore an animal that eats both plant and animal food

oncogene a cancer-initiating gene

oocyte a female sex cell in the process of a meiotic division to become an ovum

oogamy union of unlike gametes (e.g. large ovum and tiny sperm)

open reading frame (ORP) in molecular genetics – part of a reading frame that contains no 'stop' codons

opsonin type of antibody that attacks bacteria and viruses, facilitating their ingestion by phagocytic cells

order a group of related families

organ a part of an organism, consisting of a collection of tissues, having a definite form and structure, and performing one or more specialized functions

organelle a unit of cell substructure

organic compounds of carbon (except carbon dioxide and carbonates)

organism a living thing

osmolarity the osmotic concentration of a solution

osmoreceptor sense cells or organ stimulated by changes in water potential

osmoregulation control of the water balance of the blood, tissue or cytoplasm of a living organism

osmosis diffusion of free water molecules from a region where they are more concentrated (low solute concentration) to a region where they are less concentrated (high solute concentration) across a partially permeable membrane

outline give a brief account or summary

ovarian cycle the monthly changes that occur to ovarian follicles leading to ovulation and the formation of a corpus luteum

ovarian follicle spherical structures found in the mammalian ovary, containing a developing ovum with liquid surrounded by numerous follicle cells, and from which a secondary oocyte is released at ovulation

ovary female reproductive organ in which the female gametes are formed

ovulation shedding of ova from the ovary

ovule in the flowering plant flower, the structure in an ovary which, after fertilization, grows into the seed

ovum (plural, **ova**) a female gamete

oxygen dissociation curve a graph of % saturation (with oxygen) of hemoglobin against concentration of available oxygen

oxyntic cells cells in the gastric glands that secrete hydrochloric acid

P

pacemaker structure that is the origin of the myogenic heartbeat, known as the sinoatrial node

Pacinian corpuscles sensory receptors in joints

pancreas an exocrine gland discharging pancreatic juice into the duodenum, combined with endocrine glands (islets of Langerhans)

parasite an organism that lives on or in another organism (its host) for most of its lifecycle, deriving nutrients from its host

parenchyma living cells, forming the greater part of cortex and pith in primary plant growth

partial pressure the pressure exerted by each component of a gas mixture, proportional to how much of the gas is present in the mixture; the partial pressure of oxygen in air is represented by the symbol pO_2 and is expressed in kilopascals (kPa)

pathogen an organism or virus that causes a disease

pentadactyl having all four limbs (typically) terminating in five digits

pentose a 5-carbon monosaccharide sugar

peptide a chain of up to 20 amino acid residues, linked by peptide linkages

peptide linkage a covalent bonding of the amino group of one amino acid to the carboxyl group of another (with the loss of a molecule of water)

perception the mental interpretation of sense data (i.e. occurring in the brain)

pericardium a tough membrane surrounding and containing the heart

peristalsis wave of muscular contractions passing down the gut wall

pesticide a chemical that is used to kill pests

petal modified leaf, often brightly coloured, found in flowers

phagocytic cells cells that ingest bacteria, etc. (e.g. certain leucocytes, *Amoeba*)

phenotype the characteristics or appearance (structural, biochemical, etc.) of an organism

pheromone volatile chemical signal released into the air

phloem tissue that conducts elaborated food in plant stems

phosphate (P_i) phosphate ions, as involved in metabolism

phospholipid formed from a triacylglycerol in which one of the fatty acid groups is replaced by an ionized phosphate group

photoautotroph an organism that uses light energy to generate ATP and to produce organic compounds from inorganic substances

photoheterotroph an organism that uses light energy to generate ATP and obtains organic compounds from other organisms

photomorphogenesis effects on plant growth of light

photoperiodism day-length control of flowering in plants

photophosphorylation the formation of ATP, using light energy (in the light-dependent step of photosynthesis in the grana)

photosynthesis the production of sugar from carbon dioxide and water, occurring in chloroplasts and using light energy, and producing oxygen as a waste product

phototropism a tropic response of plants to light

phylogenetic classification a classification based on evolutionary relationships (rather than on appearances)

phylum a group of organisms constructed on a similar general plan, usually thought to be evolutionarily related

physiology the study of the functioning of organisms

phytoplankton photosynthetic plankton, including unicellular algae and cyanobacteria

pinocytosis uptake of a droplet of liquid into a cell involving invagination of the plasma membrane

pituitary gland the master endocrine gland, attached to the underside of the brain

placenta maternal and fetal tissue in the wall of the uterus, site of all exchanges of metabolites and waste products between fetal and maternal blood systems

plankton very small, aquatic (marine or fresh water) plants and animals, many of them unicellular, that live at or near the water's surface

plant growth substance substances produced by plants in relatively small amounts, that interact to control growth and development

plasma the liquid part of blood

plasma membrane the membrane of lipid and protein that forms the surface of cells (constructed as a fluid mosaic membrane)

plasmid small circular DNA that is independent of the chromosome in bacteria (R plasmids contain genes for resistance to antibiotics)

plasmolysis withdrawal of water from a plant cell by osmosis (incipient plasmolysis is established when about 50% of cells show some shrinkage of cytoplasm away from the walls)

plastid an organelle containing pigments (e.g. chloroplast)

platelets tiny cell fragments that lack a nucleus, found in the blood and involved in the blood clotting mechanism

pleural membrane lines lungs and thorax cavity and contains the pleural fluid

pneumocyte epithelial cell type lining the air spaces of the lungs

polarize the setting up of an electrical potential difference across a membrane

polarized light light in which rays vibrate in one plane only

pollen microspore produced in anthers (and male cones), containing male gamete(s)

pollen tube grows out of a pollen grain attached to a stigma, and down through the style tissue to the embryo sac

polygenic inheritance inheritance of phenotypic characters (such as height, eye colour in humans) that are determined by the collective effects of several different genes

polymer large organic molecules made up of repeating subunits (monomers)

polymerase chain reaction a technology in molecular biology used to amplify a single or very few pieces of DNA, generating many thousands of copies

polynucleotide a long, unbranched chain of nucleotides, as found in DNA and RNA

polypeptide a chain of amino acid residues linked by peptide linkages

polyploidy having more than two sets of chromosomes per cell

polysaccharides very high molecular mass carbohydrates, formed by condensation of vast numbers of monosaccharide units, with the removal of water

polysome an aggregation of ribosomes along a molecule of mRNA strand

population a group of organisms of the same species which live in the same area (habitat) at the same time

portal vein vein beginning and ending in a capillary network (rather than at the heart)

post-synaptic neuron neuron 'downstream' of a synapse

potential difference separation of electrical charge within or across a structure (e.g. a membrane)

potential energy stored energy

predator an organism that catches and kills other animals to eat

predict give an expected result

pre-synaptic membrane membrane of the tip of an axon at the point of the synapse

pre-synaptic neuron neuron 'upstream' of a synapse

prey–predator relationship the inter-relationship of population sizes due to predation of one species (the predator) on another (the prey)

probe a defined, labelled fragment of DNA or RNA used to identify corresponding sequences in nucleic acids

proboscis a projection from the head, used for feeding

producer an autotrophic organism

productivity the amount of biomass fixed by producers (photosynthetically)

gross productivity total amount of organic matter produced

net productivity the organic matter of organisms less the amount needed to fuel respiration

prokaryote tiny unicellular organism without a true nucleus; they have a ring of RNA or DNA as a chromosome (e.g. bacteria and cyanobacteria)

promoter region a region of DNA that initiates transcription of a particular gene

prophase first stage in nuclear division, mitotic or meiotic

proprioceptor an internal sensory receptor

prosthetic group a non-protein substance, bound to a protein as part of an enzyme, often forming part of the active site, and able to bind to other proteins

protein a long sequence of amino acid residues combined together (primary structure), and taking up a particular shape (secondary and tertiary structure)

proteome the entire set of proteins expressed by a genome

Protoctista kingdom of the eukaryotes consisting of single-celled organisms and multicellular organisms related to them (e.g. protozoa and algae)

protoplast the living contents of a plant cell, contained by the cell wall

protozoan a single-celled animal-like organism, belonging to a sub-kingdom, the Protozoa, of the kingdom Protoctista

pseudopodium a temporary extension of the body of an amoeboid cell, by which movement or feeding may occur

pulmonary circulation the circulation to the lungs in vertebrates having a double circulation

pulmonary ventilation rate breathing rate

pulse a wave of increased pressure in the arterial circulation, generated by the heartbeat

pumps proteins in plasma membranes that use energy directly to carry substances across (primary pump) or work indirectly from metabolic energy (secondary pump)

pupil central aperture in the eye through which light enters

pure breeding homozygous, at least for the gene(s) specified

Purkinje fibres fibres of the bundle of His that conduct impulses between the atria and ventricles of the heart

pyloric sphincter circular muscle at the opening of the stomach to the duodenum

pyruvic acid a 3-carbon organic acid, $CH_3.CO.COOH$; product of glycolysis

Q

quadrat a sampling area enclosed within a frame

R

radical a short-lived, intermediate product of a reaction, formed when a covalent bond breaks, with one of the two bonding electrons going to each atom

radioactive dating using the proportions of different isotopes in fossilized biological material to estimate when the original organism was alive

reaction centres protein–pigment complexes in the grana of chloroplasts, sites of the photochemical reactions of photosynthesis

receptor a cell which responds to stimuli

recessive allele an allele that has an effect on the phenotype only when present in the homozygous state

reciprocal cross a cross between the same pair of genotypes in which the sources of the gametes (male and female) are reversed

recombinant a chromosome (or cell or organism) in which the genetic information has been rearranged

recombinant DNA DNA which has been artificially changed, involving joining together genes from different sources, typically from different species

recycling of nutrients the process by which materials from dead organisms are broken down and made available for re-use in the biosphere

redox reaction reaction in which reduction and oxidation happen simultaneously

reductive division meiosis, in which the chromosome number of a diploid cell is halved

reflex a rapid unconscious response

reflex action a response automatically elicited by a stimulus

reflex arc a functional unit in the nervous system, consisting of sensory receptor, sensory neuron, (possibly relay neurons), motor neuron and effector (e.g. muscle or gland)

refractory period the period after excitation of a neuron, when a repetition of the stimulus fails to induce the same response, divided into periods known as absolute and relative

relative atomic mass the ratio of the mass of an atom of an element to the mass of a carbon atom

renal capsule the cup-shaped closed end of a nephron which, with the glomerulus, constitutes a Malpighian body

renewable energy energy that comes from exploiting wave power, wind power, tidal power, solar energy, hydroelectric power or biological sources such as biomass

replication duplication of DNA by making a copy of an existing molecule

    semi-conservative replication each strand of an existing DNA double helix acts as the template for the synthesis of a new strand

reproduction formation of new individual by sexual or asexual means

respiration the cellular process by which sugars and other substances are broken down to release chemical energy for other cellular processes

respiratory centre region of the medulla of the brain concerned with the involuntary control of breathing

respiratory pigment substance such as hemoglobin, which associates with oxygen

respiratory quotient ratio of the volume of carbon dioxide produced to the oxygen used in respiration

respiratory surface a surface adapted for gaseous exchange

respirometer apparatus for the measurement of respiratory gaseous exchange

response the outcome when a stimulus is detected by a receptor

resting potential the potential difference across the membrane of a neuron when it is not being stimulated (repolarized)

restriction enzymes enzymes, also known as endonucleases, that cut lengths of nucleic acid at specific sequences of bases

retina the light-sensitive layer at the back of the eye

retroviruses viruses which, on arrival in a host cell, have their own RNA copied into DNA which then attaches to the host DNA for a period

ribonucleic acid (RNA) a form of nucleic acid containing the pentose sugar ribose, found in nucleus and cytoplasm of eukaryotic cells (and commonly the only nucleic acid of prokaryotes), and containing the organic bases adenine, guanine, uracil and cytosine

ribosome non-membranous organelle, site of protein synthesis

ribulose bisphosphate (RuBisCo) the 5-carbon acceptor molecule for carbon dioxide, in the light-independent step of photosynthesis

rod cell one of two types of light-sensitive cell in the retina, responsible for non-colour vision

roughage indigestible matter (such as cellulose fibres) in our diet

S

saliva secretion produced by salivary glands

saltatory conduction impulse conduction 'in jumps', between nodes of Ranvier

saprotroph organism that feeds on dead organic matter (saprotrophic nutrition)

sarcolemma membranous sheath around a muscle fibre

sarcomere a unit of a skeletal (voluntary) muscle fibre, between two Z-discs

sarcoplasm cytoplasm around the myofibril of a muscle fibre

sarcoplasmic reticulum network of membranes around the myofibrils of a muscle fibre

saturated fat fat with a fully hydrogenated carbon backbone (i.e. no double bonds present)

Schwann cell cell which forms the sheath around nerve fibres

sclera the opaque, fibrous coat of the eyeball

secondary sexual characteristic sexual characteristic that develops under the influence of sex hormones (androgens and estrogens)

secondary succession a plant succession on soil already formed, from which the community had been abruptly removed

secretion material produced and released from glandular cells

sedentary organism living attached to the substratum (e.g. rock or other surface)

seed formed from a fertilized ovule, containing an embryonic plant and food store

segmentation body plan built on a repeating series of similar segments (e.g. as in annelids)

selection differential survivability or reproductive potential of different organisms of a breeding population

self-pollination transfer of pollen from the anther to the stigma of the same plant (normally the same flower)

selfing self-pollination or self-fertilization

semilunar valve half-moon shaped valves, preventing backflow in a tube (e.g. a vein)

seminiferous tubule elongated tubes in the testes, the site of sperm production

sense organ an organ of cells sensitive to external stimuli

sensory area an area of the cerebral cortex of the brain receiving impulses from the sense organs of the body

sensory neuron nerve cell carrying impulses from a sense organ or receptor to the central nervous system

sensory receptor a cell specialized to respond to stimulation by the production of an action potential (impulse)

sepal the protective outermost parts of a flower, usually green

seral stage/sere stages in a seral succession, the whole succession being known as a sere

sex chromosome a chromosome which determines sex rather than other body (soma) characteristics

sex linkage genes carried on only one of the sex chromosomes and which therefore show a different pattern of inheritance in crosses where the male carries the gene from those where the female carries the gene

sexual reproduction involves the production and fusion of gametes

shrub layer the low-level (below trees) woody perennials growing in a forest or wood, normally most numerous in clearings (e.g. where a full-grown tree has died)

sibling offspring of the same parent

sieve tube a phloem element, accompanied by a companion cell, and having perforated end walls known as sieve plates

simple sugar monosaccharide sugar such as a triose sugar (3C), pentose sugar (5C) or hexose sugar (6C)

single access key contrasting or mutually exclusive characteristics are used to divide the group of organisms into progressively smaller groupings until individual organisms (species) can be identified

sinoatrial node cells in the wall of the right atrium in which the heartbeat is initiated, also known as the pacemaker

sinus a cavity or space

sketch represent by means of a diagram or graph (labelled as appropriate). The sketch should give a general idea of the required shape or relationship, and should include relevant features

solar energy electromagnetic radiation derived from the fusion of hydrogen atoms of the Sun, reaching Earth from space

somatic cell (soma) body cell – not a cell producing gametes (sex cell)

specialization adaptation for a particular mode of life or function

speciation the evolution of new species

species a group of individuals of common ancestry that closely resemble each other and that are normally capable of interbreeding to produce fertile offspring

spermatogonia male germ cells (stem cells) which make up the inner layer of the lining of the seminiferous tubules, and give rise to spermatocytes

spermocyte cell formed in seminiferous tubules of testes; develops into sperm

sperms motile male gametes of animals

spindle structure formed from microtubules, associated with the movements of chromosomes in mitosis and meiosis

spiracle hole in the side of an insect (thorax and abdomen) by which the tracheal respiratory system connects with the atmosphere

spiral vessel protoxylem vessel with spirally arranged lignin thickening in lateral walls

spirometer apparatus for measurements of lung capacity and breathing rates

spore a small, usually unicellular reproductive structure from which a new organism arises

stamen male reproductive organ of the flower, consisting of filament and anther, containing pollen sacs where pollen is formed and released

standing crop the biomass of a particular area under study

state give a specific name, value or other brief answer without explanation or calculation

stem cell undifferentiated cell in embryo or adult that can undergo unlimited division and can give rise to one of many different cell types

steroid organic molecule formed from a complex ring of carbon atoms, of which cholesterol is a typical example

stigma part of the carpel receptive to pollen

stimulus a change in the environment (internal or external) that is detected by a receptor and leads to a response

stoma (plural, **stomata**) pore in the epidermis of a leaf, surrounded by two guard cells

stretch receptor sensory receptor in muscles

stroke volume volume of blood pumped out by the heart per minute

stroma colourless fluid contents of the chloroplast, site of the light-independent reaction in photosynthesis

style found in the female part of the flower (carpel), linking stigma to ovary

substrate a molecule that is the starting point for a biochemical reaction and that forms a complex with a specific enzyme

subthreshold stimulus a stimulus not strong enough to trigger an action potential

succession the sequences of different communities developing in a given habitat over a period of time

sugars compounds of a general formula $C_x(H_2O)_y$, where x is approximately equal to y, and containing an aldehyde or a ketone group

suggest propose a solution, hypothesis or other possible answer

summation combined effect of many nerve impulses

    spatial summation many impulses arriving from different axons

    temporal summation many impulses arriving via a single axon

surfactant compound that lowers the surface tension between liquids and a solid

suspensory ligament attaches lens to ciliary body in the vertebrate eye

symbiosis literally 'living together'; covering parasitism, commensalism and mutualism

symplast the pathway (e.g. of water) through the living contents of cells

synapse the connection between two nerve cells; functionally a tiny gap, the synaptic cleft, traversed by transmitter substances

synaptic knob the terminal swelling of a pre-synaptic neuron

synergism acting together and producing a larger effect than when acting separately

synovial fluid secreted by the synovial membrane at joints, having lubricating role

systematics the study of the diversity of living things

systemic circulation the blood circulation to the body (not the pulmonary circulation)

systemic pesticide pesticide that is absorbed and carried throughout the body

systole contraction phases in the cardiac cycle

T

target organ organ on which a hormone acts (although broadcast to all organs)

taste bud sense organ found chiefly on the upper surface of the tongue

taxis response by a motile organism (or gamete) where the direction of the response is determined by the direction of the stimulus

taxon a classificatory grouping

taxonomy the science of classification

telophase a phase in nuclear division, when the daughter nuclei form

template (DNA) the DNA of the chromosome, copied to make mRNA

tendon fibrous connective tissue connecting a muscle to bone

terminal bud bud at the apex of the stem

test cross testing a suspected heterozygote by crossing it with a known homozygous recessive

testa seed coat

testis male reproductive gland, producing sperms

testosterone a steroid hormone, the main sex hormone of male mammals

thermogenesis generation of heat by metabolism

thorax in mammals, the upper part of the body separated from the abdomen; in insects, the region between head and abdomen

threshold of stimulation the level of stimulation required to trigger an action potential (impulse)

thrombosis blood clot formation, leading to blockage of a blood vessel

thylakoid membrane system of chloroplast

thyroid gland an endocrine gland found in the neck of vertebrates, site of production of thyroxin and other hormones influencing the rate of metabolism

tidal volume volume of air normally exchanged in breathing

tight junction point where plasma membranes of adjacent cells are sealed together

tissue collection of cells of similar structure and function

tissue fluid the liquid bathing cells, formed from blood minus cells and plasma proteins

tonoplast membrane around the plant cell vacuole

total lung capacity volume of air in the lungs after maximum inhalation

toxic poisonous

toxin poison

toxoid inactivated poison

trachea windpipe

tracheal system system of tubes by which air is passed to tissues in insects

tracheole branch of the trachea in insects

trait a tendency or characteristic

transcription when the DNA sequence of bases is converted into mRNA

transcription factor a protein that binds to specific DNA sequences to control the transcription of mRNA

transect arbitrary line through a habitat, selected to sample the community

transfer RNA (tRNA) short lengths of specific RNA that combine with specific amino acids prior to protein synthesis

translation the information of mRNA is decoded into protein (amino acid sequence)

translocation transport of elaborated food via the phloem

transmitter substances substances released into the synaptic cleft on arrival of an impulse at the pre-synaptic membrane to conduct the signal across the synapse

transpiration loss of water vapour from the aerial parts of plants (leaves and stem)

tricarboxylic acid (TCA) cycle the stage in tissue respiration in which pyruvate is broken down to carbon dioxide, and hydrogen is removed for subsequent oxidation

tricuspid valve right atrioventricular valve

triglyceride fatty acid ester of the 3-carbon alcohol, glycerol – forms into globules because of its hydrophobic properties

triose a 3-carbon monosaccharide

tripeptide a peptide of three amino acid residues

trophic level a level in a food chain defined by the method of obtaining food and in which all organisms are the same number of energy transfers away from the original source of the energy (photosynthesis)

tropism a growth response of plants in which the direction of growth is determined by the direction of the stimulus

tumour abnormal proliferation of cells, either benign (if self-limiting) or malignant (if invasive)

turgid having high internal pressure

U

ultrafiltration occurs through the tiny pores in the capillaries of the glomerulus

ultrastructure fine structure of cells, determined by electron microscopy

unisexual of one or other sex

unsaturated fat lipid with double bond(s) in the hydrocarbon chain

urea NH_2CONH_2, formed from amino groups deaminated from excess amino acid

ureter tube from kidney to bladder

urethra tube from bladder to exterior

uric acid an insoluble purine, formed from the breakdown of nucleic acids and proteins

urine an excretory fluid produced by the kidneys, consisting largely of a dilute solution of urea

uterine cycle cycle of changes to the wall of the uterus (approximately 28 days)

uterus the organ in which the embryo develops in female mammals

V

vaccination conferring immunity from a disease by injecting an antigen (of attenuated microorganisms or inactivated component) so that the body acquires antibodies prior to potential infection

vacuole fluid-filled space in the cytoplasm, especially large and permanent in plant cells

vagus nerve 10th cranial nerve; supplies many internal organs, including the heart

variety a taxonomic group below the species level

vasa recta capillary loop supplying the loop of Henle

vascular bundle strands of xylem and phloem (often with fibres) separated by cambium; the site of water and elaborated food movements up and down the stem

vascular tissue xylem and phloem of plants

vasoconstriction constriction of blood supply to capillaries (of skin)

vasodilation dilation of blood supply to capillaries (of skin)

vector an organism that transmits a disease-causing organism, or a device for transferring genes during genetic engineering

vein vessel that returns blood to the heart

venous return volume of blood returning to the heart via the veins per minute

ventilation rate number of inhalations or exhalations per minute

ventral the underside

ventricle chamber, either of the centre of the brain, or of the heart

venule branch of a vein

vertebrate animal with a vertebral column

vesicle membrane-bound sac

vestibular apparatus the semicircular canals of the inner ear, concerned with balance

vestibular canal upper compartment of the cochlea

vestigial small, imperfectly developed structure

virus minute, intracellular parasite, formed of protein and nucleic acid

vitalism theory early idea that organic compounds could only be produced in living cells

vitreous humour clear jelly of inner eye

W

water table level of ground water in the Earth

wax complex form of lipid

weathering breakdown of rock

white matter nerve fibres wrapped in their myelin sheaths

X

xeromorphic modified to withstand drought

xerophyte plant showing modifications to withstand drought

xerosere succession of plants starting from dry terrain

xylem water-conducting vessels of plants

Y

yolk food stores of egg cells, rich in proteins and lipids

yolk sac membranous sac with numerous blood vessels, developed by vertebrate embryos around the yolk (e.g. in birds and reptiles) or as a component of the placenta (in mammals)

Z

zonation naturally occurring distribution of organisms in zones

zygote product of the fusion of gametes

zymogenic cells cells of gastric glands, secreting pepsinogen

Index